[redacted]

Compendium of Grape Diseases, Disorders, and Pests

SECOND EDITION

AF863740

Edited by

Wayne F. Wilcox
Cornell University
New York State Agricultural Experiment Station, Geneva

Walter D. Gubler
University of California, Davis

Jerry K. Uyemoto
United States Department of Agriculture–Agricultural Research Service (retired)
University of California, Davis

APS PRESS
The American Phytopathological Society

Front cover photograph: Andrew Hagen/Shutterstock.com

Back cover photographs (center, then clockwise from upper left):
C. N. Austin, R. C. Pearson, W. F. Wilcox, T. B. Sutton, T. J. Burr,
W. F. Wilcox, M. Fuchs, M. Sosnowski, and W. F. Wilcox

Reference in this publication to a trademark, proprietary product, or company name by personnel of the U.S. Department of Agriculture or anyone else is intended for explicit description only and does not imply approval or recommendation to the exclusion of others that may be suitable.

Library of Congress Control Number: 2015935134
International Standard Book Number: 978-0-89054-479-2

© 1988, 2015 by The American Phytopathological Society
First edition published 1988
Second edition published 2015

All rights reserved.
No portion of this book may be reproduced in any form, including photocopy, microfilm, information storage and retrieval system, computer database, or software, or by any means, including electronic or mechanical, without written permission from the publisher.

Copyright is not claimed in any portion of this work written by U.S. government employees as a part of their official duties.

Copyright is not claimed in portions of this work (Figures 39, 89, 135, 175, 176, 177, 294, and 345, and the section on Sour Rot) by an employee of Ontario Ministry of Agriculture and Food. © Queen's Printer for Ontario.

Printed in the United States of America on acid-free paper

The American Phytopathological Society
3340 Pilot Knob Road
St. Paul, Minnesota 55121, U.S.A.

Preface

The first edition of the *Compendium of Grape Diseases* was published in 1988. It was a groundbreaking book at the time and has remained a major reference for grape growers and advisers ever since. However, the volume and breadth of grape disease research conducted in the quarter century following its publication, and the degree to which the results of these efforts have been utilized to improve vineyard disease management programs, is remarkable. Thus, our intention in compiling the second edition of this volume has been to preserve the basic structure and value of the original while reflecting the many advances that have been made in understanding the subject matter since that publication's release. In doing so, we have maintained a primary focus on multiple infectious diseases of grapes that are caused by fungi, oomycetes, bacteria, phytoplasmas, viruses, and viruslike agents, including the symptoms and diagnostic features for each of the diseases; the etiological agents and environmental conditions that interact with the host to cause them; and techniques for their management.

As with the first edition, we also have included a chapter on the parasitic nematodes that affect grapes and have provided photographs and text to help diagnose and understand the symptoms of various abiotic disorders. Similarly, we again have provided text and photographs to aid in diagnosing the symptoms that are caused by certain arthropod pests but can be mistaken for symptoms of infectious diseases, along with relevant information on the biology and management of those pests. To highlight its primary focus on infectious diseases while also recognizing this additional subject matter, the new volume has been titled *Compendium of Grape Diseases, Disorders, and Pests*. For more inclusive treatments of abiotic disorders and/or pests, the interested reader is encouraged to consult additional sources that examine these topics in greater detail.

Due to the significant international value of grapes, the wide range of diseases that afflict the crop, and the critical role that disease management plays in its production, this new edition is published in an expanded format that allows additional pages of text and figures beyond those accommodated in the original; included are 375 photographs and drawings, including 280 that are new to this edition. The book is intended to be a general reference that is accessible to individuals with limited formal training in plant pathology and its associated disciplines, while also providing sufficient technical detail to be of value to subject matter specialists. Nevertheless, it remains oriented toward a primary readership of informed grape growers, their advisers in the public and private sectors of the economy, and individuals in various support industries who wish to diagnose grape diseases and disorders and are seeking science-based, practical information regarding their cause and control. Two relatively extensive sections that were not included in the first edition, Grapevine Fungicides and Spray Technology for Grapevines, attest to this particular focus on grape producers and disease management practitioners. And in an effort to provide additional context pertaining to the host plant itself, two other sections have been expanded significantly relative to their counterparts in the first edition, (i) The Genus *Vitis*, Its Species, and Its Rootstocks; and (ii) Grapevine Structure and Growth Stages.

This book is a collaborative effort of 79 authors and coauthors who are recognized authorities in their fields. These authors represent 12 different countries spread across five continents, thereby providing a deliberate international scope and perspective to the information presented. Many of the individual authors contributed photographs for their own sections and sometimes for sections written by others as well. Photographs and illustrations new to this volume were provided by 29 additional individuals who did not author sections of the text, whom we thank and gratefully acknowledge below.

The *Compendium of Grape Diseases, Disorders, and Pests* rests squarely on the shoulders the first edition, *Compendium of Grape Diseases*. That original volume provided the fundamental framework for this current one, and we should like to recognize and remember its editors, the late Drs. Roger C. Pearson (who was taken from us far too soon) and Austin C. Goheen. Many of the individual sections in the original edition provided a basis for the revised versions in this new one, and we should like to recognize and thank the previous contributors, who are listed below. Additionally, we should also like to thank Gary Grove and Heather Scheck for critically reviewing the entire manuscript; Marc Fuchs and Naidu Rayapati for their through reviews and pre-publication updates of the chapter on virus diseases; and Bob Martin and the staff of APS Press for their editorial assistance in getting the book completed.

Contributors to the first edition:

M. Bisiach, Universita degli Studi della Tuscia, Viterbo, Italy
J. Bulit, Institut National de la Recherche Agronomique, Centre de Recherches de Bordeaux, Pont-de-la-Maye, France
T. J. Burr, New York State Agricultural Experiment Station, Cornell University, Geneva, New York, U.S.A.
M. V. Carter, The University of Adelaide, Waite Agricultural Research Institute, Glen Osmond, South Australia
A. Caudwell, Institut National de la Recherche Agronomique, Dijon, France
L. P. Christensen, University of California, Kearney Agricultural Center, Parlier, California, U.S.A.
M. Clerjeau, Institut National de la Recherche Agronomique, Centre de Recherches de Bordeaux, Pont-de-la-Maye, France
B. Dubos, Institut National de la Recherche Agronomique, Centre de Recherches de Bordeaux, Pont-de-la-Maye, France
D. L. Flaherty, University of California Cooperative Extension, Visalia, California, U.S.A.
J. A. Foster, USDA-APHIS-PPQ, U.S. Plant Quarantine Facility, Glenn Dale, Maryland, U.S.A.
P. Galet, Ecole Nationale Supérieure Agronomique de Montpellier, France
W. Gärtel, Biologische Bundesanstalt für Land- and Forstwirtschaft, Institut für Pflanzenschutz im Weinbau, Bernkastel-Kues, Federal Republic of Germany
A. C. Goheen, University of California, Davis, California, U.S.A.

D. Gonsalves, New York State Agricultural Experiment Station, Cornell University, Geneva, New York, U.S.A.
J. Granett, University of California, Davis, California, U.S.A.
T. Herrera, Centro de Investigaciones Agricolas del Norte, Campo Agricola Experimental de la Laguna, Torreon, Coah., Mexico
W. B. Hewitt, University of California, Davis, California, U.S.A.
D. L. Hopkins, University of Florida, Leesburg, Florida, U.S.A.
G. L. Jubb, Jr., University of Maryland, Keedysville, U.S.A.
A. N. Kasimatis, University of California, Davis, California, U.S.A.
R. Lafon, Institut National de la Recherche Agronomique, Centre de Recherches de Bordeaux, Pont-de-la-Maye, France
P. Larignon, Institut National de la Recherche Agronomique, Centre de Recherches de Bordeaux, Pont-de-la-Maye, France
J. Lehoczky, Plant Protection Institute, Hungarian Academy of Sciences, Budapest, Hungary
L. S. Leu, Taiwan Agricultural Chemicals and Toxic Substances Research Institute, Taichung Hsien Taiwan, Republic of China
S. D. Lyda, Texas A&M University, College Station, Texas, U.S.A.
G. P. Martelli, Universita Degli Studi di Bari, Bari, Italy
J. R. McGrew, Hanover, Pennsylvania, U.S.A.
R. D. Milholland, North Carolina State University, Raleigh, North Carolina, U.S.A.
S. M. Mircetich, USDA, University of California, Davis, California, U.S.A.
I. I. Mirică, Staţiunea de Cercetare şi Producţie Viti-vinicolă, Argeş, Rumania
L. T. Morton, Broad Run, Virginia, U.S.A.
S. Nelson-Kluk, Foundation Seed and Plant Materials Service, University of California, Davis, California, U.S.A.
C. G. Panagopoulos, Athens College of Agricultural Sciences, Athens, Greece
R. C. Pearson, New York State Agricultural Experiment Station, Cornell University, Geneva, New York, U.S.A.
F. G. Pollack, Naples, Florida, U.S.A.
R. M. Pool, New York State Agricultural Experiment Station, Cornell University, Geneva, New York, U.S.A.
C. Pratt, New York State Agricultural Experiment Station, Cornell University, Geneva, New York, U.S.A.
R. D. Raabe, University of California, Berkeley, California, U.S.A.
D. C. Ramsdell, Michigan State University, East Lansing, Michigan, U.S.A.
D. J. Raski, University of California, Davis, California, U.S.A.
V. Savino, Universita Degli Studi di Bari, Bari, Italy
W. C. Schnathorst, University of California, Davis, California, U.S.A.
H. Schöffling, Landes- Lehr- und Versuchsanstalt für Landwirtschaft, Weinbau und Gartenbau, Trier, Federal Republic of Germany
H. Schüepp, Eidgenössische Forschungsanstalt für Obst-, Wein- und Gartenbau, Wädenswil, Switzerland
G. Stellmach, Biologische Bundesanstalt für Land- und Forstwirtschaft, Institut für Pflanzenschutz im Weinbau, Bernkastel-Kues, Federal Republic of Germany
L. H. Weinstein, Boyce Thompson Institute for Plant Research at Cornell University, Ithaca, New York, U.S.A.
W. F. Wilcox, New York State Agricultural Experiment Station, Cornell University, Geneva, New York, U.S.A.
R. N. Williams, Ohio State University, Ohio Agricultural Research and Development Center, Wooster, Ohio, U.S.A.
L. T. Wilson, University of California, Davis, California, U.S.A.

New photographs or illustrations provided by:
Craig N. Austin, DuPont Crop Protection, Newark, Delaware, U.S.A.
Terrence Bates, Cornell University, Portland, New York, U.S.A.
M. Bisiach (deceased), University of Milan, Milan, Italy
Lailang Cheng, Cornell University, Ithaca, New York, U.S.A.
Andrew M. Daly, CSIRO, Darwin, Northern Territory, Australia
Richard M. Dunst, Cornell University, Portland, New York, U.S.A.
Connie L. Fisk, North Carolina State University, Raleigh, North Carolina, U.S.A.
Sobhi M. Girgis, National Agricultural Research Foundation, Athens, Greece
Bradley D. Hanson, University of California, Davis, California, U.S.A.
Stephen Hessler, Cornell University, Geneva, New York, U.S.A.
Harvey C. Hoch, Cornell University, Geneva, New York, U.S.A.
James S. Kamas, Texas A & M University, College Station, Texas, U.S.A.
Markus Keller, Washington State University, Prosser, Washington, U.S.A.
L. R. Krake, CSIRO, Adelaide, South Australia, Australia
George M. Leavitt, University of California Cooperative Extension, Madera, California, U.S.A.
Pascal Lecomte, INRA, Villenave d'Ornon, France
Timothy E. Martinson, Cornell University, Geneva, New York, U.S.A.
Maxwell Norton, University of California Cooperative Extension, Merced, California, U.S.A.
Satoshi T. Ohki, Osaka Prefecture University, Sakai, Osaka, Japan
Roger C. Pearson (deceased), Cornell University, Geneva, New York, U.S.A.
Cristina Pisani, University of California, Davis, California, U.S.A.
John Roncoroni, University of California Cooperative Extension, Napa, California, U.S.A.
Annemiek Schilder, Michigan State University, East Lansing, Michigan, U.S.A.
Nadja Schoepe, Staatliches Weinbauinstitut, Freiburg, Germany
Andrew Senesac, Cornell University, Riverhead, New York, U.S.A.
Michael White, Iowa State University, Ames, Iowa, U.S.A.
Mary Jean Welser, Cornell University, Geneva, New York, U.S.A.
Alice Wise, Cornell University Cooperative Extension, Riverhead, New York, U.S.A.
Stella M. Zitter, Cornell University, Geneva, New York, U.S.A.

Finally, we should also like to recognize the contributions of countless researchers, technicians, students, laborers, grape growers, and vineyard advisers whose work and observations have contributed to the body of knowledge reflected in these pages, but whose names do not appear within.

Wayne F. Wilcox
W. Douglas Gubler
Jerry K. Uyemoto

Contributors

We gratefully acknowledge those who prepared new or revised sections for the 2nd Edition of this Compendium.

David N. Appel
Texas A&M University
College Station, Texas, U.S.A.

Apostolos D. Avgelis
Plant Protection Institute
Heraklion, Crete, Greece

Kendra Baumgartner
USDA-ARS
University of California
Davis, California, U.S.A.

Donato Boscia
Institute of Plant Virology
National Research Council
Turin, Italy

Elisabeth Boudon-Padieu
INRA CNRS
Université de Bourgogne
Dijon, France

Thomas J. Burr
Cornell University
Geneva, New York, U.S.A.

Fiona E. Constable
Department of Environment and Primary Industries
Melbourne, Victoria, Australia

Paolo Cortesi
University of Milan
Milan, Italy

Peter Cousins
E. & J. Gallo Winery
Modesto, California, U.S.A.

Kent M. Daane
University of California
Berkeley, California, U.S.A.

Jacqueline Edwards
Department of Environment and Primary Industries
Melbourne, Victoria, Australia

Mike A. Ellis
The Ohio State University
Wooster, Ohio, U.S.A.

Akif Eskalen
University of California
Riverside, California, U.S.A.

Howard Ferris
University of California
Davis, California, U.S.A.

Joseph A. Foster
USDA-APHIS-PPQ
Beltsville, Maryland, U.S.A.

Marc Fuchs
Cornell University
Geneva, New York, U.S.A.

David M. Gadoury
Cornell University
Geneva, New York, U.S.A.

R. Garau
University of Sassari
Sassari, Italy

Carmen Gispert
University of California Cooperative Extension
Indio, California, U.S.A.

Martin C. Goffinet
Cornell University
Geneva, New York, U.S.A.

Deborah A. Golino
Foundation Plant Services
University of California
Davis, California, U.S.A.

Jeffrey Granett
University of California
Davis, California, U.S.A.

W. Douglas Gubler
University of California
Davis, California, U.S.A.

Paul Gugerli
Agroscope Changins-Wädenswil ACW
Nyon, Switzerland

David R. Haviland
University of California Cooperative Extension
Bakersfield, California, U.S.A.

Georg K. Hill
DLR-Rheinhessen-Nahe-Hunsrück
Oppenheim, Germany

Lisa E. Hoffman
DuPont Crop Protection
Newark, Delaware, U.S.A.

Raymond Johnson
Agriculture and Agri-Food Canada
Sidney, British Columbia, Canada

Hanns-Heinz Kassemeyer
Staatliches Weinbauinstitut
Freiburg, Germany

Bruce C. Kirkpatrick
University of California
Davis, California, U.S.A.

Andrew J. Landers
Cornell University
Geneva, New York, U.S.A.

Bernardo Latorre
P. Universidad Católica de Chile
Santiago, Chile

Alan Little
University of Adelaide
Adelaide, Australia

Gregory Loeb
Cornell University
Geneva, New York, U.S.A.

Walter F. Mahaffee
USDA-ARS
Oregon State University
Corvallis, Oregon, U.S.A.

Michael Maixner
Julius Kühn-Institut
FRI Cultivated Plants
Siebeldingen, Germany

Giovanni P. Martelli
Università degli Studi di Bari
Bari, Italy

Sheila A. McBride
Texas A&M University
College Station, Texas, U.S.A.

Wendy McFadden-Smith
Ontario Ministry of Agriculture, Food, and Rural Affairs
Vineland Station, Ontario, Canada

Michael V. McKenry
University of California
Parlier, California, U.S.A.

Angelantonio Minafra
Instituto di Virologia Vegetale del CNR
Bari, Italy

Laura Mugnai
University of Florence
Florence, Italy

Francesca Peduto Hand
The Ohio State University
Columbus, Ohio, U.S.A.

Elsa Petit
Université Paris-Sud
Orsay, France

Alan J. L. Phillips
Universidade Nova de Lisboa
Caparica, Portugal

Charlotte Pratt (deceased)
Cornell University
Geneva, New York, U.S.A.

Ulisse Prota
University of Sassari
Sassari, Italy

Jay W. Pscheidt
Oregon State University
Corvallis, Oregon, U.S.A.

Belinda Rawnsley
Peracto SA
Adelaide, South Australia, Australia

M. Ali Rezaian
CSIRO
Adelaide, South Australia, Australia

David M. Rizzo
University of California
Davis, California, U.S.A.

Philippe E. Rolshausen
University of California
Riverside, California, U.S.A.

Amy Y. Rossman
USDA-ARS
Beltsville, Maryland, U.S.A.

Joachim Rumbolz
GAB Consulting
Lamstedt, Germany

Sead Sabanadzovic
Mississippi State University
Mississippi State, Mississippi, U.S.A.

Pasquale Saldarelli
Department of Biology Agriculture and Food Science
National Research Council
Oristano, Italy

Vito Savino
Università degli Studi di Bari
Bari, Italy

Mark R. Sosnowski
South Australian Research and Development Institute
Adelaide, South Australia, Australia

Mysore R. Sudarshana
University of California
Davis, California, U.S.A.

Giuseppe Surico
University of Florence
Florence, Italy

Turner B. Sutton
North Carolina State University
Raleigh, North Carolina, U.S.A.

Yasuo Terai
Yamanashi Prefectural Agricultural Academy
Yamanashi, Japan

Tarlochen S. Thind
Punjab Agricultural University
Ludhiana, India

Florent Pierre Trouillas
University of California
Parlier, California, U.S.A.

Jose Ramon Urbez Torres
Agriculture and Agri-Food Canada
Summerland, British Columbia, Canada

Jerry K. Uyemoto
USDA-ARS
University of California
Davis, California, U.S.A.

Justine Vanden Heuvel
Cornell University
Ithaca, New York, U.S.A.

M. Andrew Walker
University of California
Davis, California, U.S.A.

Vaughn Walton
Oregon State University
Corvallis, Oregon, U.S.A.

Wayne F. Wilcox
Cornell University
Geneva, New York, U.S.A.

Tony K. Wolf
Virginia Polytechnic Institute
Winchester, Virginia, U.S.A.

Francis P. Wong
Bayer CropScience LP
Research Triangle Park, North Carolina, U.S.A.

Frank G. Zalom
University of California
Davis, California, U.S.A.

Inga A. Zasada
USDA-ARS
Oregon State University
Corvallis, Oregon, U.S.A.

Contents

Part IV. Grapevine Disease Management

Appendices

Introduction

The Grape, Its Diseases, and Its Pathogens

The grape is the most widely planted fruit crop in the world, covering a reported area of approximately 7.5 million hectares in 2012. It grows from temperate to tropical regions, although most vineyards are planted in areas with temperate climates. The most concentrated cultures are in Europe, with countries in the EU accounting for approximately 56% of the world's planted vineyard area and 39% of worldwide production. With over 407,000 bearing hectares in 2012, the United States ranked sixth worldwide in planted area after Spain, France, Italy, China, and Turkey, in that order, and accounted for 10% of the total world production. The leading Southern Hemisphere grape-growing countries are Argentina, Chile, Australia, and South Africa (ranked ninth, 10th, 12th, and 13th in cultivated area, respectively). In terms of volume, China had risen to become the world's leading producer by 2012, accounting for 14% of all grapes harvested. Between 2000 and 2012, the proportion of worldwide production increased from 19 to 31% in Asia and from 12 to 21% in the Americas, but fell from 62 to 39% in Europe.

The grape is a crop plant of many uses. Over half of worldwide production is fermented into wine or distilled into brandy. Much is eaten fresh and, through the use of cold storage and production in the "opposite" hemisphere, is available to consumers in many countries throughout the year. Large quantities of fresh fruit also are dried into raisins. Unfermented juice, juice concentrate, and preserves are common uses for the fruit in North America and other markets. In 2010, the commodity value of grapes for fresh and processed uses in the United States was $3.63 billion, a figure that was increased several fold by the value-added nature of many products developed from these fruit.

Disease management is a critical component of grape production in many, if not most, regions. The importance of disease management derives from not only the relatively high value of this crop, but also the cost of its establishment, the perennial nature of its plantings, and its exposure to exotic pathogens, particularly since the mid-nineteenth century, as a result of human travel and activity.

Diseases in grapes, as in other crop plants, can result in substantial losses in production. In most cases, disease is the result of an interaction between a susceptible host and a living pathogenic organism. Causal agents that are living organisms, such as fungi, bacteria, viruses, and nematodes (Fig. 1), are called biotic, infectious, or parasitic pathogens. Disorders are maladies that appear similar to diseases, but are caused by nonliving entities such as a nutrient imbalance, environmental stress, or chemical toxicity; these disorders are often referred to as abiotic, noninfectious, or nonparasitic diseases. The disease complex in each viticultural region depends on the array of particular causal agents present, the susceptibility of the particular grape cultivars to the causal agents, and the climate of the region.

Pathogens

Fungi and stramenopiles. Fungi are small, frequently microscopic, multicellular or filamentous organisms that lack chlorophyll. Lacking chlorophyll, they must derive their energy either from dead animal or plant materials as saprophytes, or from living animals or plants as parasites or pathogens. Some important fungal pathogens can derive their nutrition only from their living host and, therefore, cannot persist in a growing (vegetative) state in plants or plant parts that have died. Such organisms are referred to as obligate parasites and include several large groups of important pathogens such as powdery mildews, downy mildews, and rusts, which are composed of multiple individual species that have become specialized to feed upon a single host species or group of closely related host species.

Some fungi may infect plants through direct penetration of tissue, whereas others lack the necessary enzymes to penetrate the outer protective cuticle and therefore require an injury or natural openings such as stomata, hydathodes, and lenticels. The vast majority (some prominent exceptions include powdery mildews and various soilborne pathogens) require free moisture on their host surface to initiate infection, which often determines the geographical range and temporal spectrum of their activity.

Some funguslike organisms, once referred to as "lower fungi," are now considered so biologically distinct that they have been assigned to a separate kingdom, known as the Stramenopiles. Most important plant pathogens within this classification belong to a group known as oomycetes or water molds. On grapes, the important representatives are *Phytophthora* spp. that cause root and crown rot and *Plasmopara viticola*, the cause of downy mildew. One practical consequence of the significant biological differences between fungi and Stramenopiles is that many "fungicides" developed to control one group are ineffective against the other.

Bacteria and mollicutes. Bacteria are microscopic, rod shaped or spherical, prokaryotic organisms. They survive from one season to the next in or on soil, plant debris, seeds, insects, or infected plants, and they are spread by splashing rain or running water, wind, insects and other animals, and by humans through commercial and agricultural activities. They require wounds or natural openings to enter plants. Some bacteria are confined to certain tissues in the plant, such as xylem vessels, and are transmitted by leafhoppers or by grafting. Because of their small size, they are difficult to identify on a morphological basis using light microscopy. Thus, identification is typically based on physiological, serological, or molecular characters. Although bacteria cause numerous important diseases of humans and animals, they are much less important causes of plant disease than fungi, and only a few species are important on grape, albeit significantly so.

Mollicutes, also called phytoplasmas (and originally referred to as mycoplasmalike organisms or MLOs), are small, sometimes submicroscopic organisms similar to bacteria except that they have a single, three-layered cell membrane and lack a true cell wall. They vary in shape and are usually nonmotile. As

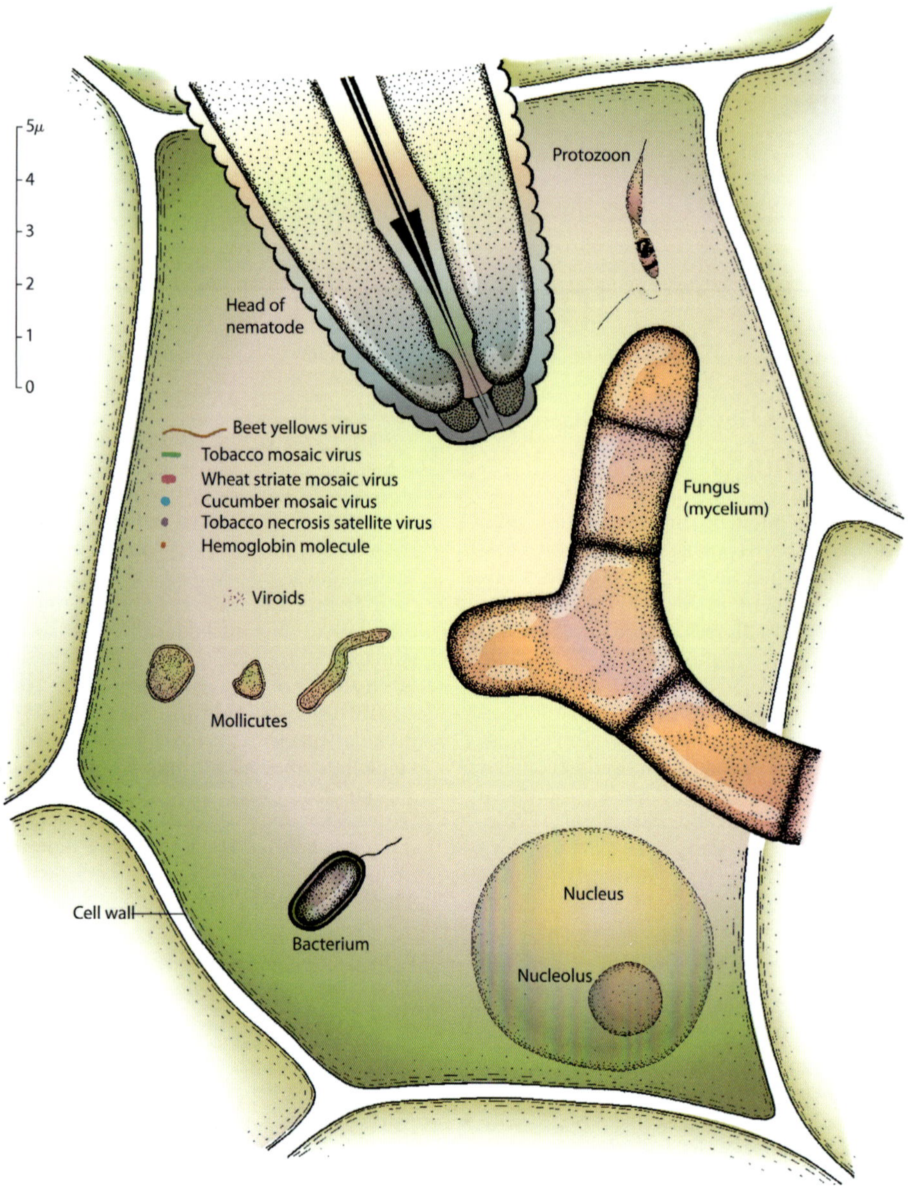

Fig. 1. Schematic diagram of the relative sizes of various plant pathogens compared with a plant cell. (Reprinted from *Plant Pathology, 5th Edition*, G. N. Agrios, 2005, with permission from Elsevier)

plant pathogens, they generally cause "yellows" type diseases and are found in the phloem where they induce dysfunction. Because of the very small size of mollicutes and the fact that the diseases they cause can be transmitted through grafting, such diseases were thought to be caused by viruses until the identity of the agents was first established in 1967. They are transmitted by leafhopper and planthopper vectors in addition to grafting.

Viruses. Plant viruses are submicroscopic particles usually consisting of ribonucleic acid (RNA) wrapped in a protein coat. Virus particles are polyhedral or rod shaped (from short and rigid to long and flexuous). Viruses do not reproduce themselves; instead, they induce the host to produce more virus particles. All plant viruses can be transmitted through vegetative propagation techniques such as grafting and budding with infected source tissue. In the propagation of grafted grapevines, either the rootstock or the scion may be the source of a virus that subsequently infects the new vine systemically. Depending upon their individual nature, some plant viruses also can be transmitted by insect or nematode vectors, by mechanical contact through wounds, or through pollen and the resulting seed.

Nematodes. Nematodes are small (15–35 × 300–1,000 μm), wormlike, root- or soil-inhabiting animals. They are round in cross section and have smooth, unsegmented bodies without appendages. In some species, the female swells at maturity, becoming pear shaped or spheroid. Parasitic nematodes have stylets for piercing host cells during feeding. They may feed only on cells near the root surface and not enter the roots (ectoparasites), or they may enter the roots and feed from within (endoparasites). A few important nematode species also cause damage to grapes by serving as vectors of specific detrimental plant viruses. Nematodes may be sedentary or may migrate short distances in the soil; they can be disseminated more widely by wind, water, and the movement of infested soil, plants, or machinery. Because they are animals, nematodes typically are not affected by standard rates of common fungicides, whereas they are affected by some broad toxicants that control insects as well as by general biocides used for soil fumigation.

Parasitic higher plants. Parasitic higher plants in the families Convolvulaceae (e.g., dodder), Orobanchaceae (e.g., broomrape), and Santalaceae have been observed on grapevines occasionally. They are not addressed in this compendium because none are reported to cause problems in commercial grape production.

Selected References

Agrios, G. N. 2005. Plant Pathology. 5th ed. Elsevier Academic Press, New York.

(Prepared by W. F. Wilcox)

The Genus *Vitis*, Its Species, and Its Rootstocks

Vitis is the most important genus within the family Vitaceae. This genus consists of perennial woody and herbaceous vines with more or less cordate, entire to lobed leaves; coiled tendrils; 5-merous flowers with petals united at the top and separating at the base to form a calyptra ("cap"); minute, nondescript sepals; and clusters that are modified tendrils forming a thryse with soft, pulpy berries. *Vitis* species are dioecious and highly heterozygous as a result of enforced outcrossing; however, the cultivated forms of *V. vinifera* and interspecific hybrids involving this species are almost all hermaphroditic, with the exception of a few pistillate cultivars. The species are diploid with a basic chromosome complement of $n = 19$ and $2n = 38$. Species of the genus *Vitis* are often divided into two sections, *Vitis* (syn. *Euvitis*) and *Muscadinia*, to distinguish the very different species that are placed in this latter taxon. All species within the *Vitis* section are freely interfertile; geographic and phenological barriers isolate them reproductively.

Species within the *Muscadinia* section are quite different and separable from those in the *Vitis* section by a number of unique morphological characters, including nonbranched tendrils; a sequential blooming habit; adhering versus shedding bark; rounded, nonbeaked seeds; the lack of a diaphragm at the nodes; and the presence of lenticels on the bark. At the cellular level, *Muscadinia* has an additional chromosome ($2n = 40$), leading to relatively poor chromosomal pairing with *Vitis* species and a very high degree of sterility in *Vitis* × *Muscadinia* hybrids ($2n = 39$). Some believe these differences warrant separation of the *Muscadinia* species into their own genus.

The most widely utilized of the *Vitis* species is *V. vinifera* L., the cultivated European bunch grape, which is used for wine, table grape, and raisin production across the temperate zones of the globe and constitutes the world's most important horticultural crop. There are thousands of *V. vinifera* cultivars and many have been in cultivation in multiple regions for hundreds of years, which has frequently led to the development of different regional names for a single cultivar (synonyms).

In the middle of the 1800s, European viticulture began to change markedly with the introduction of three very serious North American fungal pathogens: *Erysiphe* [*Uncinula*] *necator* (powdery mildew), *Plasmopara viticola* (downy mildew), and *Guignardia bidwellii* (black rot). Prophylactic sulfur and copper treatments were developed to control the mildews within a decade after their introduction and remain in use to this day. A fourth import, the aphidlike insect pest grape phylloxera (*Daktulosphaira vitifoliae*), came to Europe during this same period and virtually destroyed European viticulture over a 30-year timeframe. Grape phylloxera was finally controlled by the development of resistant rootstocks bred from resistant North American *Vitis* species. Breeders also produced hundreds of hybrids by crossing *V. vinifera* cultivars with phylloxera resistant North American species. These cultivars became known as the French-American hybrids, French hybrids, or hybrid direct producers, and although they did not possess sufficient phylloxera resistance to grow on their own roots, breeders did gain fungal-disease resistance and cold tolerance from the American species. These hybrids are still used in cold climates and often possess substantial resistance to many foliar fungi. However, their fruit and wine quality is generally considered inferior to that from *V. vinifera* cultivars.

The *Vitis* species have three centers of diversification: the Middle East, Asia, and North America. *Vitis vinifera* is thought to have originated in the Middle East and spread by birds and human migration throughout Europe and Asia. The wild forms are in a separate subspecies, *V. vinifera* ssp. *sylvestris*, and are now quite rare because of their high degree of susceptibility to downy mildew, powdery mildew, and grape phylloxera, in addition to most of the diseases discussed in this compendium. *Vitis vinifera* is tolerant of calcareous soils and has been bred with other species in efforts to incorporate lime tolerance and phylloxera resistance into an easily propagated rootstock (e.g., Fercal, 41B, and AXR#1). However, rootstocks with *V. vinifera* parentage have not proved to be adequately resistant to grape phylloxera.

The Asian *Vitis* species (e.g., *V. amurensis* Rupr.) recently have undergone taxonomic revision, and there may be as many as 37 species distributed from Korea to southern China. These species are generally susceptible to grape phylloxera, but have varying levels of resistance to the fungi that plague *V. vinifera*. *Vitis amurensis* is found in the northeast of China and has been used to hybridize with *V. vinifera* to produce cold-hardy and relatively disease-resistant wine grapes.

The North American *Vitis* species have been extensively utilized for the production of rootstocks and disease-resistant wine and table grapes. These species range from Canada to the northern parts of South America. They are generally found near sources of water or in mesic habitats with ample growing-season rainfall. The North American *Vitis* species likely originated in Central America and moved northward with glaciation events; they now occupy most of the United States, with the exception of the northwestern states and neighboring Canadian provinces. They also are found in south-central and southeastern Canada, most of Mexico where water is available, Central America, and the Caribbean. There may be over 30 species in North America; however, the ranges of many of these species overlap and, as a result of their interfertile nature, hybrid zones occur where species intergrade and create what appear to be additional "species." The result is great genetic and morphological diversity, which blurs the dividing lines between species.

In response to the inadvertent importation of grape phylloxera to Europe in the mid-1850s, resistant rootstocks were bred to combat the root feeding and vineyard destruction of this North American pest. Although many of the North American species possess strong resistance to the root feeding of grape phylloxera, few of these species propagate well from woody cuttings, a trait that distinguishes them from the easily rooted *V. vinifera*. Only two North American grape species—*V. riparia* and *V. rupestris*—root well from woody cuttings, and they were first developed to create resistant rootstocks. Selections of these species are still used as rootstocks: *V. riparia* cv. Gloire de Montpellier (syn. Riparia Gloire) and *V. rupestris* cv. St. George (syn. Rupestris du Lot). However, rootstocks of these species were found to provide poor and moderate lime tolerance, respectively, on the typically calcareous soils of Europe; therefore, a new round of rootstock breeding commenced, incorporating strong phylloxera resistance, ease of propagation, and lime tolerance. The North American *Vitis* species used in rootstock and fruit breeding are discussed below, as are the important groupings of rootstocks, listed by their species parentages.

Vitis Species and Uses

Vitis riparia Michx. is found in moist soils and near streams and rivers east of the Rocky Mountains from Montana to Texas and east to the Atlantic Ocean. It is very common in the central and northeastern United States. Dormant cuttings of this species root and graft easily, and possess strong resistance to the root feeding of phylloxera, although the leaves are heavily galled by the foliar form of the insect. The roots of *V. riparia* tend to be very shallow because of its moist alluvial soil habitat. Because of its excellent rooting ability and high phylloxera resistance, the species has been used to produce many rootstock hybrids. Its selection Riparia Gloire is used where low-vigor vines are desired because of high rainfall, deep vigor-inducing soils, or the desire for close vineyard spacings. *Vitis riparia* has also been used widely in fruit breeding because of its excellent cold tolerance and fungal resistance. Although it produces fruit with very high acidity, its berries have relatively neutral flavors, and the vines are highly productive.

Vitis cordifolia Michx. is now considered to be *V. vulpina* L. and has a long history of confusion with *V. riparia. Vitis vulpina* is found across the midwestern and southern United States, and is distinguished from *V. riparia* by its cluster architecture, open shoot tip, waxy stems, and woody cuttings that root poorly. This species was used in early hybrid combinations to produce rootstocks because of its strong phylloxera resistance. One such rootstock, 44-53 Malègue, is still used to a limited extent on soils that are relatively high in magnesium.

Vitis rupestris Scheele was once commonly found in rocky creek beds from Tennessee to Texas. However, its unusual shrubby growth habit has led to its near extinction over much of this range as a result of cattle grazing, although it is still relatively common in southern Missouri. This species has better lime tolerance than *V. riparia*, good phylloxera resistance, and propagates easily from woody cuttings. Because it has deep plunging roots serving to anchor it in erosive creek beds, it induces high vigor to its grafted scions. The rootstock cv. St. George is well suited to dry hillsides, but it is susceptible to all detrimental nematode species that infest vineyards. *Vitis rupestris* has also been used in scion breeding efforts. Its fruit has a peppery herbaceous character, but the juice has intense red-purple color. Thus, it has been used to produce red-juice varieties (teinturiers) and it is in the parentage of many of the French-American hybrids.

Vitis berlandieri Planch. is now considered to be *V. cinerea* var. *helleri* (L. H. Bailey) Moore, and is native from central Texas south to northern Mexico. Collections of this species from the limestone soils of the Edward's Plateau, Texas, were used in Europe to develop rootstocks for calcareous soils. However, it roots poorly from woody cuttings and therefore was used in hybrid combinations with the easily rooted *V. riparia* and *V. rupestris*. It has also been crossed with *V. vinifera* to produce rootstocks for very chalky soils. These rootstocks (333 E. M., 41B, and Fercal) do not possess high levels of phylloxera resistance and should be used cautiously. *Vitis cinerea* (Engelm.) Engelm. ex Millardet is common in the southern and southeastern United States and variants are found in Mexico. The clusters of *V. cinerea* are loose and open, in comparison with the compact clusters of *V. riparia* and *V. rupestris*, and because of this trait, it has been used in fruit breeding despite its herbaceous and astringent flavors.

Vitis aestivalis Michx. and its variants are found throughout the eastern United States and west to central Texas and mesic areas of Mexico. It was included in many of the French-American hybrids and has been used extensively in the United States to produce table and wine grape cultivars in hybrid combinations with *V. vinifera* and other American species. Norton (Cynthiana) is a good example of a *V. aestivalis* × *V. vinifera* hybrid grown in the eastern United States for its excellent fungal resistance and good fruit quality. Lenoir (Jacquez, Black Spanish) is a similar hybrid with excellent resistance to Pierce's disease. It has also been used as a rootstock, although its phylloxera resistance is limited.

Vitis labrusca L. is found across the eastern United States but is widespread only in the northeast. The commonly grown Concord grape is a natural hybrid of this species and *V. vinifera*, with the cultigen name *V.* × *labruscana*. Hundreds of American hybrids have been produced with *V.* × *labruscana* types and in hybrid combination with *V. riparia* and *V. aestivalis. Vitis labrusca* brings cold tolerance and good fungal resistance to its progeny, but also brings methyl anthranilate, the "Concord grape juice" character, to their fruit. Grapes with *V. labrusca* and *V. aestivalis* parentage usually have "slip-skin" berries, with a skin that readily separates from the pulpy flesh. This fruit character is also found in *V. candicans* and its forms. *Vitis labrusca* has not been used to produce rootstocks.

Vitis candicans Engelm. ex A. Gray is now known as *V. mustangensis* Buckley, and is found across central and eastern Texas and south into Mexico. It is closely related to both *V. labrusca* and *V. shuttleworthii* House, which is found scattered across central and southern Florida, most abundantly in the Tampa area. In addition to producing large, "slip-skin" berries with thick skins, all three species have dense tomentum on their leaves and shoots and exhibit strong resistance to a number of fungal pathogens. Unlike *V. labrusca*, both *V. candicans* and *V. shuttleworthii* also possess strong resistance to Pierce's disease. Unfortunately, their fruit is very astringent until it is overripe. *Vitis candicans* hybridizes freely with a number of species to create natural hybrids, which include *V. acerifolia* Raf. and *V. solonis* Munson (with *V. riparia*); *V. doaniana* Munson ex Viala (with *V. acerifolia*); and *V. champinii* Planch. (with *V. rupestris*). There are also frequently other forms of species found in Texas and the southwestern United States that appear to be hybrids between *V. candicans* and *V. monticola*, *V. berlandieri*, and *V. arizonica* Engelm., many of which appear very similar to *V. champinii*.

Vitis rotundifolia Michx. (*Muscadinia rotundifolia* Small) grows across the southern United States from Virginia to Florida, west to Tennessee and northern Arkansas, and south through eastern Texas and southern Mexico. The smaller-berried forms in central and southern Florida are called *V. rotundifolia* var. *munsoniana* (Simpson ex Munson) Moore and the tropical forms in the Caribbean and Central America are called *V. popenoii* Fennel. *Vitis rotundifolia* has very high levels of resistance to most of the fungal diseases that afflict *V. vinifera*, as well as to Pierce's disease, phylloxera, and nematodes. Hybrids between *Vitis* and *Muscadinia* species are normally sterile as a consequence of differences in chromosome number, in addition to a lack of homology between chromosomes. Nevertheless, there has been some success in producing *V. vinifera* × *V. rotundifolia* hybrids, one of which, O39-16, is the only rootstock that resists fanleaf degeneration. O39-16 has strong resistance to *Xiphinema index*, the dagger nematode vector of *Grapevine fanleaf virus* (GFLV), but the nematode's probing for feeding sites injects GFLV into the root system and the virus moves up into the scion. Once in the scion, GFLV replicates and spreads, but it does not induce disease symptoms when O39-16 is used.

Grape Rootstocks

The ***V. riparia* × *V. rupestris*** group of rootstocks lack tolerance to calcareous soils, and in general moderately restrict vigor in scions grafted to them. They are well suited to relatively deep, fertile soils where vigor control in scion canopies is important. 3309C, 101-14 Mgt, and Schwarzmann are commonly used rootstocks in this group, and all three root and graft easily. Schwarzmann and 101-14 Mgt have good resistance to root-knot nematodes, but 3309C is susceptible to root-knot and other nematodes. 3309C is also noted for its sensitivity to a number of viruses that cause graft incompatibility. Many other rootstocks outside this group also have been found to react poorly with viruses, including Freedom, 110R, and Kober 5BB. The related rootstock 1616C [*V. solonis* (*V. acerifolia*) × *V. riparia*] induces low vigor, and has very good and relatively broad resistance to nematodes.

The ***V. berlandieri* × *V. riparia*** group of rootstocks was developed to incorporate the lime tolerance of *V. berlandieri* with the ease of propagation, strong phylloxera resistance, and relatively shallow root system of *V. riparia*. *Vitis riparia* also produces canes with long internodes and few laterals, both excellent nursery production traits. This group of rootstocks is widely used around the world on both calcareous and noncalcareous soils. They typically induce low to moderate vigor in scions and propagate less well than the *V. riparia* × *V. rupestris* group. Common rootstocks in this group include: Kober 5BB, Teleki 5C, SO4, and 420A Mgt, with 5BB inducing the most vigor and 420A inducing much less vigor. 5BB, 5C, and SO4 have moderate resistance to root-knot and dagger nematodes.

The ***V. berlandieri* × *V. rupestris*** group of rootstocks combines the lime tolerance of *V. berlandieri* with the ease of propagation, strong phylloxera resistance, and relatively shallow root system of *V. rupestris*. Several of these rootstocks inherited the brushy growth habit of *V. rupestris*, which makes them less favorable for nursery production. Because they have deeper root systems, they are drought tolerant and perform much better than the rootstocks in the above two groups on dry, shallow, or droughty soils. They are also very tolerant of calcareous soils and their larger root systems allow them to adapt well to soils with nutrient deficiencies and imbalances. They are not recommended for deep, fertile soils with ample moisture because of the high vigor they would induce in scions under these conditions. The most common rootstocks in this group are 110R, 1103P, and 140Ru.

Grape rootstock breeders have used *V. champinii* for its excellent resistance to nematodes and ability to induce high levels of vigor. Two *V. champinii* selections, Dog Ridge and Ramsey (often incorrectly called Salt Creek, which is a selection of *V. doaniana*) have been used as rootstocks—Ramsey to the greatest extent on droughty, saline soils in Australia. *Vitis champinii* provides the primary source of nematode resistance in several interspecific hybrid rootstocks, including Freedom and Harmony, and nine recently released selections: RS-3, RS-9, GRN-2, GRN-3, GRN-4, GRN-5, Matador, Minotaur, and Kingfisher. The first six of these recently released rootstocks have been tested against a broad array of aggressive nematodes and the last three have been tested against aggressive root-knot nematodes, but data on field performance are limited. Freedom and Harmony both have a small percentage of *V. vinifera* in their genetic background, and therefore are best suited for sandy soils that typically do not encourage phylloxera development. Ramsey and Dog Ridge induce very high vigor in scions, but produce much smaller canopies in cool, wet soils. Freedom is less vigorous than either of these two rootstocks but is more vigorous than Harmony; however, both Freedom and Harmony usually induce higher vigor than rootstocks based on *V. berlandieri*, *V. riparia*, or *V. rupestris*.

Vitis vinifera has been used intentionally in rootstock breeding for its excellent lime tolerance and propagation ability. The *V. berlandieri* × *V. vinifera* group (Fercal, 41B Mgt, and 333 E. M.) is commonly used on chalky soils with very high lime content, which suppress grape phylloxera. However, these rootstocks are not reliably resistant on soils that encourage phylloxera development. The most notable failure of a rootstock containing *V. vinifera* parentage was the collapse of AXR#1 (*V. vinifera* × *V. rupestris*). This rootstock was used in the large scale replanting of California's north coastal counties in the 1970s and began failing from phylloxera by the mid-1980s. Tens of thousands of hectares were replanted as a result, at an estimated cost of over $1 billion.

Rootstock breeding programs are still active around the world and focus on developing selections with broad and durable nematode resistance, resistance to fanleaf degeneration, drought and salinity resistance, and better adaptability to soils with various physical and chemical problems (e.g., calcareous, saline, high magnesium, and acidic soils). There also is interest in developing a better understanding of the role of rootstock in virus-induced graft incompatibility and in genetic resistance to this phenomenon, as well as in resistance to a wide range of soilborne fungi, including *Phytophthora*, *Armillaria*, and those causing various forms of vine decline. Scion breeding efforts also continue with a focus on fruit and wine quality, and on disease and pest resistance. Recent years have seen a dramatic increase in molecular efforts directed at grape improvement. The future holds great promise for the development of grape cultivars with resistance genes to address a number of serious grape pests and diseases, thereby reducing both the costs of control and environmental and social concerns over pesticide use.

Selected References

Bailey, L. H. 1934. The species of grapes peculiar to North America. Gentes Herbarum 3:151-244.

Galet, P. 1998. Grape Varieties and Rootstock Varieties. Oenoplurimédia, Chaintré, France.

Galet, P. 1988. Cépages et Vignobles de France. Tome 1. Les Vignes Américaines, 2nd ed. Charles Déhan Press, Montpellier, France.

Moore, M. O. 1991. Classification and systematics of eastern North American *Vitis* L. (Vitaceae) north of Mexico. Sida 14:339-367.

Olmo, H. P. 1986. Grapes. Pages 294-298 in: Evolution of Crop Plants. N. W. Simmonds, ed. Longham Press, London.

USDA-ARS, National Genetic Resources Program. Species records of *Vitis*. Germplasm Resources Information Network (GRIN). National Germplasm Resources Laboratory, Beltsville, MD. http://www.ars-grin.gov/cgi-bin/npgs/html/splist.pl?12735.

(Prepared by M. A. Walker)

Grapevine Structure and Growth Stages

In this discussion, the structure and development of the grapevine are presented with reference to Section *Vitis* of the genus *Vitis*. Those wanting information on the structure of muscadine grape, Section *Muscadinia*, should consult the reference by Goffinet et al. (2001).

Grapevines are perennials grown from rooted hardwood or softwood cuttings (own-rooted), or from grafts of a scion cultivar onto a rootstock so produced. Winter dormancy separates the two growing seasons involved in the production of the inflorescence (cluster), flowers, and fruit (Fig. 2). Grapevine shoots emerge from the overwintering buds situated at the nodes of leafless canes. The shoot develops the leaves and inflorescences of the current growing season while also producing complex lateral buds that will develop the shoots and clusters for the next season.

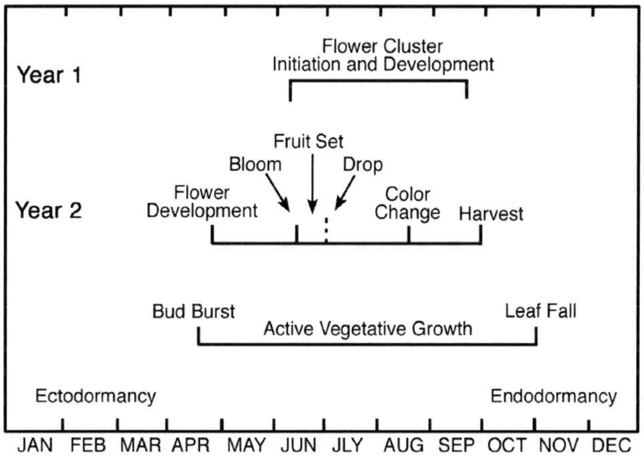

Fig. 2. Chronological relationships of reproductive and vegetative developmental stages of cv. Concord grape in New York. (Adapted from Shaulis and Pratt, 1965)

Grapevine Shoot Morphology

In spring, the shoot emerges from a bud and then develops two rows of leaves along its elongating stem. These leaves, attached to the plant stem at the nodes, alternate from side to side and may or may not be associated with a cluster or tendril at a given node. The phyllotaxy (leaf arrangement pattern) is repeated for the duration of shoot growth. *V. vinifera* and its interspecific hybrid cultivars have a discontinuous pattern of nonfoliar appendages (Fig. 3). Two nodes with a leaf opposite a cluster or tendril are followed by a node having a leaf only. Cultivars derived from *V. labrusca* typically have a tendril opposite any leaf above the fruiting zone. With either pattern, clusters (preformed in the overwintering bud) are produced at the more basal nodes of the green shoot. Once the change to production of tendrils occurs along the growing shoot, no additional clusters are produced. Tendrils usually branch once. At a given developmental stage, the tendril tips become sensitive to touch and begin to grasp nearby objects. Upon such contact, the whole tendril coils, bringing the shoot closer to that object. In highly fruitful shoots, the lowest-most tendril may be replaced by an inflorescence.

All leaves, tendrils, and clusters result from cellular activity at the extreme tip of the shoot, termed the shoot apex (Fig. 4). The shoot apex is a small hemispherical mass of rapidly dividing cells that generates both leaf and tendril (or inflorescence) primordia on its flanks. Very soon after leaf initiation, a lateral bud arises in the leaf axil. These axillary buds later give rise to any leafy summer lateral shoots on the main shoot, and also to the buds that remain over winter to produce the main shoots of the following year (see Structure of Grapevine Buds). The appendages at each node subtending the apical meristem rapidly become separated from one another by internode tissue produced through the activity of an intercalary meristem at the base of each internode. The incipient pith tissue of the stem is interrupted at the nodes by the production of tightly compacted parenchyma cells that develop into the nodal diaphragm of the mature shoot. Vines in Section *Muscadinia* lack such nodal diaphragms. The primordial vascular system develops as procambium extremely early in all elongating organs in the shoot tip. The procambium is the template for the development of all primary xylem and phloem in elongating leaves, tendrils, flower clusters, succulent and woody stems, and roots.

Anatomy of the Stem

The structure of the stem internode at a given position basal to the shoot apex continuously changes over the course of the season (Fig. 5). When the internode is still elongating rapidly near the shoot tip, the vascular system comprises only a small ring of bundles within it (Fig. 5A). Each bundle develops from a procambial strand and consists of primary xylem and phloem on the inner and outer flanks, respectively. The function of primary xylem is the conduction of water and mineral nutrients in the upward direction and the mechanical support of young, flexible organs. The primary phloem functions to bidirectionally move organic nutrients and growth regulators. The ring of vascular bundles surrounds the stem's central pith. The cells of the cortex, between the bundles and epidermis, contain chloro-

Fig. 3. Morphology of a *Vitis vinifera* cv. Chardonnay shoot at the end of bloom; leaves are in two subopposite ranks and, above the fruiting zone, nodes lack tendrils at every third node. **AB,** Axillary bud; **Bl,** blade (lamina); **CB,** compound bud; **Cl,** cluster (inflorescence); **In,** internode; **Lat,** lateral branch (wing) of cluster; **N,** node; **Ped,** peduncle (hypoclade); **Pet,** petiole; **Ra,** cluster rachis; **SL,** summer lateral shoot; **ST,** shoot tip (apical meristem); **T,** tendril. (Courtesy M. C. Goffinet)

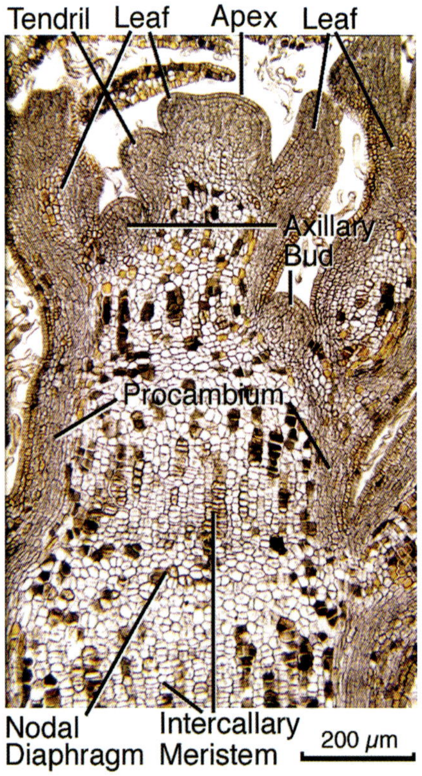

Fig. 4. Longitudinal section through the apical meristem of a rapidly elongating shoot of cv. Concord (*Vitis* x *labruscana*) grapevine. Note precocious development of buds in the leaf axils, the development of an elementary vascular system (procambium), and the rapid elongation of the stem through activity of an intercalary meristem at the base of each internode. Tendrils and leaf petioles also elongate via such activity. The diaphragm at nodes contains tightly compacted pith cells. (Courtesy M. C. Goffinet)

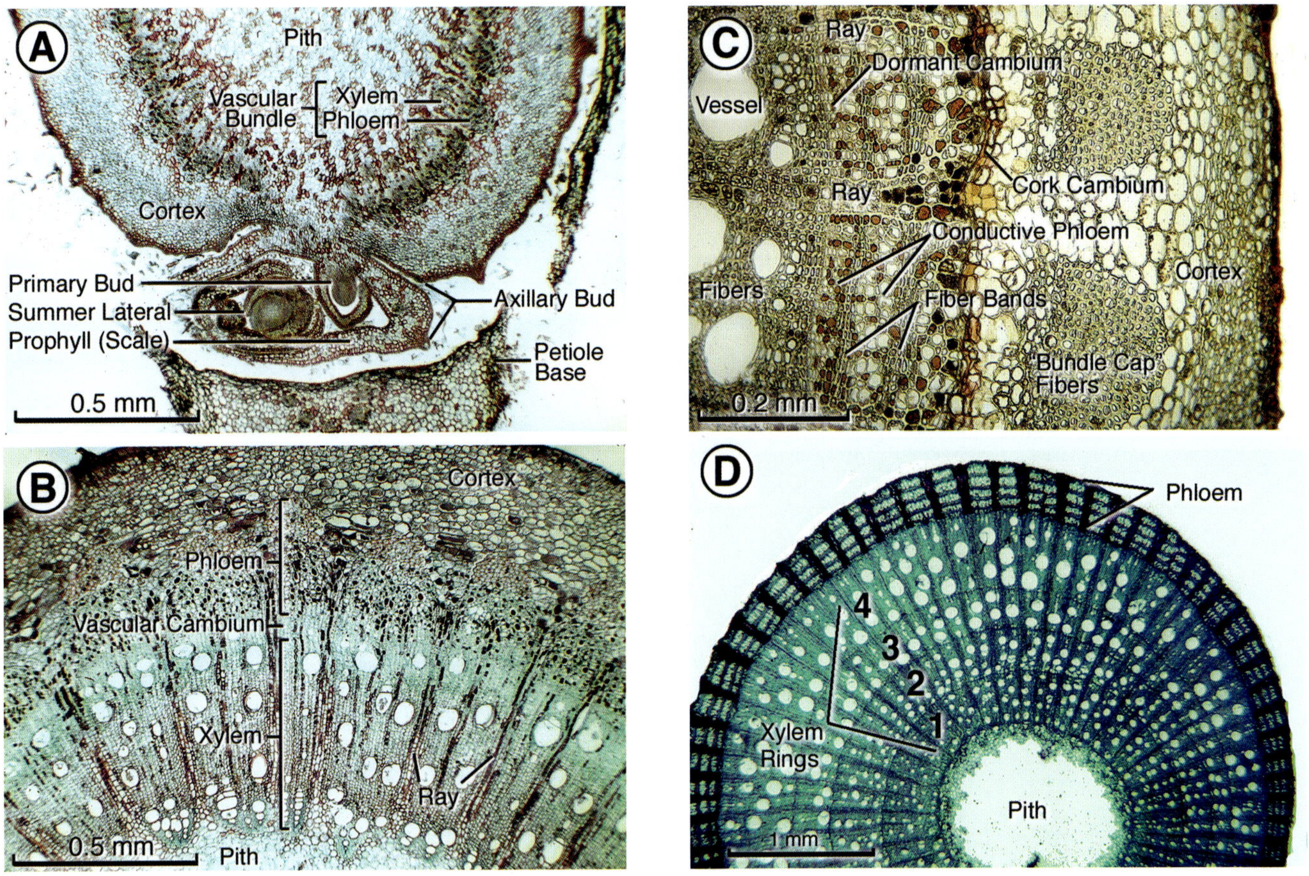

Fig. 5. Developmental anatomy of the grapevine stem. **A,** Cross section of an internode just above insertion of a leaf and associated axillary bud, about two nodes below the lowest node seen in Fig. 4. The procambial traces have developed into a ring of vascular bundles that each contain primary xylem and phloem tissue. The axillary bud already contains meristems for a precociously emergent summer lateral (prompt bud) and the primary bud, which overwinters to produce the shoot in year two. **B,** Internode section after initiation of a vascular cambium, whose cells mature into secondary xylem and phloem. **C,** Late-season internode section with senescence of cortex and outer (phloem) bundle cap fibers, cessation of vascular cambium activity, and initiation of a cork cambium and cork band just outside the secondary phloem. Phloem has fiber bands ("hard phloem") alternating with conductive ("soft") phloem groups. **D,** Cross section of a 4-year-old cordon, showing annual xylem increments; only the most recent annual increment of hard and soft phloem is discernable. Dead fibrous outer bark was removed prior to sectioning. (Courtesy M. C. Goffinet)

plasts and provide both photosynthates and some mechanical support.

As internode elongation ceases and associated leaves reach maximum size, the primary vascular tissue is insufficient for the water- and food-conducting needs of the shoot, or for the stem's mechanical support. Secondary vascular tissues develop to fulfill these functions. In the residual procambium of the internode, a sheetlike cylinder (the vascular cambium) arises, the cells of which divide to produce secondary xylem and phloem to the inner and outer side, respectively (Fig. 5B). This process crushes the older, primary phloem against the cortex while embedding the primary xylem within an enlarging ring of new, secondary xylem with its very large water-conducting vessel elements embedded in a matrix of wood fibers. Both secondary xylem and phloem are traversed radially by ray tissue. All tissues external to the cambium are termed "bark."

In late summer, the main shoots grow at a slower rate and by frost the shoot tip dies back to internodes having brown bark. The stem's vascular cambium also becomes inactive in late summer, causing the last-formed xylem and phloem cells to abut upon one another (Fig. 5C). The annual increment of secondary phloem consists of bands of fibers ("hard phloem") alternating with bands of conductive cells ("soft phloem"). The cortex senesces and a cork layer develops between the cortex and the cylinder of secondary phloem. Cork cells are dead at maturity and their walls are suberized, providing a barrier to movement of water in either liquid or vapor form. The browning of the shoot results from the senescence and death of the cortex and outermost phloem. In viticultural terms, this is known as periderm formation and is associated with "ripening" of the shoot into a cold-tolerant cane. The xylem fibers, vascular rays, and most cells of the phloem store starch and other forms of carbohydrate over the winter season.

In the second season and those following, the cambium is reactivated in spring to produce a new ring of secondary xylem (wood) and a new phloem ring (Fig. 5D). However, at the end of each summer a new ring of cork isolates all phloem of the preceding year. This older phloem senesces and becomes part of the outer, stringy bark seen in older woody stems. Of note is that even 2- or 3-year-old stems of vines in Section *Muscadinia* have a smooth bark with lenticels, long-lived cortex, and a periderm that develops just beneath the epidermis.

The anatomy of secondary vascular tissues reflects their function. The xylem (wood) serves as the water-conducting, food-storing, and supporting tissue of the plant. It is composed of vessels, parenchyma cells, and wood fibers (Fig. 6). Vessel elements are dead and empty of cytoplasm at maturity. The elements are stacked in vertical series, with adjoining elements separated by open perforation plates (Fig. 6A). Thus, once water enters a vessel, this fluid can be pulled upward by the cohesion of water and the tension on this water "column" via the active transpiration of water from leaves. Pathogens in the xylem vessels, such as the Pierce's disease bacterium, also can move or proliferate within these sap-filled conduits, at least as far as open perforation plates allow. Parenchyma cells surround each vessel, while parenchyma cells of the wood rays support radial transport of nutrients and also store carbohydrate. Mechanical injury, insect feeding, and disease organisms can stimulate the blockage of vessel elements. The vessels may become plugged with gums and mucilage as the tissue senesces, or the cell walls of the paratracheal parenchyma cells may balloon into the vessel cavity as tyloses, thereby blocking water flow (Fig. 6B).

Secondary phloem is a food-conducting tissue composed of sieve tubes, companion cells, parenchyma cells, and fibers. Sieve tubes consist of a stacked series of cells known as sieve tube elements (Fig. 7A). These cells are living at maturity, so all transported materials must move through living cytoplasm, even at the junctures between elements (the sieve plates). The last-formed phloem of the season must also serve the need for early food conduction in the next season. The sieve plates develop callose deposits in fall (Fig. 7B) that are thought to terminate phloem conduction. The deposits are enzymatically dissolved the next year. Viruses and phytoplasmas are known to travel in the sieve elements, depending on living cytoplasm for

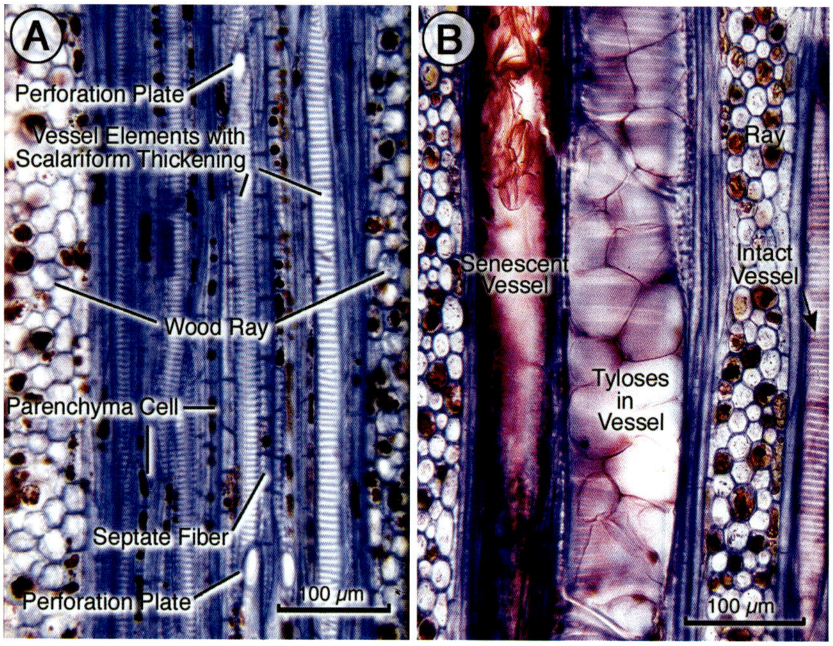

Fig. 6. Longitudinal tangential sections of secondary xylem in cane internodes. **A,** Conductive vessel elements are connected vertically in series to construct a vessel; common end walls are open due to formation of a perforation plate. **B,** Injured xylem with an intact vessel and vessels containing gum and mucilage or tyloses (invasive portions of paratracheal parenchyma cells). (Courtesy M. C. Goffinet)

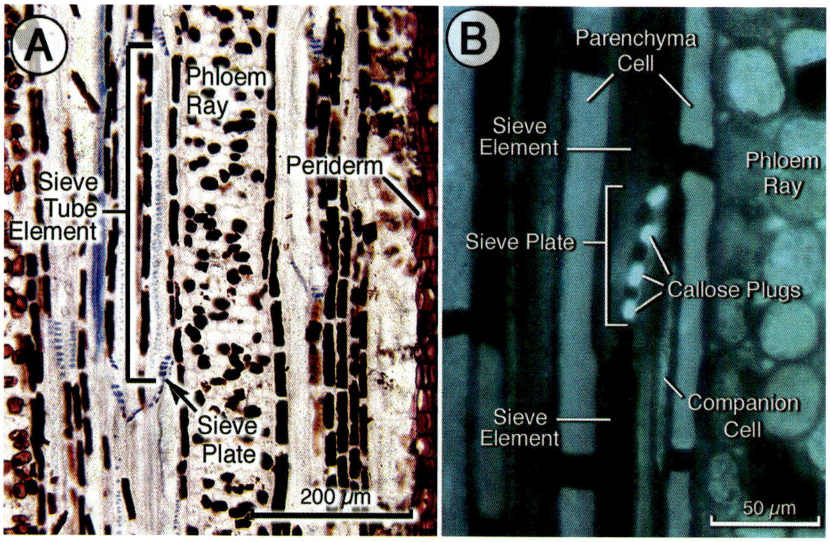

Fig. 7. Longitudinal tangential sections of secondary phloem in cane internodes, bordered by cork cells of periderm tissue. **A,** Conductive sieve-tube elements are connected vertically in series to construct a sieve tube. Common end walls are closed by a sieve plate; cytoplasmic strands that penetrate the plate connect sieve elements. Dark cells are phloem parenchyma. **B,** Fluorescence microscopy of callose plugs occluding a sieve plate between sieve tube elements in late fall; these must be digested in spring to renew phloem function. (Courtesy M. C. Goffinet)

their function, multiplication, and movement in the vine. Such pathogens may be small enough to move through the cytoplasmic strands within the sieve plates. Anatomical symptoms of the phytoplasma diseases grapevine yellows and *flavescence dorée* include poor development of phloem fiber bands and production of disorganized soft phloem groups.

Anatomy of the Root

The young root does not contain a cylinder of vascular bundles surrounding pith, as seen in stems, but instead possess a central vascular cylinder (Fig. 8). The so-called "pith" seen in some roots is actually composed of xylem parenchyma. All the innermost root tissue is primary xylem, which is surrounded by small groups of primary phloem tissue (Fig. 8A). The pericycle comprises the outermost vascular tissue. Pericycle cells contribute to the development of branch roots, the deep origin of cork tissue, and to the initiation of some sectors of the vascular cambium. The innermost cortex consists of a single layer of suberized cells known as the endodermis, which provides some regulatory function to water entry into the central cylinder. It may also serve as a barrier to pathogen entry. The cortex outside of the endodermis eventually senesces at the time secondary vascular development is initiated. The development of the vascular cambium between xylem and phloem groups eventually leads to the production of a core of secondary xylem surrounded by secondary phloem (Fig. 8B). Both tissues contain very broad parenchymatous rays. Because roots develop under a fairly constant environment, there is little differentiation of annual growth rings in the xylem (Fig. 8C). Also, both xylem and phloem differentiate many fewer fibers than they do in stems, and roots contain a high proportion of nonlignified storage parenchyma. New periderms develop as girth increases and as localized breaks occur in outermost tissues.

Leaf Structure

The grapevine leaf is "simple" and consists of a palmately lobed lamina (blade), a petiole (leaf stalk), and a pair of ephemeral stipules at the petiole base (Fig. 9A). Leaves are considered mature when expanded to full size relative to adjacent leaves on the same shoot. Stipules at the shoot tip enclose all younger leaves and the shoot apex, and as each leaf grows a time is reached when the stipules senesce and shrivel. Each heavy scale covering an overwintering bud is actually a modified leaf that consists of the stipule pair with no or minimal contribution from the petiole or blade. The leaf blade contains a palmate framework of major and lateral veins that connect with the vascular bundles of the petiole. The blade is perfused by a ramifying network of progressively smaller veins and veinlets that ultimately serve very small areas of the leaf. Thus, the food- and water-conducting vasculature is never more than a few cells away from the photosynthetically active cells. Larger veins consist of well-developed xylem tissue with large vessels and adjacent phloem-conductive elements (Fig. 9B). Such veins are typically enlarged with both parenchyma cells and a flexible aggregate of collenchyma tissue, which aids in leaf support. The lamina consists of an upper layer of vertically oriented cells—the palisade layer—and a subtending, loosely structured, spongy layer (Figs. 9B and C). These mesophyll tissues provide the leaf's photosynthetic and gas-exchange functions. These layers are sandwiched between upper and lower epidermis. Small veinlets bring water and nutrients to the leaf and remove sugars produced via photosynthesis. Stomata, also called stomates (epidermal pores surrounded by guard cells), occur more frequently on the abaxial (lower) leaf surface than on the adaxial (upper) surface. Stomata serve in CO2 and O2 exchange and in water vapor loss via transpiration, which aids in leaf cooling. Trichomes (surface hairs) may develop, especially on the lower leaf surface. Trichome number, size, and shape depend on species or cultivar and environmental conditions. At leaf drop, the petiole is severed from the stem after an abscission layer is produced at the petiole base. In some situations, the blade may abscise at the blade-petiole juncture.

Texts of grapevine ampelography distinguish species and cultivars on the basis of the organization of the shoot tips and the architecture of stems and roots, with heavy emphasis on

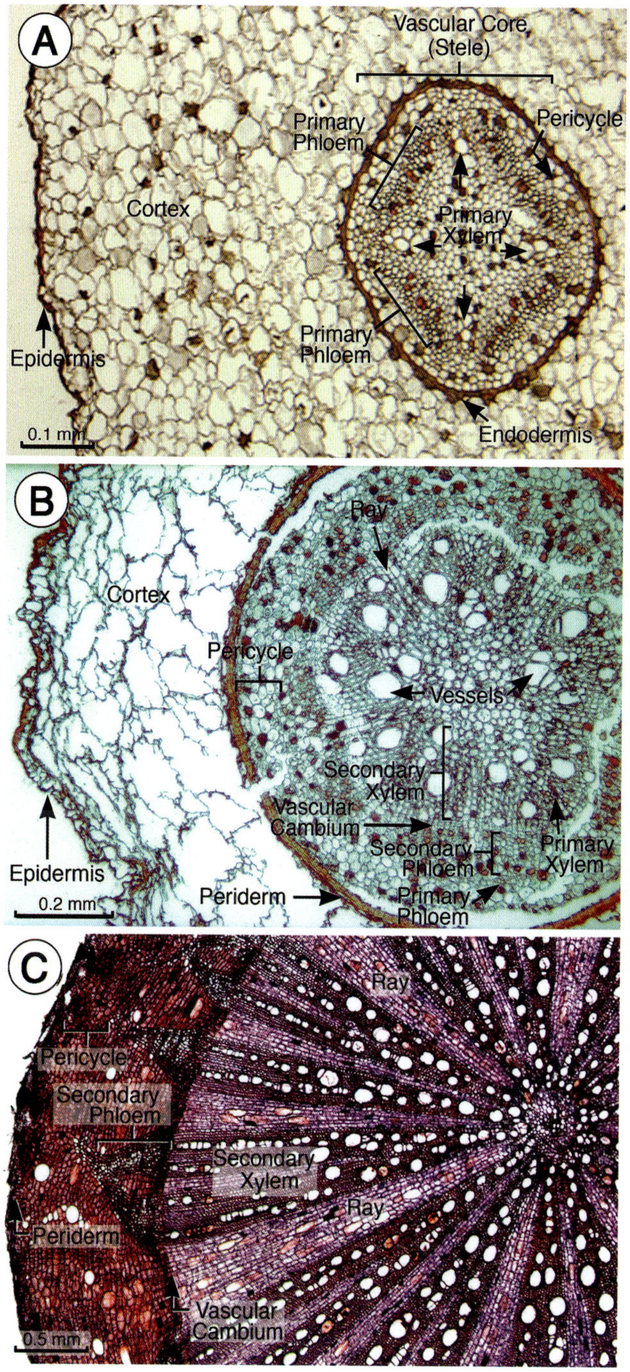

Fig. 8. Developmental anatomy of the grapevine root. **A,** Cross section of a 6-week-old root just after initiation of the vascular cambium between xylem and phloem groups. All vascular tissue is surrounded by pericycle, in which future branch roots and the periderm arise. **B,** Early development of secondary xylem and phloem via activity of a recently initiated vascular cambium. Xylem vessels are large and more secondary xylem than phloem is formed. Cortex degenerates as a periderm is initiated by pericycle tissue. **C,** Section through an older root with heavily thickened xylem and a smaller amount of phloem. The secondary xylem and phloem contain wide, radially continuous rays. Inner tissues are protected by the rind of periderm; the cortex has sloughed away. (Courtesy M. C. Goffinet)

leaf structure (lobing, venation pattern, leaf teeth, leaf hair number, type, and arrangement, etc.).

Structure of Grapevine Buds

The summer axillary bud. The bud in the axil of each leaf is called an axillary bud or lateral bud (Figs. 3, 4, and 5A). This bud is initiated within a few leaves of the shoot apex during active shoot growth. The major component of this axillary bud grows immediately into a lateral shoot and has thus been called the "prompt bud" or "summer lateral" (Figs. 3 and 5A). The summer lateral may sometimes grow vigorously, may branch, and may develop periderm, thus becoming a "persistent lateral." Flowers may develop on the summer lateral, and some cultivars are noted for this tendency. The summer lateral shoot bears a scale (prophyll) at its first node (Figs. 5A and 10B). A compound bud arises in the axil of this scale. The summer lateral shoot usually senesces and then abscises at its base just distal to the insertion of the compound bud (Fig. 10).

The overwintering bud. The bud that remains at the node after the abscission of the summer lateral shoot is termed the compound bud or "eye." A compact system of buds is contained in this eye (Fig. 10). The primary bud is the centralmost and largest of these buds. The primary bud usually bears at its basalmost nodes two prophylls, each in turn having its own axillary bud. Although they are similar, the more basal and larger of the two is often called the "secondary bud" and the smaller is called the "tertiary bud." These buds lie in a line roughly parallel to the longitudinal axis of the cane that produced the summer lateral. In Section *Muscadinia*, the "summer lateral" may not emerge the first year and subsequently emerges as the major fruitful shoot at bud break the next season.

After winter dormancy, usually only the primary bud emerges as the main shoot of the second year. All inflorescences are initiated during the first year's growing season, often completing their annual growth in the lowest 10–20 nodes by veraison. Six

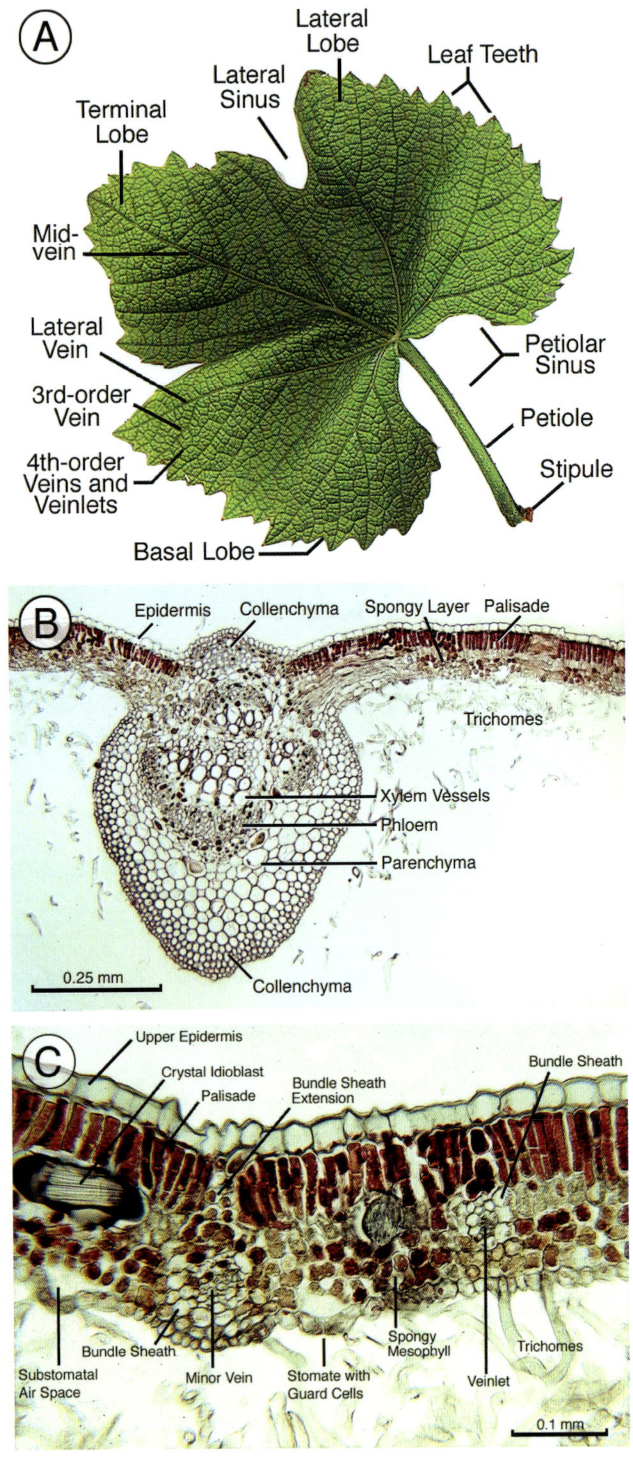

Fig. 9. Leaf morphology and anatomy. **A,** Expanding Concord leaf. **B,** Cross section of lateral vein region. The vein is supported by mechanical cells in the xylem and elastic collenchyma tissue, giving the leaf flexibility with rigidity. Photosynthetic cells occur in the thin interveinal regions. **C,** Cross section of minor vein region. A sheath of parenchyma cells encloses minor veins. Cells of the palisade and spongy layers are in close proximity to minor veinlets. (Courtesy M. C. Goffinet)

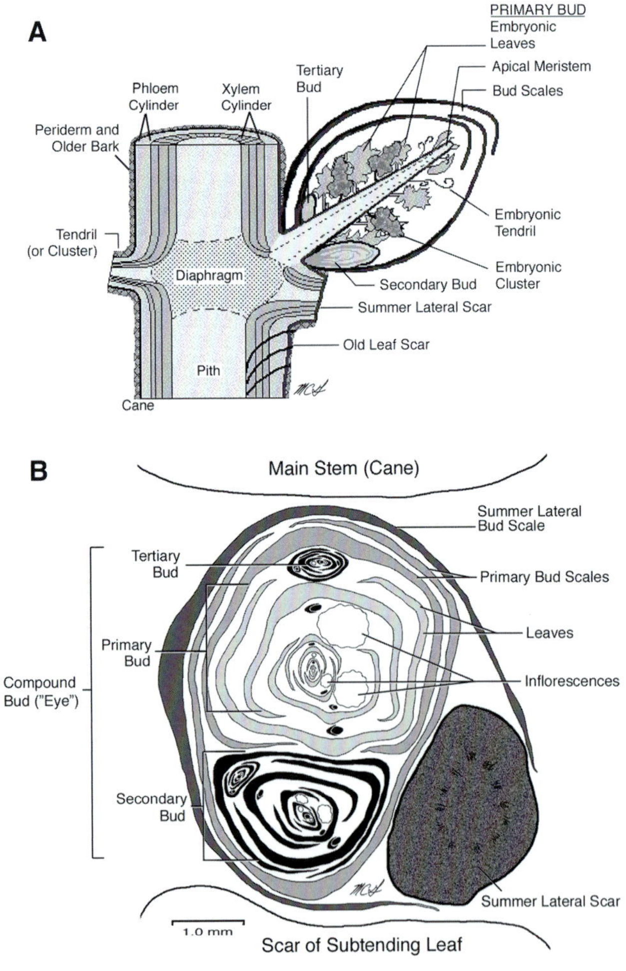

Fig. 10. Construction of the overwintering compound bud of grapevine. **A,** Longitudinal view of the node of the winter cane. **B,** Cross sectional view of the compound bud with many axillary buds (in black) on the primary axis. Embryonic clusters (inflorescence primordia) are in white. (Courtesy M. C. Goffinet)

TABLE 1. Stages in Shoot Development and Flowering/Fruiting in the Grapevine[a]

Eichhorn-Lorenz Stages[b]	Baggiolini Stages[c]
01 – Winter dormancy: winter bud scales more or less closed	A – Winter bud: bud nearly completely covered by two brownish scales
02 – Bud swelling: buds expand inside the bud scales	
03 – Wool (doeskin stage): brownish wool clearly visible	B – Bud swell
05 – Bud burst: green shoot first clearly visible	C – Green shoot
07 – First leaf unfolded and spread away from shoot	D – Leaf emergence: tips of leaves visible, bases still protected by wool
09 – Two to three leaves unfolded	E – Leaves unfolded; first leaves spread away from shoot: internodes visible
12 – Five to six leaves unfolded; inflorescences clearly visible	F – Four to six leaves unfolded; all inflorescences visible
15 – Inflorescences elongating; flowers closely pressed together	G – Inflorescences separated and spaced along the shoot
17 – Inflorescences fully developed; flowers separating	H – Flowers separated
19 – Beginning of flowering; first caps falling	
21 – Early flowering: 25% of caps fallen	
23 – Full flowering: 50% of caps fallen	I – Flowering
25 – Late flowering: 80% of caps fallen	
27 – Fruit set: young fruits beginning to swell, remains of flowers lost	J – Fruit set
29 – Berries small; bunches begin to hang	
31 – Berries pea-sized; bunches hang	
33 – Beginning of berry touch	
35 – Beginning of berry ripening; beginning of loss of green color (*véraison*)	
38 – Berries ripe for harvest	
41 – After harvest, end of wood maturation	
43 – Beginning of leaf fall	
47 – End of leaf fall	

[a] Adapted from OEPP/EPPO,1984; used by permission.
[b] Data from Eichhorn and Lorenz, 1977. See also Fig. 11.
[c] Data from Baggiolini, 1952.

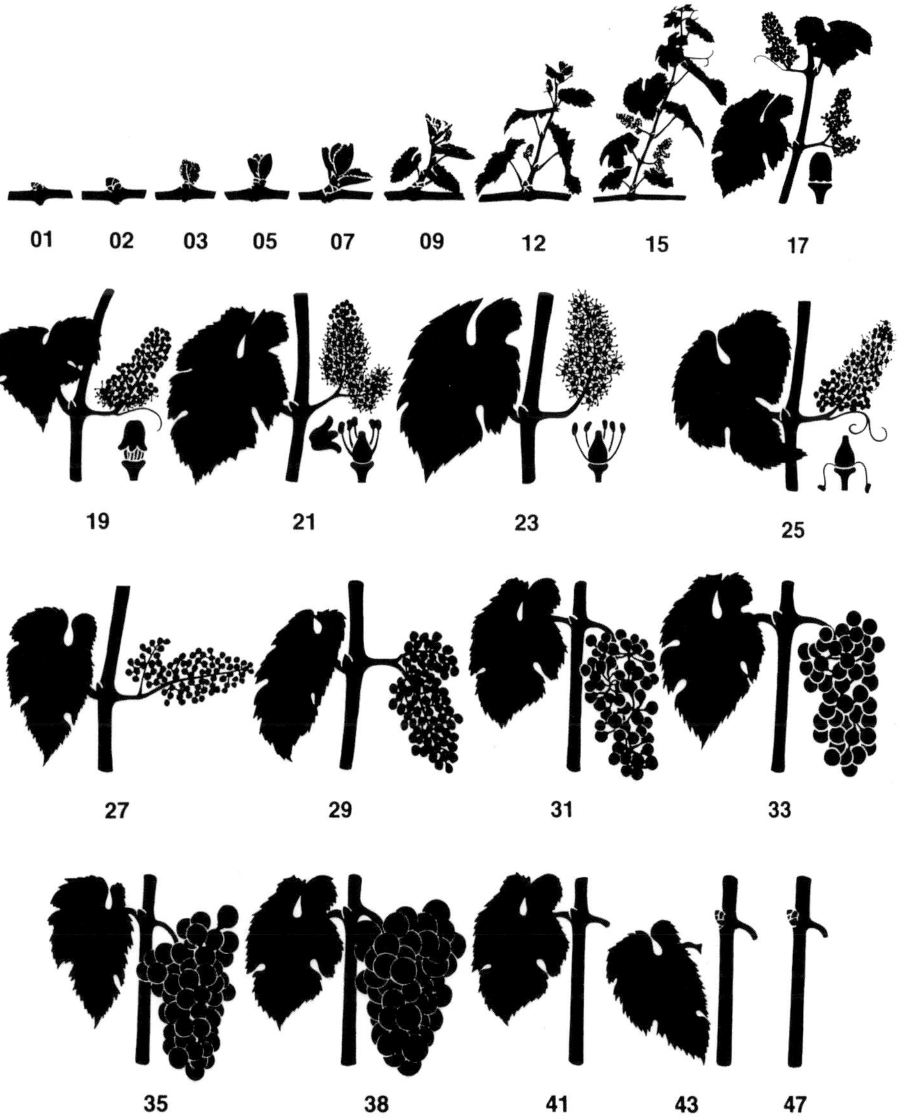

Fig. 11. Stages in grapevine shoot development from bud dormancy to leaf fall in the autumn. (Reprinted, by permission, from Eichhorn and Lorenz, 1977)

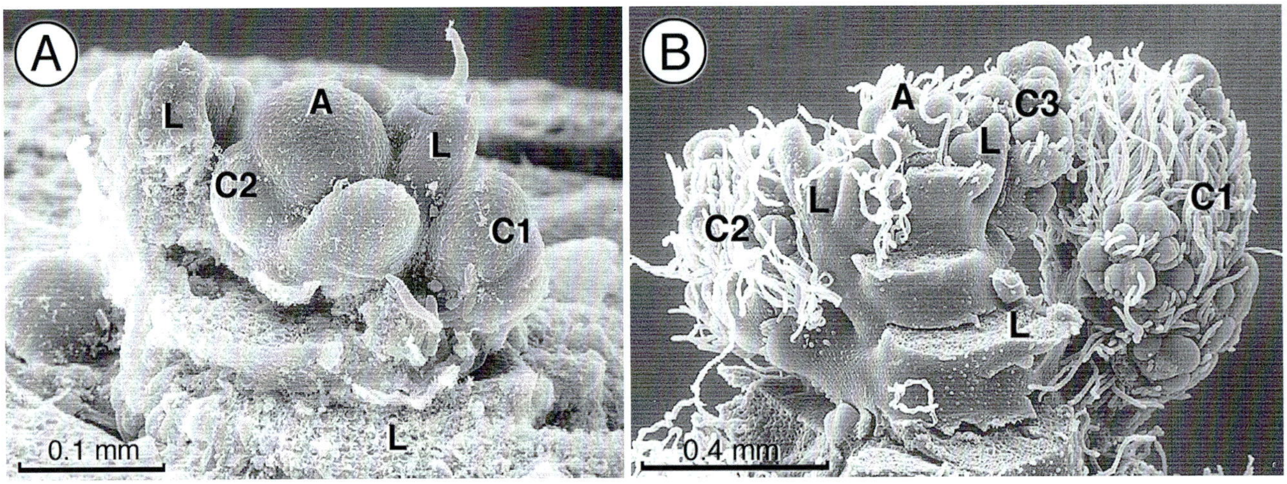

Fig. 12. Scanning electron micrographs showing year-one cluster development in primary buds of a cv. Concord grapevine. **A,** Apical meristem of primary bud ca. 3 weeks post-bloom, with clusters 1 and 2 already having been initiated. Cluster 1 has already formed a lateral bulge that will become the outer arm or wing of the mature cluster. **B,** Apical meristem of primary bud in mid-September, with three clusters and dissected or entire leaf primordia. The ultimate lobes on the cluster primordia enter winter in this state, then further ramify at bud break to produce multiple florets. **A,** shoot apex; **C1–C3,** 1st through 3rd cluster primordia; **L,** leaf primordium. (Courtesy M. C. Goffinet)

to a dozen or more embryonic leaves also may be initiated in primary buds before winter. The most developed inflorescence primordia are found in the primary bud, although the secondary bud may have large primordia, too, especially in climates with long growing seasons. The secondary and tertiary buds of year one often do not grow into shoots in year two but lie quiescent as "latent buds," from which shoots may eventually arise many years after their formation. Shoots derived from such buds on a trunk, cordon, or arm are commonly called water sprouts; if from the crown region, they are referred to as suckers. The accumulation of latent buds accounts for the remarkable ability of grapevines to sprout from older woody tissues after severe pruning, low-temperature injury, etc.

Reactivation of Buds, Phloem, Cambium, and Roots

As soon as the soil thaws in spring, sap rises and drips ("bleeds") from pruning cuts. When air temperatures reach 5°C, the previous year's phloem within the cane at the base of each swelling bud resumes its transportation of assimilates to the growing tissues. The vascular cambium begins to renew cell divisions, also beginning in the cane immediately below each emergent shoot. This reactivation spreads down the cane from each emergent shoot and gradually down the cordons and trunk into the roots. Cells of the vascular cambium divide to form new xylem on the inside and new phloem toward the outside of the cane, as previously described.

After shoots begin to emerge, some existing root tips resume elongation growth while other new roots are initiated within the vascular cylinder of parent roots, resulting in the proliferation of fine, white roots and strong "sinker" roots. Both the rejuvenated and newly formed root tips absorb water and nutrients from the soil.

Grapevine Phenological Stages

The development of the overwintering buds into fruitbearing shoots has been divided into stages (Table 1 and Fig. 11) for convenience in communicating phenologically-based pest control and other management practices. The two phenological classification systems presented in Table 1 are those most recognized by viticulturists, although several amendments and alternative systems also have been proposed. During bud swell, the hard outer bud scales separate, exposing the hairy outer surfaces of the stipules enveloping the young leaves. The shoot elongates and the preformed leaves enlarge slowly as the shoot enters a period of slow growth. The growth rate depends primarily on temperature. As temperature increases and the leaves mature, the shoot grows rapidly. This vegetative growth includes both the expansion of preformed leaves and internodes and the production of new leaves and internodes. The apex of a growing shoot tip alternately produces new leaf and tendril primordia, as previously described.

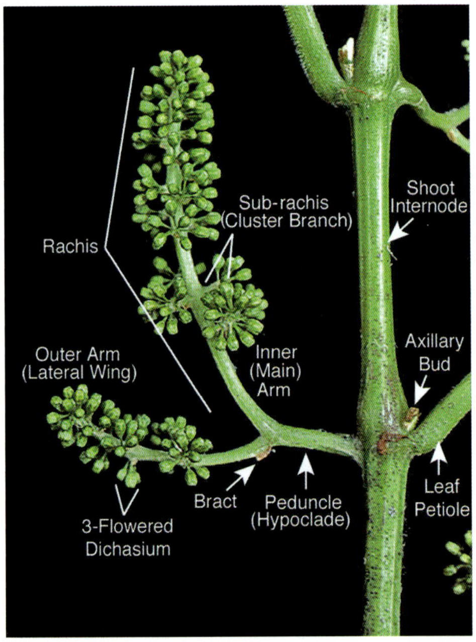

Fig. 13. Morphology of the grapevine cluster (inflorescence) shortly before bloom. The basic arrangement of florets is that of a dichasium, a three-flowered structure with a central and two subjacent flowers, each subtended by a bract. Depending on position in the cluster this arrangement is either intensified in the larger sub-rachises or reduced to a single floret near the tip of a rachis or sub-rachis. The most exaggerated and complex sub-rachis is the outer arm, or wing. (Courtesy M. C. Goffinet)

Cluster Initiation and Development

First-year events. During the postbloom period of a shoot, the primary buds in its leaf axils are already initiating and developing the inflorescences for the next season (Figs. 2 and 12).

For example, the shoot apex of the primary bud of Concord vines in New York State begins initiation of these inflorescences by early July (Fig. 12A). By mid-September, all clusters in buds are at maximum size and developmental complexity for year one (Fig. 12B). Cluster size and degree of branching decrease distally along the primary bud axis. There is some discussion in the literature as to whether the ultimate lobes of the cluster generate flowers by the end of year one or whether florigenesis begins only upon bud activity in spring of year two. In New York State, the ultimate lobes of the clusters are not individual flowers during the fall-winter period. Thus, there is cluster initiation in year one and flower initiation in year two.

Second-year events. Upon bud swell, the lobes of the formerly dormant cluster typically branch and ultimately initiate individual florets. The florets become obvious as the cluster rachis elongates in the prebloom period (Fig. 13). Florets may

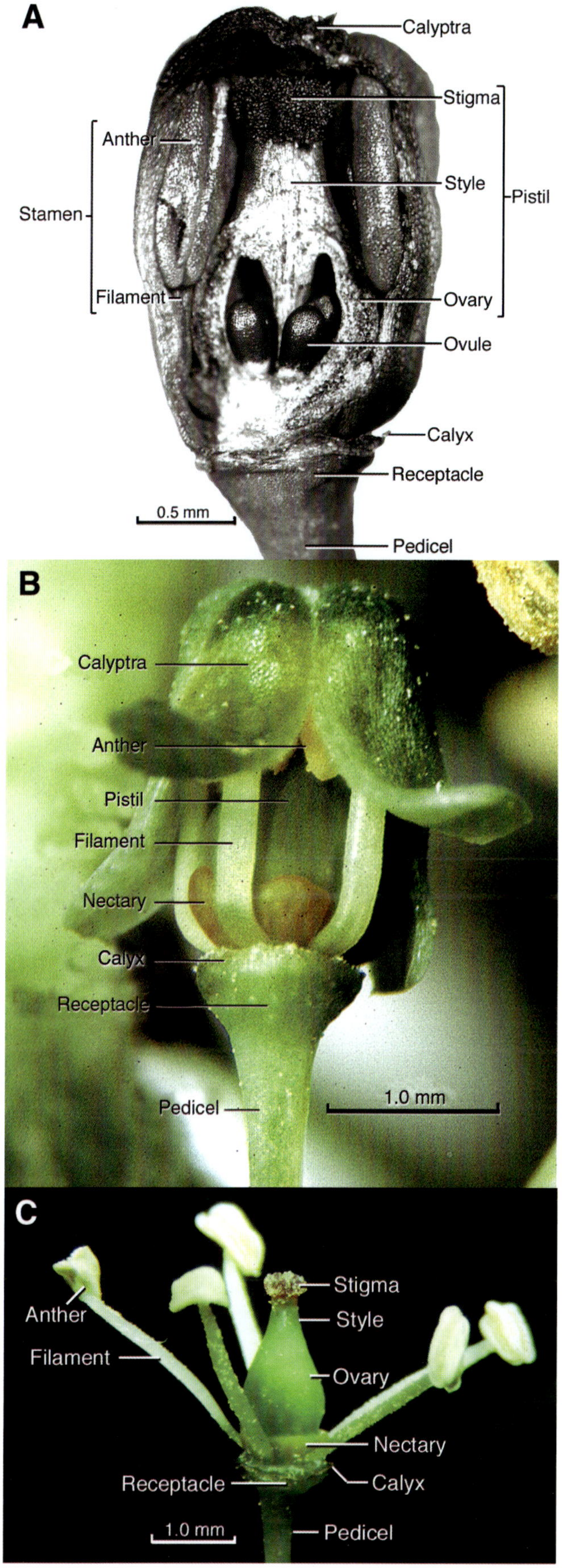

Fig. 14. Floral morphology and anthesis. **A,** Dissection of a floret 2 days before flower opening (anthesis). The mature berry results from the growth of the ovary wall tissue, while the seeds mature from the ovules. **B,** Shedding of the calyptra at beginning of anthesis, ("cap fall"). **C,** Floret at full anthesis. (Courtesy M. C. Goffinet)

Fig. 15. 'Concord' clusters at fruit set, about 10 days after bloom. **A,** Cluster morphology, with only a portion of the flowers at anthesis growing into fruits. **B,** Close-up of a small sub-rachis showing how the parts of the flower contribute to fruit development. (Courtesy M. C. Goffinet)

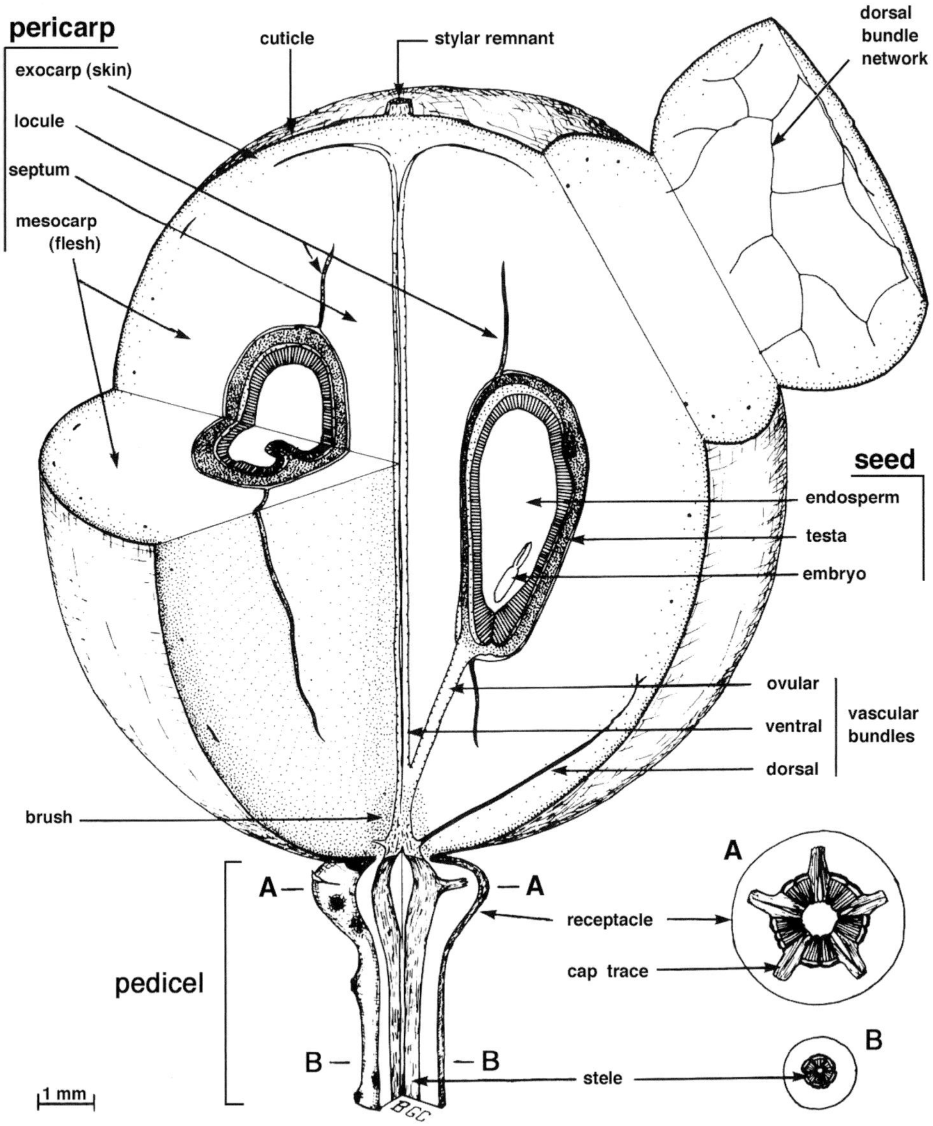

Fig. 16. Diagrammatic representation of the anatomy of a mature berry. (Adapted from Coombe, 1987. Used by permission.)

occur as singlets, but they are usually arranged in small groups on inflorescence branches (variously termed subrachises, rachillae, or dichasia, depending on their position within the cluster). A large branch, or wing, is generally evident at the shoulder of the cluster, and may be termed an outer arm because of the nature of its origin on the cluster the previous year. The main inflorescence axis, or rachis, may thus be termed the inner arm. The wing may have very poor or even no flower development, especially if on the most distal cluster and if vines were overcropped the previous season. In such cases, the outer arm becomes tendril-like. There is much evidence demonstrating that tendrils and inflorescences are homologous structures. The most basal region of the rachis, between the outer arm and the parent shoot, is termed the cluster stalk (peduncle or hypoclade).

Each flower immediately before anthesis (bloom) is covered by five to six interlocking green petals, which compose the "cap" or "calyptra" (Fig. 14A and B). At anthesis, the cap separates from the flower's receptacle and falls off (Fig. 14B). The calyx in *Vitis* is reduced to a small ring on the rim of the receptacle. The five to six stamens on each flower spread out (Fig. 14C) and their anthers rupture, thus shedding pollen, which is deposited on the stigma of the pistil. This is the bloom period, which typically lasts for just a few days in regions with cold winters to a week or longer in climates with relatively mild winter temperatures. Excessively long bloom periods (even within individual clusters or between clusters on the same vine) can complicate disease management programs when specific spray timings are dependent on floral or fruit developmental stage.

Pollen germinates on the stigma soon after bloom, in some situations perhaps even before calyptras are shed. A pollen tube emerges from each grain and grows at a temperature-dependent rate down through the style of the pistil and toward one of the four ovules in the ovary (Fig. 14A). Each ovule contains an embryo sac with one egg cell. Upon entering the embryo sac of an ovule, the pollen tube ruptures and releases a sperm nucleus that fuses with the egg nucleus within the sac (fertilization). Not all ovules become fertilized, as certain diseases or deleterious genetic or environmental factors may increase the number of immature or aborted ovules or decrease the frequency of pollen-grain germination.

With some exceptions, flowers of most cultivated grapevine species are hermaphroditic ("perfect") in that each flower develops the central female structure, the pistil, surrounded by the male structures, the stamens (Figs. 14A and C). Wild grapevines are often dioecious (i.e., either male plants or female plants). The female vines have reflexed stamens with nonfunctional pollen and male vines have reduced and nonfunctional pistils.

Fruit Set

After fertilization, the stamens are shed and the ovary containing the fertilized ovules begins to enlarge. By 1–2 weeks after bloom, depending on temperature, it is normal in all grape cultivars for many small berries to have fallen off the cluster (Fig. 15). The presence of at least one developing berry is critical to the continued development of the rachis of the cluster.

If drop is excessive and the cluster dies or consists of only a few berries, the condition is called *coulure* in French and sometimes referred to as "shatter" in English. Some cultivars are prone to *coulure*. It may result from genetic sterility or from nutritional, environmental, or pathological conditions unfavorable to ovule or pollen maturation.

Another disorder of fruit set, called *millerandage* in French or "shot berries" in English, is the production of many small, seedless berries. It appears to be related to genetic, nutritional, and environmental factors, although infection by *Tomato ringspot virus* also has occasionally been implicated.

Berry Growth and Maturation

The grapevine berry matures from the ovary region of the flower. Berry growth to maturity takes about 100 days for Concord grapes (seeded) in the Great Lakes region of the United States. Other cultivars vary from this, and climate and other factors impose additional variation. After the drop period, the increase in weight of seeded berries is generally divided into three phases: a period of rapid growth until the seed reaches its mature size; a period of slow growth ending with the beginning of loss of green color (*veraison*); and a second period of rapid growth ending in maturity, as indicated by color and the ratio of soluble solids to acid. The duration and distinctiveness of these individual periods vary according to genetic and environmental factors.

The mature seeded berry is shown in Fig. 16. The epidermis (outer layer of the skin) of a young berry usually has a few stomata, which become corky lenticels as the berry matures. Grape cultivars may be classified into two groups on the basis of the seeded or "seedless" nature of their berries. In a given seeded cultivar, mature berry size is related to the number of seeds. Berries of seedless cultivars grow more uniformly than do cultivars with seeded berries, and seedless berries also have a much shorter period of slow growth than those with seeds. Seedless cultivars fall into one of two classes with respect to the time of seed abortion: those in which the ovules do not develop past bloom (e.g., Black Corinth); and those in which the developing seeds abort (seedless table cultivars). It is usually possible to find traces of aborted seeds in the latter class. Grape seeds contain embryos that may provide seedlings or tissue-cultured plantlets with desirable traits of both parents for use in grape breeding programs.

Selected References

Baggiolini, M. 1952. Les stades repères dans le développement annuel de la vigne et leur utilisation pratique. Stn. Fed. Essais Agric. (Lausanne) Publ. 12 (MC).

Coombe, B. G. 1987. Distribution of solutes within the developing grape berry in relation to its morphology. Am. J. Enol. Vitic. 38:120-127.

De la Harpe, A. C., Swanepoel, J. J., and Swart, J. P. J. 1982. The anatomy of the genus *Vitis*: An annotated bibliography. S. Afr. J. Enol. Vitic. Suppl. 3:2.

Eichhorn, K. W., and Lorenz, D. H. 1977. Phänologische Entwicklungsstadien der Rebe. Nachrichtenbl. Dtsch. Pflanzenschutzdienstes (Braunschweig) 29:119-120.

Fournioux, J.-C., and Adrian, M. 2011. Morphologie et Anatomie de la Vigne. Éditions Féret, Bordeaux.

Goffinet, M. C., Welser, M. J., and Himelrick, D. G. 2001. Anatomy and morphology. Pages 15-50 in: Muscadine Grapes. F. M. Basiouny and D. G. Himelrick, eds. Am. Soc. Hortic. Sci., Alexandria, VA.

May, P. 2004. Flowering and Fruitset in Grapevines. Lythrum Press, Adelaide, South Australia.

Mullins, M. G., Bouquet, A., and Williams, L. E. 1992. Biology of the Grapevine. Cambridge University Press, Cambridge, UK.

OEPP/EPPO (European and Mediterranean Plant Protection Organization). 1984. WPPO crop growth key stages. Bull. OEPP/EPPO Bull. 14:295-298.

Pratt, C. 1971. Reproductive anatomy in cultivated grapes—A review. Am. J. Enol. Vitic. 22:92-109.

Pratt, C. 1974. Vegetative anatomy of cultivated grapes—A review. Am. J. Enol. Vitic. 25:131-150.

Shaulis, N. J., and Pratt, C. 1965. Grapes: Their growth and development. Farm Research, Jan.–Mar., Vol. 30, No. 4. New York State Agricultural Experiment Station, Geneva, NY, and College of Agriculture, Cornell University.

(Prepared by M. C. Goffinet and C. Pratt)

Historical Significance of Grapevine Diseases

Throughout most of recorded history, plant diseases were mysterious and feared, and the destruction they caused was felt at a very personal level. The perspective of the present day shapes our view of plant diseases, but in a broader historical context, the *science* of plant pathology is actually comparatively new. If a peasant farmer of the medieval period could visit a farm during your great-grandfather's childhood, that farmer would carry sufficient knowledge to use virtually every tool and technology at hand with little or no instruction. The medieval peasant and your own recent ancestor would likely share a similar, dark view and understanding of plant diseases. However, in the middle of the nineteenth century, great changes began to take place in the state of knowledge about plant diseases. Grapevines were at the center of those changes.

Viticulture had its beginnings in Asia and Europe and spread to other continents with colonists who brought the European grape *Vitis vinifera* with them. Plantings of *V. vinifera* in eastern North America were blighted from above and below, as unsuspecting growers placed their vineyards in the center of origin of several of this host's most virulent microbial pathogens (including the now-cosmopolitan *Erysiphe necator* and *Plasmopara viticola*) and an endemic home of the grape phylloxera. However, Spanish colonists who planted vines in the Mediterranean climate of the western coast of North America were more successful, and their vineyards thrived. Nevertheless, eastern North America was also a center of speciation for the genus *Vitis*, and colonists quickly adopted and domesticated native species such as *Vitis labrusca* and *V. muscadinia*, both of which grew well and yielded healthy fruit. Amateur botanists and expatriate members of farming families were soon sending samples of the New World vines back to Europe. These exotic species found their way into backyard gardens, larger farms, and into many of the most prestigious conservatories and research institutes of Europe. Unfortunately, some of the vines travelled in the company of their equally exotic plant pathogens and insect pests.

In 1845, Edward Tucker, a gardener employed at Margate (Kent county) in England found what appeared to be mildew growing on *V. vinfera* grapevines in a glasshouse. He sent the leaves to Rev. M. J. Berkeley, who identified the pathogen as a powdery mildew, and named the pathogen *Oidium tuckeri* in the discoverer's honor. Margate is located on the southeastern coast of England near the narrows of the English Channel, and by 1847, *Oidium tuckeri* had crossed the channel and was reported in France. Although mildews were widely known on other crops, this one was different—it was uncharacteristically virulent on grapevine. If weather was dry, the grapes split and dried into rocklike mummies; if it was wet, the berries softened and rotted on the vine. Vines in full leaf were rapidly sickened and defoliated. Before the disease was brought under control in

1854 through the widespread use of sulfur dust, grape yields in France and Germany declined by nearly 70%. Sulfur had long been employed to a minor degree in various forms against a variety of fungal plant pathogens; Edward Tucker himself was familiar with its use against powdery mildew on peach and employed it immediately against the grapevine disease in Margate upon its discovery. By the mid-1850s, however, grape powdery mildew had made dusting with sulfur a dominant and necessary activity in viticulture across Western Europe. This war upon mildew was declared at a time when "germ theory" was neither understood among the general populace nor even generally accepted within the scientific community: Louis Pasteur had only reported the microbial nature of fermentation a few years earlier and it would be another 30 years before Robert Koch would publish the first proofs of the microbial nature of disease in animals. Yet in 1854, a substantial proportion of France's 6 million grape growers were now involved in the routine application of sulfur to suppress powdery mildew on nearly 2.5 million hectares of vineyards.

In 1878, while European vineyardists were still reeling from the destruction caused by introduction of the grape phylloxera 13 years earlier, downy mildew (*Plasmopara viticola*) was first noted in southwestern France. By 1882 the disease had spread to all of France, and soon after to vineyards throughout most other parts of the continent. The disease seemed less destructive in regions with drier climates but could develop explosively in regions with relatively high and frequent rainfall, causing severe defoliation and fruit infection. Efforts to suppress it with sulfur were ineffective, but in 1885, P. M. A. Millardet reported the successful use of copper sulfate, lime, and water (Bordeaux mixture) to control downy mildew. Bordeaux mixture remained one of the most important fungicides in the world for the next half century, as its use was expanded to address a number of fungal and bacterial diseases of other crops; numerous commercial formulations based on it are still used today.

Therefore, between 1847 and 1878, vineyards in Europe were subjected to three successive plagues that nearly destroyed the culture of *Vitis vinifera*. These events had far-reaching consequences of significance to agriculture in general and to the cultivation of grapes and the discipline of plant pathology in particular. The existence of present day quarantine programs, plant certification programs, and available fungicide products is partly the result of losses caused by grapevine powdery and downy mildew in the nineteenth century. Activity against *Erysiphe necator* and *Plasmopara viticola* are often important considerations for agricultural chemical companies when deciding which new products to bring to market.

In addition to fungi and oomycetes, all other major pathogen groups are represented in the historical record of destructive epidemics on grapevines. A lethal malady known as the California vine disease was first noted near Anaheim in 1884 in a region where vines had been cultured on mission grounds for some 300 years with no previous record of serious illness. The mysterious disease was next observed in the Napa Valley in 1887 and later in other locations, and by 1906 it had destroyed approximately 16,000 ha of vineyards. The disease was renamed Pierce's disease in the 1930s after the scientist who had studied it most extensively, but the causal bacterium (*Xylella fastidiosa*) was not identified until the 1970s. In the 1990s, problems with Pierce's disease became more acute with the introduction and spread of a highly efficient vector, the glassy-winged sharpshooter (*Homalodisca vitripennis*), into southern California. Presently, the disease is limited to warmer climates by the cold sensitivity of *X. fastidiosa*. The endemic presence of both the causal bacterium and suitable vectors limit viticulture in the southern United States, Mexico, and Central America.

Production in vineyards subjected to epidemics of mildews and phylloxera eventually recovered following the development of effective management strategies (fungicides and resistant rootstocks, respectively); however, a new, insidious, deforming, and degenerating disease was first noted in Europe in 1841, and often developed among grafted vines as these became widely planted. Known in different countries as *roncet*, *court-noué*, *panachure*, and *Reisigkrankheit*, the disease had a developmental period of 12–15 years following planting, after which production declined rapidly and vines degenerated. Reclaiming soils required approximately 10 years of fallow from vines. By the mid-1930s, the disease was known to be transmitted by grafting and to be soilborne, and it was named, in some places, infectious degeneration. It was not until the late 1950s that transmission by the ectoparasitic nematode *Xiphinema index* was confirmed, representing the first example of virus transmission by a nematode. The causal agent was finally identified as *Grapevine fanleaf virus* (GFLV) in the early 1960s. Replanting vineyards with virus-free planting stock in vector-free soils has increased production by 50–100%.

The level of technology and the degree of education in pest management issues among viticulturists today is considerable when viewed in the context of the historical record. Losses of the scope experienced between 1850 and 1900 in Europe following the establishment of exotic pathogens have not been repeated. However, such potential still exists, as evidenced by current efforts to contain spread of the glassy-winged sharpshooter in California and programs to prevent movement of *X. fastidiosa* outside its current geographic range. Chronic losses resulting from intractable issues have been more typical. Catastrophic losses still occur, but tend to be focal.

Selected References

Campbell, C. 2006. The Botanist and the Vintner. Algonquin Books, Chapel Hill, NC.

Gardner, M. W., and Hewitt, W. B. 1974. Pierce's Disease of the Grapevine: The Anaheim Disease and the California Vine Disease. Departments of Plant Pathology, University of California, Berkeley and Davis.

Hewitt, W. B., Goheen, A. C., Raski, J. D., and Gooding, G. V., Jr. 1962. Studies of virus diseases of the grape in California. Vitis 3:57-83.

Large, E. C. 1940. The Advance of the Fungi. Henry Holt and Co., New York.

Millardet, P. M. A. 1885. Traitement du Mildiou et du Rot. J. Agric. Prat. 2:513-516. Pages 7-11 in: The Discovery of Bordeaux Mixture. F. J. Schneiderhan, trans. (1933) Phytopathological Classics 3. American Phytopathological Society, St. Paul, MN.

Parris, G. K. 1968. A Chronology of Plant Pathology. Johnson and Sons, Starkville, MS.

Schumann, G. 1991. Plant Diseases: Their Biology and Social Impact. American Phytopathological Society, St. Paul, MN.

(Prepared by D. M. Gadoury)

Part I. Diseases Caused by Biotic Factors

Diseases Caused By Fungi and Oomycetes

Anthracnose

Anthracnose, also known as black spot, bird's-eye rot, and bird's-eye spot, is a disease of European origin that is now found in most grape-growing regions of the world. When uncontrolled, anthracnose can reduce fruit quality and yield, and weaken the vine via significant destruction of new shoots and leaves. It was probably introduced into other parts of the world on vines imported from Europe, where it was among the most serious disease problems before the introduction of powdery and downy mildews. Most *Vitis* species are susceptible to anthracnose. It was first detected in the United States in the mid-1800s and quickly became a significant disease in the warm and humid southeastern regions of the country, where weather conditions during much of the growing season are conducive for its development. In humid grape-growing regions of South and Southeast Asia, anthracnose is well established, making some grape cultivars extremely difficult to grow. In the absence of adequate control programs, it can also cause damage on highly susceptible cultivars following early-season rains in otherwise dry climates such as the grape-growing regions of southern Australia.

Once established in vineyards, anthracnose can be very destructive and difficult to manage. However, the overall incidence and severity of anthracnose in European vineyards has now been reduced to a negligible level, owing to the widespread use of fungicides such as copper compounds, dithiocarbamates, and phthalimides to control downy mildew and other diseases. In addition to the regular use of such fungicides, the disease is also limited by the general lack of rainfall throughout much of the growing season in the Mediterranean to arid climates where many of the world's grapes are cultivated. In the humid climate of eastern North America, anthracnose is sometimes problematic on certain highly susceptible interspecific hybrid cultivars, which typically receive only modest fungicide programs to control other diseases. Such outbreaks are most severe in the warm southeastern and lower midwestern portions of the United States, but also occur sporadically in more northerly regions, especially in years with high rainfall.

Symptoms

Anthracnose, which literally means "like coal" in reference to blackening of affected tissues, attacks the aerial, succulent parts of the vine, including young shoots, leaves, petioles, tendrils, and clusters; however, lesions on shoots and berries are most common and distinctive.

The first symptoms on young shoots appear as isolated, small, reddish, circular spots that enlarge to become brownish, sunken lesions with gray centers and dark, round or angular margins (Fig. 17). A slightly raised area may form around the margins of the lesions, whose centers may extend to the pith of the shoots (Fig. 18). Individual lesions may coalesce, causing affected shoots to blight. Infected areas may crack, causing the shoots to become brittle. Affected shoots typically remain stunted in growth. Small, pink masses of spores may appear in the lesions during humid weather or following incubation of suspect tissues under warm, humid conditions. Following severe attacks, the disease causes deterioration of older canes,

Fig. 17. Early development of anthracnose on internodes, petioles, and leaves. (Courtesy M. White)

Fig. 18. Anthracnose lesions extending into the pith on a developing shoot. (Courtesy W. F. Wilcox)

resulting in reduced yield. Anthracnose lesions on shoots may be confused with hail injury; however, unlike hail damage, the edges of the lesions caused by the anthracnose fungus are raised and dark brown. In addition, hail damage usually appears on only one side of the shoot, whereas anthracnose lesions are more generally distributed. Symptoms on petioles and tendrils develop similarly to those on shoots.

On leaves, numerous small (1–5 mm in diameter), circular, brown spots appear, which later turn gray in their centers and develop dark-brown to black margins. Older lesions may have round or angular edges and the necrotic centers usually drop out, creating a "shot-hole" and tattered appearance (Fig. 19). Lesions may cover the entire blade or appear mainly along the veins. In severe attacks, lesions may coalesce and cause distortion of the leaf blade and eventually death of the entire leaf. Young leaves are more susceptible to infection than older leaves. When the veins are affected, especially on young leaves, the lesions prevent normal development and result in malformation or complete drying of the leaves; this is most obvious at the tips of the shoots, giving them a burned appearance.

Infection may also occur on clusters before flowering. The rachis and pedicels develop lesions similar to those on shoots and petioles. If the rachis is girdled, the distal portion of the cluster shrivels and dries. On berries, small, reddish-brown, circular spots develop initially, which then enlarge to an average diameter of approximately 5–7 mm and may become slightly sunken. The centers of the spots turn whitish gray and are surrounded by reddish-brown to black margins, producing an appearance that sometimes resembles a bird's eye (Fig. 20). During prolonged wet weather, a pinkish mass of fungal spores oozes from the fruiting bodies (acervuli) that develop in these lesions. Lesions on berries may extend into the pulp, which induces cracking and facilitates infection by secondary microorganisms, resulting in sour rot and poor juice quality. Sometimes, the severely affected berries shrivel and dry into mummies, retaining the outlines of the original spots.

Causal Organism

Grape anthracnose is caused by the ascomycete *Elsinöe ampelina* (de Bary) Shear (syn. *E. viticola* Raciborski). De Bary reported the disease for the first time in 1874 and named the causal fungus *Sphaceloma ampelinum*, giving the description of its conidial form. Changes in the anamorphic name to *Ramularia ampelophaga* Pass. and *Gloeosporium ampelophagum* (Pass.) Sacc. were proposed 2 and 4 years later, respectively.

The fungus produces acervuli on the exterior of lesions. Acervuli contain short, cylindrical, crowded conidiophores bearing small, ovoid, hyaline conidia (3–7 × 2–4 µm) with mucilaginous walls and one or two refringent spots. Morphological and pathological variability has been found to exist among isolates of the anamorph *G. ampelophagum*. Perithecia containing asci and ascospores may also form on infected canes and berries left on the ground or in the trellis from the previous year. Asci (80–100 × 11–23 µm) are formed in pyriform locules of ascostromata and each ascus contains eight brown to black, four-celled ascospores (29–35 × 4.5–7 µm).

Diseases caused by *Colletotrichum* spp. on many other crops are given the common name "anthracnose." This potentially ambiguous situation has sometimes resulted in grapevine diseases caused by *Colletotrichum* spp. being referenced as "anthracnose" in the literature, especially in Asia, rather than as "ripe rot," the accepted common name of the disease caused by these pathogens. On grapevines in India and the Philippines, for example, the *C. gloeosporioides* state of *Glomerella cingulata* (Stonem.) Spauld. & Schrenk can cause light-brown, irregularly scattered spots with ashy gray centers, mostly on older leaves and weaker canes during the rainy season. In addition to being less virulent than *G. ampelophagum*, *C. gloeosporioides* is also distinguished by the much larger conidia (12–21 × 3.5–6 µm) relative to those of the former fungus, and the distinctly different symptoms that it causes on infected fruit (see Ripe Rot chapter).

Disease Cycle and Epidemiology

Infected canes are the main source of pathogen survival between seasons. Sclerotia or mycelium surviving in cane lesions become active in spring and produce abundant conidia when they are wet with rain or dew for 24 h or more at temperatures above 2°C. These conidia are then splashed by rain (2 mm or more) to new growing tissues, where they germinate and cause primary infection at temperatures from 2 to 32°C if tissues remain wet for a sufficient length of time. Warm weather reduces both the wetting time needed for initial infections to occur and the subsequent incubation period prior to symptom expression. Optimum spore germination occurs at 25–30°C, with a minimum of 3–4 h of leaf wetness required for infection; the incubation period is also shortest within this temperature range (3–4 days under ideally wet conditions). By contrast, after infection at an incubation temperature of 10°C, disease symptoms do not start developing until approximately 14 days. Conidia or ascospores formed on infected berries that overwinter on the vine or vineyard floor also may cause primary infections.

Fig. 19. Foliar anthracnose symptoms, including individual and coalescing lesions and shot-holing. (Courtesy W. F. Wilcox)

Fig. 20. Anthracnose lesions on green berries exhibiting classical "birds-eye" appearance. (Courtesy M. White)

Spores infect new leaves, shoots, tendrils, and young berries, producing lesions upon which acervuli form and produce abundant conidia during periods of humid weather. These conidia serve as secondary inoculum and are responsible for further spread of the disease during the growing season. The mucilaginous conidia are dislodged from the acervuli by raindrops and are dispersed in the splashed droplets. Although most of these travel relatively short distances, conidia have been detected in windborne droplets up to 7 m away from the source. The disease develops intensely as plant growth begins, but in the latter half of summer, spread is often stopped by dry weather, depending on region and year. On the Indian subcontinent, maximum disease development and spread occurs during the monsoon season, when weather conditions become most conducive.

Epidemic development of anthracnose is governed by susceptibility of the grapevine tissue, dispersal of conidia, and prevailing weather conditions. The most important factor in determining the severity of epidemics is the frequency of rainy periods. Long dry intervals between rainy periods enable the plant to escape infection, and once foliage and fruit attain maturity, they are less susceptible to the disease. Berries become highly resistant about 50 days after anthesis, when soluble solids rise above 5–7%.

Management

Primary inoculum comes almost entirely from within the vineyard (infected canes and fallen mummified berries), so sanitation measures to reduce these sources are very important and influential. Diseased canes should be pruned during the dormant season and removed from the vineyard or destroyed. If numerous infected berries remain on the vineyard floor, the spores originating from them can be largely negated by covering the berries with soil through cultivation or, if practical, covering them with mulch.

Grape cultivars vary considerably in their susceptibility to anthracnose, and the disease is a consistent threat or problem primarily on cultivars that show the greatest level of susceptibility, including Thompson Seedless (Sultanina), Cardinal, Delicia, Italia, Citronella, and Queen of the Vineyards. In eastern North America, the interspecific hybrid cultivar Vidal Blanc and several minor seedless table grape varieties (especially Reliance) traditionally have been most affected. Several cold-tolerant hybrid cultivars derived from *V. riparia* (especially Marquette) also have been problematic. Highly susceptible cultivars should be avoided in heavy soils with poor drainage, as such sites are particularly conducive to the development of this disease.

As with most fungal diseases, canopy management also helps in anthracnose control. Shoots likely to trail near the ground should be removed and those remaining properly trained to facilitate air circulation and spray penetration. Because natural dispersal of the pathogen is limited, exclusion should be the first line of defense, and new vineyards (particularly those of highly susceptible cultivars) should be established with vines from nurseries known to be free of anthracnose. Removal of wild grapes in hedgerows or wooded areas immediately adjacent to vineyards may also help in reducing sources of initial infection.

Fungicide applications targeted against anthracnose are recommended where the disease is established in commercial vineyards. Late-dormant-season sprays (shortly before bud break) of DNOC preparations, lime sulfur, ferrous sulfate, or copper-containing products are applied in some viticultural regions to inhibit sclerotial germination and the production of primary conidial masses from the cankers of diseased canes retained after dormant pruning. In the lower midwestern region of the United States, such a "delayed dormant" application of lime sulfur is considered a critical component of anthracnose management programs in vineyards where the disease has become established. However, this material must be used with caution as it is caustic and can cause vine damage if applied after bud break.

Where disease pressure warrants it, foliar applications of fungicides are recommended at 7–14-day intervals from the time that shoots are 5–10 cm long until the berries are half grown, depending on the occurrence of potential infection events initiated by rainfall or sprinkler irrigation. Many of the fungicides used in management programs for other diseases during this period also are effective in controlling anthracnose. Bordeaux mixture and copper oxychloride are the traditional fungicides used to protect against anthracnose, but these have gradually been replaced by organic fungicides that provide good preventive action against *E. ampelina*, including mancozeb, chlorothalonil, dithianon, propineb, ziram, daconil, captan, folpet, dichlofluanid, and fluazinam. Locally systemic fungicides, such as carbendazim, thiophanate methyl, and some triazole compounds used against powdery mildew and other diseases, can control anthracnose more effectively than the contact fungicides. However, following repeated use, the fungus can develop resistance to these materials because of their specificity in mode of action; therefore, applications of such materials should be alternated with multisite, nonsystemic fungicides.

Selected References

Brook, P. J. 1973. Epidemiology of grapevine anthracnose, caused by *Elsinoe ampelina*. N. Z. J. Agric. Res. 16:332-342.

Kumar, S., Thind, T. S., and Mohan, C. 1995. Morphogenic and pathogenic variations in *Gloeosporium ampelophagum*. Indian Phytopath. 48:331-334.

Magarey, R. D., Coffey, B. E., and Emmett, R. W. 1993. Anthracnose of grapevines, a review. Plant Prot. Q. 8:106-110.

Magarey, R. D., Emmett, R. W., Magarey, P. A., and Franz, P. R. 1993. Evaluation of control of grapevine anthracnose caused by *Elsinoe ampelina* by pre-infection fungicides. Australas. Plant Pathol. 22:48-52.

Quimio, T. H., and Quimio, A. J. 1975. Notes on Philippine grape and guava anthracnose. Plant Dis. Rep. 59:221-224.

Shear, C. L. 1929. The life history of *Sphaceloma ampelinum* de Bary. Phytopathology 19:673-679.

Sivanesan, A., and Critchett, C. 1974. *Elsinöe ampelina*. Descriptions of Pathogenic Fungi and Bacteria, no. 439. C.A.B. International Mycological Institute, Kew, England.

Sutton, B. C., and Pollack, F. G. 1973. *Gloeosporium cercocarpi* and *Sphaceloma cercocarpi*. Mycologia 65:1125-1134.

(Prepared by T. S. Thind)

Armillaria Root Rot

Armillaria root rot is a fatal disease of grapevines reported from California, the southeastern United States, Europe, South America, and Australia. The common occurrence of Armillaria root rot on grapevines and orchard trees planted on sites previously inhabited by *Quercus* species (oaks) in California has been noted since the disease was first described in the 1880s, and has sometimes resulted in the common name "oak root rot."

Symptoms

The pathogen infects grapevine roots and the root crown region, killing the cambium and decaying underlying xylem tissue. The resulting symptoms include stunted shoots, dwarfed leaves, wilting, premature defoliation, and shriveling of berries (Fig. 21). The most common course of symptom development is a gradual, multiyear reduction in pruning weight and yield caused by destruction of the root system's vasculature. Less common is a rapid desiccation of leaves and berries followed by vine death, occurring between veraison and harvest on grapevines previously exhibiting moderate symptoms. Grapevines with severely stunted shoots that do not succumb to Armillaria root rot before harvest usually die during the dormant season or soon after budbreak. The disease typically develops in localized, circular sections of the vineyard (disease centers). These disease centers appear to expand radially over time and typically include a combination of dead, symptomatic, and healthy vines.

For diagnostic purposes, subterranean signs of the pathogen are much more indicative of Armillaria root rot than are aerial symptoms of the disease. *Armillaria* spp. form thick, white mats of vegetative fungal tissue (mycelial fans) beneath the bark of infected roots (Fig. 22). Armillaria root rot may be diagnosed in a vineyard by digging approximately 30 cm below the soil line to expose the base of the vine trunk where the main roots originate (the root collar). The bark is then removed with a knife, and the inner surface of the bark pieces and subtending portion of the woody root collar are examined for the presence of mycelial fans. Wood beneath a mycelial fan may be rotted and, thus, very easy to cut through. It is most useful to examine the root collars of dead and severely symptomatic grapevines, as the mycelial fans are closer to the soil surface than on grapevines with only slightly stunted shoots.

Armillaria spp. also form rhizomorphs, which are black, shoestringlike structures containing mycelium (Fig. 23). Rhizomorphs are occasionally found within the root bark covering mycelial fans and also may grow on the surface of infected roots.

Clusters of *Armillaria* basidiocarps (mushrooms) may form at or near the base of infected grapevines during the rainy season. However, the basidiocarps are not a reliable diagnostic feature as they may not form every year and they decompose within 2 weeks of emergence. Furthermore, basidiocarps of *Armillaria* spp. are not distinctive and, to the untrained eye, resemble mushrooms of a number of other fungi.

Causal Organism

Several *Armillaria* species are native to California and to other parts of western North America, but of these, only *A. mellea* (Vahl:Fr.) P. Kumm. kills grapevines. The pathogen, which also is the cause of Armillaria root rot on grape in continental Europe, has a broad host range, infecting over 500 species of woody plants. It produces mushroom fruiting bodies that are light yellowish brown, resulting in their common name of "honey mushrooms" (Fig. 24). Caps are slightly darker in their

Fig. 21. Foliar symptoms of Armillaria root rot. Shoots on the diseased vine in the foreground are severely stunted, without fruit, and in a premature state of senescence. (Courtesy K. Baumgartner)

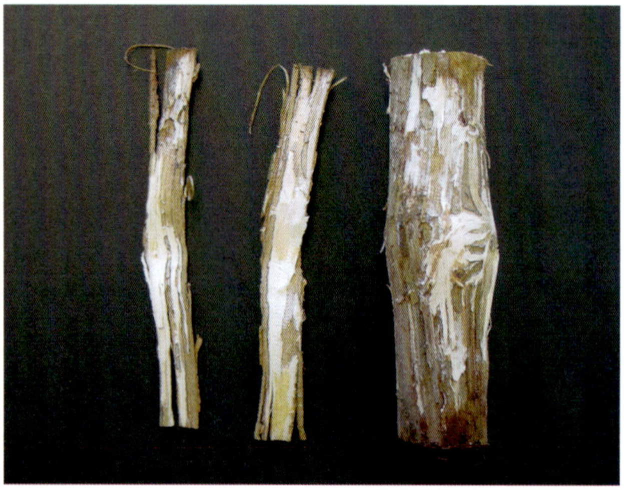

Fig. 22. Mycelial fans of *Armillaria mellea* on the underside of the bark (left and at center) and contacting surface of the wood (right), from a segment of the trunk base (root collar) that was removed below the soil line. (Courtesy K. Baumgartner)

Fig. 23. Shoestring-like rhizomorph of *Armillaria mellea* (arrows) beneath the bark of an infected grapevine. (Courtesy K. Baumgartner)

Fig. 24. A cluster of *Armillaria mellea* basidiocarps arising from the roots and root crown of an infected vine. (Courtesy K. Baumgartner)

centers, ranging broadly in size from approximately 3 to 28 cm diameter. Scales are sometimes present on the cap. Gills are pale yellow to pink. Stipes are pale yellow to pink at the top and light yellowish brown at the base, usually have a fleshy, pale-yellow to pink annulus, and range from 5 to 20 cm long and 0.5 to 3 cm thick. Clusters of *A. mellea* may contain up to 50 mushrooms, although this number can be highly variable. The hyaline, cylindric basidiospores are white in mass and are sometimes deposited by the tallest mushrooms onto the caps of shorter ones beneath them.

Reports of Armillaria root rot on grapevines outside western North America and Europe include symptom descriptions identical to those caused by *A. mellea*, although different *Armillaria* species sometimes are identified as causal agents. In the southeastern United States and the U.K., grapevine infections have been associated with *A. tabescens* (Scop.) Emel. In Australia, the disease reportedly is caused by *A. luteobubalina* Watling & Kile and, possibly, *A. novae-zealandiae* (G. Stev.) Herink. Based on current knowledge of *Armillaria* species in the Southern Hemisphere, *A. novae-zealandiae* appears to be the likely causal species in South America. However, the present unresolved taxonomy of Southern Hemisphere *Armillaria* species complicates interpretation of disease reports from these regions.

The basidiocarps of the aforementioned species are very similar in size, shape, and color to those of *A. mellea*, with the exception of *A. tabescens*, which has no annulus. Because differences in morphological features are relatively minor among individual *Armillaria* species (e.g., basidiospore length), identification to the species level is most accurate when the DNA sequences of field collections are compared with those of known species. However, regardless of the particular *Armillaria* species involved, disease symptoms, development, epidemiology, and control are the same.

Armillaria spp. are cultured easily from mycelial fans, decayed wood, surface-sterilized rhizomorphs, or the stipes of mushrooms by the use of water agar amended with a benzimidazole fungicide and streptomycin. *Armillaria* mycelium cultured from these sources is distinguished from other basidiomycete fungi by the absence of clamp connections. In culture, *Armillaria* spp. do not produce spores or any other reproductive structures.

Disease Cycle and Epidemiology

Armillaria species differ slightly in their host range and fruiting-body morphology, but they share the same life cycle and infection cycle. The pathogen's vegetative mycelium can survive for several years in the partially decayed, woody roots of a previously infected host that have remained in the soil after land clearing (residual roots). Following vineyard establishment, infection can occur when any woody part of a grapevine's root system comes into direct contact with *Armillaria* spp. mycelium from residual roots. Once infected, woody grapevine roots (whether still living or dead) serve as an additional source of inoculum for replants and, to a lesser extent, for secondary spread to neighboring vines. Disease centers appear to expand radially over time, but the capacity for subterranean expansion of *A. mellea* via rhizomorphs is extremely limited because of their infrequent occurrence and limited length. Instead, the apparent expansion of disease centers is due to a succession of primary infections. Vines planted within zones containing the highest inoculum densities become infected first, followed by adjacent vines in soil volumes with lower inoculum densities. Although *Armillaria* spp. mushrooms release windblown spores, they are not a significant source of disease spread among horticultural crops hosts.

Once *Armillaria* spp. encounter a root, they secrete lytic enzymes that dissolve the root bark. Beneath the bark, *Armillaria* spp. colonize and kill a small section of vascular cambium, then form a mycelial fan. The mycelial fan expands beneath the root bark, colonizing and killing more cambial tissue. After the cambial tissue is destroyed, the fungi decay the underlying woody root tissue by decomposing cellulose and other cell wall components. Grapevines eventually succumb to the disease after the rot expands to girdle approximately 50% of the root collar and the remaining functional vascular tissue can no longer accommodate the demand for water from the shoots. Foliar symptoms soon follow. The presence of a mycelial fan at the root collar signifies that an *Armillaria* sp. has destroyed some of the underlying cambium and is in the process of girdling the host's trunk.

Management

Once established in a planting, Armillaria root rot is relatively unaffected by conventional pesticide applications, including postplant soil fumigation and trunk injections with fungicides. Because of the pathogens' broad host range, replanting infected sites with another fruit crop is unlikely to eliminate an Armillaria root rot problem completely. Therefore, the best controls are preventive measures. Rhizomorph growth and growth of mycelium through living roots were once thought to be the two main modes of infection and disease spread; however, such assumptions originated from studies of Armillaria root rot in forest stands, and newer research has shown them to be inaccurate in vineyards. Vineyard disease incidence is best minimized by removing residual roots before planting, not by preventing postplant dispersal of rhizomorphs. When clearing an orchard, vineyard, or wooded site prior to planting a new vineyard, foliar symptoms (most obvious just before harvest) and disease centers should be looked for, and the root collars of dead and dying plants should be examined for mycelial fans. Residual roots should be removed after clearing the site, with particular attention paid to any identified disease centers. General removal of residual roots is recommended, especially for sites cleared of forest trees, as mycelial fans on the root collars of forest versus fruit trees are more difficult to locate and more likely to escape detection. Deep-ripping the soil in several directions brings large residual roots to the surface where they should be collected and removed from the site. Although it is not possible to remove all residual roots, simply moving them closer to the soil surface, where conditions are hotter and drier, lowers the probability of *Armillaria* spp. survival. In some regions, soil solarization may achieve high soil temperatures quickly, but its beneficial effects are restricted to small-diameter roots at a very shallow soil depth.

Preplant soil fumigation with materials such as methyl bromide and sodium tetrathiocarbonate may improve control of Armillaria root rot if the soil is properly prepared beforehand. Preplant soil fumigation is effective when adequate fumigant reaches the majority of residual roots, but the soil depth to which the fumigant penetrates is significantly diminished by soil moisture and various edaphic properties such as low porosity and high organic matter content. In desert and Mediterranean climates, depth of penetration is improved significantly by fumigating at the driest time of the season and by tarping after application of the material. Soil moisture should be depleted further by planting a vigorous cover crop such as *Sorghum vulgare* Pers. (sudangrass) several months before the dry season begins. Injecting sodium tetrathiocarbonate into the soil is more effective than surface application, but neither approach is as effective as applying methyl bromide followed by tarping.

The success of a curative treatment depends on the degree of damage already present in addition to the treatment's ability to decrease further colonization of healthy root tissue by *Armillaria* spp. Therefore, curative treatments are recommended only for slightly to moderately symptomatic vines. One such treatment is root collar excavation, involving permanent removal of soil from the base of the vine in an effort to protect this critical portion of the plant from further pathogen colonization (Fig. 25). This causes the mycelial fan at the root collar to recede, thereby preventing further girdling of the trunk, and

Fig. 25. Root collar excavation for control of Armillaria root rot. (Courtesy K. Baumgartner)

has improved yields on moderately symptomatic vines relative to their untreated counterparts.

Severity of Armillaria root rot is exacerbated by wet soil conditions. Therefore, in vineyards where drip irrigation is practiced, dripline emitters should be moved away from the base of the trunk as soon as possible following vineyard establishment. In Mediterranean climates, a relationship has been documented between the imposition of frequent, shallow irrigations during the dry season and development of Armillaria root rot on landscape and orchard trees. The mechanism underlying this relationship is not known, but possibilities include a decreased defense response of root crown tissues in the waterlogged portion of the soil and/or provision of a more favorable environment for mycelial growth in this frequently wet region.

Grape rootstocks exhibit a range of tolerance to infection by *Armillaria* spp. infection, and it is likely that resistant rootstocks will be recommended for control of Armillaria root rot in the future. More research is needed to identify sources of resistance.

Selected References

Baumgartner, K., Coetzee, M. P. A., and Hoffmeister, D. 2011. Secrets of the subterranean pathosystem of *Armillaria*. Mol. Plant Pathol. 12:515-534.

Baumgartner, K. 2004. Root collar excavation for postinfection control of Armillaria root disease of grapevine. Plant Dis. 88:1235-1240.

Baumgartner, K., and Rizzo, D. M. 2002. Spread of Armillaria root disease in a California vineyard. Am. J. Enol. Vitic. 53:197-203.

Baumgartner, K., and Rizzo, D. M. 2006. Relative resistance of grapevine rootstocks to Armillaria root disease. Am. J. Enol. Vitic. 57:408-414.

Baumgartner, K., and Warnock, A. E. 2006. A soil inoculant inhibits *Armillaria mellea* in vitro and improves productivity of grapevines with root disease. Plant Dis. 90:439-444.

Raabe, R. D. 1962. Host list of the root rot fungus *Armillaria mellea*. Hilgardia 33:25-88.

(Prepared by K. Baumgartner and D. M. Rizzo)

Miscellaneous Berry Rots and Raisin Molds

Berry rots and bunch rots are common viticultural problems the world over. Many rot organisms infect individual berries and seldom spread to others in the cluster, whereas other rot organisms typically spread to many or most of the berries in the cluster, causing bunch rots. Some organisms cause storage and transit rots that, although initiated primarily in the vineyard, develop after harvest.

Annual losses to berry rots are typically 3–5% but in unfavorable seasons may range from 15 to 80%. With few exceptions, rotting berries are not suitable for making wine or other products such as vinegar because they impart odd and distasteful flavors.

Various species of fungi, including yeasts, and some bacteria contribute to the rotting of berries. Some 70 species of fungi in 30 genera and a few bacteria, including acetic acid-producing species (see Sour Rot), have been found on berries, and many of these organisms are directly or indirectly involved in rot. Some infect green berries, but infections do not necessarily advance into rot. Others produce unsightly spots or scars on green berries, and still others become latent and later advance into rot as berries mature. However, many of these 70 species of fungi seem to be transient, occurring in abundance at specific times but not necessarily building up and continuing to spread.

In general, the fruit-rotting organisms may be divided into two major groups: those that infect berries directly, or "primary invaders"; and those that enter berries through wounds or follow a primary invader, known as "wound and secondary invaders." Wounds commonly found on berries are cracks in the skin caused by internal pressure associated with rain or by diseases such as powdery mildew and black measles. Hail, bird pecks, and insect feeding also cause wounds that permit entry of secondary invaders.

Evidence of rot problems in vineyards includes the odor of vinegar, juice dripping over berries of a cluster, the presence of adults and larvae of the vinegar fly (*Drosophila melanogaster* Meigen), soft spots in the skin of berries ("slip-skin"), and fungal fructifications on the surface of berries.

Some specific rot diseases are described in the following subsections under the genus name of the causal organism, which is commonly used to name a specific rot.

Primary invaders. Alternaria rot caused by *Alternaria alternata* (Fr.) Keissl. (syn. *A. tenuis* Nees) often occurs near the cap stem. The rot area is tan initially and becomes brown with age. Under humid conditions, fluffy gray tufts of mycelium and conidiophores bearing conidia develop through cracks in the brown skin over rot lesions. Infections may occur through the skin in a small drop of water or at 98–100% relative humidity. Rot at the cap stem may also develop from an infected pedicel. Infections of pedicels occur during most seasons. The fungus has been cultured from pedicels from the time of blossom shatter to berry maturity.

Cladosporium rot caused by *Cladosporium herbarum* (Pers.:Fr.) Link is characterized by a well-defined, black, soft, circular area ranging from 5–7 mm in diameter up to as much as two-thirds of a berry. At room temperature in a humid atmosphere, the surface of the rot area becomes a velvet olivaceous color due to the presence of conidiophores and conidia. Cladosporium rot is typically a storage disease that develops on fruit harvested late in the season after a rain, although it may also develop in the field under prolonged wet or humid conditions before harvest. Infection takes place through the skin either in the vineyard or in storage and may occur at temperatures from 4 to 30°C (optimum 24–26°C).

Other primary invaders include *Botrytis cinerea* Pers.; *Colletotrichum gloeosporioides* and *C. acutatum* J. H. Simmonds; *Diplodia natalensis* Pole Evans; *Elsinoë ampelina* (de Bary) Shear; *Guignardia bidwellii* (Ellis) Viala & Ravaz; *Greeneria uvicola* (Berk. & Curt.) Punithalingam; and *Phomopsis viticola* (Sacc.) Sacc. They are covered elsewhere in this compendium.

Wound and secondary invaders. Aspergillus rot caused by *Aspergillus niger* van Tiegh. is common in warm to hot climates. This rot is usually associated with a wound and is at first tan to brown, but the area is soon covered with a dusty mass of brown or black spores. These rot areas are initially soft but become firm and leathery. Under warm conditions (26–32°C) and in the presence of free water, the fungus may infect mature fruit directly through the skin.

Blue mold or Penicillium rot caused by *Penicillium* spp. is common in wounded berries (Fig. 26). The fungus produces large, dusty masses of blue spores.

Rhizopus rot is a wet, juicy rot caused by *Rhizopus arrhizus* Tesher or *R. stolonifer* (Ehrenb.:Fr.) Lind in warm, humid climates. Rotted areas on berries are soft and brown, drip juice, and under humid conditions may be covered with a cobwebby mycelium. Numerous sporangiophores with small, spherical, dark sporangia emerge through cracks in the skin of diseased berries or through the borders of a wound. In humid weather the fungus may spread to other berries in a cluster, causing a bunch rot (Fig. 27). The disease is often abundant in vineyards near plum and peach trees with decaying fruit on the ground, and also near fields of sugar beets.

Fruit rots caused by *Ascochyta* sp., *Fusarium moniliforme* Sheldon, *Hormiscium* sp., *Stemphylium botryosum* Walker, and *Torula* sp. are indiscrete rots or include a mix of organisms. Rot caused by *Helminthosporium* sp. alone appears much like that caused by *Cladosporium herbarum*.

Alternaria geophila Deszew., *Aspergillus niger*, *R. arrhizus*, and *R. stolonifer* can become primary invaders when provided free water or grape juice in the temperature range of 18–30°C. Water on the skin absorbs sugars and amino acids from the berry and provides a good food base for the pathogens.

Other wound and secondary invaders include fungi such as *Aspergillus aculeatus* Iizuka, *A. flavus* Link, *A. ochraceous* Wilhelm, *A. wentii* Wehmer, *Botryosphaeria dothidea* (G. & P.) Arx & Müller, *Candida* sp., *Chaetomium elatum* Kunze, *Cladosporium cladosporioides* (Fres.) de Vries, *C. oxysporium* Berk. & Curt., *Monilia* sp., *Penicillium brevicompactum* Dierck., *P. cyclopium* West., *P. frequentaus* West., *P. stoloniferum* Thom, *Saccharomyces cerevisiae* Kreger-van Rij, and *Sclerotinia sclerotiorum* (Lib.) de Bary.

Sour bunch rots, which can vary significantly among geographical zones according to climate, are discussed in detail elsewhere in this compendium.

Disease cycle and epidemiology. Many fruit-rotting organisms are fungi that produce large numbers of spores that are windblown, carried in dust, spread in rain, or transported on the feet and mouthparts of insects. These fungi occur in vineyards on mummified berries, dead vine parts, the bark of canes and spurs, vineyard debris, and other decaying vegetation.

Species of *Alternaria*, *Cladosporium*, and *Stemphylium* grow in bark of 1-year-old canes and sporulate there in the spring. These three fungi, like *Botrytis* and *Helminthosporium*, grow under humid conditions and sporulate on castoff flower parts that remain in the cluster.

Berry rots and bunch rots are associated with wetness. They are favored by high relative humidity, rain, and sprinkler irrigation during the time berries are maturing and approaching harvest. In general, the degree of rot development is proportional to the frequency and duration of wetting events. Rots become increasingly abundant each year in a sequence of disease-conducive seasons but may stop abruptly during one dry season. Some cultivars are more susceptible to rot than others.

Control. Several viticultural practices may aid in reducing rot. Prune to adjust the crop and prevent clumping of clusters. Thin if necessary to loosen clusters. Clean the vineyard and surrounding area of debris, prunings, and unnecessary vegetation. Adjust time and duration of sprinkler irrigation so that berries are not kept wet more than 18 h, and do not use sprinklers after fruit reach 15°Brix. Keep the vine canopy open so that fruit is well aerated. Control biological agents that injure fruit, such as the powdery mildew fungus, birds, and insects.

Raisin Molds and Rots

Molds and rots of raisins occur periodically in association with wet weather. The amount of mold and rot varies with the vineyard, the time of harvest, the district, and the season. Losses range from 0 to 2% in a normal season and may be as high as 70% in a rainy season. These diseases reduce production, increase the cost of making and processing raisins, and reduce quality. The amount of preharvest berry rot in a vineyard grown for raisins is an indication of the potential for mold and rot of raisins while drying.

Symptoms. Molds are mostly contaminants. Mold fungi grow on the surface of the raisin, where they produce conidial spore masses. They can be cleaned (brushed or washed) from the surface. Rots are caused by fungi that colonize the inside of the

Fig. 26. Bluish sporulation of *Penicillium* spp., the cause of blue mold. (Courtesy R. C. Pearson)

Fig. 27. Symptoms and signs of Rhizopus rot. (Courtesy R. C. Pearson)

berry before harvest or during drying. Fungi growing inside the raisin form mats of mycelium and sporulation tufts on the surface. The internal mycelial mats can be observed after infected raisins are soaked in a relatively large volume of warm water or are boiled in water until clear. Raisins with rot are a total loss because the rot cannot be removed by washing.

Causal organisms. The same fungi found normally in vineyards are also found on berries when harvested and laid to dry on trays in the vineyard or on racks or drying pads. Species of *Alternaria*, *Aspergillus*, *Botrytis*, *Chaetomium*, *Cladosporium*, *Helminthosporiurn*, *Hormiscium*, *Hormodendrum*, *Penicillium*, *Rhizopus*, and *Stemphylium* are common raisin mold and rot fungi.

Disease cycle and epidemiology. At harvest, berries have on their skin a representative sample of spores of fungi that occur naturally in the vineyard. When berries are moistened by dew or rain, sugars and amino acids of the berries move into surface water. Conidia germinate and grow in this surface medium. New colonies sporulate after a short period of time, and a new crop of conidia may be blown by wind or splashed by raindrops to spread the organisms.

Berries that have rot at the time of harvest and are laid on trays to dry also support fungal sporulation. Conidia on these berries are also spread by wind or rain to other berries. Furthermore, when water remains on the surface of berries or on drying trays in contact with berries for 24–36 h or more, some fungi, such as species of *Alternaria*, *Aspergillus*, *Cladosporium*, and *Rhizopus*, penetrate the skin and colonize the interiors of the berries or raisins and cause rot. These fungi may also grow on the paper trays, causing the raisins to stick to the paper. Rains or showers that keep the fruit wet for periods of 24–30 h or longer at 15–23°C produce ideal conditions for the development of molds and rots.

Control. Two approaches may be taken to control raisin molds and rots: prevention and chemical treatments. Although molds and rots on raisins generally occur after harvest while the fruit are on drying trays in the vineyard, preventive control begins early in the season and continues through drying and curing. Chemical control is a late choice and often a futile effort.

To prevent preharvest rot in the vineyard, suggested controls for berry and bunch rots should be followed. The crop should be adjusted to obtain high levels of soluble solids early in the harvest and drying season to maximize quality and permit flexibility in choice of harvesting time. Harvest dates should be chosen in accordance with the best weather forecast for satisfactory drying. Debris can be reduced by cultivating between vine rows. A good tray base should be prepared by forming soil to a firm slope faced to receive maximum sun exposure and good drainage in case of rain. Care should be taken during harvest to minimize damage to fruit—that is, by *cutting* clusters from the vine and *placing* them in only one layer on the drying tray. The fruit should be turned over when they are partially dry to shorten the drying time.

The effectiveness of a fungicide to prevent mold and rot of raisins depends on the amount of inoculum, the amount and degree of damage to berries at harvest, the fungicide used, the timing and thoroughness of fungicide application, and the duration of the wetting period. Fungicides have given reasonable protection when applied to the fruit before harvest or to fruit on drying trays soon after harvest but before rain.

Selected References

Barbe, G. D., and Hewitt, W. B. 1965. The principal fungus in the summer bunch rot of grapes. Phytopathology 55:815-816.

Bisiach, M., Minervini, G., and Salomone, M. C. 1982. Recherches expérimentales sur la pourriture acide de la grappe et sur ses rapports avec la pourriture grise. Bull. OEPP/EPPO Bull. 12:15-27.

Delp, C. J., Hewitt, W. B., and Nelson, K. E. 1951. Cladosporium rot of grapes in storage. (Abstr.) Phytopathology 41:937-938.

Harvey, J. M., and Pentzer, W. T. 1960. Market diseases of grapes and other small fruits. U. S. Dep. Agric. Handb. 189.

Hewitt, W. B. 1974. Rots and bunch rots of grapes. Calif. Agric. Exp. Stn. Bull. 868.

Martini, L. P. 1966. The mold complex of Napa Valley grapes. Am. J. Enol. Vitic. 17:87-94.

Nelson, K. E., and Ough, C. S. 1966. Chemical and sensory effects of microorganisms on grape musts and wine. Am. J. Enol. Vitic. 17:38-47.

(Prepared by W. B. Hewitt)

Bitter Rot

Bitter rot is a common disease in the southern two-thirds of the eastern United States, although it sometimes causes problems as far north as Long Island and the New England states. It appears to be relatively widespread internationally in regions subject to warm rains during the fruiting season, and has been reported from Australia, Brazil, China, Costa Rica, Greece, India, Japan, New Zealand, South Africa, and Taiwan. As rotten fruit begin to soften they have a distinct bitter taste, which is carried through the winemaking process to give the finished product an unpleasant, burnt or bitter flavor.

Symptoms

The bitter rot pathogen infects all aboveground vegetative parts of the vine. Leaf infections are more common on muscadine grapes (*Muscadinia rotundifolia*) than bunch grapes and appear as tiny, sunken, reddish-brown flecks with yellow halos. Lesions on stems and petioles are round to elliptical, reddish brown to black, and may be slightly raised or sunken. The fungus has been reported to cause girdling of the shoots of several cultivars of *V. vinifera* in Greece. Flecking of the sepals and blighting of the flower buds can also occur (Fig. 28). Typically, the bitter rot fungus first invades the berry from the pedicel (berry stem) after veraison. Diseased regions of light-skinned berries turn brown and form multiple, prominent fruiting bodies of the fungus (acervuli) after the berries have reached full size (Fig. 29), causing the epidermises and cuticles to rupture. Acervuli form in concentric rings as the rot expands through infected berries, but appear more uniformly distributed once fruit are completely rotted (Fig. 30). These symptoms are less

Fig. 28. Unopened flowers of muscadine grapes blighted by bitter rot (background leaf lesions are caused by black rot). (Courtesy C. Fisk)

apparent on dark-skinned berries, which appear roughened and sometimes iridescent when acervuli develop (Fig. 31). Eventually, infected grapes soften and begin to shrivel, becoming completely covered with massive numbers of acervuli. Infected berries may abscise, or may dry into mummies and remain firmly attached to the pedicel.

Bitter rot is often confused with ripe rot (caused by *Colletotrichum* spp.), as both are favored by the same weather conditions and symptoms on ripening fruit can be similar. Unlike ripe rot, fruit with bitter rot are characterized by their prominent masses of acervuli and often leave hands black if they are handled when wet or shortly after damp weather has occurred. In contrast, fruit with severe ripe rot infections are often covered with a slimy mass during moist weather or a dried crust of pink to salmon-colored spores, whereas those with bitter rot are not.

Fig. 29. Bitter rot on white-skinned grapes. (Courtesy T. B. Sutton)

Fig. 30. Density and distribution of *Greeneria uvicola* acervuli on grape berries affected with bitter rot. (Courtesy M. A. Ellis)

Causal Organism

Greeneria uvicola (Berk. & M. A. Curtis) Punith. (syn. *Melanconium fuligineum* (Scribn. & Viala) Cav.), the cause of bitter rot, forms acervuli (varying in diameter up to 250 µm) that are separate to confluent and subepidermal, with irregular dehiscence (Fig. 32). Conidiophores (30 × 3 µm) are hyaline, septate, and irregularly branched. Conidiogenous cells are enteroblastic, phialidic, discrete or more frequently integrated, determinate, cylindrical, tapered to the apexes, hyaline, and smooth; the collarette and channel are minute. Conidia (7.5–10 × 3–4 µm) are black, smooth, thin walled, nonseptate, cylindrical, and fusiform to oval, with truncate base and obtuse apex (Fig. 33). A perfect stage of the organism has not been described.

Disease Cycle and Epidemiology

G. uvicola overwinters as a saprophyte on fallen fruit, cold-damaged shoot tips, and necrotic bark of the trunk and cordons. In the spring, gelatinous masses of conidia are exuded

Fig. 31. Bitter rot on dark-skinned grapes. (Courtesy T. B. Sutton)

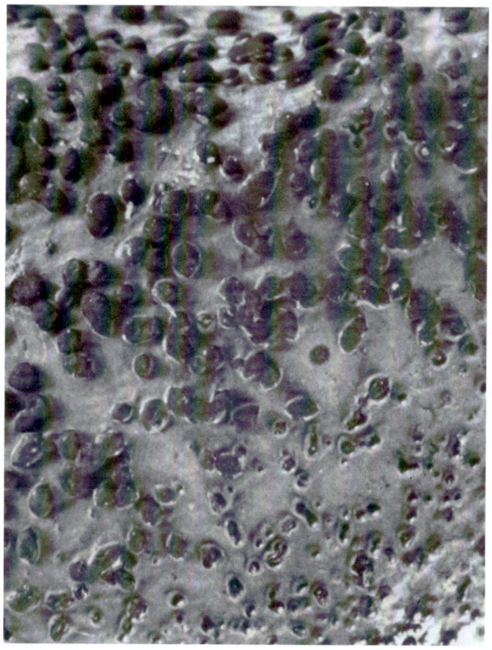

Fig. 32. Acervuli of the bitter rot fungus (*Greeneria uvicola*) fruiting on the surface of a grape berry. (Courtesy J. R. McGrew)

from acervuli and washed by rain to all green parts of the vine, including the pedicels. Infection can occur during wet periods across a temperature range of approximately 12–30°C, although temperatures of 22.4–24.6°C and wetness durations of 6–12 h are optimal. The need for relatively long and warm rains accounts for the predominant occurrence of the disease in regions with warm, humid summers.

The pathogen invades the pedicels and becomes latent until fruit mature. Then, the pathogen resumes growth and moves from the infected pedicels into the ripening fruit, causing them to rot. Secondary infections occur when conidia from infected berries are rain splashed onto other ripening fruit. Berries are susceptible to direct infection from conidia at all times between bloom and harvest but are most susceptible near veraison. Infections often occur in fruit that have been mechanically wounded by birds, insects, or hail, or is cracked owing to heavy rains.

Management

It is important to remove mummied fruit, pedicels, and other dead tissues from the vines while pruning to reduce the inoculum of *G. uvicola* for the next growing season. Training systems that minimize the retention of old wood also appear to help to reduce disease incidence. Fungicide sprays applied every 2 weeks from just after bloom until harvest are necessary to control the disease where it is a common problem, unless rainfall is absent. If late-season or preharvest sprays are omitted, the potential for bitter rot development is greatly increased. Captan, mancozeb, benzimidazole, and QoI fungicides provide good control of bitter rot, whereas copper fungicides appear to be relatively ineffective.

Grape cultivars vary in susceptibility to this disease within and among species. Bronze muscadine cultivars tend to be more susceptible than black ones. The *V. aestavalis* cultivar Cynthiana (Norton) is highly resistant. French-American hybrids are generally more resistant than most cultivars of *V. vinifera*. Within *V. vinifera*, Merlot, Riesling, and Sauvignon Blanc are moderately resistant; Cabernet Sauvignon, Cabernet Franc, and Tempranillo are slightly susceptible; Chardonnay, Sangiovese, Syrah, and Viognier are moderately susceptible; and Mourvédre and Petit Verdot are highly susceptible.

Selected References

Critopolous, P. D. 1961. Girdling of grapevine canes by *Melanconium fuligineum*. Phytopathology 51:524-528.

Kummuang, N., Smith, B. J., Diehl, S. V., and Graves, C. H., Jr. 1996. Muscadine grape berry rot diseases in Mississippi: Disease identification and incidence. Plant Dis. 80:238-243.

Longland, J. M., and Sutton, T. B. 2008. Factors affecting the infection of fruit of *Vitis vinifera* by the bitter rot pathogen *Greeneria uvicola*. Phytopathology 98:580-584.

Luttrell, E. S. 1953. Melanconium leaf and stem fleck of grapes. Phytopathology 43:347-347.

Ridings, W. H., and Clayton, C. N. 1970. *Melanconium fuligineum* and the bitter rot disease of grape. Phytopathology 60:1203-1211.

Sutton, B. C., and Gigson, I. A. S. 1977. *Greeneria uvicola*. Descriptions of Pathogenic Fungi and Bacteria, no. 538. Commonwealth Mycological Institute, Kew, Surry, England.

(Prepared by T. B. Sutton)

Black Foot Disease

First reported in France in 1961, black foot disease occurs in nearly all major viticulture regions throughout the world. It primarily affects young vines up to 8 years of age with incurable and almost invariably fatal infections, although older vines occasionally succumb as well. When the disease develops, it causes substantial economic losses resulting from reduced productivity and replanting costs. In California, black foot disease is most widespread in the North Coast region, perhaps owing to its cooler climate relative to other production regions of the state.

Symptoms

Leaves of infected vines appear chlorotic and scorched as if by water stress, with the entire vine becoming stunted and often dying. In cross section, the base of the trunk appears necrotic and xylem vessels are plugged with black inclusions and tyloses, reminiscent of those associated with Petri Disease (*Phaeomoniella chlamydospora*, *Phaeoacremonium* spp). Roots of symptomatic vines show black, sunken, necrotic lesions (Fig. 34). Often, the root system appears normal, i.e., it

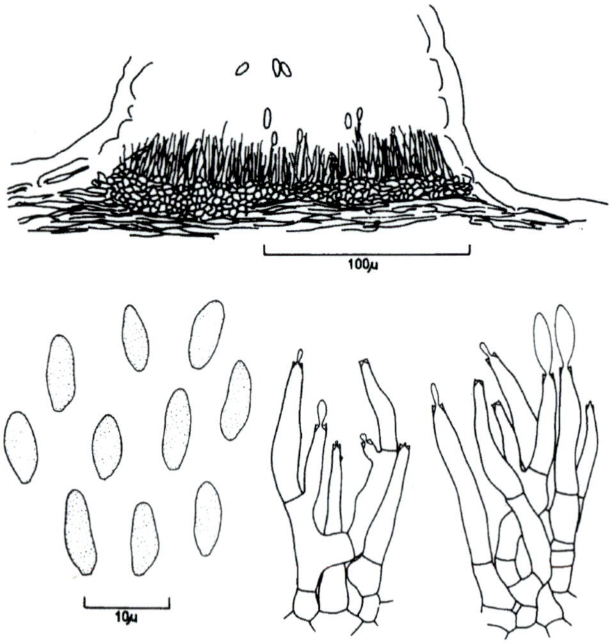

Fig. 33. Acervulus (top), conidia (bottom left), and conidiophores (bottom right and center) of *Greeneria uvicola*. (Reprinted, by permission, from Sutton and Gibson, 1977)

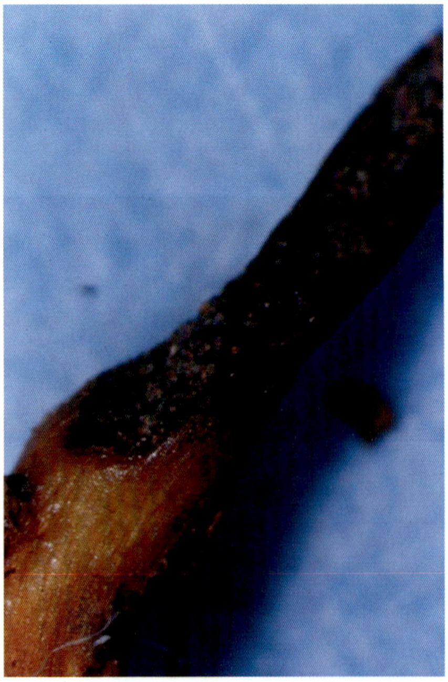

Fig. 34. Sunken root lesion associated with black foot disease. (Courtesy Elsa Petit)

is well spread but frequently shallow, developing parallel to the ground. Sometimes, an adventitious root system develops from the rootstock portion of the trunk close to the soil surface, and this can support the young plant for a while if the primary root system fails. When symptomatic vines are removed from the ground, their root systems sometimes present a so-called "J-root" configuration, typically the result of the bottom portion of the root system being bent upward at the time of planting and a failure of new roots to develop normally from these tissues thereafter (Fig. 35).

Causal Organisms

Campylocarpon fasciculare Schroers, Halleen and Crous; *Campylocarpon pseudofasciculare* Halleen, Schroers and Crous; *Cylindrocarpon obtusisporum* (Cooke & Harness); *Cylindrocarpon destructans* (Zinssm) Sholten; *Cylindrocarpon macrodidymum* Halleen, Schroers and Crous; and *Cylindrocarpon liriodendri* (Cooke & Harness) Wollew. have all been described as causal agents of black foot disease. Since *C. destructans* was first reported in 1961 in France, it has been isolated from diseased vines in Sicily, Portugal, Australia, New Zealand, South Africa, California, and southeastern Canada. *C. macrodidymum* has been reported recently in Australia, New Zealand, South Africa, California, and southeastern Canada. *C. obtusisporum* was detected in Sicily in 1975. *Campylocarpon fasciculare* and *Campylocarpon pseudofasciculare* were recently shown to cause black foot disease in South Africa.

The anamorph *C. destructans* produces microconidia and macroconidia that are mostly triseptate (22–31 × 4.5–6 µm), hyaline, cylindrical with obtuse apices, straight (rarely curved), and often having a distinct, protuberant, flat, basal abscission scar (Fig. 36). Distinctive mycelial chlamydospores, 8–25 µm in diameter, either single or in chains or clusters with thickened, golden-brown walls are profuse in older cultures (Fig. 37). The teleomorph of *C. destructans*, *Neonectria radicicola* (Gerlach & Nilsson) Mantiri & Samuels, is characterized by the presence of red, globose to subglobose perithecia, 0.17–0.35 × 0.15–0.32 mm, with a broadly conical papillum and a scarly to warted wall. These fructifications contain asci, which in turn contain eight smooth and hyaline ellipsoidal ascospores, 8–16 × 2.2–4.4 µm, each with a single septum.

C. macrodidymum produces microconidia and mostly triseptate macroconidia (31–41 × 6–8 µm), predominantly straight and minutely widening toward the tip, with an apical cell slightly bent to one side (Fig. 36). *C. macrodidymum* does not always produce chlamydospores. Its sexual stage, *Neonectria macrodidyma*, is characterized by the presence of ovoid to obpyriform, dark-red, smooth to finely warted perithecia (200–250 µm diameter) without recognizable stroma. These fructifications contain clavate to narrowly clavate asci, each bearing eight ellipsoid, uniseptate, smooth to finely warted, hyaline ascospores measuring 12–18 × 3.5–4.5 µm.

C. liriodendri is morphologically similar to *C. destructans*, distinguishable from the latter primarily on the basis of conidial shape (fusiform and gently curved) and by various molecular characters. The sexual stage of *C. liriodendri* is not known.

Both *Campylocarpon* species produce brownish colonies on potato dextrose agar and brownish hyphal strands that form in the aerial mycelium. *C. fasciculare* produces macroconidia with three or four septa (38–44.5 × 6.5–8 µm or 47–51.5 × 7.5–8.5 µm, respectively) on single or aggregated conidiophores. *C. pseudofasciculare* produces macroconidia with three, four, or five septa (37.5–48.5 × 6.5–7.5 µm, 46.5–53.5 × 7–8.5 µm, or 51–59 × 7.5–8.9 µm, respectively). It also produces sparse chlamydospores (8–11× 7–8.5 µm), whereas *C. fasciculare* does not produce chlamydospores. Neither species forms microconidia, and no *Campylocarpon* species has a known sexual stage.

Disease Cycle and Epidemiology

Cylindrocarpon species are ubiquitous soilborne fungi, causing root rot on a wide range of host plants. Based on what is known of their behavior on other hosts, it is assumed that they persist in a latent state in the soil and attack weakened plants opportunistically under certain poorly defined conditions. For example, black foot disease occurs most frequently in vineyards or sections thereof with compacted, poorly drained soils. Although it has been linked to the planting of infected nursery stock, vineyard soils naturally infested with the pathogen are

Fig. 35. "J-root" configuration on a young vine caused by the root system being bent upwards at the time of planting, showing symptoms of black foot disease that developed subsequently. (Courtesy E. Petit)

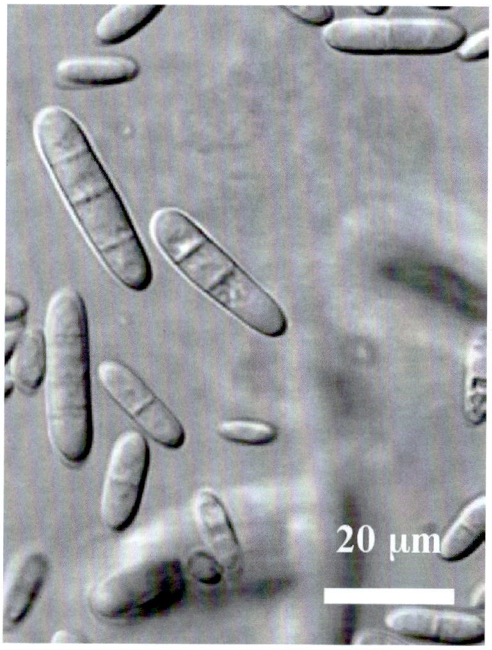

Fig. 36. Conidia of *Cylindrocarpon destructans* at 400× magnification. (Courtesy E. Petit)

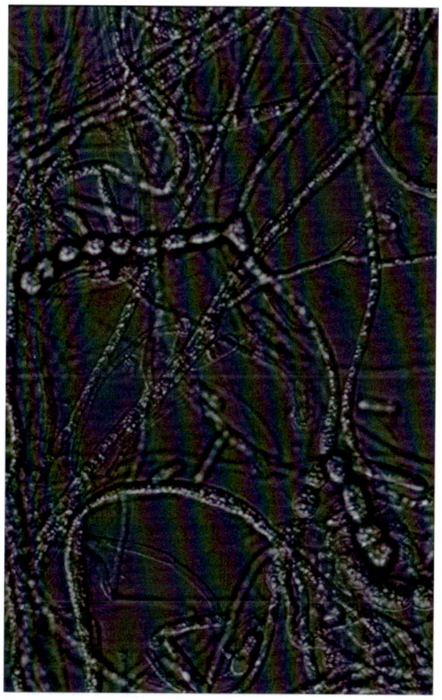

Fig. 37. Chlamydospores of *Cylindrocarpon destructans.* (Courtesy E. Petit)

believed to be a more important source of inoculum. The fungus infects through natural openings or wounds on the roots or through the crown of the rootstock. Over time, it invades lignified tissues of the plant, leading to the development of root lesions and plugging of the xylem by fungal tissue, gums, and tyloses. The attendant reduction in mineral and water uptake results in plant dieback. Diseased plants die very rapidly when young, but as vines age, infection results in a steadier decline and death may not occur for more than 1 year after the initial onset of symptoms. Gelatinous masses of conidia are probably spread through the soil by water. The production of chlamydospores allows the pathogen to survive in the soil for extended periods of time in the absence of suitable host tissue. The disease cycle of *Campylocarpon* spp. on grapevines is not known.

Management

No specific fungicide treatments against black foot disease have been developed; therefore, host resistance, cultural practices, and biological control are the primary tools available to suppress pathogenesis. Uncontaminated nursery material should be planted in soils having adequate drainage. In heavy soils, drainage in the root zone can be facilitated by planting on berms, installing tile drains, and moving drip irrigation emitters away from the vine, as appropriate. All phylloxera-resistant rootstocks show some level of susceptibility to black foot disease; greenhouse inoculations with *C. macrodydimum* and *C. liriodendri* showed 110R to be the most susceptible and 161-49C the least susceptible of those examined. In additional greenhouse studies, colonization of roots by the mycorrhizal fungus *Glomus intraradices* prior to infection by the pathogen provided good disease control.

Selected References

Alaniz, S., León, M., Vicent, A., García-Jiménez, J., Abad-Campos, P., and Armengol, J. 2007. Characterization of *Cylindrocarpon* species associated with black foot disease of grapevine in Spain. Plant Dis. 91:1187-1193.

Booth, C. D. 1966. The genus *Cylindrocarpon*. Mycol. Pap. (CMI) 104:1-56.

Gubler, W. D., Baumgartner, K., Browne, G. T., Eskalen, A., Latham, S. R., Petit, E., and Bayramian, L. A. 2004. Root diseases of grapevines in California and their control. Australas. Plant Pathol. 33:157-165.

Halleen, F., Schroers, H.-J., Groenewald, J. Z., and Crous, P. W. 2004. Novel species of *Cylindrocarpon* (Neonectria) and *Campylocarpon* gen. nov. associated with black foot disease of grapevines (*Vitis* spp.). Stud. Mycol. 50:431-455.

Petit, E., Barriault, E., Baumgartner, K., Wilcox, W. F., and Rolshausen, P. E. 2011. *Cylindrocarpon* species associated with black-foot of grapevine in northeastern United States and southeastern Canada. Am. J. Enol. Vitic. 62:177-183.

Petit, E., and Gubler, W. D. 2005. Characterization of *Cylindrocarpon* species, the cause of black foot disease of grapevine in California. Plant Dis. 89:1051-1059.

Petit, E., and Gubler, W. D. 2006. Influence of *Glomus intraradices* on black foot disease caused by *Cylindrocarpon macrodidymum* on *Vitis rupestris* under controlled conditions. Plant Dis. 90:1481-1484.

Samuels, G. J., and Brayford, D. 1990. Variation in *Nectria-Radicicola* and its anamorph, *Cylindrocarpon destructans*. Mycol. Res. 94:433-442.

(Prepared by E. Petit)

Black Rot

Black rot, caused by the pathogen *Guignardia bidwellii* (Ellis) Viala and Ravaz (anamorph *Phyllosticta ampelicida* [Engleman] Van der Aa) is an economically important disease in most viticultural regions with humid growing seasons. Although favored by warm weather, the disease is nevertheless a serious threat even in regions where relatively moderate temperatures accompany rainfall in the late spring and summer (e.g., the Great Lakes region in North America, and Germany and Switzerland in Europe). Although relatively easy to control with several groups of modern fungicides, black rot can cause virtually complete crop loss if not effectively managed. The disease was one of the original impediments to the introduction of *V. vinifera* into North America, as noted in an 1888 USDA bulletin: "Black rot is the most serious and important disease of the vine in the [eastern] United States ... there is no disease ... that causes in a few days such great losses ... after numerous trials the culture of European vines [*V. vinifera*] has been abandoned ...". To this day, the disease remains one of the major challenges to organic production of susceptible grape cultivars in regions where it occurs regularly. Symptoms of black rot were first recorded in a Kentucky vineyard in 1804 and the first specimens were collected in Alabama in 1855. The pathogen was introduced into Europe (France) in 1885, leading to the first detailed account of the disease by Viala and Ravaz the following year. Although *Guignardia bidwellii* is indigenous to eastern North America, it is also established now in some regions of Europe, South America, and Asia. It is unknown in Chile and is a quarantined pathogen in Australia.

All cultivars of *V. vinifera* appear to be highly susceptible to black rot. Most interspecific hybrid cultivars of primarily *V. labrusca* origin are susceptible to highly susceptible; those primarily of *V. aestavalis* background are moderately resistant; and those of complex origin vary widely in susceptibility, depending on their individual parentage and the criteria applied in selecting them for cultivation.

Symptoms

All new growth is susceptible to infection during the growing season, although the most diagnostic symptoms occur on the leaves and fruit. Leaf lesions occur primarily on the adaxial surface, and are roughly circular to irregularly shaped, 2–10 mm in diameter, and characteristically delineated by a narrow, dark-brown band at the margin (Fig. 38). Lesions are initially cream colored, becoming tan and finally reddish brown as they mature. Pycnidia, which appear as small black pimples, develop within the leaf lesions, often in a ring (Fig. 39). When the

Fig. 38. Leaf lesions caused by *Guignardia bidwellii*. (Courtesy M. Sosnowski)

Fig. 39. Pycnidia of *Guignardia bidwellii* within a leaf lesion. (Photograph by W. McFadden-Smith, Ontario Ministry of Agriculture and Food; Queen's Printer of Ontario, © 2009. Reproduced with permission.)

lesions fall on a main vein, the vein may turn black; blackened veins may extend beyond the border of the lesion.

On berries, the first symptom is a small cream-colored dot, which very quickly becomes surrounded by a zone of chocolate-brown necrotic tissue (Fig. 40). This necrosis continues to expand until the entire berry becomes rotted, whereupon it quickly turns dark and begins to shrivel into a hard, blue-black mummy that remains firmly attached to the pedicel (Fig. 41). Pycnidia first appear on the berry at the point of the initial infection and quickly spread over the entire surface, sometimes in concentric rings.

Lesions can also develop on the petioles, pedicels, shoots, and tendrils. On the petioles and pedicels, they appear as small, darkened depressions which soon turn black. Occasionally, lesions girdle the petiole and kill the entire leaf. Shoot infections appear as elongated black cankers ranging in length from 1 mm to 2 cm (Fig. 42). Shoot and petiole lesions often contain abundant pycnidia (Fig. 43).

Symptoms are different on the fruit of muscadine grapes (*V. rotundifolia*). Lesions are small, black, superficial, and scabby (Fig. 44). They do not cause the decay of the entire berry as occurs on bunch grapes. Multiple infections on the same berry may coalesce and form a brown to black crust on a

Fig. 40. Initial symptoms of black rot on an infected berry. (Courtesy M. Sosnowski)

Fig. 42. Severe shoot and petiole infections caused by *Guignardia bidwellii.* The small group of overwintered mummies attached to the training wire served as the source of inoculum. (Courtesy W. F. Wilcox)

Fig. 41. Various stages of black rot development on berries. (Courtesy W. F. Wilcox)

Fig. 43. Abundant formation of pycnidia in a petiole lesion caused by *Guignardia bidwellii.* (Courtesy M. Sosnowski)

portion of the berry surface, which may ultimately split. Pycnidia form in the surface of these scabby lesions.

Causal organism

G. bidwellii produces pseudothecia in a stroma on overwintered mummies. Pseudothecia are separate, black, and spherical (61–199 µm in diameter), with a flat or papillate ostiole at the apex. The centrum is pseudoparenchymatous and without paraphyses. Asci (36–56 × 12–17 µm) are fasciculate, cylindrical to clavate, short stipitate, and eight spored. The ascus wall is thick and bitunicate. Ascospores (10.6–18.4 × 4.8–8.9 µm) are hyaline, nonseptate, oval or oblong, rounded at the ends, and often present a characteristic footprint shape (Fig. 45). They contain one to four nuclei and have finely granular cytoplasm with two large refractive droplets at either end.

Spermagonia are also produced toward the end of the growing season on mummified berries in association with ascogonial stromata. They are black, spherical (45–78 µm in diameter), erumpent, and ostiolate at the apex. Bacilliform spermatia are hyaline, nonseptate, and measure 2.5 × 1 µm.

Black, erumpent pycnidia are produced from all infected tissues during the growing season. They are spherical (59–196 µm in diameter), solitary, and ostiolate at the apex. Under wet or highly humid conditions, hyaline conidia exude from the neck of the pycnidium in a white, mucilaginous cirrus. Conidia are hyaline, nonseptate, ovoid to oblong, and rounded at the ends, measure 7.1–14.6 × 5.3–9.3 µm in diameter, and are either uni- or binucleate, but occasionally trinucleate.

Three *formae speciales* of *G. bidwellii* were described by Luttrell. *G. bidwellii* f. sp. *euvitis* is pathogenic to the American bunch grape species of the section *Vitis*, and to *V. vinifera*. *G. bidwellii* f. sp. *muscadinii* is pathogenic to *V. rotundifolia* and *V. vinifera*. A third *forma specialis*, G. *bidwellii* f. sp. *parthenocissi*, is pathogenic only to species of *Parthenocissus*, a genus of climbing plants that belongs to the family Vitaceae. Besides differing in pathogenicity, *G. bidwellii muscadinii* also differs from *G. bidwellii euvitis* in appearance, growth rate of colonies in culture, and the size of pseudothecia, ascospores, and conidia.

Disease cycle and epidemiology

The disease cycle for black rot is presented diagrammatically in Fig. 46. *G. bidwellii* overwinters in mummified berries either retained on the vine or fallen to the ground, and in cane lesions. Ascospores released from mummified berries serve as the most common form of primary inoculum in commercial vineyards. They are ejected from pseudothecia after approximately 0.3 mm or more of rain, and may be discharged for up to 8 h after a rainfall. Ascospores from mummies on the ground can be discharged from budbreak through mid summer, although in both laboratory discharge tests and spore trapping studies in the field, the vast majority have been recorded by the end of the bloom period. However, drought during the prebloom and bloom period can delay ascospore depletion until rains resume. In contrast, viable ascospores continue to be produced into late summer from mummies that are retained in the canopy. Once discharged from pseudothecia, ascospores are dispersed by wind currents. Although their density is diluted with distance, trap plants 100 m or more from the nearest inoculum source (e.g., wild grapevines in adjacent woodlands) regularly become infected. Similarly, black rot is becoming an increasing problem in some densely planted regions of Europe where sporadic foci of abandoned vines serve as intense sources of primary inoculum for neighboring cultivated vineyards.

Conidia from mummified berries and from overwintered cane lesions also can function as primary inoculum. Release of conidia from cane lesions peaks near bud break and declines progressively thereafter; release may be complete by bloom or continue through harvest, depending on the vineyard and year. Viable lesions capable of producing conidia can persist in wood for at least 2 years; older lesions have not been investigated. Conidia are splash-dispersed short distances by raindrops, and increased numbers of leaf infections have been recorded beneath canes with overwintered black rot lesions. Conidia from mummies on the ground are probably of relatively minor importance, as (i) their production declines precipitously during the early growing season and is finished by bloom, and (ii) splash dispersal from the vineyard floor to host tissue within the canopy is limited. In contrast, mummies retained within the canopy between seasons pose an extreme risk, as they produce large numbers of conidia from bud break through veraison—which includes the entire period of berry susceptibility—and they are situated immediately adjacent to young berries and other susceptible tissues, facilitating their dispersal to the host.

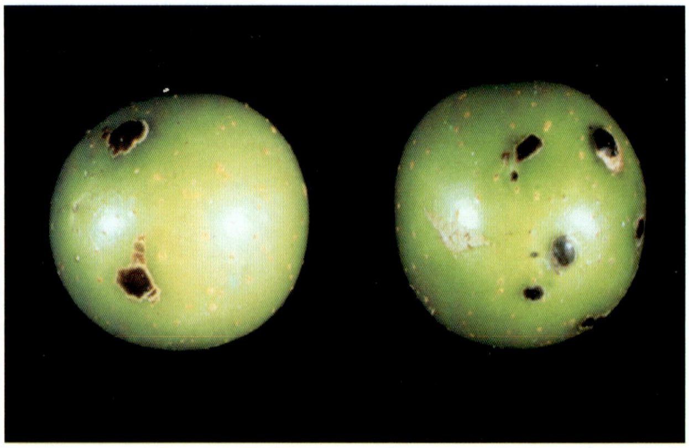

Fig. 44. Scabby lesions of black rot on muscadine berries. (Courtesy P. Bertrand)

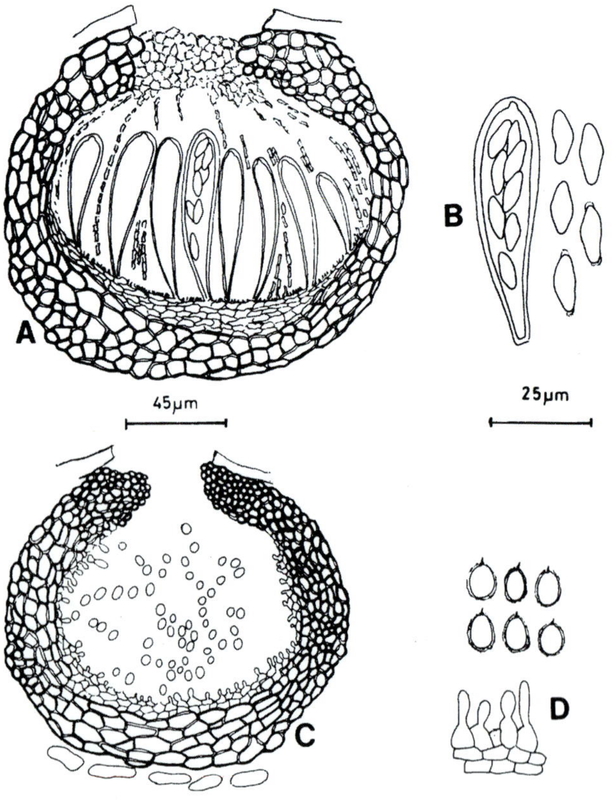

Fig. 45. Fruiting bodies and spores of *Guignardia bidwellii*. **A,** Cross section through a pseudothecium, with asci within a locule. **B,** Ascus and ascospores. **C,** Cross section through a pycnidium. **D,** Conidiogenous cells and conidia. (Modified and reprinted, by permission, from Sivanesan and Holliday, 1981)

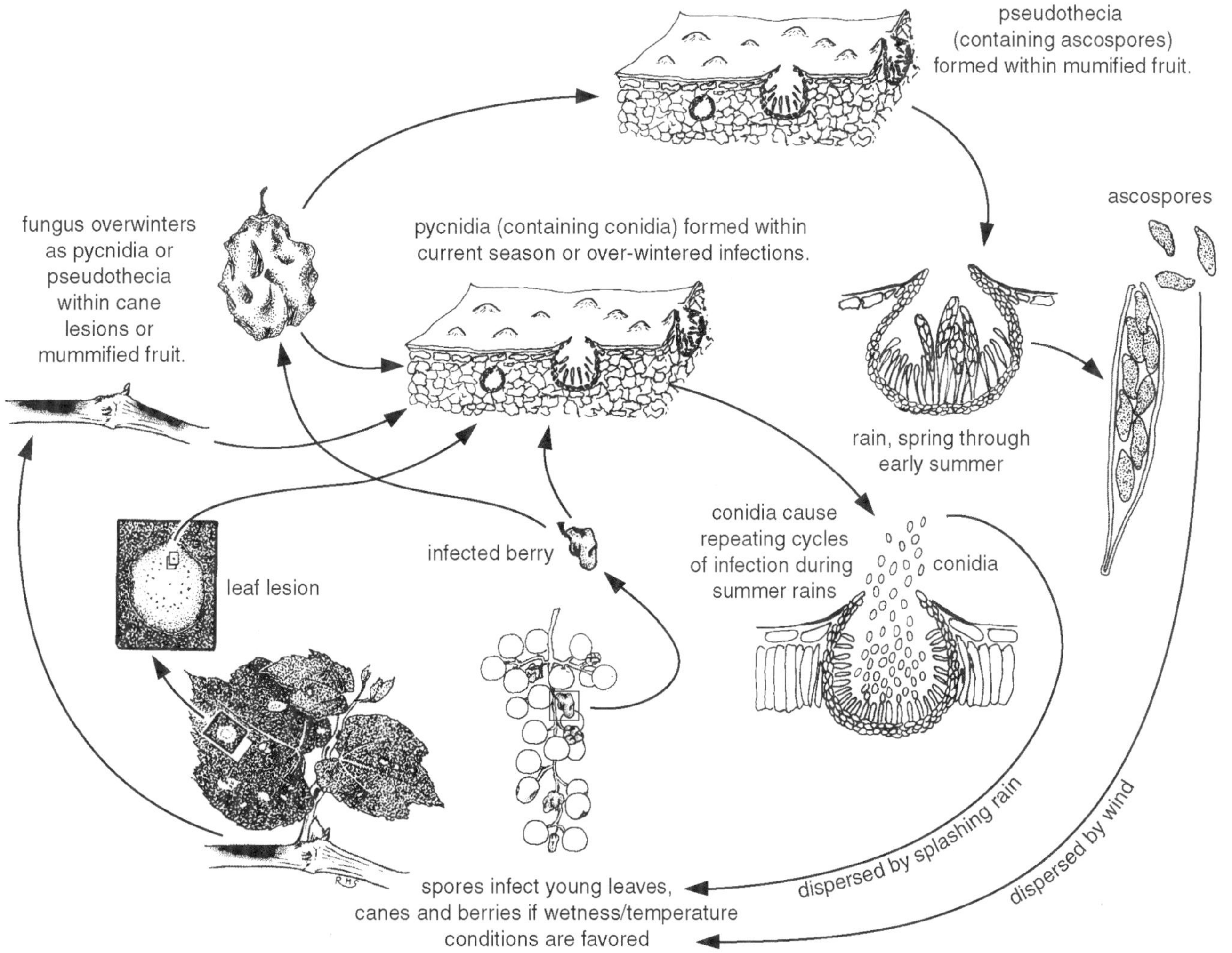

Fig. 46. Disease cycle of black rot. (Drawing by R. Sticht and E. Gotham; reproduced by permission of Cornell University and the New York State IPM Program)

Just a few mummies in the canopy can result in significant disease development in their vicinity (Fig. 42), with "hot spots" of activity often observed beneath mummies that were inadvertently retained during dormant pruning.

Following the release of ascospores or conidia during rain events, continued wetness of leaves, shoots, or berries is required for spore germination and subsequent infection. The approximate wetness duration necessary for infection to occur ranges from 24 h at a temperature of 10°C to only 6 or 7 h at temperatures of 21–26.5°C (Table 2). These data can be used to help determine the need for fungicide sprays in response to putative or anticipated infection events.

Conidia produced from pycnidia that form in new infections can serve as secondary inoculum throughout the season. Because of the copious number of conidia produced and their proximity to other fruit in the cluster, just one or two infected berries can serve as a source of inoculum to spread the disease throughout the entire bunch, particularly if they begin sporulating while the remaining fruit are still young and highly susceptible. In contrast, the epidemiological role and economic significance of leaf infections is not clear. The only detailed study of parallel disease progress on leaves versus fruit suggested that most foliar disease was caused by secondary inoculum originating from recently infected berries, rather than the converse. Nevertheless, if significant foliar infection occurs prior to bloom, the pycnidia produced within these lesions could serve as an important source of secondary inoculum threatening berries after they set. Such a scenario would also suggest a direct danger to berries from high levels of primary inoculum within or near the vine.

TABLE 2. Duration of Continuous Wetness Necessary for Infection by *G. bidwellii* at Different Temperatures (from Spotts, 1977)

(°C)	Hours of Leaf Wetness Required for Infection
7.0	No infection
10.0	24
13.0	12
15.5	9
18.5	8
21.0	7
24.0	7
26.5	6
29.0	9
32.0	12

Early researchers noted that only the young, tender portions of the vine were susceptible to infection, but it was not until recently that this ontogenic, or age-related, resistance in grape leaves was described in detail. Conidia are able to germinate and penetrate grape leaves of all ages, but subsequently are not able to colonize or cause visible symptoms on those that have become fully expanded. On such older leaves, germlings remain quiescent but are able to resume growth saprobically if the tissue is killed.

The phenomenon and practical importance of ontogenic resistance in fruit has also received new study. In the Finger Lakes region of New York, berries of *Vitis* × *labruscana* 'Concord' exhibit a period of maximum susceptibility from bloom until 2–4 weeks later, and do not become diseased when inoculated after 5 weeks postbloom. Berries of *V. vinifera* cultivars exhibit a longer period of susceptibility: maximum levels of susceptibility persist until 3–5 weeks after bloom and a small proportion of the berries remain susceptible until 6–7 weeks postbloom. These latter data agree with reports of black rot development on *V. vinifera* vines in Switzerland. The incubation period also increases markedly as berries age and lose their susceptibility to colonization by the fungus. Whereas symptoms first appear 10–14 days after inoculation of young *V. vinifera* fruit (depending largely on temperature), the minimum incubation period is approximately 21 days on those inoculated 6 weeks after bloom, with some berries in the same cluster not becoming symptomatic until a full month after inoculation. These extended incubation periods should be considered when trying to determine the infection event(s) leading to a disease outbreak in the latter part of the summer.

On muscadine grapes, *G. bidwellii* survives the winter as pseudothecia in senesced leaves and as pycnidia in infected canes. Asci mature in late winter and early spring, and ascospores are discharged for 4–5 weeks in April and May. Both ascospores and conidia can serve as primary inoculum. Immature leaves can be infected throughout the growing season, whereas berries may be infected from set until they reach full size. Conidia produced from primary infections can initiate secondary cycles of infection, as described above.

Management

Management of black rot is most effective with an integration of chemical and cultural components. Sanitation is the first line of defense and can greatly impact the efficacy of a fungicide program, depending on the extent to which it is implemented. When highly effective fungicides are not employed (e.g., organic production systems), rigorous sanitation efforts are essential for sustained control on susceptible cultivars in problematic regions. Retention of mummies in the canopy, which is particularly likely with mechanical pruning systems, provides greatly increased disease pressure relative to the same number of mummies on the ground, owing to the additional number of spores produced per mummy, the temporal shift in spore release to coincide with the period of peak berry susceptibility, and the proximity of these inoculum sources to the new fruit. Thus, removal of mummies to the ground during dormant pruning is a minimal sanitation practice, which typically is adequate in vineyards where the disease is well controlled routinely. However, in vineyards where black rot control is problematic, additional procedures may be warranted. These include cultivating and/or mulching the vineyard floor to physically cover mummies by or soon after bud break, and even physical removal of mummies from the vineyard if practical. Diseased canes should be pruned from the vine and, ideally, removed from the vineyard or destroyed.

Traditionally, fungicide applications were recommended to begin when shoots were 10–15 cm in length and to continue until the berries reached veraison. Recent research has shown the value of concentrating fungicidal protection during the period of maximum berry susceptibility, from bloom until approximately 4–6 weeks later, depending on cultivar and weather. In vineyards where sanitation is good and the level of overwintering inoculum is modest, a program of limited fungicide applications extending from the immediate prebloom period through 4 weeks postbloom has provided excellent control in several different regions where black rot is active. In vineyards where disease historically has been difficult to control and inoculum levels are higher, fungicide applications may need to begin 2 or more weeks prebloom, depending on rainfall and temperature. If more than a trace of disease develops on young clusters, fungicidal protection should continue until fruit age to the point that they are no longer susceptible to infection.

Several fungicide classes provide excellent control of black rot. Among the traditional protectants, the dithiocarbamates (mancozeb, ferbam, and ziram) are very effective, whereas captan is only moderately so. A number of triazole fungicides (including difenoconazole, flusilazole, flutriafol, myclobutanil, tebuconazole, tetraconazole, and triadimefon) are highly effective, providing moderate residual protective activity plus significant postinfection activity, thereby enabling their use in curative programs or as an emergency response following unprotected infection events. In contrast, some other sterol demethylation inhibitor fungicides (e.g., fenarimol and triflumizol) provide only modest control of this disease. The strobilurin fungicides (azoxystrobin, kresoxim-methyl, pyraclostrobin, and trifloxystrobin) also provide excellent control of black rot, primarily in a protectant mode but with modest curative activity as well. Copper provides the best control among materials approved for organic production systems, although it is only moderately effective and must be used at relatively short spray intervals under significant disease pressure. Sulfur is poor. Because black rot spores require hydrophobic surfaces for germination, surfactants that improve the wettability of waxy leaf and berry surfaces may offer some promise as alternative control materials; however, such materials have proven more effective in greenhouse settings than in limited field tests. Lime sulfur applied shortly before bud break should reduce pycnidial inoculum from diseased canes retained during pruning. This treatment may be useful where cane lesions are important sources of primary inoculum, particularly in circumstances where highly effective fungicides may not be available.

Selected References

Becker, C. M., and Pearson, R. C. 1996. Black rot lesions on overwintered canes of *Euvitis* supply conidia of *Guignardia bidwellii* for primary inoculum in spring. Plant Dis. 80:24-27.

Ferrin, D. M., and Ramsdell, D. C. 1977. Ascospore dispersal and infection of grapes by *Guignardia bidwellii*, the causal agent of grape black rot disease. Phytopathology 67:1501-1505.

Ferrin, D. M., and Ramsdell, D. C. 1978. Influence of conidia dispersal and environment on infection of grape by *Guignardia bidwellii*. Phytopathology 68:892-895.

Hoffman, L. E., Wilcox, W. F., Gadoury, D. M., and Seem, R. C. 2002. Influence of grape berry age on susceptibility to *Guignardia bidwellii* and its incubation period length. Phytophathology 92:1068-1076.

Hoffman, L. E., and Wilcox, W. F. 2003. Factors influencing the efficacy of myclobutanil and azoxystrobin for control of grape black rot. Plant Dis. 87:273-281.

Hoffman, L. E., Wilcox, W. F., Gadoury, D. M., Seem, R. C., and Riegel, D. G. 2004. Integrated control of grape black rot: Influence of host phenology, inoculum availability, sanitation, and spray timing. Phytophathology 94:641-650.

Jermini, M., and Gessler, C. 1996. Epidemiology and control of grape black rot in southern Switzerland. Plant Dis. 80:322-325.

Kuo, K., and Hoch, H. C. 1996. Germination of *Phyllosticta ampelicida* pycnidiospores: Prerequisite of adhesion to the substratum and the relationship of substratum wettability. Fungal Genet. Biol. 20:18-29.

Kuo, K., and Hoch, H. C. 1996. The parasitic relationship between *Phyllosticta ampelicida* and *Vitis vinifera*. Mycologia 88:626-634.

Luttrell, E. S. 1946. Black rot of muscadine grapes. Phytopathology 36:905-924.

Luttrell, E. S. 1948. Physiologic specialization in *Guignardia bidwellii*, cause of black rot of Vitis and Parthenocissus species. Phytopathology 38:716-723.

Molitor, D., and Beyer, M. 2014. Epidemiology, identification and disease management of grape black rot and potentially useful metabolites of black rot pathogens for industrial applications—a review. Ann. Appl. Biol. 165:305-317.

Scribner, F. L., and Viala, P. 1888. Black Rot (*Laestadia bidwellii*). Dept. of Agric. Bot. Div. Bull. 7.
Sivanesan, A., and Holliday, P. 1981. *Guignardia bidwellii*. Descriptions of Pathogenic Fungi and Bacteria, no. 710. Commonwealth Mycological Institute, Kew, Surrey, England.
Spotts, R. A. 1977. Effect of leaf wetness duration and temperature on the infectivity of *Guignardia bidwellii* on grape leaves. Phytopathology 76:1378-1381.
Spotts, R. A. 1980. Infection of grape by *Guignardia bidwellii*—factors affecting lesion development, conidia dispersal and conidial populations on leaves. Phytopathology 70:252-255.

(Prepared by W. F. Wilcox and L. E. Hoffman)

Botryosphaeria Dieback

Species in the family Botryosphaeriaceae Theiss. & P. Syd. 1918 have a cosmopolitan distibution and are found as endophytes, parasites, and/or saprophytes in and on a wide range of annual and perennial crops, ornamental plants, and forest trees. Several species are recognized as significant plant pathogens that cause a variety of symptoms in economically important perennial crops, including leaf spots, fruit rots, shoot blight, perennial cankers, and dieback syndromes that eventually culminate in host death. However, the association between species within this family and dieback of grapevines has been overlooked for many years, as such organisms largely have been considered saprophytes, secondary colonizers and/or weak parasites of this host. Recognition of grapevine diseases caused by Botryosphaeriaceae species also has remained problematic owing to misdiagnoses due to the similarity of these symptoms to those of other trunk diseases, including esca, Eutypa dieback, and Phomopsis cane and leaf spot.

The pathogenicity of members of the Botryosphaeriaceae that parasitize grapevine and the epidemiology of their diseases also have been obscured by conflicting reports of virulence and symptoms among cultivars and grape-growing regions. Furthermore, the taxonomy of species within the Botryosphaeriaceae has been subject to numerous revisions, complicating efforts to accurately identify individual isolates or to compare results among different reports, where different names may have been ascribed to common taxa and vice versa. However, research conducted in several countries since the late 1990s has clarified the status of these fungi as important grapevine trunk pathogens worldwide, resulting in advances in the understanding of their epidemiology and control.

Since a Botryosphaeriaceae species was first associated with grapevine dieback in the mid-1960s, considerable controversy and debate have ensued regarding the cause and name of trunk diseases caused by fungi within this family. Until recently, different common names, including Black dead arm, Botryosphaeria canker, and Diplodia cane dieback have been used to refer to the same symptoms (e.g., wedge-shaped cankers, dark streaking, and/or light-brown discoloration of the wood), sometimes caused by a single organism. Furthermore, multiple trunk diseases can occur within individual vineyards, and these names have sometimes been applied erroneously in the diagnosis of symptoms attributable to accompanying diseases (e.g., esca) that are caused by unrelated pathogens, and this has contributed significantly to the confusion regarding their etiology. Recent studies have shown that various Botryosphaeriaceae spp. can be isolated from grapevines exhibiting the same vascular symptoms but diverse foliar symptomatology, either alone or in combination with other trunk disease pathogens. Thus, the use of multiple unique names to denote individual diseases with well-characterized symptoms and specific causal agents does not appear to be appropriate. Consequently, the name Botryosphaeria dieback has been accepted recently by workers in this field to encompass the broad range of symptoms caused by multiple species within the Botryosphaeriaceae that infect grapevine wood.

Botryosphaeriaceae spp. have been isolated from diseased grapevine wood in viticultural regions worldwide. Resultant yield losses of 25–30% have been reported in the susceptible cultivars Cabernet Sauvignon, Cabernet Franc, and Sauvignon Blanc in France. An economic study of the California wine grape industry conducted in the late 1990s estimated losses from canker diseases (shown subsequently to be primarily Eutypa dieback and Botryosphaeria dieback) to be over $260 million per year at that time. Botryosphaeria dieback is currently recognized as one of the most prevalent diseases of grapevines worldwide, limiting both vineyard longevity and productivity. In Mexico and California, Botryosphaeria dieback is considered endemic in the arid and warmest grape-growing regions. Disease incidence up to 100% was reported in vineyards ≥10 years old in the southern California region of Coachella Valley, with several cankers per vine.

Botryosphaeria dieback decreases vineyard longevity, reduces yields, and increases production costs through necessary management programs, involving both prophylactic efforts and the additional pruning and training practices required to replace lost fruiting wood. Although the disease usually affects matures vines (at least 6 years old), it sometimes causes decline and eventual death of young vines as well. Native American and interspecific hybrid cultivars are affected in addition to those of *Vitis vinifera*.

Symptoms

A wide range of vascular symptoms associated with Botryosphaeria dieback have been characterized, but foliar symptomatology remains unclear. It now seems likely that some symptoms previously described as caused by Botryosphaeriaceae spp. (e.g., chlorosis and colored spots on the leaf margins or blade that coalesce to form large necrotic areas) were confused with one of the many manifestations of esca, or that they may have resulted from a complex combination of several fungi, including those causing esca, in a specific stage of vine development or decline. The most widely reported external symptom associated with Botryosphaeria dieback has been the total absence of spring growth from single or multiple spur positions of infected vines (Fig. 47), which has helped to differentiate this disease from the characteristic foliar symptomatology observed on grapevines affected by Eutypa dieback, esca, and/or Phomopsis cane and leaf spot. However, grapevines infected by the fungi that cause Botryosphaeria dieback can also exhibit normal and healthy shoot development so long as sufficient healthy vascular tissue remains present to support it (Fig. 48).

Fig. 47. Absence of spring growth from spurs killed as a result of Botrytosphaeria dieback. (Courtesy J. R. Úrbez-Torres)

Vascular symptoms associated with Botryosphaeria dieback include perennial cankers (Fig. 49); internal wood necrosis (Fig. 50); dark- and/or light-brown streaking of the wood (Fig. 51); and bud and pith necrosis. Additionally, several Botryosphaeriaceae spp. have been isolated from bleached canes, pruning debris, asymptomatic wood, necrotic nursery material (scion and/or rootstock), and from necrotic areas of the graft union. Among all these symptoms, perennial cankers are the most prevalent and widespread, as reflected by the former disease name of Botryosphaeria canker.

The vascular symptoms associated with Botryosphaeria dieback appear to initiate primarily from pruning wounds and to expand basipetally from that point of infection (Fig. 52). Perennial cankers also expand laterally in the spurs, cordons, and trunks for several years until only a small wedge of live tissue remains, culminating in death of the vine portions above the canker once complete girdling occurs. Although typically observed on mature vines, especially where large pruning wounds have been made while retraining them, wedge-shaped cankers caused by Botryosphaeriaceae spp. also have been observed on young vines. Cankers have been shown to develop from cracks and natural openings on the framework of vines as well. Black, ostiolate pycnidia can be found frequently beneath the bark on cankered areas in spurs, cordons and/or the vine trunk (Fig. 53).

Causal Organisms

To date, the following 23 Botryosphaeriaceae spp. have been identified on grapevines affected by Botryosphaeria dieback: *Botryosphaeria dothidea* (Moug. : Fr.), Ces. & De Not.; *Diplodia corticola* A. J. L. Phillips, A. Alves & J. Luque; *Diplodia mutila* (Fr. : Fr.) Fr.; *Diplodia seriata* De Not.; *Dothiorella americana* J. R. Úrbez-Torres, F. Peduto & W. D. Gubler; *Dothiorella iberica* A. J. L. Phillips, J. Luque & A. Alves; *Dothio-*

Fig. 48. Dead wood infected by a Botryosphaeriaceae species as revealed by removing the vine bark, with the remaining live wood supporting healthy shoot growth apical to it. (Courtesy G. M. Leavitt)

Fig. 50. Internal wood necrosis symptom of Botryosphaeria dieback. (Courtesy J. R. Úrbez-Torres)

Fig. 49. Cross-sectional view through an expanding perennial canker associated with Botryosphaeria dieback. (Courtesy J. R. Úrbez-Torres)

Fig. 51. Internal wood streaking symptom of Botryosphaeria dieback. (Courtesy J. R. Úrbez-Torres)

rella neclivorem W. M. Pitt & J. R. Úrbez-Torres; *Dothiorella vidmadera* W. M. Pitt, J. R. Úrbez-Torres & F. P. Trouillas; *Dothiorella vinea-gemmas* W. M. Pitt & J. R. Úrbez-Torres; *Lasiodiplodia crassispora* T. I. Burgess & Barber; *Lasiodiplodia missouriana* J. R. Úrbez-Torres, F. Peduto & W. D. Gubler; *Lasiodiplodia theobromae* (Pat.) Griff. & Maubl.; *Lasiodiplodia viticola* J. R. Úrbez-Torres, F. Peduto & W. D. Gubler; *Neofusicoccum australe* (Slippers, Crous & M. J. Wingf.) Crous, Slippers & A. J. L. Phillips; *Neofusicoccum luteum* (Pennycook & Samuels) Crous, Slippers & A. J. L. Phillips; *Neofusicoccum macroclavatum* T. I. Burgess, Barber & Hardy; *Neofusicoccum mediterraneum* Crous, M. J. Wingf. & A. J. L. Phillips; *Neofusicoccum parvum* (Pennycook & Samuels) Crous, Slippers & A. J. L. Phillips; *Neofusicoccum viticlavatum* (Van Niekerk & Crous) Crous, Slippers & A. J. L. Phillips; *Neofusicoccum vitifusiforme* (Van Niekerk & Crous) Crous, Slippers & A. J. L. Phillips; *Spencermartinsia viticola* (A. J. L. Phillips & J. Luque) A. J. L. Phillips, A. Alves & Crous; *Spencermartinsia westrale* W. M. Pitt, J. R. Úrbez-Torres & F. P. Trouillas; and *Sphaeropsis porosa* (Van Niekerk & Crous) A. J. L. Phillips & A. Alves.

It is only in recent years that species definitions have been clarified and for this reason some earlier reports of "*Botryosphaeria*" spp. on grapevines may not be entirely accurate. This is particularly true for species with *Fusicoccum* anamorphs, where the name of *B. dothidea* (anam. *Fusicoccum aesculi* Corda) was often used in a broad sense that encompassed what are now considered to be several distinct species. On the basis of several pathogenicity studies conducted during the last decade, species of Botryosphaeriaceae infecting grapevines are divided into three virulence rankings: those that are highly virulent (*Lasiodiplodia* spp. and *Neofusicoccum* spp.); moderately virulent (*B. dothidea* and *Diplodia* spp.); and slightly virulent (*Dothiorella* spp. and *S. viticola*).

Among the 23 botryosphaeriaceous taxa listed above, *B. dothidea*, *D. mutila*, *D. seriata*, *L. theobromae*, and *N. parvum* have been isolated most frequently from symptomatic wood. Therefore, descriptions of these five species are provided below.

Ascomata of *B. dothidea* (200–500 µm in diameter) are erumpent, brown to black, open through an ostiole, and form in botryose aggregates of up to 100, although sometimes singly. Asci are bitunicate with a thick endotunica, clavate, and stipitate. Ascospores (17–)19–24(–32) × (6–)7–8(–10) µm are ellipsoid to broadly fusoid, hyaline, smooth, thin walled, and one celled. Pycnidia are morphologically indistinguishable from the ascomata. Conidiophores are hyaline, cylindrical, and smooth, 23–35 × 4–5 µm, and are formed from the cells lining the locule wall. Conidia (17–)19.5–30(–34) × 4–6(7.5) µm are hyaline, thin walled, smooth, and narrowly or irregularly fusiform, with a subtruncate to bluntly rounded base exhibiting a minute marginal frill (Fig. 54). Colonies on potato dextrose agar initially are white, becoming green to dark green after 3 or 4 days with copious aerial mycelium, ultimately becoming dark gray (Fig. 55). Numerous conidiomata can be observed in cultures after 1 week of incubation at room temperature. Optimum colony growth occurs between 25 and 30°C.

Ascomata of *D. mutila* are dark brown, immersed, and uniloculate with the wall composed of dark-brown, thick-walled *textura angularis*. Asci are bitunicate with a thick inner wall, clavate, and stipitate. Ascospores (25–36 × 9.5–13.6 µm) are hyaline and aseptate, sometimes becoming translucent brown and one or biseptate with age (Fig. 56). Pycnidia (130–195 µm in diameter) occur individually or in groups in dead cortical tissue. They are globose, dark brown to black, immersed, thick walled, and with a central ostiole. Conidia (21–)25–28 × (10–)13–14.6 µm are hyaline; one celled and cylindrical with broadly rounded ends, some with a large central guttule; and

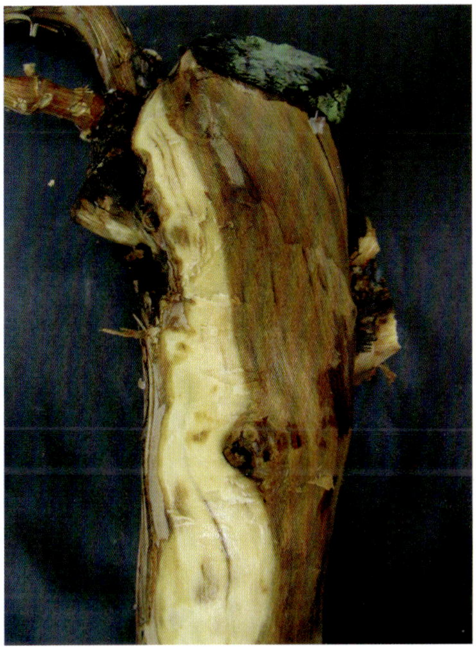

Fig. 52. Vascular symptoms of Botryosphaeria dieback expanding basipetally from one or more pruning wounds, the apparent point(s) of infection. (Courtesy J. R. Úrbez-Torres)

Fig. 53. Pycnidia of a Botryospaheriaceae species beneath the bark of an infected vine. (Courtesy J. R. Úrbez-Torres)

Fig. 54. Conidia of *Botryosphaeria dothidea*. (Courtesy J. R. Úrbez-Torres)

smooth with a thick, glassy wall. Some conidia also become pale, translucent brown, and develop a single septum after discharge (Fig. 57). Colonies on potato dextrose agar are dark green with moderate amounts of aerial mycelium (Fig. 58). Conidiomata, 1–2 mm in diameter, start to form after 1 week

Fig. 55. Colony of *Botryosphaeria dothidea* on potato dextrose agar. (Courtesy J. R. Úrbez-Torres)

Fig. 56. Bitunicate asci of *Diplodia mutila*, containing eight ascospores. (Courtesy A. J. L. Phillips)

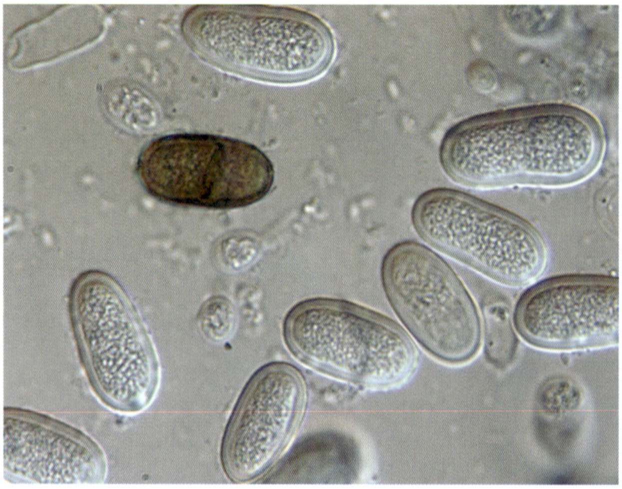

Fig. 57. Conidia of *Diplodia mutila*, representing pre- (hyaline) and post-discharged (brown) morphologies. (Courtesy J. R. Úrbez-Torres)

on water agar at room temperature. Optimum temperature for colony growth is 25°C.

Ascomata of *D. seriata* are dark brown, immersed, and uniloculate with the wall composed of dark-brown and thick-walled *textura angularis*. Asci are bitunicate with a thick inner wall, clavate, and stipitate. Ascospores (23–38 × 7–3 µm) are hyaline and one celled. Pycnidia are dark brown, unilocular with a central ostiole, immersed in young tissue, and appear individually or in groups under exfoliating bark of spurs, cordons, and trunks. Cylindrical, one-celled conidia, (15–)22.5–26 × (8)9.5–14 µm, are hyaline when immature, becoming light brown when mature and before discharge from the pycnidia, with both ends rounded or sometimes truncate at the base (Fig. 59). Colonies on potato dextrose agar are dark gray with moderate aerial mycelium (Fig. 60) and start to form numerous small (0.3–1.2 mm diameter) pycnidia after 5 days at room temperature. The optimum temperature for colony growth is 27°C.

Ascomata of *L. theobromae* (250–399 µm) are dark brown to black and aggregated. Asci (90–120 µm) are bitunicate, clavate, and contain eight ascospores. Ascospores (24–)30–35 × (7)11–14(–17) µm are hyaline and aseptate. It should be noted

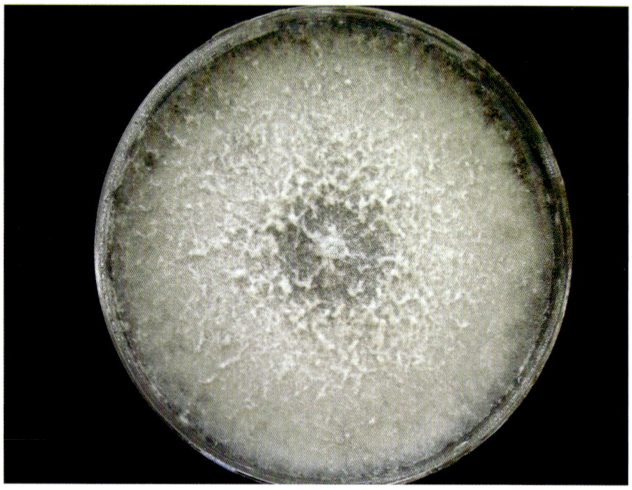

Fig. 58. Colony of *Diplodia mutila* on potato dextrose agar. (Courtesy J. R. Úrbez-Torres)

Fig. 59. Conidia of *Diplodia seriata*. (Courtesy A. J. L. Phillips)

that this teleomorph has not been unambiguously associated with *L. theobromae*. Pycnidia are single or grouped, black, globose, ostiolate, and subcutaneous to erumpent. Pycnidia vary from 93 to 625 µm in diameter, occur in or on a stroma, and may be hirsute or glabrous. Conidia are subovoid to ellipsoid-ovoid with a broadly rounded apex, (19.0–)21.0–31.0(–32.5) × (12.0–)13.0–15.5(–18.5) µm, taper to a truncate base, are thick walled and widest in the middle to upper third, and have granular contents. They are hyaline and aseptate initially, becoming dark brown with a single septum only long after discharge from the pycnidia, with melanin deposits on the inner surface of the wall arranged longitudinally to give a striate appearance (Fig. 61). Colonies on potato dextrose agar develop a mycelium appearing gray in white light, later turning dark gray with dense aerial growth (Fig. 62). The optimum temperature for colony growth is typically 30–35°C, with some strains showing optimum growth up to 40°C.

Ascomata of *N. parvum* are virtually indistinguishable from those of *B. dothidea* but are less frequently aggregated into large clusters. Asci (75–145 × 17–20 µm) are bitunicate with a thick endotunica, clavate, and stipitate. Smooth and thin-walled hyaline ascospores, (15–)18–27(–29.5) × (6–)8–11 µm, are broadly ellipsoid to fusoid and widest in the middle to upper third. Pycnidia, externally indistinguishable from ascomata, are uni- or multilocular, occurring individually or in the same stromata as the ascomata. Pycnidia are globose and nonpapillate to pyriform, opening through a nonperiphysate ostiole. Conidia are hyaline, smooth, and thin walled; unicellular, (13.5–)15–22.5(–28.5) × 4–7.5 µm; and ellipsoid with a subobtuse apex and truncate base (Fig. 63). Conidia frequently become olivaceous or light brown with time, developing 1 or 2 septa with a darker middle cell (Fig. 64). Microconidia, which

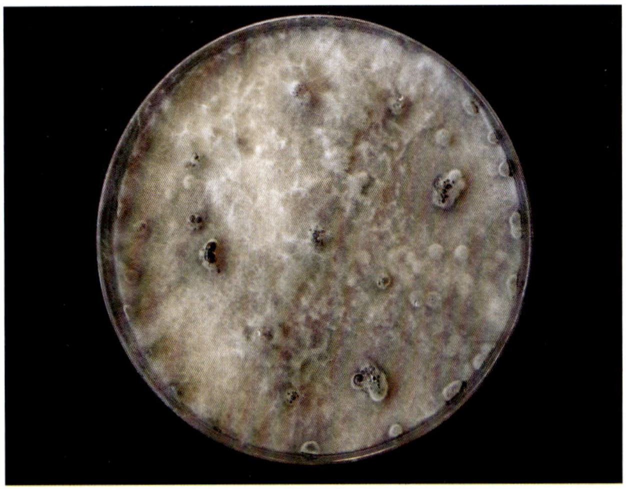

Fig. 62. Colony of *Lasiodiplodia theobromae* on potato dextrose agar. (Courtesy J. R. Úrbez-Torres)

Fig. 60. Colony of *Diplodia seriata* on potato dextrose agar. (Courtesy J. R. Úrbez-Torres)

Fig. 63. Young, hyaline, aseptate conidia of *Neofusicoccum parvum*. (Courtesy A. J. L. Phillips)

Fig. 61. Post-discharged appearance of *Lasiodiplodia theobromae* conidia. (Courtesy J. R. Úrbez-Torres)

Fig. 64. Older, darkened, multicellular conidia of *Neofusicoccum parvum*. (Courtesy A. J. L. Phillips)

are seen in some isolates, are hyaline, smooth, rod shaped, truncate at either end, and 3–5 × 1–1.5 µm. Colonies on potato dextrose agar appear similar to those of *B. dothidea*. The optimum temperature for colony growth is 28°C.

Disease Cycle and Epidemiology

The sexual stage of several Botryosphaeriaceae spp. has been described, but teleomorphs are rarely found on grapevines or other substrates. Therefore, asexual pycnidia appear to be the principal source of primary inoculum. Pycnidia of the species responsible for Botryosphaeria dieback are commonly observed embedded in the bark of diseased canes, spurs, cordons, and trunks, and can be found on old prunings left in the vineyards. Pycnidia of *B. dothidea*, *D. mutila*, *D. seriata*, *L. theobromae*, and *N. parvum* in particular have been reported from a broad range of hosts, many of which are commonly found surrounding vineyards. However, because dissemination of pycnidiospores occurs primarily via water splashing over relatively short distances, it is generally accepted that most infections result from inoculum that is produced within an infested vineyard. However, the role that windblown rain might play in long-distance dispersal of conidia into previously uninfested vineyards is currently unknown. Several Botryosphaeriaceae spp., especially *D. seriata* and *N. parvum*, have also been isolated from necrotic wood of rootstocks, scions, and/or graft unions in young vines, suggesting that infected propagation material also can play a role in the initiation and/or development of epidemics. Because some Botryosphaeriaceae species are known to act as latent pathogens, there is a potential for asymptomatic material from infested vineyards to be used for propagation material and ultimately infest new vineyards when they are planted.

Spore trapping studies conducted in both Northern and Southern Hemisphere grape-growing regions have somewhat elucidated the climatic conditions that promote spore release. Conidia of Botryosphaeriaceae spp. are primarily released from pycnidia during rain events providing at least 0.2 mm of precipitation. It has also been confirmed that overhead sprinkler irrigation can induce spore release. Seasonal spore dispersal patterns have been shown to vary among regions as a result of their greatly variable climatic conditions. For example, in California most Botryosphaeriaceae spores were trapped following rain events during the winter months, whereas in France and New Zealand spores were trapped throughout the year and the highest numbers were detected during summer months. Thus, in Mediterranean climates such as California, the dormant pruning of grapevines takes place during the primary period of rainfall (and, consequently, spore release), thereby exposing highly susceptible infection sites to inoculum of the botryosphaeriaceous fungi in these regions. For example, grapevine pruning wounds in California are susceptible to infection by *L. theobromae* and *N. parvum* to varying extents throughout the entire dormant season, but susceptibility is greatest immediately after pruning and decreases significantly over time thereafter. Thus, the significance of high spore numbers detected in some regions during the growing season, when pruning wounds appear to be much less susceptible to infection by these fungi, is still not well understood.

Management

Progress has been made in developing and implementing novel control techniques against Botryosphaeria dieback but management is still extremely difficult, due in part to the wide range of different species causing disease. Nevertheless, minimizing inoculum in the vineyard by removing and destroying all diseased wood at pruning time remains one of the most effective means to reduce the risk of new infections, regardless of the particular species involved.

Management of Botryosphaeria dieback traditionally has been achieved by remedial surgery, i.e., the excision of diseased wood, which prolongs the life of grapevines infected by various canker-forming fungi. When the disease has not progressed throughout the vine, production can be at least partly restored by pruning cankered portions a minimum of 15 cm below any discolored wood. New fruiting wood is then developed on a replacement arm derived from a shoot arising from the remaining healthy tissue. Application of wound protectants to large cuts resulting from surgery or other retraining operations is highly recommended, particularly if these are made during periods of high inoculum concentration.

Because species in the Botryosphaeriaceae infect grapevines primarily through wounds, prophylactic strategies to control these pathogens are aimed at protecting pruning wounds. Several active ingredients, including thiophanate-methyl, pyraclostrobin, various triazole fungicides (including fenbuconazole, tebuconazole, and cyproconazole plus iodocarb), and a boron-based paste have been shown to reduce the risk of infection when applied as pruning wound protectants. Although large wounds are at the greatest risk of infection by species in the Botryosphaeriaceae, studies have shown that both 1- and 2-year-old wounds also are susceptible. Ideally, protective treatments should be applied to all pruning wounds. However, the high cost of manually treating all pruning cuts, and the possibility that repeat application(s) might be required to maintain protection through the period of wound susceptibility, make such treatments economically viable only in high cash-return vineyards. Pruning wound protection through tractor-applied fungicide sprays can reduce cost significantly and facilitate treatment of large vineyards, but such applications traditionally have provided limited efficacy. However, recent research with several of the fungicides listed above, using spray application technology that maximizes deposition efficiency, has shown new promise for this approach.

Late pruning and/or double pruning, cultural practices used widely in California to reduce disease caused by *E. lata*, also have been effective in reducing infections by Botryosphaeriaceae spp. there and in other regions with similar Mediterranean climates. These techniques are designed to produce final wounds near or after the end of the rainy season, when inoculum is limited and wounds may lose susceptibility more quickly as temperatures increase. Wound protection by fungicide sprays may complement late pruning by reducing the likelihood that the limited inoculum still present by that time is able to colonize the host.

Complete control of Botryosphaeria dieback and other grapevine trunk diseases is extremely difficult owing to the high number of wounds made on individual vines each year, the long time that wounds remain susceptible, the difficulty in controlling a wide spectrum of different pathogens with a single fungicide active ingredient, and the technical and logistical difficulties in delivering fungicide sprays to multiple targets efficiently and effectively.

Selected References

Díaz, G. A., and Latorre, B. A. 2013. Efficacy of paste and liquid fungicide formulations to protect pruning wounds against pathogens associated with grapevine trunk diseases in Chile. Crop Prot. 46:106-112.

Kuntzmann, P., Villaume, S., and Bertsch, C. 2009. Conidia dispersal of *Diplodia* species in a French vineyard. Phytopathol. Mediterr. 48:150-154.

Phillips, A. J. L. 2002. *Botryosphaeria* species associated with diseases of grapevines in Portugal. Phytopathol. Mediterr. 41:3-18.

Phillips, A. J. L., Alves, A., Abdollahzadeh, J., Slippers, B., Wingfield, M. J., Groenewald, J. Z., and Crous, P. W. 2013. The *Botryosphaeriaceae*: Genera and species known from culture. Stud. Mycol. 76:51-167.

Pitt, W. M., Sosnowski, M. R., Huang, R., Qiu, Y., Steel, C. C., and Savocchia, S. 2012. Evaluation of fungicides for the management of Botryosphaeria canker of grapevines. Plant Dis. 96:1303-1308.

Rolshausen, P. E., Úrbez-Torres, J. R., Rooney-Latham, S., Eskalen, A., Smith R. J., and Gubler, W. D. 2010. Evaluation of pruning wound susceptibility and protection against fungi associated with grapevine trunk diseases. Am. J. Enol. Vitic. 61:113-119.

Úrbez-Torres, J. R. 2011. The status of Botryosphaeriaceae spp. infecting grapevines. Phytopathol. Mediterr. 50:S5-S45.

Úrbez-Torres, J. R., and Gubler, W. D. 2011. Susceptibility of grapevine pruning wounds to infection by *Lasiodiplodia theobromae* and *Neofusicoccum parvum*. Plant Pathol. 60:261-270.

Úrbez-Torres, J. R., Battany, M., Bettiga, L. J., Gispert, C., McGourty, G., Roncoroni, J., Smith, R. J., Verdegaal, P., and Gubler, W. D. 2010. *Botryosphaeriaceae* species spore trapping studies in California vineyards. Plant Dis. 94:717-724.

Úrbez-Torres, J. R., and Gubler, W. D. 2009. Pathogenicity of Botryosphaeriaceae species isolated from grapevine cankers in California. Plant Dis. 93:584-592.

(Prepared by J. R. Úrbez-Torres, A. J. L. Phillips, and W. D. Gubler)

Botrytis Bunch Rot and Blight

Botrytis bunch rot or gray mold occurs in vineyards throughout the world, although it is most severe in regions experiencing moderate temperatures and rainfall or extended periods of high humidity between veraison and harvest. The disease can cause serious economic losses by reducing both crop yield and quality. Yield reductions may be associated with damaged flower clusters early in the season or, more commonly, berry rot pre- and/or postharvest. Economic losses also may accrue from the costs associated with manually sorting diseased clusters pre- or postharvest, and from the lower prices received for diseased fruit or buyers rejecting it outright for failure to meet quality standards. In wine production, quality losses result from the modified chemical composition of diseased berries. For example, *B. cinerea* converts simple sugars (glucose and fructose) to glycerol and gluconic acid and secretes polysaccharides such as β-glucan, which hinder the clarification of wine. It also produces an enzyme (laccase) that catalyzes the oxidation of phenolic compounds, resulting in a browning of white wines and color instability in red wines. In table grape production, the most serious damage is the loss of fruit quality due to berry rot in the field, during transit or in storage. Postharvest losses are highly problematic for producers and shippers of table grapes owing to the pathogen's propensity to remain quiescent and undetected within clusters until tissues senesce, its ability to grow at near-freezing temperatures, the extended periods of fruit storage and/or transit that may be imposed, and the extremely low tolerance for diseased fruit in the wholesale and retail marketplace.

In certain cultivars and under conducive preharvest weather conditions (ideally, foggy evenings and mornings, sunny afternoons), Botrytis infections free of secondary contaminants can develop in the ripening clusters, giving rise to the so-called "noble rot." This form of the disease is desirable for some wine styles and is an important component of some exceptional and highly priced sweet white wines.

Symptoms

Although rotting berries are the most frequent and problematic symptom of infection by *B. cinerea*, blights (sudden, severe necrosis of tissue) and rots (decomposition and putrefaction of tissue) can occur on all herbaceous tissues of the vine. Early in the growing season, infected buds and young shoots may turn brown and desiccate (Fig. 65). Later in the preflowering period, particularly following prolonged periods of rain, fog, and/or dew, large irregular, necrotic patches appear on scattered leaves, often localized at the leaf base or on the edge of the lamina (Fig. 66) where water accumulates. Infection of young clusters can occur that appear as brown to black spots on rachises (Fig. 67), calyptra (Fig. 68), and pedicels of inflorescences. Under humid or wet conditions, infections can blight the entire cluster, with infected tissues yielding the characteristic gray, cottony mycelia and conidia of the fungus (Fig. 69). Young, immature berries may develop a soft brown rot with characteristic sporulation if the early summer is extremely wet. From approximately veraison onward, berries can begin exhibiting symptoms from previously quiescent infections that occurred earlier in the season, typically near flowering. Small, water-soaked lesions appear on the fruit, or a symptom known as "slip skin," in which the berry skin appears shiny (Fig. 70) and is separated easily from the flesh by rubbing, can develop. Tufts of gray, fungal mycelia and characteristic conidia appear on the surface of infected berries (Fig. 71). Often, rotted berries first develop near the center of the cluster (Fig. 72); such infection loci can spread quickly from berry to berry in a process known as "nesting" (Fig. 73), and infection may eventually engulf the entire cluster (Fig. 74). Additionally, other fungi and

Fig. 65. Browning and desiccation of young shoots infected with *Botrytis cinerea*. (Courtesy A. Wise)

Fig. 66. Sporulating *Botrytis cinerea* on a leaf of cv. Müller-Thurgau. (Courtesy R. C. Pearson)

bacteria can invade the rotting berries, producing a variety of colors, smells, and tastes (see Sour Rot). Infected table grapes in cold storage often develop a wet rot of the rachis, laterals, and pedicels, which can become covered by a mycelial mat, sometimes with sporulation. Individual infected berries develop circular, brown lesions that gradually cover the whole fruit, culminating in the "slip-skin" symptom. In severe cases,

Fig. 67. Dark lesion caused by *Botrytis cinerea* on young rachis. (Courtesy W. D. Gubler)

Fig. 68. Necrosis of young florets due to infection by *Botrytis cinerea.* (Courtesy W. Mahaffee)

Fig. 69. Blighting of young cluster by *Botrytis cinerea*, with characteristic sporulation of the fungus. Infection has progressed into the nodal region of the shoot. (Courtesy A. Wise)

Fig. 70. Shiny "slip skin" appearance of a 'Pinot Noir' berry infected with *Botrytis cinerea.* (Courtesy W. Mahaffee)

Fig. 71. Characteristic tufts of gray conidia of *Botrytis cinerea* arising from infected grape berries. (Courtesy W. F. Wilcox)

Fig. 72. Initial disease locus of *Botrytis cinerea* established within the center of a cv. Pinot Noir cluster, radiating outward. (Courtesy W. Mahaffee)

infections can spread among tightly packed berries and clusters to cause extensive loss (Fig. 75).

Late in the growing season, poorly hardened shoots may have patches of bleached bark where black sclerotia or grayish, sporulating mycelia frequently are present. The infected part mostly surrounds a node (Fig. 76), suggesting entry of the fungus through a diseased leaf or petiole or through a leaf scar.

Newly grafted grapevine cuttings incubated in callusing boxes at temperatures near 30°C and at high humidity may be infected and destroyed by rapid growth of the pathogen. The fungus also can develop beneath the film of paraffin used to seal the junction of grafted vines, thereby inhibiting development of the union between rootstock and scion.

Causal Organism

The causal fungus is the ascomycete *Botryotinia fuckeliana* (de Bary) Whetzel, although only the anamorph (asexual form), *Botrytis cinerea* Pers., is commonly observed in vineyards. *B. cinerea* will grow on most media suitable for the general cultivation of fungi. Hyphae are branched, septate, and hyaline but can appear tan to black when melonization occurs during moisture-limiting conditions. Conidiophores develop directly from hyphae and are generally long (1–5 mm), stout, dark brown, irregularly branched near the apex, with enlarged apical cells bearing clusters of conidia on short sterigmata (Fig. 77). Conidia are unicellular (8–14 × 6–9 μm), generally multinucleate, ovate, and individually hyaline but appearing grayish in mass with a smooth surface.

The mycelium can differentiate into two types of survival structures: chlamydospores and sclerotia. Chlamydospores are

Fig. 73. "Nesting" pattern of spread of *Botrytis cinerea* through a cv. Chardonnay cluster, resulting from physical contact of healthy with diseased berries. (Courtesy W. F. Wilcox)

Fig. 74. Culmination of berry-to-berry spread of *Botrytis cinerea* throughout an infected cv. Chardonnay cluster. (Courtesy R. C. Pearson)

Fig. 75. Storage rot caused by *Botrytis cinerea.* (Courtesy R. C. Pearson)

Fig. 76. Bleached nodal regions of one-year-old cane infected with *Botrytis cinerea*. (Courtesy W. D. Gubler)

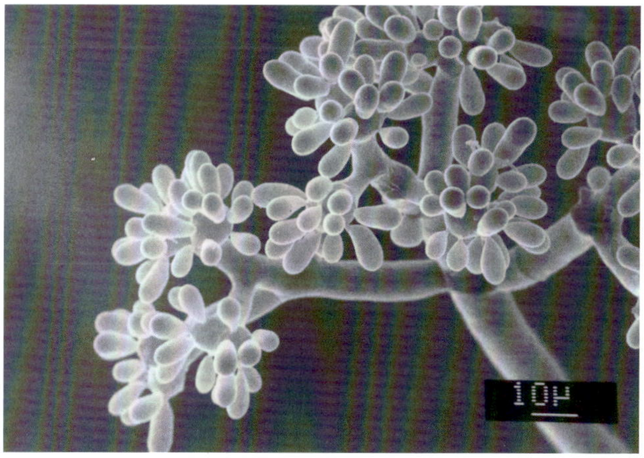

Fig. 77. Conidia on conidiophore of *Botrytis cinerea.* (Courtesy R. Wind)

hyaline cells of extremely variable shape and size, with terminal or intercalary placement. Chlamydospores can germinate to produce microconidia, hyphae, or macroconidia. Sclerotia can be of both a macro- and micro- form. Macrosclerotia are brown to black, irregularly shaped, up to 5 mm in diameter, and firmly attached to a substrate. They have a rind of thick-walled cells and a medulla of filamentous hyphae loosely arranged in a gelatinous matrix. Microsclerotia are individual or small groups of hyphal cells that have become thickened and are sometimes melonized. Microsclerotia germinate to produce mycelia, whereas macrosclerotia produce either conidia or, following sexual recombination, apothecia. Apothecia are cupulate, stalked, and brownish, with a stipe about 4–5 mm long. Ascospores (7 × 5.5 µm) are hyaline, one-celled, ovoid-ellipsoid, and smooth.

B. cinerea may also produce microconidia on phialides. The phialides usually arise freely from single hyphal cells on old aerial mycelium. They also can develop from germ tubes produced by macroconidia, mature hyphae, or inside empty hyphal cells, and from appressoria and sclerotia. The microconidia (2–3 µm in diameter) are hyaline and one celled, and are formed in chains or in clusters enclosed within a protective covering (hülle).

Disease Cycle and Epidemiology

Botrytis bunch rot has a complex disease cycle, influenced by multiple pathogen-host-environment interactions, many of which are not well understood. This complexity and the myriad possible combinations of the factors involved likely underlie the diversity of conclusions drawn from studies of the disease and its management in variable climates and production systems around the world.

B. cinerea is a "weak" pathogen that requires an exogenous nutrient source or a compromised host for successful infection and colonization. Thus, highly succulent, necrotic, injured, or senescing tissues are most susceptible. Old flower parts and ripe fruit are particularly important senescing tissues with respect to the epidemiology and economic impact of the pathogen on grapes. The fungus thrives under conditions of high humidity and still air. It has the capacity to remain quiescent, or latent, for extended periods in infected tissues, contributing to the complexity and occasional unpredictability of the disease. All of these characteristics influence disease development and management.

B. cinerea is not specific to grapevines; it can infect almost all dicots, many monocots, and some Pteridophyta plant species under the proper conditions. It also can live as a normal resident in various dead plant tissues, providing the capacity to survive indefinitely as a saprophyte in climates to which it is adapted. In vineyards, it typically overwinters as sclerotia, mycelia, or chlamydospores in infected grape tissues on the vine or vineyard floor. Old cluster components (rachides and mummified berries) and leaf tissues (petioles and blades) have been documented as particularly important sources of vineyard inoculum in several regions. The pathogen also can overwinter in asymptomatic weeds, becoming active and sporulating once these plants senesce naturally the following season or are killed by herbicide applications. In spring, conidia (or, rarely, ascospores) are produced from these various sources (Fig. 78) and function as primary inoculum. The production of conidia is favored by successive interrupted periods of surface wetness, moderate temperature, and changes in the atmospheric humidity. Conidia are released largely by air movement, especially during periods of rapidly changing humidity, as well as by splashing water, mechanical disturbances, or vertebrates and insects. Although the latter animals can serve as vectors, dispersal of conidia is primarily on wind currents, with concentrations in the air a function of the prevailing environment. High wind velocities and accompanying turbulence facilitate conidia escaping the canopy zone and becoming entrained in the free-moving air currents above it, effecting longer-range dispersal. In general, however, conidia produced from sources on the ground and within the canopy are transported <1 m from their origin.

Once deposited on a suitable substrate, conidia can germinate if conditions are favorable, but can otherwise remain dormant for long periods of time. Germination and subsequent infection can occur at temperatures between 0 and 30°C, with 15–25°C being broadly optimal and 18–20°C being ideal. These processes are favored by the presence of free water (e.g., rain or dew), but also can occur in its absence when relative humidity is above 93% for an extended period. Conidial germination requires an exogenous nutrient source; thus, it and subsequent infection are favored by the presence of damaged plant tissues. At any time during the infection process, *B. cinerea* can enter a state of quiescence, in which the pathogen remains alive but no longer grows. The cause of quiescence is poorly understood but appears to be plant mediated in response to changing environmental conditions. Such infections are increasingly likely to become active again as tissues progressively senesce or should they become moribund.

Young leaves, clusters, and shoots can become infected in the spring when daytime temperatures regularly exceed 15°C and moisture conditions are sufficient to induce formation of spores from overwintering survival structures and their subsequent germination and penetration of host tissues. Young leaves may carry high levels of asymptomatic latent infections, which can become active and sporulate if the leaves are killed through natural causes or viticultural practices. Lesions on intact leaves and blighting of young, green shoots and inflorescences may become apparent after several days of rain or fog if temperatures are favorable. It appears likely that disease outbreaks following such long wet periods reflect an increase in host susceptibility under these conditions, as conidia require only 15 h of wetting to germinate and infect susceptible berries at 15–20°C and *B. cinerea* preferentially colonizes numerous host tissues when their water content is high. Although direct losses from early infection of vegetative tissues are minimal, these can provide an important source of secondary inoculum adjacent to healthy inflorescences at the time of flowering.

From bloom to the beginning of berry touch, inflorescences and young clusters are highly susceptible to infection by *B. cinerea*. Furthermore, the structural parts of the clusters are dusted with pollen then, which can serve as an exogenous food source to support the pathogen's growth. Berry infections that occur during the period of bloom through early postbloom typically remain quiescent, but sometimes become active as fruit become ripe; such infections serve as sources for secondary spread of bunch rot should activation occur preharvest, or they can initiate storage rot decay if they remain quiescent until postharvest. Opinions and experiences differ with respect to the primacy of infections initiated during the bloom and early postbloom period. Some authorities believe the period to be

Fig. 78. Conidia of *Botrytis cinerea* produced from sclerotia on overwintered grape cane. (Courtesy W. D. Gubler)

the singlemost important stage in the progress of an epidemic, whereas others contend its importance to be rivaled or even superseded by events that occur during the preharvest period. To some extent, such perspectives appear to be influenced by the specific contexts in which they are formed, including the likelihood of significant preharvest rainfall occurring in the region. Nevertheless, all workers agree that this is a potentially critical phase in the disease cycle.

Several infection pathways from the floral organs to the berries have been demonstrated. Early investigations of the phenomenon suggested that *B. cinerea* first infects the style during bloom, and it grows from there into the ovule and remains quiescent until the berry ripens; however, subsequent investigators were largely unable to reproduce this finding. In contrast, several investigators have demonstrated infection of the receptacle area, both directly through wounds exposed as the calyptra are shed ("cap scars") and via growth of fungal germlings through the stamens and into the receptacle at their point of attachment. Dead floral tissues also function as important sources of berry infection when they become infected during flowering (Fig. 79) and then adhere to the cluster after they are shed, becoming trapped within as the bunches close. These food sources support saprophytic growth of the fungus until it is able to move into the subtending berries as their resistance breaks down during ripening. Such infections via blossom debris likely contribute to the common phenomenon of bunch rot originating near the center of the cluster, as does similar saprophytic growth from unpollinated, aborted fruitlets that become infected near flowering (Fig. 80) and remain trapped within the bunches after they close.

In addition to floral organs, *B. cinerea* also readily establishes quiescent infections within the structural cluster parts (rachises and pedicels) while they are young. Although these tissues develop moderate resistance as they age, conidia and germlings can survive on their surfaces for extended periods of time. In contrast, healthy grape skins are relatively resistant to infection and colonization early in their development. However, damage to the berry skin by other pathogens (e.g., powdery mildew), insects, sun scald, or mechanical factors increases the likelihood of its infection by *B. cinerea* and eventual rot of the entire berry as it ages.

Although young, green berries may become rotted by *B. cinerea* if the postbloom weather is extremely wet, the frequency of diseased green berries typically is very low. Rather, in most regions and years, berry rot symptoms are expressed from ripening onward, when the senescing fruit apparently are no longer able to maintain quiescent infections in a latent state and the newly active fungus colonizes the fruit. Studies have shown that the majority of quiescent infections typically remain so through harvest, but the factors that favor their activation versus continued quiescence are poorly understood. Environmental factors that have been shown to promote the development of rot from previously quiescent infections include persistently high soil moisture during the postveraison period, high atmospheric humidity during the immediate preharvest period, and high levels of nitrogen in the berries. Various forms of stress, particularly wounding of the berries at the point of their attachment to the pedicel, as may occur from high winds or cluster compaction as the fruit swell before harvest, also have been shown to activate quiescent infections. Injuries from compaction and the resulting activation of quiescent infections are another likely reason that bunch rot often begins in the center of the cluster.

Berries that become symptomatic as a result of activated quiescent infections are usually limited in number. Nevertheless, their economic impact can be vastly disproportionate to their actual numbers in table grape production, where commercial tolerance for diseased fruit is virtually nonexistent and symptoms may not appear until after bunches have been packed and shipped or placed into storage. In wine grape production, significant losses from Botrytis bunch rot are usually the result of extensive secondary spread during the postveraison period, with activated quiescent infections serving an important role as primary disease foci that initiate the epidemic. Such spread is by far most severe in tightly compressed bunches, which facilitate rapid spread of the fungus from berry to berry through mycelial growth at the point of contact, resulting in the common "nesting" pattern of spread. For example, it has been shown that preharvest disease can progress virtually throughout a compacted cluster from a single berry that develops rot postveraison, with only minor spread developing in loose clusters on the same vine. High berry nitrogen content and high soil water content (presumably leading to elevated berry water content) also have been shown to increase the rate of secondary disease development in addition to increasing the activation of quiescent, primary infections.

Conidia produced on symptomatic berries also can be responsible for secondary disease spread. Infection may occur via direct penetration of the skins as their resistance decreases near maturity or through wounds caused by various factors including rain cracking, bird claws and feeding, and insects. Insects implicated as initiators of wounds subsequently colonized by *B. cinerea* and/or as vectors of fungal spores to these wounds include the grape berry moth (*Lobesia botrana*), the New Zealand flower thrips (*Thrips obscuratus*), the Mediterranean fruit fly (*Ceratitis capitata*), the light-brown apple moth (*Epiphyas postvittana*), and the vinegar fly (*Drosophila melanogaster*). Following inoculation of fresh wounds, berries

Fig. 79. Grape floral tissue infected with sporulating *Botrytis cinerea.* (Courtesy S. M. Zitter)

Fig. 80. Aborted fruitlets infected with *Botrytis cinerea* and sporulating during the preharvest period. (Courtesy S. M. Zitter)

typically become necrotic and sporulate within approximately 1 week, providing further opportunity for epidemic spread through vegetative and conidial inoculum.

Management

Perhaps more than for any other major fungal disease of the grape fruit and foliage, control of Botrytis bunch rot is dependent upon a conscientious integration of feasible cultural practices with targeted chemical and/or biological treatments appropriate to climatic conditions and market demands. Although many considerations go into the initial choice of cultivar, clone, and planting site, growers in regions prone to Botrytis should remember that such choices can have a profound effect on their ability to manage this disease for the lifetime of the vineyard. Cultivars vary considerably in their susceptibility to Botrytis bunch rot because of differences in berry chemistry, active responses to attempted infection, and cluster architecture. A factor sometimes as important as cultivar, but one whose choice entails fewer commercial implications, is that of clone; loose-clustered clones of any given cultivar typically have fewer problems with Botrytis bunch rot than those with tight clusters, and such differences can be pronounced. Planting sites, row spacings, and row orientations that favor good air movement through the canopy similarly aid in the overall management of the disease. Rootstocks matched to the site to control vine vigor make necessary canopy management far easier.

In climates conducive to Botrytis development, canopy management is a critical component of a control program. Through canopy management, the microclimate around the berry clusters can be manipulated to reduce relative humidity and the duration of berry wetness by increasing airflow, light penetration, and temperature. This alteration of the berry microclimate can be done through the choice of training system and the subsequent imposition of shoot positioning, hedging, and leaf removal. Shoots should be positioned to develop an evenly spaced canopy, which should be maintained throughout the season so that light and air can continue to move through it. Leaf removal further increases light penetration and air flow, as well as the deposition upon clusters of spray materials applied to provide bunch rot control, and has proven to be a valuable practice in multiple regions where it has been implemented. For best results, leaf removal should be conducted as soon as possible after fruit set. Earlier removal may result in poor set or return bloom (although such is not always true and some studies have indicated benefits from this practice), and later removal may diminish some of the potential benefits in addition to making the fruit more susceptible to subsequent sun scald. In regions and years where dry weather prevails throughout most of the summer, particularly in the preharvest period, such cultural practices alone may provide adequate levels of bunch rot control.

In arid regions, irrigation should be managed to balance canopy vigor and avoid excessively succulent berries with tightly compacted clusters. The judicious use of nitrogen fertilizers will limit the potential for infection by reducing excessive canopy growth in addition to maintaining berry tissues in a physiological state less conducive for fungal colonization.

There are multiple reports of reducing disease severity through the use of chemical growth regulators to loosen cluster compactness and thereby limit the spread of disease through berry-to-berry contact. Although common in some table grape production systems, this technique has several potential pitfalls on wine grape cultivars that must be addressed, including reduced fruit set and/or return bloom, in addition to local product registration issues. Nevertheless, the technique offers significant opportunity when these pitfalls can be addressed successfully. Loosening clusters through a mechanical reduction of flower numbers during bloom, e.g., with a plastic brush as is practiced in some table grape production regions, also has been shown to limit berry-to-berry spread but is labor intensive and costly. Experimental programs utilizing leaf removal at or before bloom have successfully reduced cluster compaction but have sometimes resulted in poor yield or reduced bloom the following year.

In regions where economic loss from Botrytis bunch rot is a regular concern, effective fungicides must usually be applied in addition to implementing appropriate cultural practices. Fungicide efficacy is a function not only of the inherent product efficacy and application timing but also of good spray coverage, which can be difficult to achieve late in the season without good canopy management. Most fungicides traditionally used to control other diseases have limited to no activity against *B. cinerea*, at least at the standard rates employed. Therefore, several classes of fungicides have been developed which are active primarily against this fungus and a few related species, including the dicarboxamides, the anilinopyrmidines, and the hydroxyanilides. Some more recent QoI (strobilurin) and SDHI (carboxin) fungicides also have shown good activity. *B. cinerea* is a pathogen at high risk for fungicide resistance development, owing to its high reproductive potential, the ready dispersal of its conidia, and the multiple sprays sometimes applied for its control; implementing basic resistance management practices is critical for the sustainable use of these products.

On the basis of the disease epidemiology, potential fungicide applications are traditionally focused on four phenological growth stages: (i) mid-bloom through the beginning of fruit set, to prevent establishment of latent infections within the cluster and limit the colonization of eventual blossom debris; (ii) just before berry touch, the last opportunity to fully cover all berry surfaces; (iii) the beginning of veraison, to protect healthy berries as they become susceptible to new infections and (as provided by some systemic fungicides) to suppress the activation of existing latent infections; and (iv) one or more applications beginning 2–4 weeks after the veraison application, to control secondary spread. The importance of each timing varies among regions and years, depending largely on cultivar, cultural practices, and the weather conditions during that growth stage. In regions with wet springs but dry summer and preharvest conditions, the early applications are often considered most effective. In New York State, where rainfall throughout the growing season varies greatly among years, a series of spray timing trials conducted across three decades showed the critical number and phenological timing of sprays to be inconsistent and weather dependent. Disease forecasting systems are currently used in some regions to predict the occurrence or probability of infection events based upon environmental factors, although the complex set of factors regulating bunch rot development has restricted the utility of this approach in many others. In some cases, models may be more useful for predicting the onset of symptom expression than the potential severity of disease at harvest or following cold storage.

Applications of several registered biological control products based upon antagonistic microorganisms (e.g., species of the fungal genera *Aureobasidium*, *Trichoderma*, and *Ulocladium*, fermentation products of *Bacillus* spp. bacteria) have provided a measure of control in some regions but have been ineffective in others, particularly when disease pressure has been significant. For fresh market grapes, postharvest treatment or exposure to volatile compounds that inhibit growth of *B. cinerea*, such as sulfur dioxide, are typically necessary to insure against decay in storage and extend shelf life.

Selected References

Broome, J. C., English, J. T., Marois, J. J., Latorre, B. A., and Aviles, J. C. 1995. Development of an infection model for Botrytis bunch rot of grapes based on wetness duration and temperature. Phytopathology 85:97-102.

Coley-Smith, J. R., Verhoeff, K., and Jarvis, W. R. 1980. The Biology of *Botrytis*. Academic Press, London, UK.

Elmer, P. A. G., and Michailides, T. J. 2004. Epidemiology of *Botrytis cinerea* in orchard and vine crops. Pages 243-272 in: *Botrytis*: Biology, Pathology, and Control. Y. Elad, B. Williamson, P. Tudzynski,

and N. Delen, eds. Kluwer Academic Publishers, Dordrecht, The Netherlands.
Gubler, W. D., Marois, J. J., Bledsoe, A. M., and Bettiga, L. J. 1987. Control of Botrytis bunch rot of grape with canopy management. Plant Dis. 71:599-601.
Hill, G. K., Stellwaag-Kittler, F., Huth, G., and Schlösser, E. 1981. Resistance of grapes in different developmental stages to *Botrytis cinerea*. Phytopathol. Z. 102:328-338.
Holz, G., Gütschow, M., Coertze, S., and Calitz, F. J. 2003. Occurrence of *Botrytis cinerea* and subsequent disease expression at different positions on leaves and bunches of grape. Plant Dis. 87:351-358.
Keller, M., Viret, O., and Cole, F. M. 2003. *Botrytis cinerea* infection in grape flowers: Defense reaction, latency, and disease expression. Phytopathology 93:316-322.
Mlikota Gabler, F., Smilanick, J. L., Mansour, M., Ramming, D. W., and Mackey, B. E. 2003. Correlations of morphological, anatomical, and chemical features of grape berries with resistance to *Botrytis cinerea*. Phytopathology 93:1263-1273.
Nair, N. G., Guilbaud-Oulton, S., Barchia, I., and Emmett, R. W. 1995. Significance of carry over inoculum, flower infection and latency on the incidence of *Botrytis cinerea* in berries of grapevines at harvest in New South Wales. Aust. J. Exp. Agric. 35:1177-1180.
Thomas, A. C., Matthee, F. N., and Kotze, J. M. 1981. Survival of *Botrytis cinerea* from table grapevines in South Africa. Phytophylactica 13:157-160.
Vail, M. E., and Marois, J. J. 1991. Grape cluster architecture and the susceptibility of berries to *Botrytis cinerea*. Phytopathology 81:188-191.

(Prepared by W. F. Wilcox, W. Mahaffee, and W D. Gubler)

Dematophora Root Rot

Dematophora root rot is a serious disease on many herbaceous and woody plants throughout much of the temperate world. It is most common on deciduous fruit trees and grapes. On grape it is commonly known as *pourridié* as well as *aubernage, bianco, blanc, blanc des racines, blanquet, champignon blanc, malbianco, mal nero, marciume radicale bianco, morbo bianco, pourridié de la vigne, pourriture*, Rosellinia root rot, *Weinstockfaule*, white root rot, and *Wurzelpilze*. The disease has been reported mainly from European countries but also on grape in North and South America, Africa, Australia, New Zealand, India, the former Soviet Union, and Japan.

Symptoms

Symptoms of Dematophora root rot are not very diagnostic. Diseased plants may die rapidly or slowly throughout one season, or may linger into a second year before succumbing. Vines that linger frequently bear an excessively large crop the year before they die. Diseased plants tend to be aggregated and to occur in discrete areas of the vineyard.

The leaves remain attached on vines that die rapidly. On vines that decline gradually, tendrils and leaves are weak and stunted, wilting is common, and shoots may arise from the base. Dead vines can be pulled easily from the soil because of severe deterioration of the roots. Frequently, vines break off at the soil line, where the fungus has weakened the wood. The bark below the soil line is darkened and sloughs easily. At the root crown, there may be a black, gummy ooze.

Under moist conditions, the fungus produces abundant hyphae on the surface of infected roots, giving them a white, fluffy appearance (Fig. 81). The hyphae tend to grow along the smaller roots and often form flattened strands in soil cavities around the roots. As these fungal strands age, they darken and may take on a tan or brown cast.

The fungus grows rapidly in the infected vine and produces small, white plaques scattered throughout the wood (Fig. 82). These plaques differ considerably from those of *Amillaria mellea*, which are confined to the area between the bark and the wood (see Armillaria Root Rot).

Pieces of invaded roots or stems, when put in a moist chamber, are rapidly covered with abundant growth of white hyphae. The fungus also may produce sclerotialike masses on the surface of infected tissues.

Causal Organism

The causal organism, *Rosellinia necatrix* Prill. (anamorph *Dematophora necatrix* Hartig), produces nearly spherical, brown to black perithecia, aggregated and embedded in a brown web of hyphae on the host surface. Perithecia are about 1–2 mm in diameter. Young perithecia have a distinct protrusion around the ostiole that is visible with a dissecting microscope,

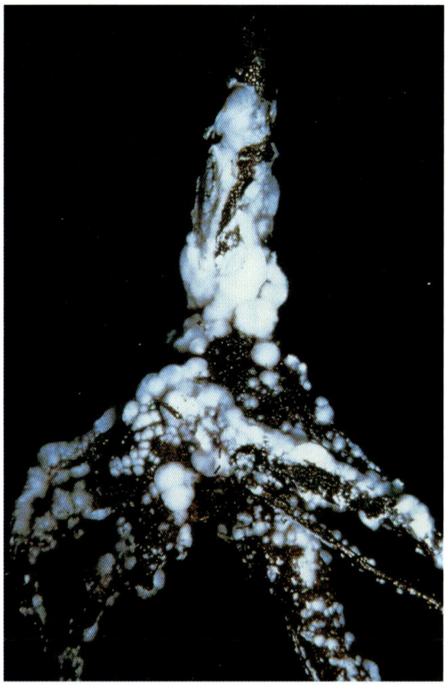

Fig. 81. Hyphae of *Dematophora necatrix* growing from an infected plant under moist conditions. (Courtesy A. H. McCain)

Fig. 82. Small, discontinuous plaques of *Dematophora necatrix* on an infected root with bark removed. (Courtesy R. D. Raabe)

but ostioles on old perithecia are difficult to find. The perithecial stage takes several years to develop and is rarely found.

Asci are cylindrical (8–12 × 250–380 µm), long-stalked, and unitunicate. The eight ascospores are one celled, cymbiform, straight or curved, and dark brown. They measure 5–8 × 3–50 µm and have a longitudinal slit running parallel to the long axis of the spore for about one-third of its length.

The conidial stage is composed of brown, rigid synnemata up to 1–5 mm high. The stipe is 40–300 µm thick and is often dichotomously branched toward the apex. Conidia (2.5 × 3–4.5 µm) are produced in large numbers.

Microscopic signs of the pathogen include distinct enlargements of the ends of hyphal cells next to the septa. This occurs most frequently in older hyphae and is helpful in identifying the fungus.

Disease Cycle and Epidemiology

The fungus grows through soil, using roots it has killed as a food base. It is favored by the presence of moisture and organic material, which also can serve as a food base, and is frequently found in soils high in clay content. The pathogen is spread by means of infested soil and infected planting material. Although the *Dematophora* stage may produce many conidia, most researchers have not been able to germinate them. Therefore, the role of spores in the dissemination of the fungus is unclear. The fungus withstands drying and can remain viable in air-dried pieces of wood in the laboratory for several years. It grows optimally at 22–28°C but does not grow at 31°C.

Management

Controlling Dematophora root rot is extremely difficult. Many fumigants, such as allyl bromide, ammonium hydrosulfide, bromopicrin, carbon tetrachloride, chloroform, chloropicrin, ethylene dibromide, and formalin pentachloroethane have failed to control the disease in small-scale experiments. In field experiments, methyl bromide was ineffective in California but was effective in Israel. Carbendazim and dazomet also have been reported as providing control.

The use of resistant rootstocks is a logical approach to control. Resistance has been reported in *V. cinerea*, *V. vinfera* 'Carignane', and in Solonis, a complex hybrid. In an experiment established in an infested field in California, some survivors included Iona, Red Malaga, Palomino, Dog Ridge, Salt Creek, and St. George, *V. arizonica*, *V. flexosa*, and several hybrid selections. Further testing of rootstocks is warranted.

Selected References

Berlese, A. N. 1892. Rapporti tra *Dematophora* e *Rosellinia*. Riv. Patol. Veg. 1:1, 5-17, 33-46.

Hansen, H. N., Thomas, H. E., and Thomas, H. E. 1937. The connection between *Dematophora necatrix* and *Rosellinia necatrix*. Hilgardia 10:561-565.

Khan, A. H. 1949. The root disease caused by *Rosellinia necatrix* (Hart.) Berl. Ph.D. thesis. University of California, Berkeley.

Sivanesan, A., and Holliday, P. 1972. *Rosellinia necatrix*. Descriptions of Pathogenic Fungi and Bacteria, no. 352. Commonwealth Mycological Institute, Kew, England.

Sztejnberg, A., Omary, N., and Pinkas, Y. 1983. Control of *Rosellinia necatrix* by deep placement and hot treatment with methyl bromide. Bull. OEPP/EPPO Bull. 13:483-485.

Viala, P. 1891. Monographie du Pourridié. Libraire de l'Académie de Médecine, Paris. C. Coulet, Montpellier, France.

(Prepared by R. D. Raabe)

Downy Mildew

Downy mildew, caused by *Plasmopara viticola*, is perhaps the most serious disease of grapevines in climates with relatively warm and humid summers. When highly susceptible cultivars are grown in such climates the potential exists for complete defoliation and crop destruction in the absence of effective control programs. The disease was first described in 1834 by Schweinitz on *Vitis aestivalis* in the southeastern United States, and was considered to be of minor importance on this and other moderately to highly resistant North American grape species that co-evolved with the pathogen. However, downy mildew was a principal reason (along with phylloxera) for the death of early *V. vinifera* plantings established by colonists in eastern North America. The pathogen was presumably introduced to Europe via infected planting material and/or infested soil, as there were numerous introductions of New World vines to European farms and research institutions for at least 30 years prior to the discovery of the disease in the Bordeaux region in 1878. The reproductive capacity of the pathogen, the near-universal susceptibility of *V. vinifera* cultivars, and very conducive weather allowed downy mildew to spread rapidly across the continent, and the disease was present in viticultural regions throughout Europe within a few years of its original discovery. Downy mildew is now a serious threat in virtually all viticultural regions worldwide where rainfall occurs regularly during the late spring and summer. Conversely, it is not a serious annual problem in regions where low rainfall coupled with hot, dry weather in the spring and summer limits its spread (e.g., much of Argentina, Australia, California and western North America, Chile, southern Peru, Spain, and much of the Mediterranean region). However, serious losses can be experienced in some of these areas when sporadic or atypical wet weather results in conditions favorable for downy mildew.

Unlike in *Vitis vinifera*, there is a wide range of susceptibility to downy mildew among *Vitis* interspecific hybrids. Among North American grapevine species, *V. aestivalis* and *V. labrusca* are moderately susceptible, whereas *V. cordifolia*, *V. rupestris*, and *V. rotundifolia* are relatively resistant. Certain *Vitis* interspecific hybrid cultivars exhibit organ-specific resistance: leaves may be moderately to highly susceptible while fruit are highly resistant (e.g., Aurore and Delaware) or leaves may exhibit moderate levels of resistance while clusters, tendrils, and shoot tips are highly susceptible (e.g., Chancellor). Several new European cultivars (e.g., Regent) have been developed from progeny of crosses between *V. vinifera* and resistant North American species in an effort to incorporate the most desirable qualities of both parental branches. These progeny were selected for resistance to downy mildew, and resistant clones were then repeatedly backcrossed to *V. vinifera* for multiple generations, producing vines with the phenotype and fruit quality characteristics of *V. vinifera* while maintaining downy mildew resistance introduced from the North American species.

Symptoms

The disease affects all green parts of the vine, particularly leaves, inflorescences, and young berries. Lesions viewed from the upper leaf surface at first appear slightly darker and shiny, as if bruised. These spots rapidly become yellowish and circular, a symptom observed most commonly on young leaves and sometimes referred to as an "oil spot" (Fig. 83). Typically, the tissue within these lesions soon begins to turn necrotic (Fig. 84). Lesions become more angular and reddish brown as leaves age and infected tissue eventually becomes completely necrotic (Fig. 85). Leaf lesions often are delimited by veins, particularly when initiated late in the season, and late-season lesions also may be smaller than those initiated earlier (Fig. 86). Following a damp night, sporulation is prominently visible on the lower leaf surface immediately beneath these symptoms. Sporulation on young lesions appears as a dense, white, cottony mass (Fig. 87). With each repeated cycle of sporulation, the proportion of necrotic tissue increases until spore formation is limited to the lesion's margin (Fig. 88). Leaves with multiple lesions typically abscise, and severe infection can result in defoliation (Fig. 89), which limits the accumulation of sugars and flavors in fruit and decreases the vine's cold hardiness during the subsequent winter. Infected shoot tips thicken, curl, and eventually become brown and necrotic. Similar symptoms are seen on petioles, tendrils, and young inflorescences, which may ultimately abscise. Shoot tips and rachises that become infected as they are rapidly elongating may show a severe epinasty or "shepherd's crook" appearance (Fig. 90). All of the preceding vine tissues

Fig. 83. "Oil spot" symptoms of downy mildew on young cv. Delaware leaves. (Courtesy R. C. Pearson)

Fig. 84. Symptoms of downy mildew on the adaxial leaf surface in early summer; necrosis within lesions becomes progressively more severe with successive sporulation events. (Courtesy W. F. Wilcox)

Fig. 85. Old, necrotic downy mildew leaf lesions. (Courtesy W. F. Wilcox)

Fig. 86. Symptoms of downy mildew on a cv. Chardonnay leaf in the late summer, with many lesions restricted in size and delimited by leaf veins. (Courtesy W. F. Wilcox)

Fig. 87. Characteristic sporulation of *Plasmopara viticola* from the underside of an infected cv. Ugni Blanc leaf. (Courtesy W. F. Wilcox)

Fig. 88. Sporulation of *Plasmopara viticola* confined to the viable margin of an otherwise necrotic lesion. (Courtesy D. M. Gadoury; reproduced by permission from Phytopathology 95:1447)

may support a white cottony coating of sporangiophores and sporangia when conditions favor sporulation.

Young berries are highly susceptible, turning grayish when infected (gray rot). Under humid conditions, they too can be covered with a downy felt of sporulation (Fig. 91). Berries become less susceptible as they mature, eventually becoming impervious to direct infection as their stomata develop into lenticels. However, the pedicels (berry stems) remain susceptible for some time after this, and older berries can become infected if the pathogen expands into them from the pedicels, causing the so-called brown rot or leather berry symptom. Berries so infected usually will not support sporulation since they lack the functional stomata through which sporangiophores must emerge, although sporulation may occur from the pedicels or the berries themselves if a slight wound is made to permit egress of the sporangiophores. Infected older berries of white cultivars may turn dull gray green (Fig. 92), while those of black cultivars turn pinkish red (Fig. 93). Infected berries remain firm compared with healthy berries, which soften as they ripen. These infected berries drop easily, leaving a dry stem scar. Portions of the rachis or the entire cluster also may drop, especially if prebloom infections cause girdling of the rachis.

Fig. 89. Unsprayed vines severely defoliated by downy mildew, with adjacent fungicide-treated vines retaining healthy leaves. (Photograph by W. McFadden-Smith, Ontario Ministry of Agriculture and Food; Queen's Printer of Ontario, © 2014. Reproduced with permission.)

Fig. 90. Necrosis and "shepherd's crook" appearance (epinasty) of a rachis infected by *Plasmopara viticola* while expanding rapidly prior to bloom. (Courtesy W. F. Wilcox)

Causal Organism

The causal agent of grapevine downy mildew, *Plasmopara viticola* (Berk. & M. A. Curt.) Berl. & De Toni, is an Oomycete member of the family Peronosporales. Characteristics that include heterokont flagella, cell walls containing glucans, and divergent molecular phylogeny place the Peronosporales within the kingdom Chromista, taxonomically distinct from the true fungi (Mycota). However, ecologically and epidemiologically, Oomycete plant pathogens share many similarities with the true fungi, and are treated as such in the practical aspects of disease management.

P. viticola is a biotrophic pathogen attacking members of the genus *Vitis*. It develops intercellularly within the parasitized tissues of the vine in the form of tubular, coenocytic hyphae 8–10 µm in diameter, bearing globular haustoria that are 4–10 µm in diameter. Following penetration of the host cell wall, haustoria invaginate the host cell membrane.

Fig. 91. Characteristic sporulation of *Plasmopara viticola* on young, infected berries following a humid night. (Courtesy R. C. Pearson)

Fig. 92. Dull gray-green appearance of older white-fruited berries infected by *Plasmopara viticola*. (Reproduced by permission from Phytopathology 95:1447)

P. viticola is heterothallic; hence, sexual reproduction occurs only when two isolates of opposite mating types are within close proximity in the same tissue. For this reason, initiation of the sexual stage may be delayed until disease increases to a level or persists for a duration that makes the chance pairing of sexually compatible isolates possible. Antheridia and oogonia derive from hyphal tips wherein, following meiosis, the haploid nucleus of the antheridium migrates to and then fuses with that of the oogonium, forming a diploid oospore (20–120 μm in diameter). Oospores are most commonly recovered from infected leaves, often near the periphery of distinct lesions, but also throughout the lamina of leaves that exhibit numerous small colonies and abundant sporulation along the major and minor veins. Oospores are robust survival structures resistant to unfavorable temperatures and drought, and can survive more than one season in vineyard soil. The temporal limit of oospore viability has not been determined for *P. viticola*, but it has been shown to exceed 10 years for at least a fraction of the population in some other Oomycete species.

Oospores germinate via production of one or occasionally two slender germ tubes (2–3 μm in diameter and of variable length), which terminate in a pyriform primary sporangium (28 × 36 μm). Primary sporangia remain attached to the germ tube and, under saturated conditions, release approximately 30–60 zoospores at least 3 h after oospore germination begins; some literature indicates that this process takes 10 h. Portions of the oospore population may continue to germinate as long as environmental conditions remain favorable throughout the summer.

Asexual reproduction, responsible for secondary cycles of the disease, results in the formation of ellipsoid and hyaline sporangia (14 × 11 μm), which are borne upon sterigmata at the tips of dendritic sporangiophores (140–250 μm) (Fig. 94) that emerge through stomata (Fig. 95). This requirement for stomatal emergence limits the tissues and locations from which sporulation can be observed, e.g., only on the abaxial (under) side of leaves and on young berries. Detached sporangia in a water film quickly (20–30 min at optimal temperatures) cleave four to 10 biflagellate zoospores (6–8 × 4–5 μm) that are immediately released through a pore in the side of the sporangial wall. Zoospores are generally uninucleate, rarely bi- or multinucleate. Zoospores as well the vegetative hyphae are diploid.

Fig. 93. Pinkish-red berries of a black-fruited cultivar infected by *Plasmopara viticola* near the end of their susceptible period. (Courtesy W. F. Wilcox)

Disease Cycle and Epidemiology

The disease cycle is illustrated in Fig. 96. In temperate climates, *P. viticola* overwinters as oospores in fallen leaves or, more commonly, in the surface layers of soil as infected leaves are consumed and excreted by detritivores or otherwise decompose. Epidemics can be initiated once suitable host tissue is present and these oospores are sufficiently warmed and moistened to initiate germination. The precise conditions for oospore germination have not been determined; however, based on studies conducted with other Oomycetes, it is probable that soil moisture contents between field capacity and saturation are most favorable. Upon production of primary sporangia from the germinating oospores, zoospores are released during episodes of soil saturation and are dispersed to grapevines by splashing water droplets or in windblown aerosols derived from them. Zoospores that are deposited on green tissues swim through water films to stomata on the lower leaf surfaces, shoot tips, inflorescences, and young berries by means of flagellar propulsion. Then, they shed their flagella and encyst, forming a cell wall. An infection peg emerges from each cyst and penetrates the stoma. Under optimal conditions, the elapsed time between zoospore release and their penetration of the host can be <90 min. Infections from oosporic sources are frequently associated with rain events of at least 2–3 mm coincident with temperature ≥10°C during the rain event, beginning in the early stages of vine growth (as little as 10 cm of shoot growth in some reports). Depending upon the primary inoculum load in the vineyard, the favorability of rain and temperature, and the

Fig. 94. Dendritic sporangiophore of *Plasmopara viticola*, bearing mature sporangia upon terminal sterigmata. (Courtesy H.-H. Kassemeyer)

Fig. 95. Sporangiophores of *Plasmopara viticola* emerging through stomata on the underside of a leaf. (Courtesy H.-H. Kassemeyer)

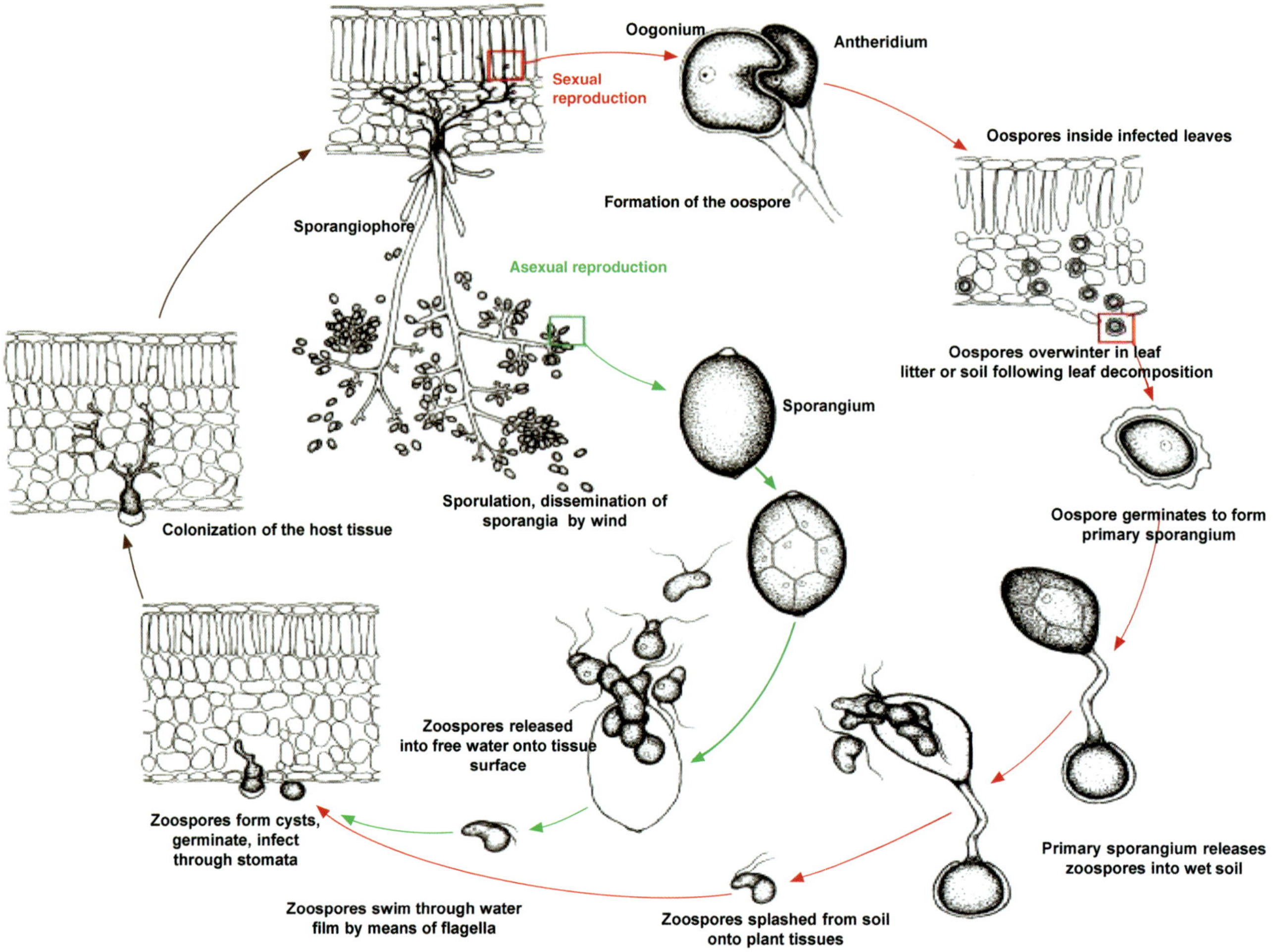

Fig. 96. Disease cycle of downy mildew of grape. (Courtesy N. Schoepe)

innate susceptibility of the cultivar, the first infection event may rarely result in an incidence of leaf or cluster infection approaching 50%. However, when one or more of these factors is suboptimal, repeated secondary cycles are much more commonly required before disease becomes severe.

Upon penetration of the stoma by zoospores (originating from either oosporic or asexual secondary sporangia), the infection peg extends into the substomatal cavity, where it dilates into a vesicle. From the substomatal vesicles, hyphae grow into the surrounding intercellular space and form numerous branches. Spread of the mycelium is limited by the veins, frequently resulting in a mosaic pattern of sporulation and necrosis in older leaves. Once colonization has developed to the extent that sporulation is possible, the subsequent length of the latent period is dependent upon the occurrence of exacting environmental conditions for this to occur: at least 4 h of darkness at temperatures >12.5°C and a relative humidity of 95–100%. Under vineyard conditions, minor temperature fluctuations and radiant cooling of plant tissues during night hours typically cause water to condense upon them once the ambient relative humidity exceeds 90%; thus, sporulation can usually be expected when the ambient relative humidity is near 90% at night unless the air temperature is rising.

Upon induction of sporulation, a secondary substomatal vesicle forms, from which develop the hyphae that protrude from the stomata and differentiate to produce the dendritic sporangiophores. Sporangia form on their tips and, once mature, are detached from the sporangiophores following the dissolution of a callose cross wall in water. This entire process, from induction of sporulation to release of sporangia, occurs within as little as 4 h at an optimal temperature of 18–22°C. Wind currents disseminate the sporangia to new host tissue where they germinate by releasing zoospores, which initiate a new round of secondary infections as per the process described above.

Infection typically requires a rain event to dislodge the sporangia from their sporagiophores and provide the surface moisture necessary for zoospore release and motility, although this can also happen with heavy dew. Following release of the sporangia, the duration of wetting required for infection to occur is relatively brief: approximately 1.5–2 h at 15–25°C, whereas 4, 6, and 8 h may be required at 10, 5, and 30°C, respectively. Symptoms generally appear within 5–6 days when postinfection temperatures remain between 18 and 26°C, although the incubation period can be as little as 4 days on highly susceptible tissues under optimal environmental conditions. Temperatures substantially lower or higher than the optimum persisting for a prolonged period can extend the incubation period significantly, e.g., 12 days at 12 or 34°C, with temperatures intermediate to the optimum producing correspondingly intermediate incubation period durations. At the coldest temperatures investigated (8–10°C), the incubation period may last 16–19 days.

Leaves are most susceptible to infection when emergent or undergoing expansion, but on susceptible cultivars they never reach a high level of resistance and once infected generally become severely diseased if conditions remain favorable. Berries acquire ontogenic (age-related) resistance to direct infections coincident with the conversion of functional stomata to lenticels, approximately 2–3 weeks after bloom. However, such berries may still become diseased via the growth of pathogen mycelium from infected rachises and pedicels, which remain susceptible to new infections for one or more weeks beyond this time.

Lesion longevity and production of sporangia are dependent upon the number of cycles of sporulation. Sporangium production is maximal through the first or second sporulation event but declines rapidly with each successive event until most of the lesion becomes necrotic and spore formation is restricted to the periphery, generally once the lesion has sporulated four to five times. In the vineyard, ungerminated sporangia die within 8–10 h after sunrise under clear skies but remain viable for at least 48 h after formation if conditions are overcast. There are some reports of survival extending up to 5–7 days; this would provide the potential for long-distance dispersal, which could be meaningful in new viticultural districts if the pathogen was not yet established.

The complex interactions between temperature, daylight, rainfall, humidity, and host development have encouraged the creation of several models to forecast disease development based upon an integration of these various factors. Many, if not most, are based upon the same foundational studies, as amended by more recent work, and thus produce similar advisories when provided the same host and environmental inputs. The most destructive epidemics occur in years with prolonged cloudy weather, frequent rain events coincident with temperatures near the optimum of 25°C, high humidity resulting in extended periods of evening leaf wetness, and the absence of extended rain-free periods with temperatures above 30°C. Regular, turbulent, evening thunderstorms particularly favor the spread of the disease throughout a vineyard.

Management

Site selection and canopy management practices that promote rapid drying of leaves, inflorescences, and clusters help to limit the potential for infection and disease spread. Mulches that impede the movement of primary zoospores from the soil to the grapevines; early-season elimination of low-lying foliage most subject to infections derived from overwintering oospores (e.g., shoots ["suckers"] arising from the basal trunk, or seedlings from uncomposted pomace used for mulch [Fig. 97]); and the destruction of nearby secondary inoculum sources (wild or feral grapevines, or abandoned vineyards) theoretically should all aid in limiting disease development, to the extent that such techniques are practical. However, because of the high susceptibility of most cultivated grapevines, some degree of fungicidal control is almost always necessary in regions and years with weather conditions conducive to downy mildew development.

A number of surface-acting fungicides (cupric salts, dithiocarbamates, and phthalimides) are efficacious as preventive treatments. The risk of *P. viticola* developing resistance to them is low, as they are fungitoxic at multiple cellular sites.

Fig. 97. Sporulation of *Plasmopara viticola* from the underleaf of an infected grape seedling growing on the vineyard floor; these plants originated from uncomposted seeds used for mulch. (Courtesy R. C. Pearson)

The duration of their effectiveness is limited by both the rate of vine growth (tissues present at the time of application are the most completely protected) and reductions of the initial deposit resulting from washing off by rain and weathering through hydrolysis and photolysis. The duration of protection provided by such fungicides can, therefore, vary greatly (e.g., from 3–14 days, typically 7–10 days) depending upon vine growth, post-application weather, the retentive properties of the fungicide, and its innate efficacy against downy mildew.

Over the past 30–40 years, many groups of fungicides active against downy mildews have been developed which exhibit variable degrees of systemicity, i.e., absorption by and translocation within the plant. These include cymoxanil, phenylamides, QoI (e.g., strobilurin) and QoO materials, carboxylic acid amide (CAA) products such as dimethomorph and mandipropamid, iprovalicarb, ametoctradin, and phosphonates. In addition to providing some residual protective activity of the treated tissues, these also have the advantages of not being washed off by rainfall and providing variable levels of postinfection (sometimes called "curative") activity when applied before symptoms appear, i.e., during the incubation period. These fungicides affect *P. viticola* most strongly within the early stages of colonization. At optimum temperatures for the pathogen, the mycelium is well developed by 3 days after the start of infection, and most of the penetrating fungicides are unable to prevent the subsequent development of lesions when applied beyond this stage. However, while still allowing lesion development to proceed, applications of these materials during the late incubation or initial-symptom stages of infection often inhibit subsequent sporulation from such lesions, thereby reducing secondary spread of the disease from them. However, the penetrating fungicides typically act at a single or very limited number of target sites in the metabolism of *P. viticola*, and there is a high risk of selecting strains of the pathogen resistant to specific classes of these materials unless they are carefully managed. Once resistance begins to develop, the great reproductive capacity of *P. viticola* (large numbers of spores widely dispersed in wind currents, a short incubation period promoting multiple generations per year) allows a few resistant individuals to multiply to a level where the fungicide no longer provides effective control, unless their reproduction is restricted some other way. Therefore, it is often recommended that penetrating fungicides with curative effects be applied in combination with at least one other fungicide active against the local population of *P. viticola*; this is typically, but not necessarily, a surface-active protective fungicide at minimal risk for resistance development itself. Fungicide classes at risk of resistance development should be applied no more than two or three times per season (all materials within a class), and applications during a severe outbreak of the disease (large populations from which to select resistant individuals) should be avoided.

Selected References

Blaeser, M., and Weltzien, H. C. 1978. Die Bedeutung von Sporangienbildung, ausbreitung und keimung fur die Epidemiebildung von *Plasmopara viticola*. Z. Pflanzenkr. Pflanzenschutz 85:155-161.

Hill, G. 1989. Effect of temperature on sporulation efficiency of oilspots caused by *Plasmopara viticola* (Berk & Curt. ex de Bary) Berl. & de Toni in vineyards. Vit. Enol. Sci. 44:86-90.

Kennelly, M. M., Gadoury, D. M., Wilcox, W. F., Magarey, P. A., and Seem, R. C. 2007. Primary infection, lesion productivity, and survival of sporangia in the grapevine downy mildew pathogen, *Plasmopara viticola*. Phytopathology 97:512-522.

Kennelly, M. M., Gadoury, D. M., Wilcox, W. F., Magarey, P. A., and Seem, R. C. 2005. Seasonal development of ontogenic resistance to downy mildew in grape berries and rachises. Phytopathology 95:1445-1452.

Kiefer, B., Riemann, M., Büche, C., Kassemeyer, H. H. and Nick P. 2002. The host guides morphogenesis and stomatal targeting in the grapevine pathogen *Plasmopara viticola*. Planta 215:387-393.

Langcake, P., and Lovell, A. 1980. Light and electron microscopical studies of the infection of *Vitis* spp. by *Plasmopara viticola*, the downy mildew pathogen. Vitis 19:321-337.
Mueller, K. 1936. Die biologischen Grundlagen für die Peronsopora-bekämpfung nach der Inkubationskalender-Methode. Z. Pflanzenkr. Pflanzenschutz 46:104-108.
Rumbolz, J., Wirtz, S., Kassemeyer, H. H., Guggenheim, R., Schäfer, E., and Büche, C. 2002. Sporulation of *Plasmopara viticola*: Differentiation and light regulation. Plant Biol. 4:413-422.
Spencer-Phillips, P. T. N., Gisi, U., Lebeda, A. 2002. Advances in downy mildew research. Kluwer Academic Publisher, Dordrecht, Boston, London.
Unger, S., Bueche, C., Boso, S., and Kassemeyer, H.-H. The course of colonization of two different *Vitis* genotypes by *Plasmopara viticola* indicates compatible and incompatible host-pathogen interactions. Phytopathology 97:780-786.
Wong, F. P., Burr, H. N., and Wilcox, W. F. 2001. Heterothallism in *Plasmopara viticola*. Plant Pathol. 50:427-432.
Wong, F. P., and Wilcox, W. F. 2001. Comparative physical modes of action of azoxystrobin, mancozeb, and metalaxyl against *Plasmopara viticola* (grapevine downy mildew). Plant Dis. 85:649-656.

(Prepared by H.-H. Kassemeyer, D. M. Gadoury, G. Hill, and W. F. Wilcox)

Esca, Petri, and Grapevine Leaf Stripe Diseases

Esca is a name given in antiquity to a disease of grapevine that appeared as a sudden, lethal wilting of the entire vine. It was first formally described on grapevine in France as folletage in 1865, and the cause of death was later attributed to wood decay (white rot) caused by basidiomycetous fungi, mainly *Phellinus igniarius* and *Stereum hirsutum*; the term *apoplexy* was substituted for that of folletage. In 1926, another symptom was linked to the same syndrome: interveinal chlorosis and necrosis on the leaf blades, later described as leaf stripes, and the term *esca* was officially assigned to the disease. This was done after it was recognized that the names *esca* and *iska* had been used since ancient times in southern Italy and a Greek region (Smirne) of the modern Turkey, respectively, to identify apoplectic grapevines showing rotted wood, albeit in the absence of leaf stripe symptoms. More recent research has attributed the foliar symptoms to ascomycetous vascular (tracheomycotic) pathogens distinct from the wood decay organisms originally described as the causal agents of esca. These vascular pathogens can produce different symptoms and syndromes at different stages in the life of the vine, variously termed as black wood streaking of rootstock, Petri disease in young vines, and leaf stripe disease (also known as "young esca").

Presently, it seems advisable to use the term *esca* in its original meaning of apoplexy and decay, and to assign the name "grapevine leaf stripe disease" (GLSD) to the vascular diseases associated with foliar symptoms. As it is very common to have wood rot and leaf stripe disease on the same vine, especially on vines greater than 8–10 years old, the term *esca proper* (EP = GLSD + white rot) has been suggested to indicate the syndrome of vascular disease and wood rot and therefore to include foliar symptoms, i.e., a complex of two different diseases. This complex of diseases is also known in the United States as black measles, because the most frequent and important symptom appears on berries in the form of small, gray to purple to brown spots on the fruit surface, making table grapes unmarketable.

Petri disease, initially called by names such as young vine decline or black goo, occurs in many of the world's production areas. It is a form of decline usually associated with vines less than 7 years old, showing fungal infection of the rootstock. It has become most common in California's North Coast production region, but also has been reported in eastern North America, Europe, Australia, New Zealand, and South Africa. An increased prevalence of the disease in California coincided with a massive replanting of grapevines in Napa, Sonoma, and Mendocino counties in the 1990s as a result of phylloxera (*Daktulosphaira vitifoliae* [Fitch]) damage to the common and formerly resistant rootstock AXR1. Although the rootstocks of new vines had greater resistance to phylloxera, they were more susceptible to the fungal causes of Petri disease. In the United States and other countries, the disease is reportedly linked to water deficit or other stress factors, which favor symptom expression in infected vines. Although widespread in occurrence, vines showing decline due to Petri disease usually constitute a minor portion (1–5%) of a newly planted vineyard.

GLSD is an important disease of grapevine worldwide, both in the presence and absence of white rot. (Henceforth, the acronym GLSD/EP is be used for simplicity, as most reports have not distinguished between the single disease and the combination.) GLSD/EP can be found in California on wine, table, or raisin grapes in years with high spring rainfall and high summer temperatures. The same disease is present across Europe in a wide range of climates, as well as in North Africa and, more rarely, in Australia and South Africa. Like Petri disease, GLSD/EP has increased in significance in California in recent years and in Europe since the end of the 1980s. Leaf symptoms that once were uncommon on vines less than 10 years old are now common on vines 1–6 years old.

The disease affects crop quality and yield, as well as the productive life of the vineyard. Losses are normally underestimated because the expression of symptoms fluctuates considerably; a vineyard may exhibit variable disease incidence among seasons, as individual infected vines may show symptoms one year but not the next. In California, losses are greatest on table grape varieties, especially older Thompson Seedless (Sultana) vineyards in the San Joaquin Valley. Affected table grape clusters are unmarketable because of their disagreeable appearance and the acrid taste that can occur; they are also more susceptible to Botrytis bunch rot. In Europe, wine grapes are also heavily affected, with Cabernet Sauvignon, Ugni Blanc, Sauvignon Blanc, Mourvèdre, and Sangiovese being only some of the most susceptible cultivars. Affected wine grapes have less sugar and greater acidity than those from healthy vines, and generally produce lower-quality wine.

Symptoms

The first symptom of infection by the vascular fungi can be noticed in nursery material as a brown to black streaking of the wood. Longitudinal sections show dark streaks, single or in bundles, appearing in cross section as spots or groups of black vessels along the wood rings. A black, gumlike substance exudes from these spots within a few minutes after cutting through affected vessels (Fig. 98). The streaking often starts

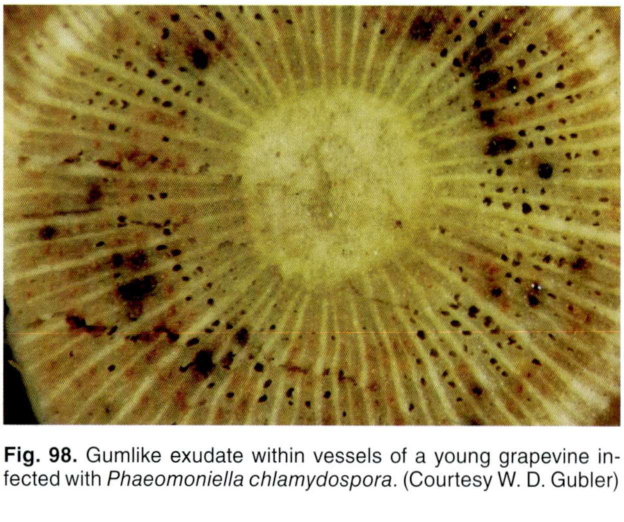

Fig. 98. Gumlike exudate within vessels of a young grapevine infected with *Phaeomoniella chlamydospora*. (Courtesy W. D. Gubler)

at the graft union in young vines or at the surface of a pruning wound on mature vines. These internal symptoms of fungal infection may be related to the decline of young vines once they are planted in the field (Petri disease).

Symptoms of Petri disease include reduced trunk diameter, shortened internodes, stunted shoot growth, shoot dieback, and chlorotic and/or necrotic leaves (Fig. 99). The most prominent internal symptom is the presence of brown to black streaks or spots in the xylem vessels of the woody cylinder, as described above (Fig. 100).

The most visible symptom of GLSD/EP in California is a peculiar berry spotting, particularly noticeable on white varieties. The skin is peppered with small, tan to purplish spots (Fig. 101). These spots may appear any time between fruit set and ripening, most commonly as berries are just attaining their full size, and may affect entire or only parts of clusters. In both severely affected vines and those showing early symptom expression, berries often crack and dry on the vine.

Shoot tip dieback, tendril dieback, and leaf collapse may occur in the spring (Fig. 102). Leaf symptoms usually develop on canes with fruit symptoms, but also can develop on those with normal fruit. Typically, affected leaves display small, chlorotic, interveinal areas that enlarge, eventually desiccating and becoming necrotic; in dark-fruited varieties, the chlorotic tissue may turn red before desiccating (Fig. 103) or red margins may surround the dead interveinal areas. A succession of

Fig. 99. Field symptoms of young vines affected with Petri disease. (Courtesy J. R. Úrbez-Torres)

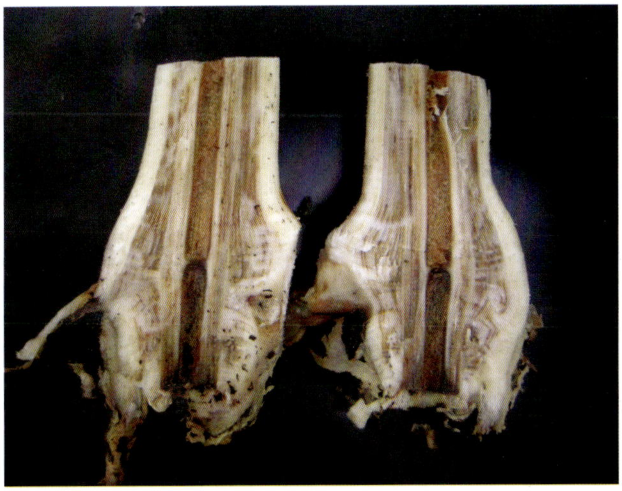

Fig. 100. Xylem necrosis and wood streaking of a young grapevine with Petri disease, as viewed in longitudinal section. (Courtesy J. R. Úrbez-Torres)

Fig. 101. "Measles" symptom of GLSD/EP. (Courtesy W. F. Wilcox)

Fig. 102. Tendril and shoot tip dieback, the first symptoms of GLSD/EP in the spring. (Courtesy J. R. Úrbez-Torres)

Fig. 103. Interveinal leaf-reddening symptom of GLSD/EP on certain dark-fruited varieties. (Courtesy J. R. Úrbez-Torres)

chlorotic/pigmented and necrotic tissues that alternate between major veins bordered by a strip of unaffected tissue produces a so-called "tiger stripe" pattern (Fig. 104). Severely affected leaves drop and the cane may die back from the tips. Foliar symptoms may appear at any time during the growing season, but they are most prevalent during mid- and late summer. Entire vines or only portions of a vine may be affected, but occasionally the symptoms appear so quickly and dramatically that the entire vine collapses and dies (apoplexy) (Fig. 105). The latter symptom is most commonly seen in mid-season in regions with higher summer temperatures. Apoplexy occurs most commonly in the Mediterranean areas of Europe, and some workers have suggested that it may be a symptom of a distinct disease that can occur in conjunction with GLSD/EP.

Internal symptoms of GLSD include the same brown to black longitudinal streaks (Fig. 106) or spots that appear in the rings of xylem vessels and often exude a black, sticky gum when cut in cross section (Fig. 107), as described above for Petri disease. A brownish-red necrosis may also develop primarily along the central part of the trunk, similar to the sectorial necrosis caused by species in the Botryosphaeriaceae, *Eutypa lata*, and *Phomopsis viticola* following their infection of the wood through wounds. However, this canker development has not been attributed to the *Phaeomoniella* or *Phaeoacremonium* spp. that are known to cause GLSD.

The peculiar discontinuity from year to year of the foliar symptoms of GLSD has not yet been explained. In California, there appears to be a close correlation between symptom expression and current year infection, with symptomatic vines sometimes appearing fully healthy the following year and again showing symptoms in an unpredictable manner after one or more additional years have passed. Foliar symptoms are reportedly linked to toxic metabolites (certain polysaccharides and polypeptides, scytalone, and isosclerone) produced by the causal fungi *Phaeomoniella chlamydospora* and *Phaeoacremonium aleophilum*. It is possible that the involvement of toxic metabolites in foliar symptomatology contributes to the complex and still-unclarified interaction between the pathogens, the host, and environmental factors with respect to variable symptom development.

White heart rot was the first internal symptom to be described as typical of esca. However, even though this concept stood for over six decades after the disease was named and foliar symptoms attributed to it, all attempts to reproduce the foliar and fruit symptoms described above failed when vines were inoculated with several basidiomycete fungi that caused such wood decay. Recent research indicates that white rot in

Fig. 104. Interveinal necrosis and "tiger-striping" symptom of GLSD/EP. (Courtesy W. F. Wilcox)

Fig. 105. Apoplexy (sudden collapse) symptom of GLSD/EP. (Courtesy P. Lecomte)

Fig. 106. Internal wood streaking in a mature grapevine affected with GLSD/EP, viewed in longitudinal section. (Courtesy J. R. Úrbez-Torres)

Fig. 107. Internal wood symptoms in a mature grapevine affected with GLSD/EP, viewed in cross section. (Courtesy J. R. Úrbez-Torres)

wood does not affect vine performance except for longevity. Extensive decay in older vines has been associated with apoplectic symptoms, but apoplexy has not been reproduced by inoculation with any pathogen.

Causal Organisms

In California, the primary pathogen consistently associated with vines showing symptoms of Petri disease has been *Phaeomoniella chlamydospora* (W. Gams, Crous, M. J. Wingf. & Mugnai) Crous & W. Gams. The pathogen more closely associated with GLSD/EP of older vines in California vineyards is *Phaeoacremonium aleophilum* W. Gams, Crous, M. J. Wingfield & L. Mugnai, although both pathogens occur together in many wood samples of cordons and spurs. However, *Pa. chlamydospora* is always the more prevalent of the two species associated with both Petri disease and GLSD/EP in Europe, Australia, and South Africa. Twenty-five different *Phaeoacremonium* spp. have been isolated from grapevines to date, but *Pm. aleophilum* is the most prevalent worldwide and the only one consistently associated with foliar symptoms. The internal wood, foliar, and fruit symptoms described for these diseases above have been reproduced following inoculation with both *Pa. chlamydospora* and *Pm. aleophilum*.

Both *Pa. chlamydospora* (Chaetothyriales, Herpotrichiellaceae) and *Pm. aleophilum* produce conidia on phialidic conidiophores. The initially yeastlike, whitish mycelium of *Pa. chlamydospora* soon becomes dark green (Fig. 108), with slender phialides bearing a typically hyaline conidiogenous cell that produces straight, 3–4 × 1–3 µm conidia. The mycelium of *Pm. aleophilum* is honey brown, frequently producing a yellow pigment on oatmeal or potato dextrose agar (Fig. 109), and supports phialides that produce ellipsoid to allantoid, 4–6 × 1–3 µm conidia. *Pa. chlamydospora* produces tiny, *Phoma*-like pycnidia on the exposed vascular tissue of grapevine cordons and spurs, which serve as sources of inoculum in infested vineyards. *Pa. chlamydospora* has been isolated as a pathogen only on grapevine, with the exception of a single isolation from *Convolvulus arvensis*, which suggest a possible, albeit rare, risk of infection from weeds.

Togninia minima (Tul. & C. Tul.) Berl. (Diaporthales, Togniniaceae), the teleomorph of *Pm. aleophilum*, spreads via ascospores produced by perithecia on infected vines. Though *Pm. aleophilum* occurs in different host species including stone fruits, olive trees, and kiwifruit vines, no perithecia have been found on these crops. Perithecia of *T. minima* have been found on grapevines exhibiting leaf stripe and black measles symptoms in vineyards located throughout California and South Africa; the perithecia are mostly aggregated, (160–)250–285(–420) µm in diameter at their basal part and (200–)285–325(–400) µm tall, with a long, sometimes double neck. Perithecia of four other *Togninia* species have been associated with GLSD on grapevines and have been found on trees such as ash (*Fraxinus* spp.) in riparian and forest ecosystems near vineyards. All have been shown to be pathogenic on grapevine; however, *T. minima* is the most prevalent of these species in vineyards.

The vascular disease caused by *Pa. chlamydospora* and *Pm. aleophilum* is, especially in older vines, typically the result of infection through pruning wounds. Later, this tissue is sometimes colonized by basidiomycetous decay fungi, resulting in wood rot. Species of the genus *Fomitiporia* appear to be among the more common of such organisms involved, including *F. mediterranea* (=*F. punctata*) in Europe, where the white wood decay symptom has been reproduced by inoculating grapevines with this species. *Fomitiporia mediterranea* forms resupinate basidiocarps of a cinnamon color on grapevine trunks and various woody host plants. Other common colonizers include *F. polymorpha* in California, *F. australiensis* in Australia, and various *Fomitiporia* and *Phellinus* species in South Africa. Some workers hypothesize that the succession of such decay organisms following attack by the vascular fungi contributes to the complexity of the symptomatology of EP/GLSD, although this has never been demonstrated experimentally.

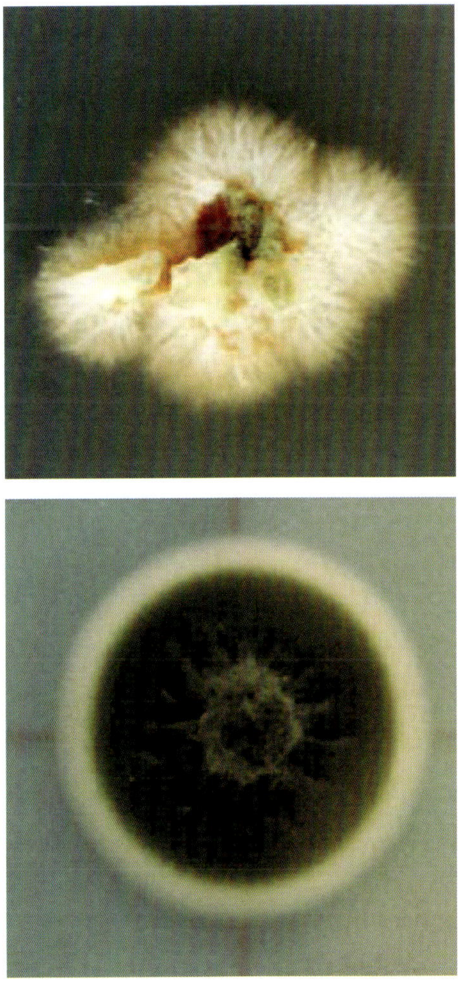

Fig. 108. Young, whitish (top) and older, dark-green (bottom) colonies of *Phaemoniella chlamydospora* growing on malt agar. (Courtesy L. Mugnai, reproduced by permission from Plant Dis. 83:411)

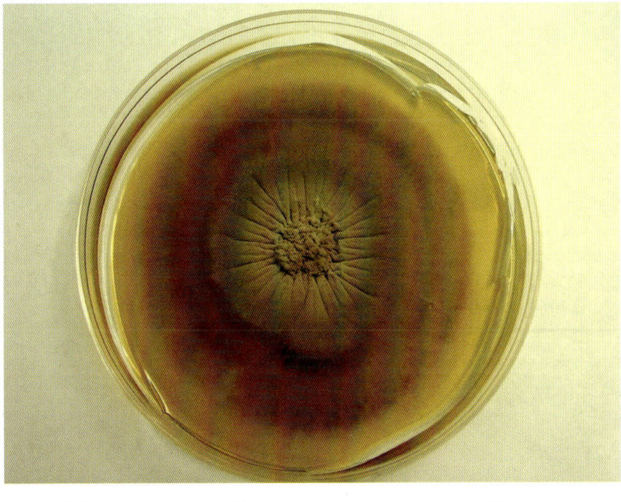

Fig. 109. Colony of *Phaeoacremonium aleophilum* producing a yellow pigment on potato dextrose agar. (Courtesy J. R. Úrbez-Torres)

Disease Cycle and Epidemiology

Phaeomoniella chlamydospora overwinters as pycnidia residing on the grapevine, most commonly in 3- to 5-year-old pruning wounds. Mycelium formed on exposed infected wood can also produce conidia via its phialides. Spores are released throughout the year, but the number of spores released is very low during the coldest winter months (below 10°C) and during hot periods. The greatest numbers of spores are usually trapped in early to late autumn and in spring to early summer, and are always linked to rainfall. New infections occur through pruning wounds, which have been shown to remain susceptible for up to 16 weeks, although their susceptibility decreases over time.

Togninia minima perithecia reside inside dead, decaying vascular tissue, e.g., old pruning wounds or vine injuries. Ascospores are released during fall and winter rainfall events, corresponding to the period of dormant pruning in many regions. Spores are forcibly ejected from the asci to a height of approximately 10 cm, whereupon they are transported by wind currents or are dispersed to pruning wounds by water droplets.

When spores of one (or both) of these fungi reach injured wood tissue, they germinate and growth occurs in the xylem vessels, causing plugging of the vessels through the production of gums and tyloses. The end result of this infection is internal discoloration of the wood, which in association with other poorly understood factors can lead to the development of typical leaf stripe symptoms in summer. In a California study, spur positions inoculated in February showed severe stunting of shoots (approximately 50%) in early summer and both leaf and fruit symptoms developed by August. The pathogens were able to colonize the vascular tissue of a 1-year-old spur and move into the basal bud tissue within 7 months of inoculation.

In addition to field spread via aerial inoculum, *Pa. chlamydospora* and *Pm. aleophilum* also are distributed via infected nursery plants. As vascular pathogens, these fungi can move in infected mother plants, resulting in asymptomatic but infected propagating material. Perhaps relatedly, routine isolations from symptomatic younger vines (less than 7–8 years old) in California yielded the pathogens primarily from the rootstock portion of the plant, suggesting an infection court other than pruning wounds for such vines.

Phaeomoniella chlamydospora and *Pm. aleophilum* appear to be ubiquitous fungi. Both have been shown to exhibit endophytic behavior in grapevines and are able to reside in the trunk xylem undetected, sporadically causing symptoms under the influence of various stress factors such as low water and high temperature conditions.

Management

Control of the various syndromes associated with esca, Petri disease, and GLSD requires a broad and integrated approach. The first step in control of the tracheomycotic fungi is prevention of infection in the nurseries, accomplished through the identification, maintenance, and use of clean scion and rootstock material. For control of Petri disease, it is also preferable to plant into sites where existing infected vines are not present. Avoiding or ameliorating vine stress is also highly advisable, e.g., planting new vineyards in a manner not to promote "J" rooting, limiting yields during the initial establishment years thereafter, and providing timely irrigation as necessary.

Reduction of surface inoculum can be obtained with dormant sprays designed to kill spores produced on the outer vine surface before their discharge. In Europe and California, thorough spraying with sodium arsenite during the dormant season was long practiced as a means of controlling GLSD/EP, and the syndrome appeared to become significantly more common after this material was banned from use. Late dormant sprays of calcium polysulfide (lime sulfur), a general oxidant that reduces surface inoculum prior to its dispersal to pruning wounds, has provided positive results in some trials. However, this product must physically contact the inoculum in order to be effective, so sprays must be applied in high water volumes to thoroughly wet the cordons and spurs, such that the material is forced through exfoliating bark and down onto the diseased tissues. Pruning grapevines before such applications is recommended to improve coverage and to reduce the amount of chemical needed.

Infections of new pruning wounds can potentially be reduced by protecting them through wound sealants containing fungicides or biological control agents specifically selected for such use. Research to identify the materials most suitable for this purpose is ongoing in several regions.

As with nearly all diseases, appropriate cultural practices are an important component of an integrated control program. Diseased wood should always be removed from the vineyard and burned if possible, as this reduces inoculum potential within the vineyard. Although infection can occur through pruning wounds of any size, the probability of infective spores being deposited upon a susceptible pruning wound is directly proportional to its exposed area; therefore, utilizing pruning and training systems that minimize large wounds is desirable in terms of disease management. When the disease is already present in the vineyard, trunk renewal or retraining cordons when symptoms first appear provides an immediate, though often temporary, reduction in the incidence of symptomatic vines. Theory does not suggest nor does experience indicate that the vascular pathogens can spread on pruning implements; therefore, it is not necessary to prune healthy and diseased vines separately.

In regions such as California, where spore release is concentrated in winter and early spring, the practice of double pruning or late pruning has been shown to reduce the establishment of new infections for various wood diseases, including GLSD/EP. Initially, all canes are pruned to five to seven buds in the early winter, a step that can easily be mechanized (e.g., with a tractor-mounted rotary saw), leaving the entire vineyard with spur positions in which new infections resulting from winter rains progress slowly under ambient seasonal temperatures. The second cut, designed to remove any such new infections before they reach the lower bud positions to be retained, is made in late winter or early spring after most spores have been released and new pruning wounds are less susceptible to infection. Ideally, this second cut should be followed by application of an effective fungicide, delivered with a sprayer with directed nozzles.

Unfortunately, this approach is not considered to be applicable in Europe, where the greatest number of spores is collected in late spring (May–June), and where there is no evidence of reduced susceptibility to infection in early spring versus winter. In these regions, choosing a careful pruning technique, selecting trellising systems that avoid large wounds, and applying a suitable and effective wound dressing continue to comprise the traditional preventive program.

Selected References

Di Marco, S., Mazzullo, A., Calzarano, F., and Cesari, A. 2000. The control of esca: Status and perspectives. Phytopathol. Mediterr. 39:232-240.

Edwards, J., and Pascoe, I. G. 2004. Occurrence of *Phaeomoniella chlamydospora* and *Phaeoacremonium aleophilum* associated with Petri disease and esca in Australian grapevines. Australas. Plant Pathol. 33:273-279.

Eskalen, A., Feliciano, A. J., and Gubler, W. D. 2007. Susceptibility of grapevine pruning wounds and symptom development in response to infection by *Phaeoacremonium aleophilum* and *Phaeomoniella chlamydospora*. Plant Dis. 91:1100-1104.

Ferreira, J. H. S., van Wyk, P. S., and Calitz, F. J. 1999. Slow dieback of grapevine in South Africa: Stress-related predisposition of young vines for infection by *Phaeoacremonium chlamydosporum*. S. Afr. J. Enol. Vitic. 20:43-46.

Fischer, M. 2002. A new wood-decaying basidiomycete species associated with esca of grapevine: *Fomitiporia mediterranea* (Hymenochaetales). Mycol. Prog. 1:315-324.

Fourie, P. H., and Halleen, F. 2004. Proactive control of Petri disease of grapevine through treatment of propagation material. Plant Dis. 88:1241-1245.

Gramaje, D., and Armengol, J. 2011. Fungal trunk pathogens in the grapevine propagation process: Potential inoculum sources, detection, identification, and management strategies. Plant Dis. 95:1040-1055.

Larignon, P., and Dubos, B. 1997. Fungi associated with esca disease in grapevine. Eur. J. Plant Pathol. 103:147-157.

Mostert, L., Halleen, F., Fourie, P., and Crous, P. W. 2006. A review of *Phaeoacremonium* species involved in Petri disease and esca of grapevines. Phytopathol. Mediterr. 45:S12-29.

Mugnai, L., Graniti, A. and Surico, G. 1999. Esca (black measles) and brown wood-streaking: Two old and elusive diseases of grapevines. Plant Dis. 83:404-418.

Rooney-Latham, S., Eskalen, A., and Gubler, W. D. 2005. Occurrence of *Togninia minima* perithecia in esca-affected vineyards in California. Plant Dis. 89:867-871.

Surico, G., Marchi, G., Braccini, P., and Mugnai, L. 2000. Epidemiology of esca in some vineyards in Tuscany (Italy). Phytopathol. Mediterr. 39:190-205.

Surico, G., Mugnai, L., and Marchi, G. 2008. The Esca Disease Complex. Pages 119-136 in: Integrated Management of Diseases Caused by Fungi, Phytoplasma and Bacteria. A. Ciancio and K. G. Mukerji, eds. Springer, Heidelberg, Germany.

Weber, E. A., Trouillas, F. P., and Gubler, W. D. 2007. Double pruning of grapevines: A cultural practice to reduce infections by *Eutypa lata*. Am. J. Enol. Vitic. 58:61-66.

(Prepared by W. D. Gubler, L. Mugnai, and G. Surico)

Eutypa Dieback

Eutypa dieback, known also as "dying arm" and formerly as "dead arm," is one of the most destructive diseases of the trunk and other woody tissues of grapevines. It is found in most countries of both hemispheres, with the frequency in any one region seemingly limited only by the incidence of rainfall. The disease occurs chronically and affects productivity, berry quality, and the lifespan of infected vines. A California study conducted in the late 1990s estimated that the loss of net income to wine grape growers attributed to this disease (an estimate now known to have included losses from Botryosphaeria dieback) was $260 million dollars per annum at that time, or 14% of the gross farm gate value of California wine grapes.

Eutypa lata, the primary causal organism, has a wide host range that includes approximately 80 species distributed across at least 27 vascular plant families. Most hosts are tree species, including some that are components of natural forests. The most severely affected horticultural hosts are grapevine, apricot, cherry, kiwifruit, and blueberry.

Symptoms

The first symptoms of Eutypa dieback usually are not seen until grapevines are at least 10 years old, with symptomatic vines typically becoming more numerous each year thereafter. Symptoms initially appear in the spring when healthy new shoots are approximately 25–50 cm long. The most readily recognized and diagnostic symptoms involve reduced vigor and deformation of the affected shoots, with young leaves appearing chlorotic, reduced in size, and cupped; these often develop small, necrotic spots and tattered margins, sometimes with larger areas of necrosis, as they age (Fig. 110). A marked dwarfing of the internodes accompanies the development of the leaf symptoms, contributing to a stunting of the shoot. Decreases in shoot length and leaf size become progressively greater over succeeding years. Clusters on affected shoots may have a mixture of large and small berries (Fig. 111).

At disease onset, mildly affected shoots are readily visible until late spring but then become obscured by healthy growth as the canopy continues to develop (Fig. 112). However, as foliar symptoms on diseased arms become increasingly severe in succeeding years, eventually only extremely dwarfed shoots remain (Fig. 113). Finally, all of the arms fail to produce new shoots in the spring and the vine dies.

Close examination of a cordon or trunk with vascular connections to symptomatic shoots usually reveals a canker surrounding a pruning wound that was made several years previously. Removal of bark is necessary to reveal the extent of the canker emanating from the point of origin (Fig. 114). In cross section, a zone of brown, necrotic sapwood typically is found, expanding from an initial small, wedge-shaped sector of symptomatic tissue to include a progressively greater area over time (Fig. 115). However, this symptom can also be produced by other canker-forming fungi and is not diagnostic for Eutypa dieback.

The fungus does not infect and colonize green current-season shoots, and therefore it cannot be isolated from these tissues. Rather, the foliar symptoms are believed to be induced by

Fig. 110. Characteristic symptoms of Eutypa dieback on affected young leaves, including size reduction, chlorosis, cupping, and necrosis. (Courtesy M. Sosnowski)

Fig. 111. Clusters from shoots exhibiting symptoms of Eutypa dieback. (Courtesy M. Sosnowski)

translocated toxins, such as eutypine, generated by mycelium of the fungus in the wood cankers. However, recent studies were unable to detect any of the known toxins in symptomatic green tissues, nor were any detected in the sap of infected grapevines. Thus, the mechanisms leading to disease expression are more complex than initially thought, and remain at best only partially understood.

Causal Organisms

Eutypa lata (Pers.:Fr.) Tul. & C. Tul. (syn. *E. armeniacae* Hansf. & Carter, anamorph *Libertella blepharis* A. L. Smith [syn. *Cytosporina* sp.]) was long believed to be the sole causal agent of Eutypa dieback. More recently, extensive disease surveys in nascent and established viticultural regions and improved molecular tools have associated multiple other species in the Diatrypaceae with wedge-shaped cankers, fruiting bodies of these fungi have been identified on diseased wood, and several of these species have produced necrotic lesions on grapevine cuttings in greenhouse bioassays. However, the foliar symptoms of Eutypa dieback have only been reproduced by inoculation with *E. lata* and a newly recognized *Eutypa* sp., which should be considered the only causal agents of this disease.

Eutypa lata is the primary causal organism. It is found worldwide, but is most common and problematic in regions with Mediterranean climates. Asci, produced within perithecia, are borne on pedicels 60–130 μm long and measure 30–60 × 5–7.5 μm, with an apical pore. The eight ascospores are pale

Fig. 112. Stunted shoot exhibiting symptoms of Eutypa dieback (center), becoming obscured by growth of healthy shoots in late spring. (Courtesy M. Sosnowski)

Fig. 113. Later-stage symptom of Eutypa dieback, with all shoots on an affected arm stunted and deformed. (Courtesy W. F. Wilcox)

Fig. 114. Large canker of Eutypa dieback emanating from an old pruning wound, revealed by removing the rough outer bark. (Reprinted, with permission, from Moller and Kasimatis, 1981)

Fig. 115. Zone of necrotic sapwood exposed in cross section through a cankered grapevine arm, revealing the extent of colonization by an expanding *Eutypa lata* infection. (Courtesy M. Sosnowski)

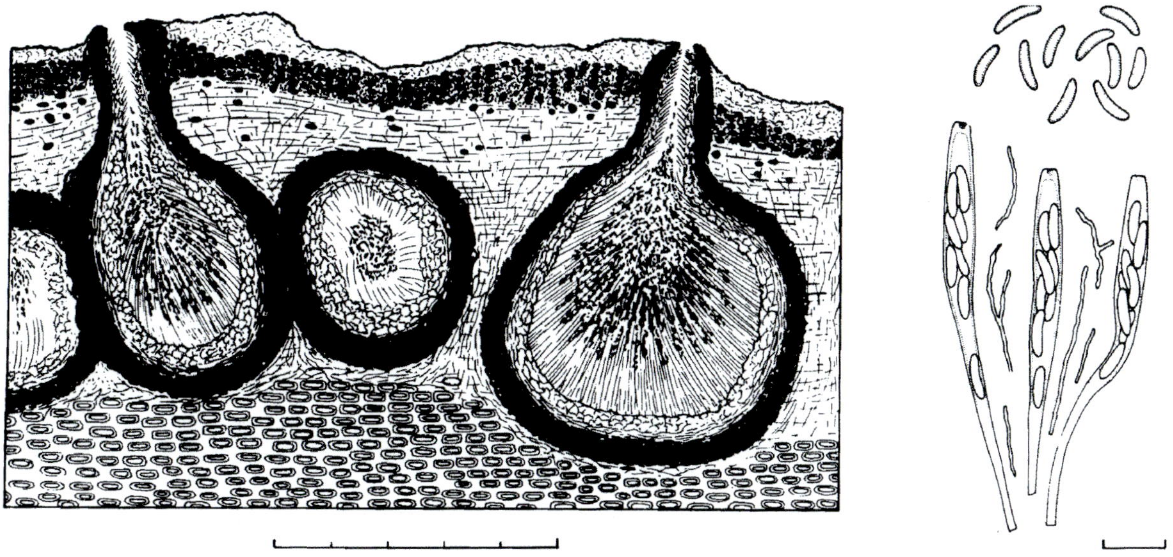

Fig. 116. Vertical section of perithecial stroma (left) and asci and ascospores (right) of *Eutypa lata*. (Reprinted, with permission, from Carter and Talbot, 1974)

Fig. 117. Perithecia of *Eutypa lata* embedded in a stroma on infected grapevine wood. (Courtesy M. Sosnowski)

Fig. 118. Cavities of perithecia of *Eutypa lata* revealed in a shallow slice cut from the stroma with a sharp blade. (Courtesy M. Sosnowski)

yellow and allantoid, measuring 6.5–11 × 1.8–2 µm (Fig. 116). On grapevines, perithecia (300–700 µm wide) are produced within stromata on diseased wood several years after the initial infection, at first in small patches surrounding the infection site, or sometimes on the wound stub that formed the original point of entry. Later, as the vine becomes more extensively invaded, larger areas of stromatic tissue may form within the outer bands of the dead wood of the trunk or cordon (Fig. 117). Stromata, visible after the loose bark has exfoliated, are black and continuous; perithecia are revealed when a shallow slice is cut from their surface with a sharp blade (Fig. 118).

The newly recognized *Eutypa* sp. has been identified only in the colder, humid climates of eastern North America. It shares several morphological features with *E. lata*, but is distinguished from the latter morphologically by its smaller perithecia width (200–400 µm) and ascospore length (5.5–9 µm).

Both Eutypa species may be isolated readily from small chips excised aseptically from the margin of discolored sapwood in diseased arms or trunks, using common laboratory growth media such as potato dextrose or malt agar. White mycelium grows from infected wood chips after 3–4 days at 20–25°C. Perithecia are not produced in culture, but conidiomata may develop after 6–8 weeks, often exuding the characteristic single-celled conidia that measure 18–45 × 0.8–1.5 µm (Fig. 119) in white to beige cirri. Conidia of *Eutypa* sp. and *E. lata* are similar in size. Exposing culture plates to an alternating 12-h light/dark regime or to near-ultraviolet radiation promotes sporulation.

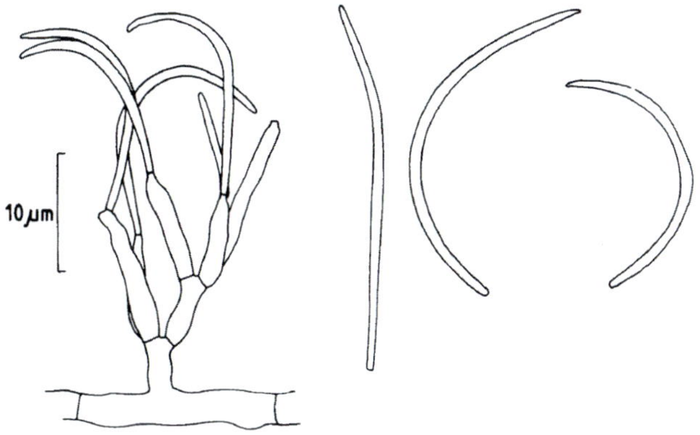

Fig. 119. Conidiogenous cells and conidiophores (left) and conidia (right) from a culture of *Libertella blepharis*. (Drawings by B. C. Sutton)

Positive identification of the causal agents of Eutypa dieback is confounded by an unresolved species concept within the family Diatrypaceae owing to the discordance between morphological descriptions of fungal type specimens and their phylogenetic analyses. The lack of sexual fruiting bodies produced in culture also has hindered the delimitation of species boundaries within this group. The presence in grapevine wood cankers of *Eutypa* species in addition to *E. lata* and *Eutypa* sp., including *E. laevata* and *E. leptoplaca*, and other closely related diatrypaceous fungi, such as species of *Diatrypella*, *Diatrype*, *Cryptovalsa*, *Cryptosphaeria*, and *Eutypella*, has further complicated disease diagnosis. Specimens of these various fungi look alike in culture, and the paucity and plasticity of associated morphological characters make them unreliable for use in identification to the species level. Molecular tools are currently being utilized to clarify the identity and role of potential etiological agents other than *E. lata* and the newly recognized *Eutypa* sp. In most diseased vineyards, the incidence of these other organisms is not as significant as that of *E. lata*, and it is possible that some or many have a restricted geographical distribution and/or a lesser ability to produce virulence factors (e.g., cell-wall degrading enzymes and phytotoxins) to colonize the plant hosts.

Disease Cycle and Epidemiology

Little is known about the biology of *Eutypa* sp., hence this discussion focuses on *E. lata*, a cosmopolitan organism that is restricted to geographical locations with at least 350 mm of annual rainfall. Although the fungus produces both conidia and ascospores in nature, only its ascospores are infectious. Ascospores are discharged with rainfalls of approximately 2 mm or greater. They are disseminated in wind currents and can travel long distances (more than 50 km), although the likelihood of infection is increased when an inoculum source is nearby. Ascospore discharge begins within 2 h of the onset of rainfall and continues for approximately 36 h, after which perithecia become depleted. At least 12 days are required for discharged asci to be replenished with new, mature ascospores. In regions where winters are temperate, ascospores are released and dispersed primarily in fall and spring. In regions where winter temperatures below 0°C prevail, dissemination of ascospores is greatest as temperatures begin to rise in late winter and early spring. In the US states of New York and Michigan, a winter snow blanket can prevent ascospore dispersal, although discharge can occur when snow is melting even in the absence of rainfall. Perithecia of *E. lata* are produced on a wide range of woody hosts, including cultivated fruit and nuts crops, ornamentals, and native trees. In California for example, apricot (*Prunus armeniaca*), cherry (*P. avium*), almond (*P. dulcis*), oleander (*Nerium oleander*), willow (*Salix* spp.), big leaf maple (*Acer macrophyllum*), and California buckeye (*Aesculus californica*) can serve as inoculum reservoirs for this pathogen. The host range for *Eutypa* sp. has not been established.

Infection is initiated when ascospores enter into the open ends of xylem vessels exposed by freshly made pruning wounds. Pruning-wound susceptibility is affected by the age of the wound and the time of pruning. Generally, the susceptibility of wounds diminishes markedly during the first 2 weeks following pruning, and after 5–6 weeks they are unlikely to become infected. Wound susceptibility declines more rapidly with high degree-day accumulation after pruning. In California, wounds made in December and January remain susceptible longer than wounds made in February and March.

Ascospores germinate within 24 h at the optimal temperatures of 20–25°C. Germination occurs within the vessels, usually 2 mm or more beneath the wound surface. The mycelia proliferate slowly, at first within the vessels and later through associated elements of the functional wood. *Eutypa lata* produces cell wall-degrading enzymes and phytotoxic secondary metabolites that contribute to the deterioration of the integrity and function of the vessel elements and in the subsequent development of the wood dieback. Cankers develop slowly on grapes, and no symptoms are seen during the first one or two growing seasons after infection. By the third or fourth season, a canker is usually visible, often accompanied by the foliar symptoms described above. Several more years may elapse before the affected arm or trunk is killed. Because of the slow progress of the disease, its full economic impact is not likely to be felt until after a vineyard reaches maturity. One of the greatest threats to vine productivity posed by Eutypa dieback is the possibility of infection of the many large wounds made when mature vines are reworked to change the pruning and trellising system, adapt the growth habit for mechanical harvesting, or change scion cultivars in the field.

Management

In regions where inoculum is produced abundantly on multiple hosts, Eutypa dieback cannot be effectively managed with sanitation methods alone, although such methods are a meaningful component of an integrated management program. In regions with vast plantings of grapes and few alternate hosts nearby, the regular excision and destruction of cankered wood to limit inoculum production may be highly beneficial in and of itself. Because early-stage infections are most apparent in spring prior to extensive canopy development, vineyards should be scouted for the presence of characteristic foliar symptoms when healthy shoots are 25–50 cm long, marking affected cordons for removal during the next dormant pruning cycle.

In some production regions, surgical removal of diseased cordons has been an effective method to control Eutypa dieback on affected vines. Usually, during the season prior to removal, a shoot arising from tissue scheduled for retention is trained to the wire to replace the cordon scheduled for removal. The following year, the diseased cordon is removed by cutting at least 10 cm below any visible canker symptoms. This remedial treatment allows individual vines to continue producing and potentially recover, although it addresses neither the prevention of new infections nor undetected nascent infections.

No grape cultivars are known to be immune, but where differences in cultivar tolerance are recognized, it is advisable to prune the least tolerant cultivars last. For example, Cabernet Sauvignon and Chardonnay are more susceptible than Merlot or Zinfandel. In California, delaying pruning until the end of the dormant season, by which time pruning wounds are healing faster, has effectively limited the spread of the disease. In large vineyards, however, this technique may not allow pruning to be completed before bud break. Therefore, a "double pruning" system was developed, whereby vine growth above the top wire is first cut in early winter (December or January) with a tractor-mounted saw, allowing a second cut to be made quickly by hand in the late dormant season (March), finally leaving two-bud spurs. Any new infections initiated after the first pruning will not have expanded to the position of the second cut by the time it is made, leaving the retained spur tissue free of the pathogen.

In many parts of eastern North America, growers of *V. vinifera* and other cold-sensitive cultivars routinely utilize a multiple-trunk training system and practice a regular program of "trunk renewal," periodically replacing older individual trunks with new ones derived from shoots that arise from basal latent buds. This practice significantly limits the severity of Eutypa dieback associated with the trunks, owing to the very slow development of the disease after infection.

Wounds are the main entrance port for the causal agents of Eutypa dieback. Although the fungus can invade 1-year-old canes after pruning, large wounds are more likely to become infected, owing to their greater surface area and attendant probability of being impacted by ascospores, and should be both minimized and protected where possible. Treatment of fresh pruning wounds with a concentrated slurry of thiophanate-methyl provides good protection against Eutypa dieback and several other important trunk diseases (e.g., Botryosphaeria dieback and esca). However, several applications may be required because the activity of a single fungicide application may not persist throughout the remaining period of host susceptibility. Similarly, a 5% (w/w) concentration of boric acid in commercial grafting or wound-protection paste also has provided a highly effective barrier against the invasion of pruning wounds by *E. lata*, as boron inhibits both ascospore germination and subsequent mycelial growth. The advantage of this paste is that it provides a physical barrier that is not washed off with rainfall, and therefore only one postpruning application is necessary. However, boron has limited activity against other trunk diseases (e.g., Botryosphaeria dieback) and cases of phytotoxicity have been reported in which the bud(s) below the treated wound fail to grow in the spring. Both treatments should be applied by hand as a wound dressing to optimize coverage of the infection court and subsequent disease control. Various international studies have reported the efficacy of other fungicides for control of *E. lata*, including members of the DMI (tebuconazole, flusilazole, and myclobutanil) and QoI (pyraclostrobin) classes. Research has also revealed the efficacy of some members of these groups against other pathogenic species of Diatrypaceae and Botryosphaeriaceae. Postpruning application of fungicides by conventional spray equipment can also be successful if good wound coverage is achieved, and ongoing research to increase spray deposition efficiency is improving the efficacy of this technique.

Selected References

Carter, M. V. 1991. The status of *Eutypa lata* as a pathogen. Phytopathological Paper 32. International Mycological Institute, Surrey, UK.

Carter, M. V., Bolay, A., and Rappaz, F. 1983. An annotated host list and bibliography of *Eutypa armeniacae*. Rev. Plant Pathol. 62:251-258.

Gramaje, D., Ayres, M. R., Trouillas, F. P., and Sosnowski, M. R. 2012. Efficacy of fungicides on mycelial growth of diatrypaceous fungi associated with grapevine trunk disease. Australas. Plant Pathol. 41:295-300.

Munkvold, G. P., Duthie, J. A., and Marois, J. J. 1994. Reductions in yield and vegetative growth of grapevines due to Eutypa dieback. Phytopathology 84:186-192.

Munkvold, G. P., and Marois, J. J. 1995. Factors associated with variation in susceptibility of grapevine pruning wounds to infection by *Eutypa lata*. Phytopathology 85:249-265.

Pearson, R. C. 1980. Discharge of ascospores of *Eutypa armeniacae* in New York. Plant Dis. 64:171-174.

Pitt, W. M., Trouillas, F. P., Gubler, W. D., Savocchia, S., and Sosnowski, M. R. 2013. Pathogenicity of diatrypaceous fungi on grapevines in Australia. Plant Dis. 97:749-756.

Ramos, D. E., Moller, W. J., and English, H. 1975. Production and dispersal of ascospores of *Eutypa armeniacae* in California. Phytopathology 65:1364-1371.

Rolshausen, P. E., and Gubler, W. D. 2005. Use of boron for the control of Eutypa dieback of grapevines. Plant Dis. 89:734-738.

Rolshausen, P. E., Mahoney, N. E., Molyneux, R. J., and Gubler, W. D. 2006. A reassessment of the species concept in *Eutypa lata*, the causal agent of Eutypa dieback of grapevine. Phytopathology 96:369-377.

Rolshausen, P. E., Greve, L. C., Labavitch, J. M., Mahoney, N. E., Molyneux, R. J., and Gubler, W. D. 2008. Pathogenesis of *Eutypa lata* in grapevine: Identification of virulence factors and biochemical characterization of cordon dieback. Phytopathology 98:222-229.

Rolshausen, P. E., Urbez-Torres, J. R., Rooney-Latham, S., Eskalen, A., Smith, R. J., and Gubler, W. D. 2010. Evaluation of pruning wound susceptibility and protection against fungi associated with grapevine trunk diseases. Am. J. Enol. Vitic. 61:113-119.

Rolshausen, P. E., Baumgartner, K., Travadon, R., Fujiyoshi, P. T., Pouzoulet, P., and Wilcox, W. F. 2014. Identification of *Eutypa* spp. causing Eutypa dieback of grapevine in eastern North America. Plant Dis. 98:483-491.

Sosnowski, M. R., Shtienberg, D., Creaser, M. L., Wicks, T. J., Lardner, R., and Scott, E. S. 2007. The influence of climate on foliar symptoms of Eutypa dieback in grapevines. Phytopathology 97:1284-1289.

Sosnowski, M. R., Wicks, T. J., and Scott, E. S. 2011. Control of Eutypa dieback in grapevines using remedial surgery. Phytopathol. Mediterr. 50:S277-S284.

Sosnowski, M. R., Loschiavo, A. P., Wicks, T. J., and Scott, E. S. 2013. Evaluating treatments and spray application for the protection of grapevine pruning wounds from infection by *Eutypa lata*. Plant Dis. 97:1599-1604.

Travadon, R., Baumgartner, K., Rolshausen, P. E., Gubler, W. D., Sosnowski, M. R, Lecomte, P., Halleen, F., and Péros, J-P. 2012. Genetic structure of the fungal grapevine pathogen *Eutypa lata* from four continents. Plant Path. 61:85-95.

Travadon, R., Rolshausen, P. E., Gubler, W. D., Cadle-Davidson, L., and Baumgartner, K. 2013. Susceptibility of cultivated and wild *Vitis* spp. to wood infection by fungal trunk pathogens. Plant Dis. 12:1529-1536.

Trese, A. T., Burton, C. L., and Ramsdell, D. C. 1980. *Eutypa armeniacae* in Michigan vineyards: Ascospore production and survival, host infection, and fungal growth at low temperatures. Phytopathology 70:788-793.

Trouillas, F. P., and Gubler, W. D. 2010. Host range, biological variation, and phylogenetic diversity of *Eutypa lata* in California. Phytopathology 100:1048-1056.

Trouillas, F. P., Pitt, W. M., Sosnowski, M. R., Huang, R. J., Peduto, F., Loschiavo, A., Savocchia, S., Scott, E. S., and Gubler, W. D. 2011. Taxonomy and DNA phylogeny of Diatrypaceae associated with *Vitis vinifera* and other woody plants in Australia. Fungal Diversity 49:203-223.

Weber, E. A., Trouillas, F. P., and Gubler, W. D. 2007. Double pruning of grapevines: A cultural practice to reduce infections by *Eutypa lata*. Am. J. Enol. Vit. 58:61-66.

(Prepared by P. Rolshausen, M. Sosnowski, F. P. Trouillas, and W. D. Gubler)

Grapevine Leaf Rust

Grapevine leaf rust occurs throughout eastern Asia, extending from East Timor and Indonesia northward to Korea and Japan. The disease also has long been established in some warm and humid regions of the Americas from Colombia, Venezuela, and Central America through the West Indies and into the southeastern United States, occasionally extending as far north as North Carolina, although it is seldom a commercial problem in the United States. It was first detected in Brazil in 2001, and has now become endemic in some tropical and subtropical regions of that country. In 2001, it also was detected in the far north of Australia (Darwin); the National Grapevine Leaf Rust Eradication Program began two years later, and involved the implementation of a quarantine zone, extensive surveys, removal of diseased vines, and monitoring of healthy plants. The disease was declared eradicated from Australia in July 2007, and this was considered to be the first successful eradication of an established rust disease in the world. Grapevine leaf rust is not known to occur in Europe.

Symptoms

Small, yellow-orange pustules of uredinia, either scattered or densely distributed, appear on the lower leaf surface (Fig. 120), usually with corresponding chlorosis on the upper leaf surface. On severely affected leaves, the chlorosis becomes necrotic, forming small, black lesions (Fig. 121). Toward the end of the growing season, dark-brown telia develop among the uredinia (Fig. 122). Severe infection causes leaf yellowing, premature defoliation, and reduced fruit quality and yield.

Causal Organism

Until 2000, the causal agent of grapevine leaf rust was considered to be a member of *Phakopsora ampelopsidis*, a species complex. In that year, the taxonomy of these fungi on vitaceous hosts was revised, dividing the pathogens among three species on the basis of morphology and host specificity. The species pathogenic on *Ampelopsis* and *Parthenocissus* spp. were designated as *Phakopsora ampelopsidis* and *P. vitis*, respectively, and that on *Vitis* spp. was described as a new species, *P. euvitis* Y. Ono. *P. euvitis* was believed to be restricted to *Vitis* spp. until 2004, when it was shown to be capable of infecting two Australian native vitaceous species, *Ampelocissus acetosa* and *A. frutescens*, under experimental conditions. *P. euvitis* is a macrocyclic rust that produces spermogonia and aecia on *Meliosma myriantha*, a deciduous tree native to Japan, and uredinia and telia on *Vitis* species. The aecial anamorph is known as *Aecidium meliosmae-myrianthae* P. Hennings & Shirai.

Uredinia of *P. euvitis* appear as minute yellow-orange pustules, 0.1–0.5 mm in diameter, primarily on the lower leaf surface but occasionally on the upper leaf surface if infection is severe. The uredinium is surrounded by cylindrical to weakly incurved paraphyses, 30–75 µm high, with moderately thickened dorsal walls (1.5–4.0 µm). Urediniospores are obovoid, obovoid-ellipsoid, or oblong-ellipsoid, 15–29 × 10–18 µm, with pale-yellow walls 1.5 µm thick, echinulate with 6 scattered or rarely 4 equatorial germ pores. Telia appear as brown to blackish-brown minute pustules, 0.1–0.2 µm in diameter, subepidermal, and crustose. Teliospores are formed in chains and are densely packed in three to five layers, with the top-layer spores being pale brown. They are oblong to oblong-ellipsoid and 13–32 × 7–13 µm. Basidiospores are kidney shaped and 8–11 × 5–8 µm.

Fig. 120. Small, yellow-orange uredial pustules of the grapevine leaf rust fungus, *Phakopsora euvitis.* (Courtesy L. S. Leu)

Other rust fungi have been recorded on *Vitis* species. *Phakopsora uva* Buriticá & Hennen is reported on unidentified *Vitis* species in Colombia and Mexico, but has not been found elsewhere. Additional reported species include *Phakopsora cronartiiformis* (Barcl.) Dietel; *Catenulopsora vitis* (Butler) Mund. & Thirum.; and *Uredo caucensis* Mayor.

Disease Cycle and Epidemiology

P. euvitis is a heteroecious, macrocyclic rust fungus (Fig. 123), the basidiospores of which infect *Meliosma myriantha*, a tree native to Japanese forests. Spermogonia and aecia are produced in sequence on infected leaves, and have only been reported in Japan. Aeciospores infect the alternate host, *Vitis*, on which uredinia and telia are produced. Outside Japan, only uredinia and telia are found.

In tropical and subtropical regions, uredinia are produced throughout the year on green leaf tissue of *Vitis* species. In temperate regions where *Vitis* spp. are dormant in winter, telia develop in autumn as the temperatures fall and host leaves senesce. Teliospores germinate at temperatures between 10–30°C, with 15–25°C representing optimum temperatures for both teliospore germination and basidiospore formation. Urediniospore germination can occur from 8–32°C, with an optimum of 24°C. High humidity enhances germination but light is detrimental. Appressoria can be observed 6 h after inoculation with urediniospores and penetrate through stomata after 12 h. Uredinial pustules appear 5–6 days after inoculation at 16–30°C and after 15–20 days at 12°C. Temperatures >35°C limit disease development. Fully expanded leaves are more susceptible to infection than young leaves, which is thought to be related to stomatal development. High humidity at night is necessary for development of epidemics.

Management

The tropical *Vitis* species *V. tiliaefolia*, *V. simpsoni*, and *V. coriacea* have been documented to be nearly immune. Over 400 grapevine genotypes representing 14 different *Vitis* spp. (principally *V. vinifera*) were screened for resistance in Australia and the majority were found to be moderately to highly susceptible. None were classified as immune.

Twenty-one fungicide products were tested for their control of *P. euvitis* infection. Fungicides effective against rust diseases on other crops, such as dithiocarbamates and triazoles, control this disease as well. Copper is also reported to be effective. Anecdotal evidence suggests that typical fungicide programs used in countries such as the United States and Japan to manage other fungal diseases also suppress grapevine leaf rust.

Fig. 121. Small, black lesions on the upper side of a grapevine leaf severely infected by *Phakopsora euvitis.* (Courtesy A. M. Daly)

Fig. 122. Dark-brown telia mixed with yellow-orange uredinia of the grapevine leaf rust fungus, *Phakopsora euvitis.* (Courtesy L. S. Leu)

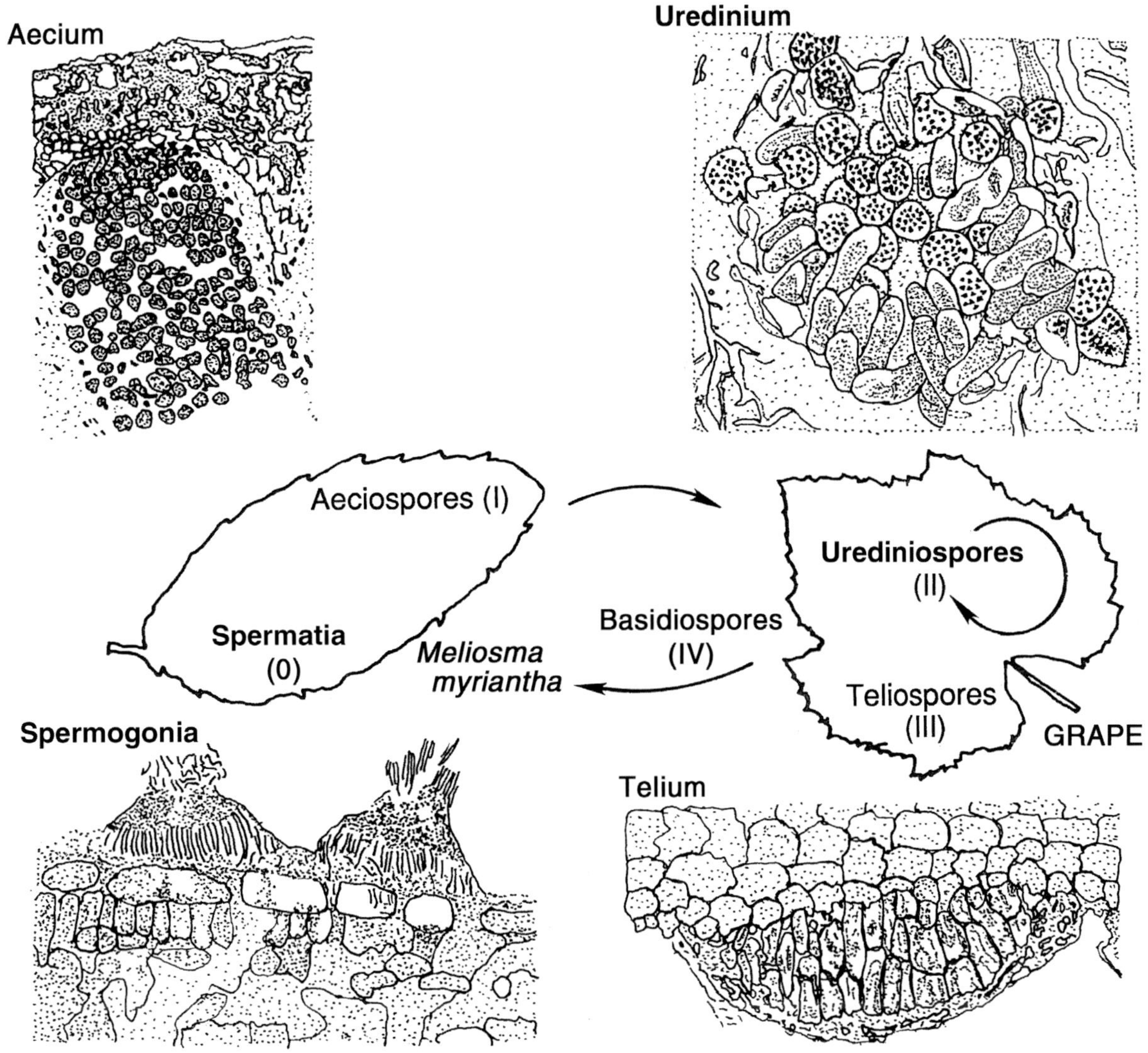

Fig. 123. Disease cycle of grapevine leaf rust, caused by *Phakopsora euvitis*. (Courtesy A. Kudo)

Selected References

Chatasiri, S., and Ono, Y. 2008. Phylogeny and taxonomy of the Asian grapevine leaf rust fungus, *Phakopsora euvitis*, and its allies (Uredinales). Mycoscience 49:66-74.

Daly, A. M., Hennessy, C. R., and Schultz, G. C. 2005. New host record for the grapevine leaf rust fungus, *Phakopsora euvitis*. Australas. Plant Pathol. 34:415-416.

Hennessy, C. R., Daly, A. M., and Hearnden, M. N. 2007. Assessment of grapevine cultivars for resistance to *Phakopsora euvitis*. Australas. Plant Pathol. 36:313-317.

Leu, L. S., and Wu, H. G. 1983. Uredospore germination, infection and colonization of the grape rust fungus, *Phakopsora ampelopsidis*. Plant Prot. Bull. (Taiwan) 25:167-175.

Naruzawa, E. S., Celoto, M. I. B., and Papa, M. F. S. 2006. Estudos epidemiologicos e controle quimico de *Phakopsora euvitis* [Epidemiological studies and chemical control of *Phakopsora euvitis*]. Fitopatol. Brasiliera 31:41-45.

Ono, Y. 2000. Taxonomy of the *Phakopsora ampelopsidis* species complex on vitaceous hosts in Asia including a new species, *P. euvitis*. Mycologia 92:154-173.

Weinert, M. P., Shivas, R. G., Pitkethley, R. N., and Daly, A. M. 2003. First record of grapevine leaf rust in the Northern Territory, Australia. Australas. Plant Pathol. 32:117-118.

(Prepared by J. Edwards)

Macrophoma Rot

Macrophoma rot affects both bunch and muscadine (*V. rotundifolia*) grapes; however, the disease is most prevalent and destructive on muscadines grown in the southeastern United States. In susceptible muscadine cultivars such as Higgins and Fry, 20–30% of the ripening berries can be lost as a result of this disease. In addition to causing fruit rot on several temperate and subtropical species, the causal fungus also causes stem blighting and wood cankers on numerous hosts, including grapes (see Botryosphaeria Dieback).

Muscadine cultivars vary in their susceptibility to the disease. Black-skin cultivars such as Noble and Nesbitt are resistant, as is the bronze cultivar Scuppernong. However, most bronze cultivars are susceptible, with Triumph and Fry being the most susceptible and Carlos being intermediate in susceptibility. There is no information on the relative susceptibility of cultivars of *V. vinifera* and other bunch grapes to Macrophoma rot.

Symptoms

In early stages, the rachis becomes soft and develops latitudinal dark lesions that are difficult to differentiate from those caused by *Phomopsis viticola*. Berry symptoms usually are not visible until fruit begin to mature, but may develop suddenly, sometimes in association with rachis blight. On muscadine berries, initial symptoms consist of circular, flat or slightly sunken lesions, 1–4 mm in diameter (Fig. 124); young lesions are black with small, tan or buff-colored centers in which pycnidia are embedded. On *V. vinifera*, initial symptoms are not distinct. On both muscadine and bunch grapes, a tan to brown, soft rot expands from the initial infection site to eventually envelop the entire fruit of susceptible cultivars (Fig. 125). Affected fruit drop from the vine, shrivel, and are finally reduced to dry, hollow shells with pycnidia scattered over the entire surface.

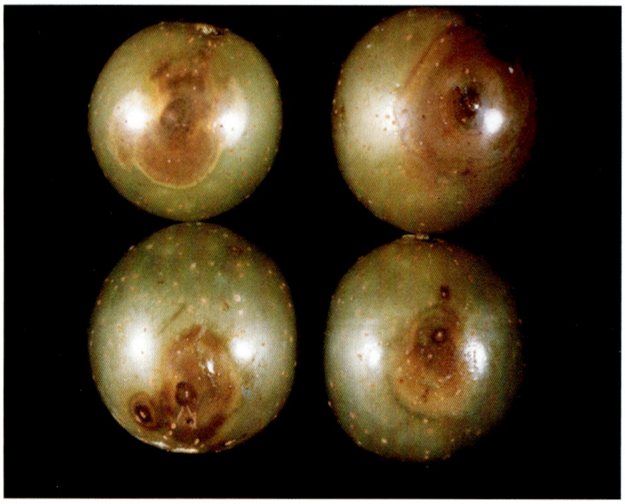

Fig. 124. Initial symptoms of Macrophoma rot on muscadine grapes. (Courtesy P. Bertrand)

Fig. 125. Muscadine grapes enveloped by light-brown soft rot due to infection by *Botryosphaeria dothidea* sensu lato, the causal agent of Macrophoma rot. (Courtesy T. B. Sutton)

Causal Organism

Botryosphaeria dothidea (Moug. ex Fr.) Ces. & de Not. (syn. *B. ribis* Grossenbacher & Duggar), anamorph: *Fusicoccum aesculi* = *Macrophoma* sp., has been reported as the causal organism of Macrophoma rot. *B. dothidea* causes cankers in grapevine spurs, trunks, and arms, but only the pycnidial stage has been found on blighted shoots and diseased fruit. Pycnidia are spherical and range from 153 to 197 μm in diameter. Conidia (14–25 × 5–8 μm) are hyaline, unicellular, narrowly elliptical to ovoid, and rounded at either end.

As described elsewhere in this volume, *B. dothidea* is a taxon that for many years was applied broadly to a variety of similar organisms, some of which are now assigned to different taxa. Thus, the identity of the species causing Macrophoma rot may benefit from a reassessment in light of current concepts of the genus *Botryosphaeria*.

Disease Cycle and Epidemiology

Very little information is available concerning the disease cycle and epidemiology of Macrophoma rot of grape. The fungus is known to overwinter as pycnidia on infected berries and stems. Pycnidia, spermagonia, and ascocarp initials are produced in stromata on stems during late fall and winter. Conidia are released during rainy periods throughout the growing season and are splashed or blown onto fruit. The optimal temperature for growth and sporulation of the causal organism is approximately 28°C.

Management

Sanitation practices that aid in the management of other fruit rot diseases by reducing primary inoculum (e.g., removal of overwintered mummies, rachis tissues, and diseased or dead wood during pruning) are important for management of Macrophoma rot as well. As with all fungal fruit rots, maintenance of an open canopy that dries rapidly, limits humidity, and facilitates spray coverage is very beneficial. Where the disease is a problem, it is necessary to apply protectant fungicides on a biweekly schedule from just after bloom until harvest except during periods of drought.

Selected References

Clayton, C. N. 1975. Diseases of muscadine and bunch grapes in North Carolina and their control. N. C. Agric. Exp. Stn. Bull. 451.

Jenkins, W. A. 1941. Diseases of muscadine grapes. Pages 19-29 in: Further Studies with Muscadine Grapes. L. Ascham, T. L. Bissell, W. L. Brown, T. A. Pickett, E. F. Savage, M. M. Murphy, Jr., and J. G. Woodroof, eds. Ga. Agric. Exp. Stn. Bull. 217.

Kummuang, N., Smith, B. J., Diehl, S. V., and Graves, C. H., Jr. 1996. Muscadine grape berry rot diseases in Mississippi: Disease identification and incidence. Plant Dis. 80:238-243.

Luttrell, E. S. 1948. *Botryosphaeria ribis*, perfect stage of the *Macrophoma* causing ripe rot of muscadine grapes. Phytopathology 38:261-263.

(Prepared by T. B. Sutton)

Minor Foliage Diseases

The literature contains numerous reports of fungi described on grape foliage. Most of these were collected from vines of the many wild species of *Vitis* in diverse climates. Many are localized or limited to a few host species, some have a wide host range that happens to include *Vitis*, and all appear under a multitude of synonyms. Minor foliage diseases typically are not a problem in commercial vineyards that receive general disease-control practices, but some may occasionally bridge the gap between wild hosts and cultivated vines when conditions particularly favor their growth, such as unusually wet growing seasons. They are reported disproportionately from nontraditional growing districts, often by newer growers who cannot identify them as one of the usual diseases described in reference manuals.

Leaf Blight

In the United States, leaf blight or Pseudocercospora leaf blight, also called Isariopsis leaf spot, occurs primarily in the Southeast and westward into Texas, where it has been particularly problematic on Cabernet Sauvignon. It has not been reported on muscadine grapes. It also occurs in Massachusetts, Connecticut, Kansas, Illinois, and California on vines of wild species. It has been reported throughout the warmer grape-growing areas of the world under various synonyms. The disease tends to appear after harvest when spraying is discontinued (Fig. 126) and may cause considerable defoliation in a wet year if not controlled.

Symptoms

Spots are irregular to angular, often numerous, 2–20 mm in diameter, and initially chlorotic. Lesions soon turn brown to black and become necrotic, and may coalesce (Fig. 127). They have clearly defined borders on the upper leaf surface and diffuse margins on the lower leaf surface. Lesions typically appear first on the lower, shaded leaves.

Fig. 127. Numerous late-season leaf infections caused by *Pseudocercospora vitis*. (Courtesy J. Kamas)

Fig. 126. Initial chlorotic lesions and subsequent necrotic, coalesced lesions caused by *Pseudocercospora vitis*. (Courtesy J. Kamas)

Causal Organism

Pseudocercospora vitis (Lév) Speg. (syn. *Isariopsis clavispora* (Berk. & Curt.) Sacc.), the imperfect stage of *Mycosphaerella personata* Higgins, is the causal agent of leaf blight. The fruiting structures are slender, black, bristlelike synnemata (200–500 µm long) bearing olive-brown, elongate conidia (25–99 × 4–8 µm) with 3–17 septa.

The teleomorph may develop on dead leaves late in the season. Perithecia are round (60–90 µm in diameter), black, embedded, and warty above. Asci are club shaped (30–40 × 6–10 µm), and ascospores (10–20 × 2.5–3.6 µm) may be forcibly discharged.

Selected References

Deighton, F. C. 1976. Studies on Cercospora and allied genera. VI. Pseudocercospora Speg., Pantospora Cif. and Cercoseptoria Petr. Mycological Paper 140. Commonwealth Mycological Institute, Kew, England.

Higgins, B. B. 1929. Morphology and life history of some ascomycetes with special reference to the presence and function of spermatia. II. Ga. Agric. Exp. Stn. J. Ser. Pap. 28:287-296.

Rhodes, A. S. 1926. Diseases of grapes in Florida. Fla. Agric. Exp. Stn. Bull. 178.

(Prepared by J. R. McGrew and F. G. Pollack)

Leaf Blotch

The fungus that causes leaf blotch is widespread throughout the eastern United States (New York to Wisconsin and south to Texas and North Carolina) and has been reported from Italy and northern Portugal as well. Lesions are most often found on the foliage of rootstock cultivars derived from native American *Vitis* spp. but sometimes appear on *V. labrusca*, *V. vinifera*, and *Vitis* interspecific hybrid cultivars. The fungus also has been found fruiting on berries that were left on the vine past normal harvest. Only fruit infection has been reported from Italy.

The leaf blotch fungus has been found in association with the larvae of the grape leaffolder, *Desmia funeralis* (Hübner), fruiting on frass inside the rolled-up portion of the leaf blade and sometimes producing a leaf lesion with synnemata in the immediate area of the infested frass.

Inoculation trials in Portugal indicate that rootstock clones are more susceptible than *V. vinifera* and interspecific hybrid fruiting cultivars. When rootstocks were grown in Maryland without fungicide applications and leaf blotch was obviously present, the overall reduction in effective leaf area was minor.

Symptoms

Leaf lesions generally appear after midseason and range in size from 5 mm in diameter to as much as one-third of a leaf. Smaller lesions may have distinct, dark margins, while larger ones show distinct, light-colored, zonate rings or arcs (Fig. 128). Synnemata are produced within 3 or 4 days of the appearance of a lesion.

Causal Organism

Briosia ampelophaga Cav., the cause of leaf blotch, forms scattered and conspicuous synnemata, more frequently on lesions on the lower leaf surface but sometimes on both surfaces. Synnemata have relatively thick, white stipes consisting of parallel, hyaline hyphae that are massed, anastomosed, and interwoven to form rigid, upright structures as long as 1 mm. The stipe is capped with a dark, spherical mass, which may be 1 mm in diameter, consisting of chains of dry, dark-brown conidia. Individual conidia are globose and measure 3–5 µm in diameter.

Fig. 128. Leaf blotch, caused by *Briosia ampelophaga*, on a leaf of the interspecific hybrid cv. Rougeon. (Courtesy J. R. McGrew)

Selected References

Cavara, F. 1888. Intorno al dissecamento dei grappoli della vite. Atti Ist. Bot. Univ. Pavia 1:293-324.

Doutel Serafim, D. J. 1955. *Coremium luteolum* S. Camara: Causa de uma doença das folhas algumas videiras. Agron. Lusit. 17:297-334.

(Prepared by J. R. McGrew and F. G. Pollack)

Zonate Leaf Spot

Zonate leaf spot, also called target spot, has been reported infrequently on wild and cultivated grapes. The fungus that causes this disease has a wide host range of both herbaceous and deciduous plants and is known to be present from Florida to Massachusetts and north into Canada as well as in Japan, China, and India. It has been reported to produce severe defoliation on several other crop species.

The disease appears very sporadically on grapes, and may be extensive in a vineyard one year yet not appear the following season. Differential susceptibility among cultivars has not been noted. Inoculum responsible for a severe outbreak in a vineyard may originate from any of several species of nearby infected trees or shrubs. Field studies have shown that the fungus moves only a limited distance (less than 50 m).

Symptoms

The generally circular lesions of zonate leaf spot may appear at any time during the growing season following several consecutive days of high humidity and accompanying rainfall, fog, or dew. They resemble lesions of leaf blotch except that the larger lesions tend to have a more concentric zonation (Fig. 129). The central portion of older lesions tends to disintegrate, leaving a hole in the leaf. Under some conditions, lesions can be numerous and probably affect vine growth.

Fig. 129. Zonate leaf spot, caused by *Cristulariella moricola.* (Courtesy J. R. McGrew)

Causal Organism

Cristulariella moricola (Hino) Redhead (syn. *C. pyramidalis* Waterman & Marshall) is the causal organism of zonate leaf spot. A sclerotial stage produced by this fungus, known as *Sclerotium cinnamomi* Sawada, may serve as an overwintering stage. The entire fungal fruiting structure, which resembles a miniature conifer, serves as a functional conidium (Fig. 130). It is pale, conical, up to 0.5 mm in length, and borne on a slender stalk that is one cell thick and four to many cells long. The stalk produces globose or lobed cells that proliferate by budding, finally forming a dense mass of cells. These cells are not freed individually as conidia. The entire structure, which can break away from the stalk, may be carried by wind. It can be found affixed horizontally to a leaf surface at the center of a newly formed lesion. The fruiting structures are formed only during periods when the relative humidity is above 96%. Lesions may resume production of these structures even after dry periods of 60 days. In culture, the conical structures produce minute phialides and globose phialospores (2.4–3.5 μm in diameter) from the globular cells.

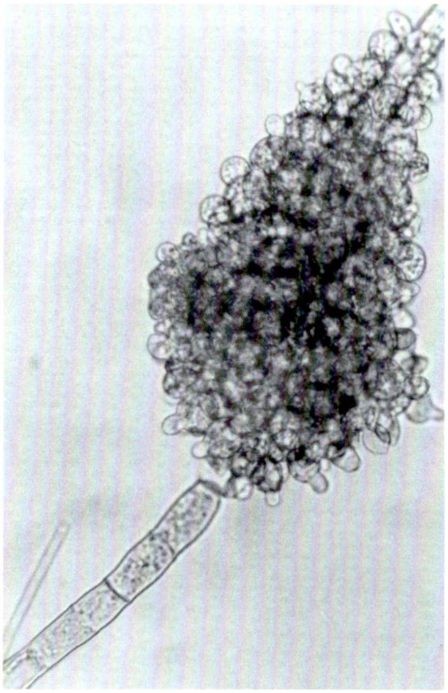

Fig. 130. Fruiting structure of *Cristulariella moricola.* (Courtesy A. J. Latham)

Selected References

French, W. J. 1972. *Cristulariella pyramidalis* in Florida: An extension of range and new hosts. Plant Dis. Rep. 56:135-138.

Pollack, F. G., and Waterworth, H. E. 1969. A leafspot disease of Kenaf in Maryland associated with *Cristulariella pyramidalis.* Plant Dis. Rep. 53:810-812.

Redhead, S. A. 1979. Mycological observations: 1. On *Cristulariella;* 2. on *Valdensinia;* 3. on *Neolecta.* Mycologia 71:1248-1253.

Trolinger, J. C., Elliott, E. S., and Young, R. J. 1978. Host range of *Cristulariella pyramidalis.* Plant Dis. Rep. 62:710-714.

(Prepared by J. R. McGrew and F. G. Pollack)

Septoria Leaf Spot

Septoria leaf spot, also called mélanose, occurs infrequently in the eastern United States and has been reported from New York to Florida and westward, from Wisconsin to Texas. It was presumably introduced into Europe during the phylloxera crisis, and has now been documented in France, Germany, Spain, Switzerland, European portions of the former Soviet Union, and Algeria. The disease is known to affect American *Vitis* spp., muscadine grapes, and some *V. labrusca* cultivars. *V. vinifera* cultivars are reported to be immune.

Symptoms

Few to many reddish-brown to black spots appear generally after midseason. They are interveinal, angular, and generally 1–2 mm in diameter, but can extend up to 2 cm in diameter on muscadine foliage. The margins of lesions are often thickened. Later, if there are several spots on a leaf, the surrounding area may become yellow.

Causal Organism

Leaf spot is caused by *Septoria ampelina* Berk. & Curt. Pycnidia are 50–60 μm in diameter and have a large, open ostiole. Conidia (40–60 × 2 μm) are hyaline and undulating, with three to six septa.

Selected Reference

Boubals, D., and Mur, G. 1983. Une autre maladie de la vigne sevit dans le Penedes (Espagne). Prog. Agric. Vitic. 100:453.

(Prepared by J. R. McGrew and F. G. Pollack)

Other Minor Foliage Diseases

Brulure

Brulure, which is caused by the fungus *Anthostomella pullulans* (de Bary) Bennett, occurs on all parts of the vine. The incitant is distributed worldwide and is generally a saprophyte.

Cladosporium Leaf Spot

Cladosporium viticola Cesati causes a leaf spot disease on older foliage of both wild and cultivated vines. The disease has been reported in Europe and in the eastern United States.

Cercospora Leaf Spot

Cercospora leaf spot (cercosporiose), caused by *Phaeoramularia dissiliens* (Duby) Deighton, produces variable yellowish to dark spots. It has been reported from Pakistan through Asia Minor, Europe, and North Africa.

Tar Spot

Tar spot, caused by *Rhytisma vitis* Schw., produces a black spot 2–4 mm in diameter, with or without a circular, brown halo up to 1 cm in diameter. It is common on wild vines in the southeastern United States.

Miscellaneous Leaf Spots

A leaf spot caused by *Asperisporium minutulum* (Sacc.) Deighton produces lesions with indefinite margins. It has been reported only on *V. californica* in California and Oregon.

Another leaf spot, caused by *Phaeoramularia heterospora* (Ell. & Gall.) Deighton, has been reported on *V. californica* and *V. girdiana* in California and on *V. vinifera* in Israel.

Fig. 131. Moderate symptoms of the physiological disorder Rupestris speckle on the interspecific hybrid cv. Chambourcin. (Courtesy T. E. Martinson)

Fig. 132. Severe symptoms of the physiological disorder Rupestris speckle on a different cv. Chambourcin vine in the same vineyard as pictured in Fig. 131. (Courtesy T. E. Martinson)

Rupestris Speckle

Rupestris speckle appears to be a physiological disorder associated only with *V. rupestris*. It may appear in a milder form on interspecific hybrids derived from this species, such as Villard Blanc and Chambourcin. Symptoms are similar to those of Septoria leaf spot or oxidant stipple but tend to be on older and shaded leaves. The necrotic areas vary from circular to angular, are generally less than 2 mm in diameter, and tend to show a yellow halo (Fig. 131). Spotting is more pronounced on weak or stressed vines (Fig. 132), and a range of severities may be found in the same vineyard. No control measures have been prescribed, and the effect on vine growth is thought to be insignificant.

Selected References

Bennett, F. T. 1928. On *Dematium pullulans* de B. and its ascigerous stage. Ann. Appl. Biol. 15:371-391.

Deighton, F. C. 1976. Three fungi on leaves of Vitis. Trans. Br. Mycol. Soc. 67:223-232.

Griffiths, D. A. 1974. The origin, structure and function of chlamydospores in fungi. Nova Hedwigia 25:503-547.

Viala, P. 1893. Les Maladies de la Vigne. 3rd ed. Coulet, Montpellier, France.

(Prepared by J. R. McGrew and F. G. Pollack)

Phomopsis Cane and Leaf Spot

Phomopsis cane and leaf spot is widely distributed throughout the viticultural regions of the world, and has been reported from every continent on which grapes are grown. The disease is especially problematic in regions or years where the climate following bud break may keep grapevines wet for extended periods of time. Although the common name of the disease is Phomopsis cane and leaf spot, economic losses in regions with humid growing seasons are more often the result of infections of the fruit and rachises (cluster stems). Fruit and rachis infections typically are sporadic but can be severe under heavy disease pressure, with yield losses up to 30% documented in vineyards with abundant inoculum if prolonged rains occur in the spring and early summer and control programs are limited. Such infections are uncommon in drier regions such as those with Mediterranean climates, where losses are generally limited to the occasional weakening and breaking of green shoots at infection sites on the basal internodes.

Information on relative susceptibility among *Vitis* species and the cultivars within them is limited. Most *V. vinifera* cultivars appear to be susceptible, although the disease is particularly common on spur-pruned varieties. Limited inoculation studies suggest that accessions of *V. aestivalis* and *V. amurensis* are relatively resistant; those of *V. riparia* and *V. yenshanensis* are highly susceptible; and interspecific hybrids vary widely in susceptibility, depending on their parentage. Most common *V. labrusca* cultivars and cultigens are susceptible, and cv. Niagara appears to be highly susceptible. In the Great Lakes region of North America, Phomopsis rachis and fruit infections cause greater consistent losses on these cultivars than does any other infectious disease, likely owing to (i) the cultivars' pendulous growth habit, which causes new, susceptible tissues to develop beneath sources of splash-dispersed inoculum; (ii) their susceptibility to this disease relative to others more problematic on *V. vinifera* (e.g., powdery mildew); and (iii) the less intensive disease management programs that they typically receive relative to *V. vinifera* and interspecific hybrid cultivars.

Symptoms

Lesions on shoots and leaves are the most common symptom of the disease. Shoot infections result in black, elongated lesions that are most numerous on the first three to six basal internodes (Fig. 133). Cracks or fissures may develop within these lesions, and eventually crust over to produce a rough, black appearance by the end of the season. The stem is weakened around these infection sites, which may cause it to break in high winds before shoots become attached to the trellis wires (Fig. 134). During the dormant season, infected canes may become bleached or grayish in appearance, and numerous tiny, black fungal fruiting bodies (pycnidia) erupt through the surface to produce a speckled appearance.

Elongated, black lesions similar to those on young shoots develop on infected petioles (Fig. 133) and may cause affected leaves to turn yellow and drop if severe. On infected leaf blades, small, light-green lesions with irregular margins develop early in the season. As the disease progresses, these spots turn dark brown to black with surrounding yellow halos (Fig. 135). The infected tissue may drop out, causing a "pin-pricked" appearance. Heavily infected leaf blades are often misshapen and crinkled. The economic impact of foliar infections is questionable and does not appear to be great.

Lesions on rachises are typically sunken and black (Fig. 136) and cause rachises to become brittle; clusters may break

Fig. 133. Black, elongated lesions caused by *Phomopsis viticola* on the basal internodes and leaf petioles of cv. Concord, with a dead pruning stub (the likely inoculum source) directly above. (Courtesy W. F. Wilcox)

Fig. 134. Wind breakage of a young shoot at the site of an internode lesion caused by *Phomosis viticola*. (Courtesy W. F. Wilcox)

Fig. 135. Numerous small, dark leaf lesions surrounded by a yellow halo, characteristic foliar symptoms of Phomopsis cane and leaf spot. (Photograph by W. McFadden-Smith, Ontario Ministry of Agriculture and Food; Queen's Printer of Ontario, © 2009. Reproduced with permission.)

at these points under the weight of the maturing crop or during harvest operations, leading to a reduction in yield. Rachis infection during the early stages of cluster development may lead to the loss of the "wing" (lateral branch of the cluster) (Fig. 137). On highly susceptible cultivars, the fungus may advance from rachis and pedicel infections into individual berries (Fig. 138), or the entire cluster may be girdled and lost (Fig. 139). After infected berries become completely necrotic, multiple black fruiting bodies (pycnidia) of the fungus erupt through the skin, giving them a rough texture (Fig. 140). Eventually, diseased berries may shrivel into black raisin-like mummies very similar to those caused by black rot or bitter rot, but they more frequently fall to the ground beforehand.

Fruit rot caused by Phomopsis is sometimes confused with black rot and bitter rot, as the same regional weather patterns are conducive to the development of all three. In general, Phomopsis fruit rot is distinguished by its late appearance (typically, just before harvest); the fact that diseased berries are easily detached from their pedicels; and the coincidence of fruit rot with the above-described symptoms on basal internodes, petioles, leaf blades, and/or rachises.

Fig. 136. Sunken, black lesions on basal internodes and cluster rachis of cv. Concord caused by *Phomosis viticola.* (Courtesy R. C. Pearson)

Fig. 137. Loss of the "wing" (lateral branch) on a cluster of cv. Concord (arrow) resulting from infection by *Phomosis viticola* during the early stage of cluster development (left) compared with a healthy cluster with wing intact (right). (Courtesy W. F. Wilcox)

Causal Organism

Phomopsis viticola (Sacc.) Sacc. (syn. *Fusicoccum viticola* Reddick) is the causal agent of Phomopsis cane and leaf spot. There has been a great deal of confusion as to whether an ascogenous stage of the fungus exists; if so, it is rarely encountered. A number of other species of *Phomopsis* are associated with grape plants, but only *P. viticola* is known to be strongly pathogenic.

Phomopsis viticola produces black pycnidia ranging in diameter from 0.2 to 0.4 mm. Each pycnidium has one or more locules, depending on the substrate. Pycnidia are globose, but

Fig. 138. Expansion of *Phomopsis viticola* from infected rachis and pedicel tissues into ripening berries of cv. Niagara. (Courtesy W. F. Wilcox)

Fig. 139. Girdling of cv. Niagara rachis with resulting loss of the cluster following infection by *Phomopsis viticola.* (Courtesy W. F. Wilcox)

may be conical at maturity. An ostiole opens at the peak of a short neck. The opening on a pycnidium is usually round and smooth but may be irregular and sometimes ragged.

Pycnidiospores often ooze through the ostiole, either in a gelatinous matrix or in long, curled, yellowish to cream-colored cirri (Fig. 141). The inner surface of a pycnidium is typically lined with straight pycnidiophores of two types. One type is elongate with a pointed apex, measures 12–20 × 2 µm, and bears nonseptate, hyaline, elliptical-fusoid pycnidiospores (or alpha conidia, 7–10 × 2–4 µm) that are acute on one or both ends and multiguttulate; guttules are sometimes grouped at the polar ends. The other type of pycnidiophore is short (5–8 × 1.5 µm) and produces scolecospores or beta conidia. Beta conidia, which are much less abundant than alpha conidia, are long and curved or threadlike, nonseptate spores measuring 18–30 × 0.5–1 µm (Fig. 142). Both alpha- and beta-conidia can be formed in the same pycnidium, although beta conidia rarely germinate and their function has not been determined. It has been reported that beta conidia are infrequently observed when cultures are incubated at 20°C, but are produced abundantly when cultures are incubated at 30°C. Spore-trapping studies from vineyards in the northcentral (Michigan) and northeastern (New York) United States have yielded primarily beta conidia in the mid-summer but far larger quantities of almost exclusively alpha conidia earlier in the season.

In green shoots, the mycelium invades mostly cortical parenchyma tissue. The mycelium is hyaline and forms pseudoparenchymatous mats among host cells. In addition to causing necrosis of infected cells, the mats may turn black, giving spots and lesions this color. Eventually, black pycnidia comprising several layers of pseudoparenchymatic cells arise from these areas and provide the conidia that serve as inoculum for new infections. In culture, the branched mycelium of *P. viticola* is hyaline, septate, and forms a slightly raised, dense mat, often in concentric rings under alternating light-dark regimes. Portions of the mat or the margins of concentric rings turn dark gray to black as the culture ages. Pycnidia form irregularly in the dark parts of the mycelial mat, either singly or in clumps.

Fig. 140. Pycnidia on necrotic berry of cv. Niagara infected with *Phomopsis viticola.* (Courtesy W. F. Wilcox)

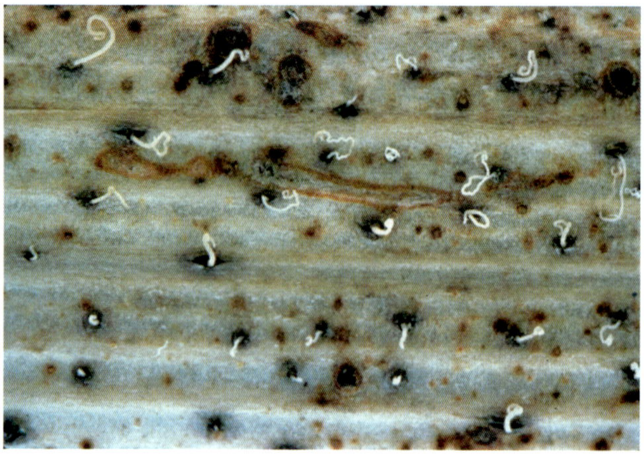

Fig. 141. Pycnidia of *Phomopsis viticola* exuding spores in cirri on the surface of a one-year-old cane. (Courtesy W. Gärtel)

Disease Cycle and Epidemiology

In the late winter and early spring, spores are produced within pycnidia that formed in previously infected woody tissues. (Although the exact time of pycnidia production in woody tissues is not clear, and may vary among climates, they generally are not observed in newly infected shoots until some time after these have lignified into canes and become dormant). Because shoot infections are most frequent on basal internodes, inoculum production is particularly favored by spur-pruning systems in which only basal cane segments are retained during dormant pruning, thereby maximizing the retention of infected wood. Infected pruning stubs, which once originated as basal internode segments, are similarly important as potential inoculum sources (Fig. 133). Mechanical hedging systems, which retain large numbers of basal cane segments, can be especially problematic for management of this disease. The fungus can continue to produce pycnidia and spores for an indeterminate number of years after infection first occurs, even after infected wood dies. Therefore, a serious disease outbreak in one season may provide inoculum for new infections for multiple years afterward.

Spores ooze out of the pycnidia in a sticky mass during humid or wet weather and are then splashed by rain onto newly devel-

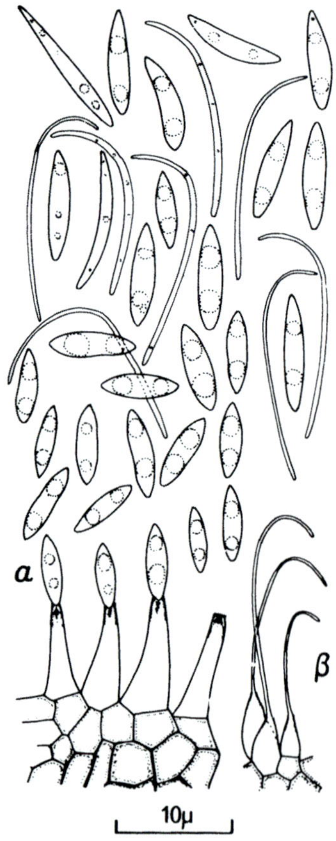

Fig. 142. Conidiogenous cells and conidia of *Phomopsis viticola*, with alpha (a) and beta (β) spores. (Reprinted, by permission, from Punithalingam, 1979)

oping tissues. Splash dispersal results in significant numbers of spores moving only a short distance, and it is not unusual to see the disease localized in "hot spots" within individual vines where inoculum is plentiful. Infection can occur during wet periods at temperatures ranging from approximately 5 to 35.5°C (optimum 16 to 20°C), with the resulting disease severity a function of both temperature and wetness duration. Significant infection frequently occurs during extended spring rains at temperatures well below the optimum.

In the north-central and northeastern regions of the United States, most inoculum is dispersed between bud break and bloom, and few infectious alpha spores are available beyond the pea-sized berry stage. With the exception of diseased berries, which sporulate shortly after these organs become necrotic during the preharvest period, conidia are seldom produced from new infections during the current growing season. Consequently, secondary spread of the disease is rare, even when summer rainfall is not limiting. Thus, the period during which new infections are possible and fungicidal protection may be necessary is restricted by both the onset of dry weather (in some regions and/or seasons) and the depletion of primary inoculum.

Shoot and leaf lesions usually appear 3–4 weeks after infection. Rachis infections may appear within a month after infection, although they typically are most pronounced during the preharvest period. Fruit infections usually are not apparent until after veraison and typically become most visible during the last 1–3 weeks before harvest. Although pycnidia can develop within newly diseased berries, the potential for secondary disease spread from these sources is limited by the timing of their appearance and the requirement for a postinfection incubation period, whose length usually exceeds the time remaining until harvest.

On shoots, the young growing tip is most susceptible and tissue becomes relatively resistant as it matures. Although shoot tissues two or three internode positions back from the growing tips are highly resistant, susceptible tissue is continually produced as the shoot keeps growing, and the timing of infection is restricted only by the availability of inoculum and favorable weather. Leaves become resistant to infection once they are fully expanded. Rachises are highly susceptible as nascent clusters first become exposed, when shoots are about 8–10 cm long (E-L stage 12). Infections during the very early period of cluster development are the most damaging and may cause serious girdling as the uninfected portions of the inflorescence expand around them, with subsequent berry loss. Rachis susceptibility declines a few weeks after bloom, coinciding with the depletion of inoculum. Berry infection apparently can occur via expansion of infections initiated in the pedicel (berry stem), beginning soon after the time that small clusters first emerge, or directly through the fruit epidermis after set. Berry infections typically remain latent (inactive) until the preharvest period, when they colonize and rot the fruit. Some studies indicate that berries become resistant to the initial establishment of infection after the pea-sized stage of development (E-L stage 31), whereas others show them to remain susceptible to new infections throughout the season. However, because few infective *P. viticola* spores are available beyond mid-summer, it appears likely that in regions where they occur, fruit infections are initiated primarily between the time of cluster emergence and the early postbloom period.

Management

Phomopsis cane and leaf spot can be managed through an integrated program of sanitation and fungicide application. Sanitation should be the first line of defense. To whatever extent possible, the introduction of *P. viticola* into the vineyard should be avoided by use of pathogen-free propagation materials when establishing a new vineyard. Once the disease has appeared, as much diseased and dead wood as practical should be removed during pruning. To reduce a major source of inoculum, debris from pruning should be destroyed by shredding and incorporating it into the soil, or by burning. Owing to the extremely local dispersal of infective spores and the monocyclic nature of this disease, serious Phomopsis outbreaks are almost always the result of high inoculum levels within a vineyard in addition to very wet weather conditions. When used with recommended sanitation practices, even a modest fungicide program (which not only protects against current losses but also limits establishment of the fungus in subsequently retained wood) typically provides adequate control of the disease.

Broad-spectrum protectant fungicides such as mancozeb, captan, folpet, ziram, chlorothalonil, and dithianon are very effective against Phomopsis and are commonly used for its control throughout the world. More modern fungicide groups, such as the DMI and QoI materials, have provided control in some regions but not in others. Sulfur is recommended in some regions but has provided little to no control under heavy disease pressure in New York trials. Copper also is relatively ineffective; thus, growers using organic production practices should focus strongly on sanitation programs, particularly in regions where this disease can be problematic. Fungicides targeted against Phomopsis should be applied during the very early part of the growing season as influenced by the occurrence of favorable weather conditions, inoculum levels in the vineyard, varietal susceptibility, and local experience. Controlling early-season infections of shoots is very important for reducing sources of primary inoculum for subsequent growing seasons. The first few weeks after cluster emergence is the most important portion of the season for controlling rachis and fruit infections in regions where they occur. Continued protection against fruit infections may be necessary through the early postbloom period if the weather remains wet in high-risk sites where they are a potential problem.

A "delayed dormant" (immediately before bud break) application of materials such as lime sulfur (calcium polysulfide) and fluazinam has often been effective in reducing, but not eliminating, disease development during the subsequent growing season, presumably by reducing the production of primary inoculum from pycnidia on the vines. However, the potential benefit of this approach must be considered with respect to its cost. A predictive model has been developed to forecast specific infection events for this disease; however, the current lack of fungicides with significant curative activity against *P. viticola* limits the application of such a predictive system within commercial disease management programs.

Selected References

Cucuzza, J. D., and Sall, M. A. 1982. Phomopsis cane and leaf spot disease of grapevine: Effects of chemical treatments on inoculum level, disease severity, and yield. Plant Dis. 66:794-797.

Erincik, O., Madden, L. V., Ferree, D. C., and Ellis, M. A. 2001. Effect of growth stage on susceptibility of grape berry and rachis tissues to infection by *Phomopsis viticola*. Plant Dis. 85:517-520.

Erincik, O., Madden, L. V., Ferree, D. C., and Ellis, M. A. 2003. Temperature and wetness-duration requirements for grape leaf and cane infection by *Phomopsis viticola*. Plant Dis.87:832-840.

Mostert, L., Crous, P. W., Kang, J.-C., and Phillips, A. J. L. 2001. Species of *Phomopsis* and a *Libertella* sp. occurring on grapevines with specific reference to South Africa: Morphological, cultural, molecular and pathological characterization. Mycologia 93:146-167.

Phillips, A. L. 2000. Excoriose, cane blight and related diseases of grapevines: A taxonomic review of the pathogens. Phytopathol. Mediterr. 39:341-356.

Pscheidt, J. W., and Pearson, R. C. 1989. Effect of grapevine training systems and pruning practices on occurrence of Phomopsis cane and leaf spot. Plant Dis. 73:825-828.

Pscheidt, J. W., and Pearson, R. C. 1989. Time of infection and control of Phomopsis fruit rot of grape. Plant Dis. 73:829-833.

(Prepared by W. F. Wilcox, M. A. Ellis, B. Rawnsley, A. Rossman, and J. Pscheidt)

Phymatotrichopsis Root Rot

Phymatotrichopsis root rot (formerly Phymatotrichum root rot), commonly called Texas root rot or cotton root rot, also has been reported in mango and avocado orchards in the semitropical states of Veracruz, Michoacan, and Sinaloa in Mexico.

Symptoms

Phymatotrichopsis root rot usually appears as patches in the vineyard. Natural spread of the pathogen is slow and it appears to be restricted to certain areas of the vineyard, although the reason for this is not understood. Affected vines may suddenly wilt and die during the early or midsummer, with their leaves becoming brown and brittle and remaining rigidly attached to the dead vine (Fig. 143). Before wilting, leaves on some infected vines may gradually turn yellow or red, with interveinal yellow spots and irregular necrotic areas on the leaf blade and margins. Some diseased vines may show discoloration and necrosis of a few leaves in early summer, with no further progression of symptoms for the remainder of the season. Shoot growth is reduced and leaves are dull green on such vines the following year.

Vines showing complete foliar discoloration or necrosis in early summer become partly or totally defoliated by midsummer, and their clusters become exposed and sunburned (raisining). Occasionally, these vines appear to recover during late summer and produce new shoots. Such vines may die during the winter or decline the following year.

Fig. 143. Sudden midsummer collapse of a vine affected with Phymatotrichopsis root rot, with characteristic adherence of brittle, brown foliage. (Courtesy J. Kamas)

Fig. 144. Mycelial strands of *Phymatotrichopsis omnivora* on the surface of dead infected roots, a reliable diagnostic indicator of Phymatotrichopsis root rot. (Courtesy Department of Plant Pathology, CAE La Laguna CIAN-INIA-SARH, Matamoros, Coah., Mexico)

Much of the grapevine root system is rotted by the time the first foliar symptoms are observed. Thick or thin mycelial strands of the pathogen are usually observed on the surfaces of infected roots (Fig. 144). Often, the root cortex is easily separated from the woody cylinder.

Causal Organism

The causal fungus, *Phymatotrichopsis omnivora* (Duggar) Hennebert (syns. *Phymatotrichum omnivorum* (Shear) Duggar, *Ozonium omnivorum* Shear), produces large, sparingly branched, leader hyphal cells. White mycelial strands, which become brown with age, are formed by the parallel growth of hyphae and lateral branching of short hyphal cells. The fungus can be easily identified with a microscope by observing these strands, which have numerous acicular and characteristically cruciform hyphae (Fig. 145).

Globose to irregularly shaped, brown sclerotia, 1–2 mm in diameter, are produced singly or in clusters from infected roots in soil. Spore mats, which consist of numerous hyphae with globose to elongated conidiophores bearing single-celled, spherical to ovate conidia (4.8–5.5 μm in diameter), are sometimes produced on the surface of the soil after several days of irrigation or rain (Fig. 145).

Disease Cycle and Epidemiology

Sclerotia are produced in soil following infection, usually 15–75 cm deep in the profile although they can be found to a depth of 2 m. They are the survival structures of the fungus, capable of persisting to serve as the source of primary inoculum

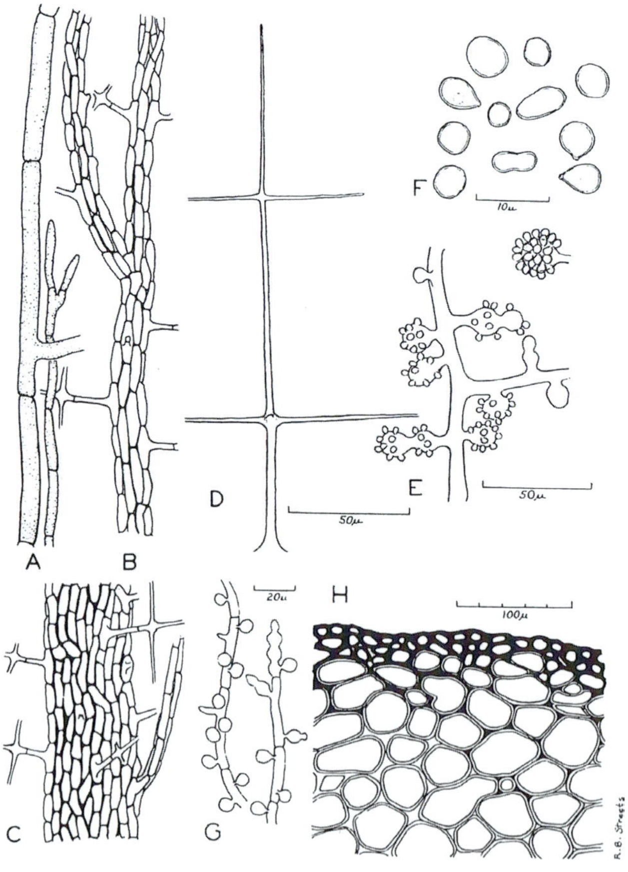

Fig. 145. Morphology of *Phymatotrichopsis omnivora*. **A,** Single hypha. **B,** Strand formed by branching hyphae surrounding a large central hypha. **C,** Mature strand with cruciform hyphae. **D,** Cruciform hypha with acicular points. **E,** Conidia on conidiophores from a sporemat. **F,** Conidia freed from conidiophore. **G,** Conidiophores before conidiation. **H,** Cross section of a sclerotium, with thick-walled rind cells and parenchyma. (Reprinted, by permission, from Streets and Bloss, 1973)

for at least 12 years. Sclerotia germinate by producing mycelial strands, which colonize root surfaces and form infection cushions at natural openings such as lenticels and cracks where lateral roots emerge.

The fungus infects the root cortex and then invades the vascular elements, impeding water translocation. Sudden wilt of the vine occurs as a result of either extensive rotting of the root system or occlusion of the vascular tissue of the main roots. The fungus moves from diseased to adjacent healthy vines by colonizing their overlapping root systems, which are typically distributed along the trellis where drip irrigation exists.

The optimum temperature for growth of *P. omnivora* is 28–30°C, which occurs in soil only during the summer months. In cotton, the disease is less severe in soils with a pH of 6 or lower, and is most serious where soil is slightly acid to alkaline (pH 6.5–8.5) and calcareous.

Management

In regions where Phymatotrichum root rot occurs, planting in infested soil can be avoided by mapping the areas where susceptible crop plants such as alfalfa and cotton show symptoms. When purchasing property intended for new vineyards, any history of cultivating cotton or alfalfa on the site should be determined. Only healthy propagation material should be used when planting in pathogen-free soil. Roots with partially rotted tissue may carry hyphal strands on their surface and mycelium in their vascular tissue, thereby introducing the pathogen.

Vines grafted onto Dog Ridge rootstock are a good option in sites to be replanted. This rootstock has good vigor and root regeneration characteristics that appear to enhance its survival compared with own-rooted *V. vinifera* cultivars. However, excess vigor and low fruitfulness may be a problem with some cultivars on this rootstock in fertile soils.

Selected References

Herrera, P. T. 1984. Investigación sobre portainjertos y su resistencia a pudrición de la raíz por *Phymatotrichum omnivorum*. Pages 83-93 in: En Memorias ler. Simposia International sobre Pudrición Texana, Biologia, y Control de *Phymatotrichum omnivorum* (Shear) Duggar. Escuela de Agricultura y Ganadeyia, Hermosillo, Son. México-Dic.

Lyda, S. D. 1978. Ecology of *Phymatotrichum omnivorum*. Annu. Rev. Phytopathol. 16:193-209.

Mortensen, E. 1938. Nursery tests with grape rootstock. Proc. Am. Soc. Hortic. Sci. 36:153-157.

Mortensen, E. 1952. Grape rootstock for southwest Texas. Tex. Agric. Exp. Stn. Prog. Rep. 1475.

Perry, R. L. 1980. Anatomy and morphology of *Vitis* roots in relation to pathogenesis caused by *Phymatotrichum omnivorum*. Ph.D. diss. Texas A&M University, College Station.

Streets, R. B., and Bloss, H. E. 1973. Phymatotrichum root rot. Monograph 8. American Phytopathological Society, St. Paul, MN.

(Prepared by S. McBride and D. Appel)

Phytophthora Crown and Root Rot

Phytophthora crown and root rots have been reported on grapevines in Australia, Chile, India, New Zealand, Spain, and the United States (California). Many vineyards and grape nurseries in South Africa have suffered serious economic loss due to the disease, but in most grape-growing regions it tends to be localized and associated with susceptible rootstocks and particularly wet sites and/or years.

Symptoms

Aboveground symptoms are similar to those of Armillaria root rot and include poor shoot growth (Fig. 146), leaf chlorosis, premature development of fall leaf colors, wilting, defoliation, and vine death. Affected vines may survive in an unthrifty state for several years or be killed within one growing season, depending on the rate and extent of disease progression.

Crown and root rot do not always occur together. Portions of the root system and/or lower trunk infected by the pathogen become necrotic. The stelar and cortical areas of rotted roots become dark brown to black, whereas infected bark and subtending vascular tissues of the root crown and trunk become reddish brown to black. Removal of the outer bark with a sharp knife reveals zones of healthy and diseased tissues that are delineated by a distinct margin (Fig. 147) until the vine dies and desiccates. Crown rot lesions may extend up the trunk above the soil surface, or the necrosis may be limited to subterranean portions of the trunk or root crown. Unlike root crown infections caused by *Armillaria*, those caused by species of

Fig. 146. Poor spring growth of several contiguous vines afflicted with Phytophthora crown and root rot (center) compared with healthy neighboring vines. (Courtesy B. A. Latorre)

Fig. 147. Reddish-brown root crown lesion characteristic of Phytophtora crown and root rot, exposed by cutting through the bark after removing soil surrounding the trunk. (Courtesy B. A. Latorre)

Phytophthora do not exhibit plaques of mycelium in or under the bark.

Grapevines of any age can be affected by *Phytophthora* spp., but the rate of disease progression and frequency of mortality are greatest in young, small vines. In affected vineyards, disease occurrence is often associated with areas that have poor water drainage for a variety of reasons, e.g., disease frequency may be greater in hillside areas with shallow soil overlying a water-impermeable layer than in lower areas with deep soil. The distribution of diseased vines may also reflect a distribution of inoculum related to both edaphic and external factors, e.g., the distribution of infested planting stock.

Causal Organisms

Species of *Phytophthora* confirmed as primary causes of the disease in pathogenicity tests include *P. cactorum* (Leb. & Cohn) Schroet., *P. cinnamomi* Rands, *P. cryptogea* Pethyb. & Laff., *P. drechsleri* Tucker, and *P. parasitica* Dastur. *Phytophthora megasperma* Drechs. has been isolated from diseased grapevines in South Africa and California, and *P. cambivora* (Petri) Buisman was isolated from diseased vines in California, but neither species has caused measurable disease consistently in pathogenicity tests. *Phytophthora citricola* Sawada was associated with declining grapevines in New Zealand, but its pathogenicity was not confirmed.

Worldwide, *P. cinnamomi* is the most prevalent and virulent species of *Phytophthora* on grape; it is an important pathogen of the crop in Australia, California, Chile, India, New Zealand, Spain, and South Africa. In South Africa, *P. cinnamomi* was more virulent and more frequently isolated than the other species causing root and crown rot in that country, which included *P. cactorum*, *P. cryptogea*, and *P. parasitica*. *Phytophthora parasitica* also was isolated from diseased grapevines in California, whereas *P. drechsleri* has only been isolated from diseased vines in Chile, where *P. cryptogea* also has been found.

Isolations of *Phytophthora* spp. from tissue recently infected by the pathogen are more successful than those from older necrotic lesions well colonized by secondary decay organisms, and are greatly facilitated by the use of semi-selective media. Some workers have incorporated the fungicide hymexazol to suppress closely related and commonly isolated *Pythium* spp., which otherwise grow on these media and typically outcompete *Phytophthora* spp., although hymexazol can also inhibit growth of some *Phytophthora* spp.

Phytophthora spp. can be isolated from infested soil or water by using parts of plants (e.g., fruits, cotyledons, and leaf disks) as bait for swimming zoospores. Soil samples must be flooded (typically for 48 h) to facilitate the production and dispersal of zoospores to the baits. Many workers have used firm, green, unblemished pear fruits as bait for *Phytophthora* spp., including those that infect grape. The *Phytophthora* spp. can then be isolated from the necrotic bait tissue onto a semi-selective medium for further identification or study.

Some key morphological characteristics of individual *Phytophthora* spp. that affect grapevines are as follows:

Phytophthora cactorum produces both sporangia and sex organs abundantly in single culture on solid media. Sporangia are broad and ellipsoidal, ovoid, or obpyriform (26–36 × 22–30 µm); conspicuously papillate; sympodial and nonproliferous; and caducous with a short pedicel (up to 4 µm long). Oogonia are smooth-walled and spherical (18–40 µm), tapering at the base. Antheridia are attached paragynously. Some isolates form chlamydospores. On cornmeal agar, cultures are slightly radiate to uniform, with very little aerial mycelium.

Phytophthora cambivora, *P. cinnamomi*, *P. cryptogea*, and *P. drechsleri* are heterothallic and produce nonpapillate, noncaducous sporangia in soil extract solution. On solid media, *P. cryptogea* and *P. drechsleri* may form a few sporangia as cultures age, but *P. cambivora* and *P. cinnamomi* form none. Sporangia are usually broadly ovoid for *P. cambivora* (44–69 × 33–48 µm); ellipsoid to ovoid or obpyriform for *P. cinnamomi* (typically 75 × 40 µm); and ovoid to obpyriform with a rounded base for *P. cryptogea* versus ovoid, obpyriform, or ellipsoid with a tapered base for *P. drechsleri* (typically about 37–40 × 23 µm for both). Hyphae of *P. cinnamomi* are uniquely coralloid. On cornmeal agar, colonies of *P. cinnamomi* have a rosette appearance whereas those of *P. cryptogea* and *P. drechsleri* are radiate.

P. parasitica Dastur produces sporangia in liquid and on solid media. Sporangia are broadly ovoid or pyriform to spherical, variable in size, papillate, sometimes intercalary, and noncaducous. Chlamydospores form on solid media. As cultures age, some isolates produce sex organs in single culture; oogonia vary widely in size, and antheridia are amphigynous. On cornmeal agar, most cultures exhibit a characteristic "bushy" and irregular rosette pattern.

Disease Cycle and Epidemiology

Species of *Phytophthora* can be introduced into vineyards via infected plant material or infested soil and water; irrigation water from rivers, canals, and ponds is frequently contaminated with them. Although species of *Phytophthora* generally are not competitive as saprophytes in decaying tissues, they can typically persist for many years once introduced into agricultural soils, presumably as thick-walled oospores and chlamydospores that facilitate survival under adverse conditions. Evidence suggests that zoospores, produced from sporangia that arise either from these resting spores or as secondary inoculum from recently infected tissues, are the principal agents of infection. Episodes of soil saturation with water stimulate the production, release, and motility of zoospores, thereby facilitating crown and root infection. Conversely, drainage to field capacity or drier states suppresses the release and motility of zoospores, thereby greatly limiting or preventing new infections while dry soil conditions prevail. Chlamydospores and oopspores produced within infected tissues are liberated into the soil as the tissues decay, thereby completing the disease cycle.

Phytophthora-induced diseases generally are favored by cool to moderate temperatures, which in Mediterranean climates such as those from which this disease is most commonly reported occur largely from late winter through early summer and again in the fall. However, responses to temperature differ among individual *Phytophthora* spp., e.g., infections by *P. cinnamomi* are skewed toward warmer conditions than are those by *P. cactorum*. Temperature ranges for sporulation, zoospore release, and zoospore motility (all being components of the infection process) are more restrictive than are temperatures for mycelial growth and survival. Local climates and viticultural practices, especially as they relate to edaphic conditions, can have large influences on the seasonal risk of Phytophthora crown and root rot development.

Genetic susceptibility to *Phytophthora* spp. varies significantly among grape rootstocks, and susceptible rootstocks have been shown to support higher population densities of *P. cinnamomi* in the soil. It is not known whether environmental or seasonal factors have important effects on the physiological susceptibility of grapevines to *Phytophthora* spp., but such effects have been shown for other hosts.

Management

Integrating cultural and genetic approaches provides the most economical strategy for managing disease caused by *Phytophthora* spp., although chemical approaches can provide valuable augmentation for high-risk areas. To minimize disease risk, vineyards should be established on well-drained soils. In some problem areas, soil water drainage can be improved by installing surface or subsurface drainage conduits. Planting vines on berms helps to avoid water accumulation around root crowns. On irrigated sites, it is advisable to minimize periods of soil saturation, avoiding those that last longer than 24 h or techniques that result in applications of excessive water near the root crown. Research has shown that foliar treatments with phosphonate can reduce the incidence of Phytophthora diseases on several crops, including root rot caused by *P. cinnamomi* on grape. Soil treatment with mefenoxam and metalaxyl also suppressed *P. cinnamomi* on grape. Pre-plant soil fumigation can improve the subsequent growth of young vines, but only provides short-term reductions in soilborne populations of *Phytophthora* spp. within infested sites.

Careful rootstock selection can reduce the incidence and severity of Phytophthora crown and root rot. Under field conditions in South Africa, cultivars of own-rooted *V. vinifera* were less susceptible to *P. cinnamomi* than several rootstocks that are hybrids between *V. vinifera* and other *Vitis* spp. Furthermore, rootstock cultivars of *V. rupestris* (St. George and Rupestris du Lot), *V. berlandieri* × *rupestris* (1045 Paulsen, 99 Richter, 110 Richter, and 1103 Paulsen), and *V. riparia* × *rupestris* (3306 C) were highly susceptible to *P. cinnamomi*, compared with several rootstocks with *V. vinifera* in their parentage (i.e., Jacquez, 3-5 USVIT, 3-6 USVIT, and 2-1 USVIT). Of the many rootstocks used for grape in California, only vines on Rupestris St. George have been known to suffer widespread loss due to infection by *P. cinnamomi*, and only this rootstock among more than 20 others was consistently susceptible to *P. cinnamomi* in replicated greenhouse tests.

Selected References

Agnihothrudu, V. 1968. A root rot of grapes in Andhra Pradesh. Curr. Sci. 37:292-294.

Erwin, D. C., and Ribeiro, O. K. 1996. Phytophthora Diseases Worldwide. American Phytopathological Society, St. Paul, MN.

Latorre, B. A., Wilcox, W. F., and Muñoz, R. 1997. Crown and root rot of table grapes caused by *Phytophthora* spp. in Chile. Vitis 36:195-197.

Marais, P. G. 1980. Fungi associated with decline and death of nursery grapevines in the Western Cape. Phytophylactica 12:9-13.

Marais, P. G. 1988. Grapevine roots and soil-borne fungi. Pages 106-137 in: The Grapevine Root and its Environment. J. L. Van Zyl, ed. Tech. Comm. 215. Republic of South Africa, Department of Agriculture and Water Supply, Pretoria.

Marais, P. G., and Hattingh, M. J. 1986. Reduction of root rot caused by *Phytophthora cinnamomi* in grapevines by chemical treatments. Plant Dis. 70:109-111.

Zentmyer, G. A. 1980. *Phytophthora cinnamomi* and The Diseases It Causes. Monogr. 10. American Phytopathological Society, St. Paul, MN.

(Prepared by G. T. Browne and B. A. Latorre)

Powdery Mildew

Perhaps because of its lack of dependence on rainfall for secondary spread, high reproductive potential, and the general susceptibility of *Vitis vinifera*, powdery mildew is the most consistently troublesome disease of grapevines worldwide. Powdery mildew, also called Oïdium, Mehltau, Echter Mehltau, Schimmel, Äscherich, Oidio, mal bianco, and cenicilla, is caused by *Erysiphe necator* Schw. The fungus was first described in eastern North America, the region to which it appears to be native, in 1834 by Schweinitz as *Erysiphe necator*, and this taxonomic assignation has recently been readopted, although *Uncinula necator* (Schw.) Burr. is encountered more commonly in literature published between 1900 and 2005. The organism was exclusively a minor pathogen of native *Vitis* species in North America until 1845, when it was first observed in a glasshouse in Margate, England by Tucker, for whom Berkeley named the anamorph *Oidium tuckeri* in 1847. By 1850, the pathogen had spread throughout Europe and caused staggering losses that often approached 100%. By 1858, however, the discovery that sulfur could be used to combat the disease and the development of methods for its large-scale application brought the disease under control in France and much of Europe. Although *O. tuckeri* was widely presumed in Europe to be the same fungus described by Schweinitz in North America, this was not proven until the discovery of fruiting bodies of the sexual stage (chasmothecia, syn. cleistothecia) in France by Couderc in 1892. The disease was first reported in California vineyards in 1859.

Symptoms

Erysiphe necator is, with the exception of the haustoria, wholly external to the host tissues, and grows upon the surface of the plant epidermis (Fig. 148). Thus, it is the actual presence of the pathogen's mycelium and conidia, i.e., the whitish, powdery-appearing growth from which the disease gets it name that is the most conspicuous feature of powdery mildew. Although these fungal structures are technically a sign of the disease, for the sake of convenience we shall use the term *symptoms* to refer to the visual manifestations of the pathogen itself in addition to the appearance of infected host tissues.

All green tissues of the vine may be infected. On leaves, symptoms vary with the source of inoculum, tissue age at the time of infection, and the age of the mildew colony. When ascospores are the principal or sole source of primary inoculum, colonies on the lower leaf surface are the predominant symptom during the earliest stages of the epidemic. They are often inconspicuous and are most frequently found on the first-formed leaves closest to the trunk or arms of the vine, where the ascospore-bearing chasmothecia are most likely to have overwintered (Fig. 149). Infection of the lower leaf surface often produces chlorotic spots on the upper surface (Fig. 150), which may sometimes be confused with the "oil spot" symptom of downy mildew. On cultivars with dense leaf hairs on the lower leaf surfaces, the pathogen may not be visible to the naked eye, but with care can be resolved by a 15–20× hand lens. On leaves with smooth lower surfaces, mildew colonies are at first white with a silver-gray or very light-brown tinge. On the upper leaf surfaces, young colonies appear whitish and show a somewhat metallic sheen. They may only be visible at a low viewing angle, by looking across the leaf; this is especially true for colonies that have not yet sporulated. Colonies are roughly circular, occur singly or in groups, and range in size from a few millimeters to a centimeter or more in diameter. Older and profusely sporulating colonies are most conspicuous, and are stark white (Fig. 151). Senescent colonies eventually turn grayish, and by mid- to late season may bear large numbers of yellow (immature) or black (mature) chasmothecia, the sexual fruiting

Fig. 148. Hyphae of *Erysiphe necator* growing on the surface of an infected grape leaf. (Courtesy D. M. Gadoury)

bodies of the pathogen (Fig. 152). Dead epidermal cells may be observed throughout the colonized area, as natural mortality in the mildew colony, use of fungicides, mycoparasites, or resistance responses in the leaf result in the deaths of segments of the mildew colony and the infected epidermal cells (Fig. 153). Severely affected leaves usually senesce, develop necrotic blotches, and fall prematurely (Fig. 154). Infection of stems initially produces symptoms similar to those on leaves, with affected areas turning black as the disease progresses and epidermal cells are killed. Eventually, the mildew colonies are killed as periderm forms in maturing canes, resulting in a dark, weblike scarring of the affected area (Fig. 155).

Berries may be infected directly by windborne conidia or ascospores, or by spread of a colony originating on the rachis or pedicel (Fig. 156). Prebloom infections of the cluster (including those of the nascent inflorescence during bud formation the previous season) or infection of berries within one week after their formation result in the most severe manifestation of the disease, and such tissues may be completely coated with whitish mildew colonies. The epidermis of such berries ceases to grow during berry expansion, often causing the skin to split

Fig. 149. Inconspicuous powdery mildew colony on the abaxial (lower) surface of a young leaf near the vine trunk, resulting from primary, ascosporic infection by *Erysiphe necator.* (Courtesy D. M. Gadoury)

Fig. 150. Chlorotic spots on the adaxial (upper) surface of a leaf that has been colonized by *Erysiphe necator* on the abaxial surface directly opposite. (Courtesy R. C. Pearson)

Fig. 151. Characteristic foliar symptom of powdery mildew, consisting of white sporulating colonies of *Erysiphe necator.* (Courtesy W. F. Wilcox)

Fig. 152. Immature (yellow) and mature (black) chasmothecia of *Erysiphe necator* on the surface of a grape leaf with powdery mildew. (Courtesy D. M. Gadoury)

Fig. 153. Patches of necrotic epidermal leaf cells resulting from powdery mildew development, surrounded by senescing (gray) and young (white) colonies of *Erysiphe necator.* (Courtesy W. F. Wilcox)

(Fig. 157), and the berry eventually desiccates and mummifies (Fig. 154) or becomes rotted by other fungi and bacteria. Berries become highly resistant to new infections as they age, but those in a transitional stage between susceptible and resistant (typically, about 3–4 weeks after anthesis, or cap fall) may develop diffuse infections characterized by a loose colony of nonsporulating hyphae. These are generally not visible to the naked eye, but can be resolved with low-power magnification. Diffuse colonies die as berries continue to mature, leaving behind a network of necrotic epidermal cells (Fig. 158).

In viticultural regions where low winter temperatures do not preclude it, overwinter survival of infected buds results in a distinctive foliar symptom when these buds give rise to shoots the following spring. In the most severe cases, such shoots are systemically infected and so heavily coated with fungal growth that they appear stark white and stand out like white flags in the vine, resulting in the term *flag shoots* (Fig. 159). More commonly, colonization of a flag shoot is less extensive and infection of a single leaf, or of leaves on one side of the shoot only, is observed (Fig. 160). The degree of colonization of the bud prophylls and leaf primordia within lateral buds may determine the macroscopic appearance of flag shoots the following season; incomplete colonization leads to the appearance of colonies on only one or two leaves and/or parts of the petioles and stem, whereas heavy colonization of the bud interior may lead to stark white, heavily colonized shoots that die soon after sporulation occurs. Breaking of infected buds is often delayed, and flag shoots are generally less vigorous than those arising from healthy buds (Fig. 161). Accordingly, their discovery in the vineyard may be delayed, and made more difficult by the obscuring growth of the healthy shoots. Leaves on flag shoots may be cupped, dwarfed, or distorted, even before conspicuous mildew colonies are seen (Fig. 162). Copious sporulation on flag

Fig. 154. Symptoms of severe powdery mildew development on an unsprayed 'Chardonnay' vine, including leaf necrosis, premature defoliation, and berry desiccation. Healthy, treated vines are in the background. (Courtesy W. F. Wilcox)

Fig. 155. Blackened tissue on shoots infected with *Erysiphe necator.* (Courtesy R. C. Pearson)

Fig. 156. Expansion of an *Erysiphe necator* colony to a young 'Concord' berry following prebloom infection of the rachis. (Courtesy D. M. Gadoury; reproduced by permission from Phytopathology 91:951)

Fig. 157. 'Chardonnay' berries infected by *Erysiphe necator* soon after their formation, coated with white mildew colonies and splitting due to poor epidermal expansion. (Courtesy C. N. Austin)

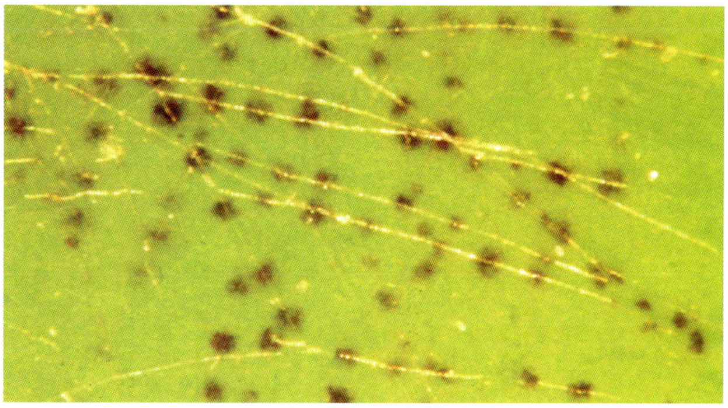

Fig. 158. Nonsporulating hyphae of *Erysiphe necator* within a diffuse powdery mildew colony on a berry inoculated approximately 3 weeks after the end of bloom, and subtending patches of necrotic epidermal cells. (Courtesy D. M. Gadoury; reproduced by permission from Phytopathology 97:1358)

shoots depends upon occurrence of favorable air temperatures and may be observed as early as 2 weeks after shoot growth begins. Even where they occur, flag shoots are usually present in low frequency. However, they provide intense point sources of inoculum early in the growing season, and this feature often pinpoints their presence in the canopy retrospectively, as numerous colonies resulting from secondary disease spread are eventually observed on the surrounding shoots and vines.

Causal Organism

Erysiphe necator Schw. (syns. *Uncinula necator* (Schw.) Burr., *E. tuckeri* Berk., *U. americana* Howe, *U. spiralis* Berk. & Curt; anamorph *Oidium tuckeri* Berk.) is obligately parasitic upon genera within the Vitaceae, including *Vitis*, *Cissus*, *Parthenocissus*, and *Ampelopsis*. Hyaline hyphae (4 to 5 μm diameter) are superficial upon epidermal cells, with the exception of globose haustoria formed at regular intervals along the hyphal strands beneath lateral, multilobed appressoria. Haustoria, the feeding organs of the fungus, penetrate the epidermal cell walls and associate with the outer surface of the subsequently invaginating epidermal cell membrane, through which they absorb nutrients. Segments of semipersistent hyphae frequently die in response to temperature extremes, ultraviolet radiation, fungicide applications, or development of resistance in host tissues, which may lead to the death of many of the epidermal cells harboring haustoria.

Conidiophores are multiseptate (10 to 400 μm in height) and form perpendicular to the prostrate hyphae, densely in rapidly

Fig. 159. Heavily colonized "flag shoot" resulting from springtime growth of a bud infected by *Erysiphe necator* the previous growing season. (Courtesy R. C. Pearson)

Fig. 160. Young shoot only partially colonized by *Erysiphe necator* following perennation of the pathogen within the bud from which it arose. (Courtesy W. D. Gubler)

Fig. 161. Flag shoot with inconspicuous powdery mildew colonies (far left), exhibiting reduced and deformed growth relative to adjacent healthy shoots. (Courtesy M. Sosnowski)

Fig. 162. Dwarfed, cupped leaves in a close-up of the powdery mildew flag shoot pictured in Fig. 161, with few signs of fungal growth yet developed. (Courtesy M. Sosnowski)

Fig. 163. Chains of hyaline conidia of *Erysiphe necator*. (Courtesy M. J. Welser)

growing colonies on susceptible tissue but less so on more resistant tissues. Conidia are hyaline, cylindro-ovoid, with one to two large water-containing vacuoles. They measure 27–47 × 14–21 µm, although their dimensions in various mountants are strongly affected by the osmotic potential of the spore cytoplasm relative to the mountant employed, and the presence of a highly elastic conidial cell wall. Individual conidiophores generally produce only one conidium per day, although under still-air conditions, conidia may accumulate in chains over a period of several days (Fig. 163), the oldest conidium at the distal end. Long conidial chains are rarely seen under field conditions, as the conidia are easily dislodged and dispersed by even low-velocity wind. Conidia germinate via a single germ tube, which terminates in a lobed appressorium.

Populations are bipolar-heterothallic, and ascocarps are initiated within 24 h of contact between hyphae of compatible mating types. Nearly all literature published before 2010 referred to the ascocarps as *cleistothecia*, but this term has been superseded by *chasmothecia* in more recent literature. Within 72 h, the hyaline ascocarp initial increases to 40 µm in diameter. Anchorage hyphae form as short outgrowths of cells of the outer ascocarp wall and become intertwined in the subtending mildew colony, but form no functional connections or appressoria. Ascocarps turn yellow after about 7 days as a yellow lipid accumulates in the centrum and wall cells. The upright and bristlelike appendages are initiated after about 3 weeks of age, and are arranged equatorially. The cells of the outer ascocarp wall eventually darken, as do the basal cells of the appendages that form the characteristic uncinate tips during the final phase of development, about 4 weeks after the ascocarp is initiated; by this time, the ascocarp wall is heavily melanized and dark brown. Anchorage hyphae and the parent hyphae die after 4 to 5 weeks under typical field temperatures during the mid- to late summer, and the chasmothecium develops a basal concavity, effectively disconnecting it from the mildew colony.

Chasmothecia contain four to six asci at maturity, each of which contains usually four and as many as seven hyaline, ovate to subglobose ascospores, 15–25 µm × 10–14 µm; however, as with conidia, exact spore dimensions are sensitive to the osmotic potential of the mountant used in their examination. Although ascocarps may reach morphological maturity in 4 weeks, physiological maturity may not be reached for several months, particularly in colder climates. Physiologically mature chasmothecia dehisce circumscissally when in contact with free water. No longer under restriction of the ascocarp wall, the asci expand greatly, push back the upper half of the chasmothecium, and discharge ascospores forcefully through a slitlike rupture of the ascus tip. Like conidia, ascospores germinate with a single germ tube, which terminates in a lobed appressorium.

Disease Cycle and Epidemiology

The general disease cycle for powdery mildew is shown in Fig. 164. In areas with relatively mild winters, mycelium with haustoria and conidia can survive the dormant period within

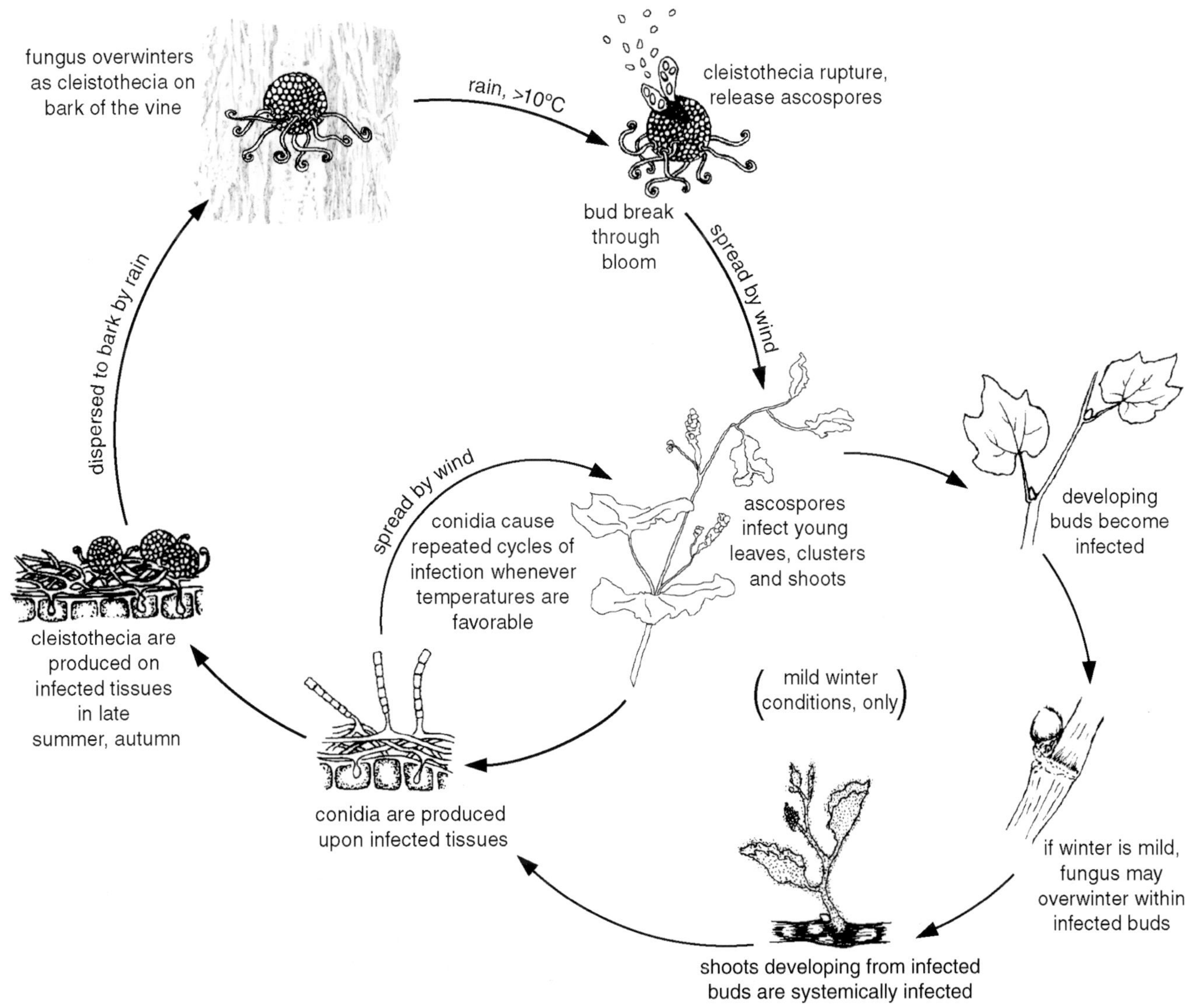

Fig. 164. Disease cycle of powdery mildew. (Drawing by E. Gotham from R. Sticht)

infected lateral buds, which give rise to conidia-bearing flag shoots the following spring. Buds that subsequently give rise to flag shoots can form soon after shoot growth begins in the spring. Surface tissues of basal bud positions are very susceptible to infection from budbreak until shoots have elongated to contain six true leaves (BBCH stages 9–16), and colonization of their surface during this period promotes ingress of the colony to the interior of the bud, as bud scales and leaf prophyls develop. With the exception of the meristem, all tissues of the dormant bud may be colonized subsequently. Susceptibility of lateral buds near the base of shoots decreases relatively quickly as they age—i.e., by BBCH stage 19 (nine true leaves present)—owing to the progressive suberization of the bud surface. However, as shoots continue to extend, new lateral buds in susceptible stages continue to be formed near the apex of the growing shoot. Whereas on spur-pruned vines the exclusive retention of basal buds necessarily restricts the origin of flag shoots to buds infected very early in the growing season, on cane-pruned vines flag shoots may be found emerging from nearly any position (hence, from infections initiated at any time during the previous season). It is presumed that infections of lateral buds reduce their winter hardiness, accounting for the relative lack to complete absence of flag shoots following the occurrence of winter temperatures well below freezing in colder viticultural regions (e.g., interior regions of Europe, and eastern Washington and New York in the United States). However, it is also possible that the pathogen itself is incapable of surviving these low temperatures as nascent infections. For example, debilitation of active *E. necator* colonies and transient resistance to new infections have been demonstrated following brief (e.g., 2 h) exposure of leaves to temperatures between 2 and 8°C, and such overnight low temperatures have been documented to occur commonly before bloom in many viticultural regions.

The pathogen also overwinters in nearly all viticultural regions as chasmothecia, providing an additional source of primary inoculum where flag shoots are common, and the principal to sole source where flag shoots are rare to absent. Chasmothecia are formed primarily on foliage once disease severity has increased to a level where pairing of compatible mating types upon the same tissue becomes possible. On heavily diseased vines, they may be found by mid-summer; however, under effective control programs, significant numbers usually do not form until the end of the season, after spray programs have eased. In areas where chasmothecia do occur by mid-summer, e.g., the central coast of California and South Australia, there can be late-season ascospore release with subsequent new infections from these sources in addition to conidia. Chasmothecia also may be found on other severely infected tissues, such as berries (Fig. 165), rachises, or shoots late in the growing season. Given sufficient time to mature morphologically (i.e., approximately 4–6 weeks after initiation, depending on temperature), chasmothecia become detached from the anchorage hyphae. In areas with sufficient rainfall, nearly all mature ascocarps are dispersed by rain to either the bark of the vine, where some survive the winter, or to the soil. They may also be dispersed by very high winds. Naked chasmothecia apparently do not survive on or in soil. However, in climates with relatively low rainfall and mild winter temperatures (e.g., California, South Australia, or Italy), they do not completely disperse from the leaves and some survive within leaf litter on the vineyard floor.

In spring, ascospores are discharged when the ascocarps are wet from rain or irrigation (Fig. 166). Fog and heavy dew may also release ascospores if chasmothecia become sufficiently wet. The period of ascospore release in New York begins near bud break and continues until the bloom or early postbloom period. Similar periods of spring ascospore release have been reported in several other regions. In both California and South Australia, some chasmothecia may also release ascospores during fall or winter rains. In general, ascosporic infection is favored by precipitation in excess of 2–3 mm coincident with temperatures above 10°C, although infection becomes increasingly likely as temperatures increase to an optimum value of approximately 25°C. Although free water is necessary to initiate discharge of the ascospores from chasmothecia, continued wetness is not required for subsequent spore germination and infection.

Where ascospores are the only propagules of primary inoculum, new leaves may be infected as they form, but foliar disease severity typically does not increase markedly until after bloom. Therefore, although disease incidence may increase rapidly beforehand, low severity usually makes it inconspicuous to the casual observer. However, without effective fungi-

Fig. 165. Small, black chasmothecia of *Erysiphe necator* forming on severely infected cv. Chardonnay berries in late summer. (Courtesy W. F. Wilcox)

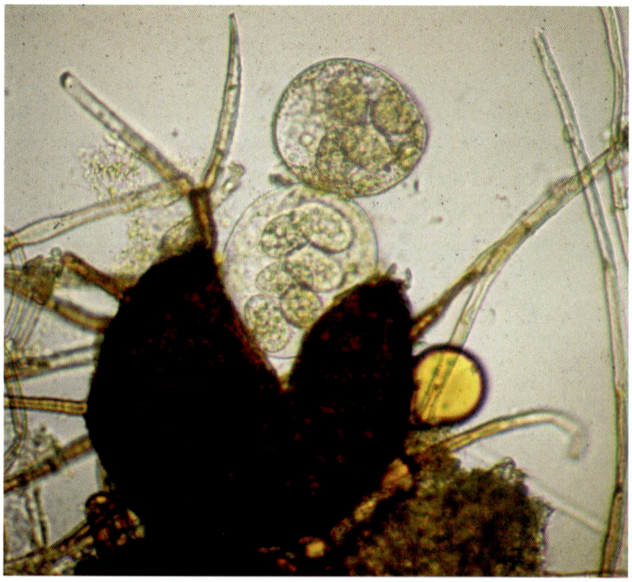

Fig. 166. Rupturing chasmothecium of *Erysiphe necator* with asci containing ascospores (× 400). (Courtesy R. C. Pearson)

cide treatment, foliar severity increases rapidly toward bloom because temperatures optimal for mildew development persist for increasing portions of the day. It is usually during this phase that observers first notice an outbreak of the disease.

Where flag shoots serve as sources of primary inoculum, an additional level of complexity may be introduced into the development of the epidemic. Coincident with the random distribution of ascosporic inoculum, a variable density of flag shoots may occur throughout the vineyard. A focus of disease develops early in the growing season around each flag shoot, as intense conidial production spreads the disease to nearby shoots and vines. Usually, these disease foci persist until after bloom, but may become less obvious as they coalesce and general disease levels increase throughout the vineyard. On varieties highly susceptible to bud infection (e.g., Carignane), flag shoots may be observed on the same vines year after year, thus making prior flag shoot location a key component of disease monitoring for advisory programs.

Temperature is the principal environmental factor that regulates growth and secondary spread of the pathogen. Mildew colonies grow and sporulate most rapidly from 23 to 30°C, and the latent period (time from the start of infection to production of new conidia from the resulting colony) can be as short as 5 to 6 days if temperatures are constantly within this range. However, the pathogen is considerably less robust and the latent period increases significantly under cooler conditions, i.e., to 11, 18, and 25 days at constant temperatures of 15, 12, and 9°C, respectively. Recent studies conducted under the fluctuating temperature conditions of the field suggest that latent periods can be approximately 30% longer than would be expected on the basis of daily average temperatures if nighttime lows drop below 8°C, particularly during the first 4–6 weeks of the growing season. The lower and upper temperature limits for disease development are 6 and 32°C, respectively. Germination of conidia is inhibited at 35°C and conidia can be killed after exposure to higher temperatures, with the lethal exposure time a function of temperature, e.g., 12, <5, and 0.5 h at temperatures of 38, 40, and 44°C, respectively, in laboratory studies. Similarly, many of the hyphae within mildew colonies may be killed by high temperatures that persist for a critical duration, e.g., 20, 6, and 1 h at 36, 40, and 44°C, respectively. However, mortality resulting from high temperatures is commonly incomplete in the field, especially within the shaded interior of the canopy, and surviving portions of colonies may resume activity if temperatures again become favorable.

Other environmental factors influencing disease development include atmospheric humidity and sunlight exposure. Although *E. necator* is unique among the fruit and foliar pathogens of grapevines in its ability to thrive in both arid and humid climates, powdery mildew develops optimally at a relative humidity of approximately 85%, and both incidence and severity decrease moderately but progressively as the air becomes drier. For example, under controlled conditions at an optimum temperature of 25°C, disease severity at 40% relative humidity was approximately one-half that at 85%, with a linear response to the humidity level occurring over that range. Germination of conidia responds similarly, but unlike mycelial growth and sporulation, germination frequency is severely inhibited at relative humidity values above approximately 87%.

Ultraviolet (UV) light also is detrimental to the fungus, which resides primarily on the surface of infected grape tissues and is without protective pigmentation. Exposure to UV radiation reduces both the frequency and rate of conidium germination, appressorium formation, and subsequent colony expansion, prolonging the latent period and reducing disease severity relative to tissues protected from UV radiation. These effects appear to be cumulative, i.e., the latent period is similarly increased by four daily 6-h exposures or a single 24-h exposure. Furthermore, surface temperatures of grapevine tissues in direct sunlight often are substantially higher than those in the shade, e.g., temperature increases ranging from 2 to 13°C have been measured for exposed versus shaded leaves in climates as diverse as New York and South Australia, depending on the intensity of the sunlight and the water status of the vines at the time of measurement. Furthermore, the detrimental effects of UV radiation are more pronounced at 28°C than at 20 or 25°C, indicating the synergistic effects of UV and heat at supraoptimal temperatures. Thus, disease severity is often several times greater on shaded tissues than on those afforded good sunlight exposure.

Free water is detrimental to *E. necator* mycelium and conidia; however, in the field, this effect may be more than offset by the favorable impacts of moderate temperatures, limited direct sunlight, and high humidity that are all likely to precede and follow rain events. Thus, the most severe mildew epidemics often accompany atypically rainy growing seasons, despite the deleterious impact upon the pathogen of discrete periods of free water.

Leaves are most susceptible to infection when half expanded, and susceptibility is reduced significantly after they have fully expanded and begun to age. However, leaves never become immune to infection, and older ones can support substantial quantities of powdery mildew. Additionally, in some environments and management systems, the indeterminate growth of shoots and forcing of laterals by summer hedging insures a continual supply of highly susceptible young leaves, and an ever-increasing supply of older, less susceptible ones.

In contrast to that of leaves, the period of fruit susceptibility is relatively brief. Individual berries of *V. vinifera* cultivars are highly susceptible to infection for the first 1–2 weeks after they have set, and infection during this period can lead to complete destruction of the fruit. Although older references usually associate sugar accumulations of 8% or more with berry resistance, a high level of ontogenic (age-related) resistance is widely observed at 3–4 weeks after the completion of bloom, at which time the concentration of soluble solids (°Brix) is still basal, fluctuating between 4 and 5%. The inconspicuous, diffuse infections that develop on berries that become infected as they transition to a resistant state at this time may result in a degradation of wine quality by both exacerbating the severity of Botrytis bunch rot and elevating populations of other microorganisms that lead to spoilage during fermentation. Diffuse infections also have been associated with increased wounding by fruit-feeding insects, particularly wasps (*Vespula* spp.), which appear to be attracted by increased levels of volatile compounds emitted by these berries even before visible symptoms of decay are seen. Berries of less susceptible *Vitis* species and interspecific hybrids appear to develop ontogenic resistance earlier than those of *V. vinifera*, e.g., berries of *Vitis* × *labruscana* 'Concord' are virtually immune within 2 weeks after fruit set.

Management

All cultivars of *V. vinifera*, a species that evolved in the absence of the pathogen, are susceptible to powdery mildew, some especially so. A number of cultivars of native North American species and interspecific hybrids show considerably more resistance, but demand for them is limited outside a few specific regions and markets. Therefore, commercial control of powdery mildew generally requires a significant use of fungicides, although cultural practices are important components of an integrated management program and can greatly impact the success and requisite number of fungicide applications. Pruning and training of the grapevine canopy should provide good ventilation and promote sunlight exposure in the cluster zone to whatever extent is possible without causing sunburn. This not only creates a microclimate around bunches that is less favorable to the pathogen (by reducing humidity and exposing them to solar radiation), but also greatly facilitates their effective coverage by the applied fungicides. Selective removal of leaves around the clusters immediately after fruit set, while the berries are still highly susceptible to infection, has been shown to aid in control of powdery mildew as well as Botrytis bunch rot. Biological control by species of the soilborne bacterium *Bacillus*, the mycoparasite *Ampelomyces quisqualis*, and the mycophageous mite *Orthotydeus lambi*, has met with limited commercial and/or experimental success, although these organisms may augment the level of suppression provided by fungicides in plantings where they can be introduced successfully or occur naturally.

Fungicide applications typically commence early in the growing season. In regions where chasmothecia are the major or sole source of primary inoculum, infections can occur once an adequate rain (2–3 mm) has fallen and temperatures reach at least 10°C; thus, fungicide programs may need to begin shortly after bud break if such conditions occur, but may be delayed by one to several weeks thereafter if cool conditions or a lack of rainfall delays the development of primary infections. In California, an application of wettable sulfur near completion of budbreak yields enhanced suppression of powdery mildew, presumably by acting against the earliest ascosporic inoculum. In areas with regular occurrences of flag shoots, these sites of primary infection are present within 1–2 weeks after bud break, and fungicide coverage should commence as early as BBCH stage 13 (three unfolded leaves) to limit secondary spread and colonization of young lateral buds at the base of the new shoots.

Several systems for determining the most efficient timing of fungicide applications throughout the remaining season have been utilized in different parts of the world. In California, a disease risk assessment index has been developed based largely upon the rapid reproduction of *E. necator* at temperatures between 21 and 30°C, the relatively slow development of disease under cooler conditions, and the deleterious effect of temperatures >35°C. In Germany, a different system has been developed, initially utilizing a combination of the previous season's disease severity and lowest winter temperature (i.e., the probability of bud perennation and subsequent flag shoot formation) plus current temperature and relative humidity conditions, assuming a minimum of 65% relative humidity for robust disease development. Recently, this system has been adapted also to account for the heightened sensitivity of clusters to infection in the prebloom through early postbloom period as discussed below. In some warmer regions of Australia, an empirical system has been developed in which sprays are restricted to the period of 2–6 weeks following bud break, after which conditions are typically too hot and sunny for significant disease development. Although all such systems are based upon certain fundamentals of pathogen biology, each tends to emphasize the influence of specific factors that are particularly determinative under local conditions.

The timing of fungicide applications and choice of materials may be influenced by the stage of crop development. Because individual berries of *V. vinifera* cultivars are most susceptible to infection from the immediate prebloom period until 1–2 weeks after their caps fall and retain significant susceptibility for approximately 2 additional weeks, management efforts should be at their peak during this critical period, placing particular emphasis on factors such as fungicide quality and rate, appropriate intervals between sprays, and quality of the spray application technique. However, the protracted, asynchronous opening of grapevine flowers that occurs in viticultural regions with relatively warm winters (e.g., California, South Australia, Chile, etc.) may extend this critical period for the youngest berries in a cluster until approximately 6 weeks after the general onset of flowering. Applications after berries have become resistant to infection are generally designed to prevent infection of cluster rachises and the foliage. Although the necessary degree of foliar control may vary among cultivar and intended market for the fruit, healthy foliage is generally required to ripen the crop and provide storage carbohydrates to the vine, which may be critical to survival in regions with low winter temperatures. Good season-long control of foliar (and bud) infections is also the best means of minimizing the level of primary inoculum available to initiate disease the following season. To maintain such control, postharvest fungicide applications may be necessary in regions where conditions are conducive to disease development for prolonged periods after the crop has been taken. In general, from 4 to 6 weeks of frost-free weather are required for late-season infections to form mature chasmothecia before the leaves sustaining their development are killed.

Sulfur is the fungicide used most widely against powdery mildew worldwide. Its efficacy, relatively low cost, and lack of pathogen resistance have made it a central component of management programs for over 160 years. Sulfur provides control through a combination of protective, curative (postinfection), and eradicative activities, although its value for eradication is limited against dense, well-developed colonies. Activity appears to accrue through both direct contact of the applied material (especially when applied in aqueous suspensions) and redistribution through the vapor phase. Due to the latter mechanism, activity has often been stated to decline at temperatures below 18°C, although controlled studies have shown little to no loss of efficacy at lower temperatures. Conversely, phytotoxicity may occur with applications made above 30°C, and is progressively more likely as temperatures increase beyond this limit. Studies in South Australia have indicated that sulfur phytotoxicity is dependent upon both temperature and relative humidity, with little to no injury at temperatures as high as 45°C if relative humidity remains below 70%, whereas phytotoxicity occurs at temperatures above 35°C when relative humidity exceeds 70%. Phytotoxicity is also genotype-dependent, being more likely on some *V. vinifera* cultivars than others. Severe phytotoxicity, regardless of temperature, precludes the use of sulfur on some cultivars of *Vitis* spp. native to North America and their interspecific hybrids, particularly red- or purple-fruited types. The utility of sulfur also is compromised in various regions by its being washed off treated vines during rains, although this limitation can be addressed to some extent by utilizing sprayable formulations with very small particle sizes, increasing the application rate, and/or adding an appropriate surfactant to the spray solution.

A number of different classes of modern organic fungicides provide excellent control of powdery mildew, including the benzimidazoles; the DMI and morpholine subgroups of the ergosterol biosynthesis inhibitors; the strobilurin members of the QoI class; various SDHI materials; and several newer powdery mildew-specific compounds that include quinoxyfen and the related the related proquinazid, metrafenone, and cyflufenamid. Relative to sulfur, these materials offer the advantages of maintaining efficacy with lower use rates and longer application intervals; being much less subject to removal by rainfall; and in some cases, providing a broader spectrum of activity against additional fungal pathogens. However, the high reproductive rate of *E. necator*, its ability to infect under a wide range of climatic conditions, and its ready dispersal by air currents collectively promote the rapid development of pathogen populations resistant to specific fungicides if these materials are used in a manner that selects for resistant biotypes. The speed with which control failures have occurred in specific commercial vineyards illustrates the need for implementing resistance management practices whenever utilizing modern synthetic fungicides, virtually all of which should be considered at risk for resistance development.

Erysiphe necator is unique in that, unlike all other fungal pathogens of grapes, its "body" (thallus) resides almost entirely on the surface of infected tissues. This feature renders it susceptible to topical applications of an assortment of materials that have little, if any, direct activity against other fungal pathogens, the thalli of which are protected from such treatments inside the tissues that they infect. Examples of effective topical materials include botanical and petroleum-based oils, inorganic salts (e.g., potassium bicarbonate, monopotassium [dihydrogen potassium] phosphate), and hydrogen peroxide (dihydrogen oxide). Such materials, which are effective primarily to exclusively when used in a curative or eradicative mode, rely on direct contact with the fungal thallus and, therefore, are highly dependent upon complete spray coverage for their activity. However, residues of many such materials have little or no effect upon spore germination and penetration processes, so they provide limited to no protective activity following an application. Copper, captan, dithiocarbamates, and several other traditional protectant fungicides provide moderate suppression of the disease, but their activity is generally inadequate on highly susceptible varieties, particularly under high disease-pressure conditions.

Selected References

Austin, C. N., Meyers, J., Grove, G. G., and Wilcox, W. F. 2011. Quantification of powdery mildew severity as a function of canopy variability and associated impacts on sunlight penetration and spray coverage within the fruit zone. Am. J. Enol. Vitic. 62:23-31.

Austin, C. N., and Wilcox, W. F. 2011. Effects of fruit zone leaf removal, training system, and variable irrigation on powdery mildew development on *Vitis vinifera* L. Chardonnay. Am. J. Enol. Vitic. 62:193-198.

Austin, C. N., and Wilcox, W. F. 2012. Effects of sunlight exposure on grapevine powdery mildew development. Phytopathology 102:857-866.

Bulit, J., and Lafon, R. 1978. Powdery mildew of the vine. Pages 525-548 in: The Powdery Mildews. D. M. Spencer, ed. Academic Press, New York.

Carroll, J. E., and Wilcox, W. F. 2003. Effects of humidity on the development of grapevine powdery mildew. Phytopathology 93:1137-1144.

Delp, C. J. 1954. Effects of temperature and humidity on the grape powdery mildew fungus. Phytopathology 44:615-626.

Doster, M. A., and Schnathorst, W. C. 1985. Effects of leaf maturity and cultivar resistance on development of the powdery mildew fungus of grapevines. Phytopathology 75:318-321.

Gadoury, D. M., Cadle-Davidson, L., Wilcox, W. F., Dry, I. B., Seem, R. C., and Milgroom, M. G. 2012 Grapevine powdery mildew (*Erysiphe necator*): A fascinating system for the study of the biology, ecology and epidemiology of an obligate biotroph. Mol. Plant Pathol. 13:1-16.

Gadoury, D. M., and Pearson, R. C. 1988. Initiation, development, dispersal, and survival of cleistothecia of *Uncinula necator* in New York vineyards. Phytopathology 78:1413-1421.

Gadoury, D. M., and Pearson, R. C. 1990. Ascocarp dehiscence and ascospore discharge in *Uncinula necator.* Phytopathology 80:393-401.

Gadoury, D. M., Seem, R. C., Ficke, A., and Wilcox, W. F. 2003. Ontogenic resistance to powdery mildew in grape berries. Phytopathology 93:547-555.

Gadoury, D. M., Seem, R. C., Wilcox, W. F., Henick-Kling, T., Conterno, L., Day, A., and Ficke, A. 2007. Effects of diffuse colonization of grape berries by *Uncinula necator* on bunch rots, berry microflora, and juice and wine quality. Phytopathology 97:1356-1365.

Gubler, W. D., Rademacher, M. R., Vasquez, S. J., and Thomas, C. S. 1999. Control of powdery mildew using the UC Davis powdery mildew risk index. APS Feature, American Phytopathological Society, St. Paul, MN. doi:10.1094/APSnetFeature-1999-0199.

Hill, G. K. 1990. The influence of annual weather patterns on epidemics of *Uncinula necator* in Rheinhessen. Wein-Wissenschaft 45:43-46.

Jarvis, W. R., Gubler, W. D., and Grove, G. G. 2002. Epidemiology of powdery mildews in agricultural pathosystems. Pages 169-199 in: The Powdery Mildews: A Comprehensive Treatise. R. R. Bélanger, W. R. Bushnell, A. J. Dik, and T. L. W. Carver, eds. American Phytopathological Society, St. Paul, MN.

Kast, W. K., and Bleyer, K. 2010. The expert system OiDiag-2.2 - a useful tool for the precise scheduling of sprays against the powdery mildew of vine (*Erysiphe necator* Schwein.). Pages 151-153 in: Proc. 6th Int. Workshop on Grapevine Downy and Powdery Mildew. INRA-INITA, Bordeaux, France.

Kast, W. K., and Bleyer, K. 2011. Efficacy of sprays applied against powdery mildew (*Erysiphe necator*) during a critical period for infections of clusters of grapevines (*Vitis vinifera*). J. Plant Pathol. 93:S1.29-S1.32.

Moyer, M. M., Gadoury, D. M., Cadle-Davidson, L., Dry, I. B., Wilcox, W. F., and Seem, R. C. 2010. Development of an advisory system for grapevine powdery mildew in eastern North America: A reassessment of epidemic progress. Plant Health Progress doi:10.1094/PHP-2010-0526-02-SY.

Moyer, M. M., Gadoury, D. M., Cadle-Davidson, L., Dry, I. B., Magarey, P. A., Wilcox, W. F., and Seem, R. C. 2010. Effects of acute low temperature events on development of *Erysiphe necator* and susceptibility of *Vitis vinifera*. Phytopathology 100:1240-1249.

Pearson, R. C., and Gadoury, D. M. 1987. Cleistothecia, the source of primary inoculum for grape powdery mildew in New York. Phytopathology 77:1509-1514.

Peduto, F., Backup, P., Hand, E. K., Janousek, C. N., and Gubler, W. D. 2013. Effect of high temperature and exposure time on *Erysiphe necator* growth and reproduction: Revisions to the UC Davis powdery mildew model risk index. Plant Dis. 97:1438-1447.

Rumbolz, J., and Gubler, W. D., 2004. Susceptibility of grapevine buds to infection by the grapevine powdery mildew fungus *Uncinula necator* (Schw.) Burr. Plant Pathol. 54:535-548.

(Prepared by D. M. Gadoury, W. F. Wilcox, J. Rumbolz, and W. D. Gubler)

Ripe Rot

As its name implies, ripe rot develops on grapes as they mature and near harvest. This disease was first reported in the United States in 1891 and has since been found in most warm and moist regions where bunch and muscadine grapes are grown. Losses from the disease are greatest during warm, wet seasons.

Symptoms

The primary symptom of this disease is rotting of the ripe fruit. Affected berries initially develop circular, reddish-brown spots of decay on their skins (Fig. 167). The spots subsequently enlarge to include entire berries, which characteristically become covered with salmon-colored masses of conidia as they decay (Fig. 168). Eventually, diseased berries shrivel and superficially resemble those in the terminal stages of several other rot diseases (black rot, bitter rot, Phomopsis), although the presence of the characteristic conidia is diagnostic for ripe rot (Fig. 169.). Rotted fruit of most grapes typically remain attached to the rachis, although those of *Vitis rotundifolia* may drop when the rot is complete. Symptoms are not common on leaves or shoots.

Fig. 167. Initial symptom of ripe rot, consisting of a circular red-brown spot of decay on the skin. (Courtesy T. B. Sutton)

Causal Organisms

Ripe rot is caused by homothallic and heterothallic strains of *Colletotrichum gloeosporioides* (Penz.) Penz. & Sacc. (teleomorph *Glomerella cingulata* (Stonem.) Spauld. & Schrenk) and *C. acutatum* J. H. Simmonds (teleomorph *G. acutata* Guerber & Correll). *Colletotrichum* spp. produce subepidermal acervuli which are arranged in circles on the surface of the fruit. A mass of sticky, salmon-colored conidia is discharged from the acervuli. Conidia of *C. gloeosporioides* are hyaline and guttulate, 12–21 × 3.5–6 µm, rounded at the ends, and are often slightly curved. Conidia of *C. acutatum* are hyaline, straight, mostly fusiform, 8–16 × 2.5–4 µm, and frequently pointed at one or both ends. Perithecia are subspherical and more or less grouped. Asci are subclavate and measure 42–60 × 10–12 µm; ascospores measure 12–24 × 4–6 µm.

Diseases caused by *Colletotrichum* spp. on many other crops are given the common name "anthracnose," whereas anthracnose on grapes is the accepted name for the disease caused by *Elsinoë ampelina*, an unrelated organism. Unfortunately, this potentially ambiguous situation has resulted in *Colletotrichum* spp. being ascribed as the causal agents of "anthracnose" on grapes in some literature and nontechnical publications, particularly from Asia. Accounts that utilize this common name should be read carefully to determine the true etiology of the disease.

Fig. 168. Characteristic salmon-colored spore masses of a *Colletotrichum* sp. on muscadine berries with ripe rot. (Courtesy T. B. Sutton)

Fig. 169. Advanced symptoms and signs of ripe rot on a *V. vinifera* cluster, including decay and shriveling of affected berries, some with pinkish spore masses of a *Colletotrichum* sp. (Courtesy W. F. Wilcox)

Disease Cycle and Epidemiology

Colletotrichum spp. and *G. cingulata* survive from one season to another on the vine as dormant mycelium in mummified fruit and infected pedicels, and in dead bark and other tissues on the cordons that have been colonized. Ascospores, dispersed from these structures by wind during rainy periods in the spring, and conidia spread short distances by splashing rain, serve as primary inoculum. Production of conidia from mummies and pedicels is greatest in early spring and decreases during the summer months; few or no primary spores are produced by mid- to late summer. Sporulation on ripe fruit near harvest provides an abundant source of secondary inoculum immediately adjacent to healthy berries, and frequent rains during this period can result in severe crop loss in the absence of adequate control measures.

Fruit are susceptible to infection by *Colletotrichum* spp. and *G. cingulata* from bloom until harvest, but do not show symptoms until they begin to ripen. Disease development is favored by warm (25–30°C), wet weather. Spores germinate in a film of water and can penetrate the cuticle within a matter of hours under favorable conditions, but the fungus remains quiescent until the fruit mature.

Management

Cultural practices, such as pruning out dead spurs, removing overwintered mummies, and removing weak or dead cordons, are important to help reduce inoculum of both ripe rot and bitter rot (*Greeneria uvicola*) in the vineyard. Converting to a cane pruning system rather than spur pruning has reduced disease development in some cases, presumably by removing certain perennial sources of inoculum. It is necessary to apply protectant fungicides on a 2-week schedule from bloom until harvest where the disease is a problem, except during periods of drought. Captan, EBDC fungicides, and the strobilurin members of the QoI group have good activity against ripe rot, whereas the DMI fungicides are not effective.

Cultivars within various grapevine species and interspecific hybrids differ in their susceptibility to ripe rot. Within *V. vinifera*, Barbera, Chardonnay, Merlot, and Syrah are moderately resistant; Cabernet Franc, Tempranillo, and Sauvignon Blanc are slightly susceptible to moderately susceptible; and Cabernet Sauvignon and Petit Verdot are highly susceptible. Dark-skinned muscadine cultivars are resistant, whereas most bronze cultivars are susceptible. Among French-American hybrids, Traminette is moderately resistant; Chambourcin and Seyval Blanc are slightly susceptible; and Vidal Blanc is highly susceptible. The *V. aestavalis* cv. Cynthiana (Norton) is resistant.

Selected References

Clayton, C. N. 1975. Diseases of muscadine and bunch grapes in North Carolina and their control. N.C. Agric. Exp. Stn. Bull. 451.

Daykin, M. E., and Milholland, R. D. 1984. Ripe rot of muscadine grape caused by *Colletotrichum gloeosporioides* and its control. Phytopathology 74:710-714.

Daykin, M. E., and Milholland, R. D. 1984. Histopathology of ripe rot caused by *Colletotrichum gloeosporioides* on muscadine grape. Phytopathology 74:1339-1341.

Kummuang, N., Smith, B. J., Diehl, S. V., and Graves, C. H. Jr. 1996. Muscadine grape berry rot diseases in Mississippi: Disease identification and incidence. Plant Dis. 80:238-243.

Quimio, T. H., and Quimio, A. J. 1975. Notes of Philippine grape and guava anthracnose. Plant Dis. Rep. 59:221-224.

(Prepared by T. B. Sutton)

Rotbrenner and Angular Leaf Scorch

Symptoms of rotbrenner were well known in Europe during the nineteenth century. Before Müller-Thurgau first described the causal agent in 1903, the disease was thought to be caused by unfavorable weather conditions, water deficiency, or excessive soil moisture. Although rotbrenner has been identified in most European countries, it is most problematic in cool viticultural regions (e.g., Austria, the Champagne and Alsace regions of France, Germany, Hungary, Luxembourg, and Switzerland) and is generally confined to certain locally restricted areas. In some of these areas the disease results in severe losses annually, whereas in others it occurs only sporadically. The disease also has been observed to develop during several succeeding years in some locations, then not to reappear for several more.

Rotbrenner affects the cultivated European grape, *V. vinifera*, as well as other *Vitis* species and interspecific hybrids. It may also occur on Virginia creeper *(Parthenocissus quinquefolia)* and Boston ivy *(P. tricuspidata)*. In the 1980s, a similar disease was noted to occur sporadically on the foliage of grapevines in the state of New York, particularly those of interspecific hybrid cultivars. Subsequent research showed that this disease was caused by a unique species within the same genus as the fungus that causes rotbrenner, and the disease was named angular leaf scorch.

Symptoms

Rotbrenner leaf lesions initially are yellow on white-fruited cultivars and bright red to reddish brown on red- and black-fruited cultivars. With time, a reddish-brown necrosis develops in the center of the lesion, leaving only a thin margin of yellow or red tissue between the necrotic and green areas of the leaf (Fig. 170). The lesions are typically confined by the major veins and the leaf margin, and may be several centimeters wide. Atypical symptoms, such as freckled spots or discoloration scattered over the leaf surface, often occur late in the season. Early infections may occur on the first to the sixth leaf of young shoots, resulting in minor losses. Later infections may attack leaves up to the 10th or 12th position on the shoot and can result in severe defoliation.

The fungus also may attack inflorescences before or during bloom, causing them to rot and dry out (Fig. 171). Unlike Botrytis blight, only the pedicels are attacked; the fungus does not seem to invade the cluster rachis. In severe cases, the berries are destroyed, leaving the rachis bearing only a few individual berries or none at all. High levels of infection during bloom can result in 80–90% crop loss.

Lesions of angular leaf scorch are similarly constricted by the major veins and laminar margin, giving rise to the disease name. Lesions usually retain a yellow margin (Fig. 172), although it may be red on some red-fruited cultivars of *V. vinifera* or absent on some native American cultivars such as Concord. Most of the leaf blade may be killed in the case of severe infection (Fig. 173). Infections of the inflorescences are rare.

Causal Organisms

Pseudopezicula tracheiphila (Müll.-Thurg.) Korf & Zhuang (syn. *Pseudopeziza tracheiphila* Müll.-Thurg, anamorph *Phialophora tracheiphila* (Sacc. & Sacc.) Korf [syn. *Botrytis tracheiphila* Sacco & Sacc.]), the cause of rotbrenner, is a discomycete with minute (up to 0.6 mm diameter), sessile, gelatinous, whitish to faintly colored, gregarious apothecia. The fungus appears to be composed of two mating types, which exhibit a bipolar heterothallic mating system. The apothecia rise from the leaf tissue and are often associated with the veins. The inoperculate, broadly clavate, eight-spored asci (115–145 × 18–28 μm) show a blue pore in iodine only after pretreatment in aqueous

Fig. 170. Symptoms of rotbrenner on leaves of a white *V. vinifera* cultivar. (Courtesy W. Gärtel)

Fig. 171. Dried inflorescence affected with rotbrenner. (Courtesy A. Bolay)

Fig. 172. Lesion of angular leaf scorch on the white-fruited interspecific cultivar Aurore. (Courtesy R. C. Pearson)

potassium hydroxide. The ellipsoid, hyaline ascospores (19–26 × 9–14 µm) are flattened on one side and produce a sporelike vesicle upon germination, which gives rise to one or more germ tubes or conidiophores. Paraphyses are branched and curved or slightly deformed at the apex, filiform, septate, and hyaline.

The anamorph may form on malt agar, producing hyaline, septate, short conidiophores that are coarser than the vegetative hyphae. Conidiogenous cells are monophialidic and lageniform, with well-defined but thin-walled collarettes. Conidia (2–3 × 1.5–2 µm) are ellipsoid, hyaline, and unicellular. The hyphae grow in a characteristic sin-wave pattern (Fig. 174) that, when observed in the vessel elements of diseased tissue, is often considered diagnostic.

Angular leaf scorch is caused by the species *P. tetraspora* Korf, Pearson, & Zhuang (anamorph *Phialophora*-type). This species is distinguished from *P. tracheiphila* by its smaller apothecia (0.1–0.3 mm in diameter) and its broadly clavate asci (80–100 × 20–22 µm) that contain only four binucleate ascospores, in contrast to the eight-spored European species. Apothecia form readily on dead, infected leaf tissue after several days incubation in a moist chamber.

Fig. 173. Severely diseased leaf of the red-fruited interspecific cultivar Rougeon infected with *Pseudopezicula tetraspora.* (Courtesy R. C. Pearson)

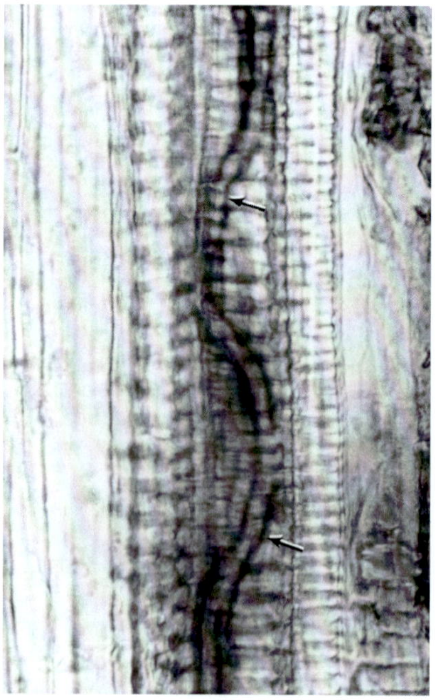

Fig. 174. Hyphae (arrows) of *Pseudopezicula tracheiphila* growing in a sine-wave pattern inside a vessel of an infected leaf. (Courtesy H. Schüepp)

Disease Cycle and Epidemiology

The incidence and severity of rotbrenner are influenced strongly by the abundance of apothecia on fallen leaves and the quantity of released ascospores. Apothecia are formed primarily on overwintered leaves in the spring, although they may develop on current-season infections in late summer or fall if provided sufficient wetness of the leaf substrate. Apothecial primordia mature as soon as temperatures rise at the end of the winter and ascospores may be released even before bud burst under wet and warm conditions, although infection appears to occur most commonly after shoots have elongated to the six-leaf stage. Heavy rainfall and prolonged wetting periods favor infection and can lead to severe disease, but little is known regarding the precise temperature and moisture conditions necessary for ascospore maturation, release, or germination and infection.

After an incubation period of 2–4 weeks, the fungus may invade the vascular elements of infected leaves, causing symptom development. If unable to invade the vessel elements, the fungus remains quiescent and can be isolated from green leaves showing no symptoms. Conditions required for the pathogen to invade the vascular system are not well understood; however, soil conditions and a water supply that place the vine under temporary stress appear to be important factors.

The epidemiology of angular leaf scorch is poorly understood, but is assumed to be similar to that of rotbrenner. The first infections can occur soon after bud break given suitable temperatures and rain, with a 3- to 4-week incubation period before the appearance of symptoms. Although secondary spread is uncommon, apothecia may form on the dead portions of diseased leaves still attached to the vine following several consecutive days of persistent rain, leading to additional infections when weather conditions favor their development.

Management

Although rotbrenner infections can occur when the first leaf reaches a width of approximately 5 cm, the most critical period for disease development is from the six-leaf stage through the end of the bloom. In locations where severe symptoms occur annually, fungicide control programs should be implemented routinely. In vineyards in which the disease occurs sporadically, fungicide treatments should be implemented in wet springs that favor the maturation of apothecia. Sprays should be applied at 7- to 10-day intervals between the six- and 12-leaf stage, depending on rainfall forecasts, the rate of plant development, and the material used. Later in the season, sprays targeting rotbrenner can be integrated with those sprays against downy mildew. Dithiocarbamates are the traditional materials used for control of this disease, although some newer fungicides also are effective, including some triazole and strobilurin products. Monitoring ascospore release by means of spore traps may provide additional information to aid in the forecasting of disease pressure.

Angular leaf scorch is seldom serious enough to warrant targeted control programs. It appears to be controlled primarily by fungicides targeted at a complex of other diseases active during the early part of the growing season. Mancozeb is known to be effective; it is assumed that other fungicides effective against rotbrenner are likely to be effective against angular leaf scorch as well, although this has not been proved.

Selected References

Korf, R. P., Pearson, R. C., Zhuang, W., and Dubos, B. 1986. *Pseudopezicula* (Helotiales, Peziculoideae), a new discomycete genus for

pathogens causing an angular leaf scorch disease of grapes ("Rotbrenner"). Mycotaxon 26:457-471.
Levadoux, M. L. 1944. Le brenner (*Pseudopeziza tracheiphila* Müll.-Thurg.). Bull. Off. Int. Vin. 17:43-54.
Müller-Thurgau, H. 1903. Der rote Brenner des Weinstockes. Centralbl. Bakteriol. Parasitenkd. Infektionskr. Abt. 2 10:1-38.
Pearson, R. C., Smith, F. D., and Dubos, B. 1988. Angular leaf scorch, a new disease of grapevines in North America caused by *Pseudopezicula tetraspora*. Plant Disease 72:796-800.
Pearson, R. C., Siegfried, W., Bodmer, M., and Schüepp, H. 1991. Ascospore discharge and survival in *Pseudopezicula tracheiphila*, causal agent of Rotbrenner of grape. J. Phytopathology 132:177-185.
Reiss, K., and Zinkernagel, V. 1994. Heterothallie bei *Pseudopezicula tracheiphila* (Müll.-Thurg.) Korf & Zhuang (syn. *Pseudopeziza tracheiphila* Müll.-Thurg.). Z. Pflanzenkrankh. Pflanzenschutz 101:212-214.
Reiss, K., Gutmann, M., Bartscherer, H.-Ch., and Zinkernagel, V. 1997. *Pseudopezicula tracheiphila* on *Vitis vinifera*: Microscopical studies of the infection mechanism. Z. Pflanzenkrankh. Pflanzenschutz 104:483-491.
Reiss, K., and Zinkernagel, V. 1997. Epidemiologische Untersuchungen zur Ascosporenfreisetzung von *Pseudopezicula tracheiphila* (Müll.-Thurg.) Korf & Zhuang, dem Erreger des Roten Brenners am Wein. Z. Pflanzenkrankh. Pflanzenschutz 104:113-112.
Schüepp, H. 1976. Verstärktes Auftreten des Rotbrenners der Rebe. Schweiz. Z. Obst Weinbau 112:379-381.
Siegfried, W., and Schüepp, H. 1983. Der Rotbrenner der Rebe, eine unberechenbare Krankheit. Schweiz. Z. Obst Weinbau 119:235-239.

(Prepared by H.-H. Kassemeyer and W. F. Wilcox)

Sour Rot

Sour rot (pourriture acide, pudrición ácida), also known in parts of the western United States as summer rot, is a poorly defined condition characterized by the rapid decomposition of ripening grape berries. The term sometimes is used to define a general decay syndrome that may involve various bacteria, yeasts, and/or filamentous fungi other than species known to be primary decay pathogens (e.g., *Botrytis cinerea*, *Greeneria uvicola*, etc.). However, berries affected with "true" sour rot characteristically exude the vinegarlike odor of acetic acid, from which the condition gets its name, and often that of ethyl acetate (nail polish remover). Berries with sour rot can contain acetic acid concentrations of 0.3–12.4 grams per liter, resulting from the oxidation of ethanol, which can be produced by associated yeasts; they also typically contain high concentrations of gluconic acid, a result of the metabolism of glucose by acetic acid bacteria, and dihydroxyacetone, produced when acetic acid bacteria oxidize glycerol formed by yeasts. Dihydroxyacetone persists through fermentation into the wine, affecting its sensory quality and sulfur dioxide binding properties. Additionally, infected berries have elevated concentrations of glycerol and tartaric acid relative to healthy berries, and higher levels of titratable acidity.

Complexes of different but overlapping groups of organisms appear to be associated with the disease in different climates. Sour rot typically develops in berries injured by primary pathogens, insects, or birds or in berries subjected to abiotic forms of physical damage such as cracking due to rain or compression within expanding bunches. Thus, tight-clustered and thin-skinned varieties or clones tend to be particularly susceptible to sour rot development. Botrytis bunch rot and sour rot are often conflated, so that sour rot is sometimes thought to be the final and most destructive stage of bunch rot. However, the two rots are caused by distinctly different etiological agents and, although they often occur together in regions or seasons with wet preharvest conditions, the occurrence of one does not always imply the occurrence of the other. Crop damage due to sour rot can be extensive because infections that begin in a single berry can spread rapidly to adjacent berries and destroy a substantial proportion of the cluster. Clusters with significant levels of sour rot can seriously reduce the quality of wines made from them. Thus, entire clusters supporting diseased berries may be excluded from harvest or discarded between harvest and crush even though many healthy berries remain, resulting in a degree of economic loss that exceeds the losses due only to rotten fruit. Economic losses also accrue when rapidly spreading disease forces growers to harvest the remaining crop prematurely in order to avoid further damage, resulting in a loss of crop quality beyond that directly attributable to decay.

Symptoms

Sour rot develops during the late stages of ripening when the berries have high sugar content. In humid climates, infected berries of white varieties turn light chestnut brown (Fig. 175), whereas those of red varieties become brick or purplish red (Fig. 176). The color change usually begins at the point of pedicel attachment in berries without obvious injuries. The flesh is rapidly and completely degraded, becoming soft and watery, typically with strong acetic acid (and sometimes ethyl acetate) odors. The skin becomes thin and fragile and breaks down, releasing juice that attracts drosophilids (fruit/vinegar flies) and other insects that can vector the causal organisms to uninfected berries. Signs of *Botrytis*, *Aspergillus*, *Penicillium*, and other filamentous fungi may be common on other parts of the same

Fig. 175. Symptoms of sour rot on cv. Riesling. (Photograph by W. McFadden-Smith, Ontario Ministry of Agriculture and Food; Queen's Printer of Ontario, © 2014. Reproduced with permission.)

Fig. 176. Purplish-red discoloration of 'Pinot Noir' berries symptomatic of sour rot, with a drosophilid fly on a healthy berry within the cluster. (Photograph by W. McFadden-Smith, Ontario Ministry of Agriculture and Food; Queen's Printer of Ontario, © 2014. Reproduced with permission.)

clusters or others nearby (Fig. 177). There is no "slip skin" symptom as found with botrytis bunch rot. In the final stages of decay, the berries frequently become hollow as a result of insect feeding and consist only of empty shells of dry skin. Numerous vinegar flies often emerge from infected clusters and their larvae are commonly observed on and in infected berries. Wasps are also attracted to the leaking juice.

Causal Organisms

The acetic acid bacteria, *Acetobacter* spp. and/or *Gluconobacter* spp., are always associated with "true" sour rot and can cause its decay symptoms on inoculated fruit in the absence of other organisms. They are obligately aerobic, Gram-negative, catalase positive, oxidase-negative, and well adapted to grow in sugar-rich environments. They can typically oxidize ethanol to acetic acid.

Gluconobacter spp. are ovoid or rod-shaped cells, 0.5–1.0 × 2.6–4.2 µm, occurring singly or in pairs, and rarely in chains. They may be motile or nonmotile; if motile, cells have 3–8 polar flagella. Colonies are white, yellow, or even brown toward the middle; circular, raised or convex; and regularly edged. Growth is optimum at 25–30°C and pH 5–6, although most strains can grow at pH as low as 3.5. They are negative for nitrate reduction, gelatin liquefaction, indole production, and H_2S production. They do not oxidize lactate or acetate to CO_2 and H_2O and do not cause a color change when incubated on Dextrose sorbitol mannitol agar. All strains produce 2-ketogluconic acid from D-glucose. *G. oxydans* (previously *Acetobacter suboxydans*) is most frequently identified in the literature as the species causing sour rot. It is the most common acetic acid bacterium on the surface of the healthy grapes, likely owing to its superior ability to metabolize sugars. *G. cerinus*, *G. albidus*, and *G. frateurii* have also been cited as causal agents of sour rot.

Acetobacter spp. are ellipsoid to rod-shaped cells, straight or slightly curved, 0.6–0.9 × 1.0–4.0 µm, occurring singly, in pairs, or chains. They can be motile or nonmotile; if motile, flagella are peritrichous. They form pale colonies and most strains produce no pigments. Their optimal temperature for growth is 30°C, with an optimal pH of 4.0–6.0. They are negative for nitrate reduction, gelatin liquefaction, indole production, and H_2S production. The *Acetobacter* species reported to cause sour rot are *A. aceti* and *A. pasteurianus*.

Although many yeast species have been isolated from berries with sour rot, only a few are able to cause the fruit disintegration and characteristic ethyl acetate odor associated with the disease in some locales. *Hanseniaspora uvarum* (Niehaus) Shehata et al. (anamorph *Kloeckera apiculata* (Reess) Janke) is the yeast most frequently identified as a causal organism of sour rot. Cells are apiculate, spheroidal to ovoidal, 1.5–5.0 × 2.5–11.5 µm, and occur singly or in pairs, with reproduction by bipolar budding. On glucose-yeast extract-peptone agar it produces white to creamy, smooth, glossy colonies that are slightly raised in the middle. On Wallerstein Laboratory nutrient agar it produces intense green, flat, smooth, opaque colonies. *H. uvarum* ferments only glucose and can produce acetic acid and its esters, including ethyl acetate, and grows in the presence of cycloheximide.

Fig. 177. Riesling clusters with (i) sour rot, characterized by chestnut-brown color and lack of sporulation on berries in the mid- and lower portion of the rightmost cluster; contrasting with (ii) Botrytis bunch rot, characterized by the fuzzy, gray sporulation on infected berries above those with sour rot and in the adjacent cluster. (Photograph by W. McFadden-Smith, Ontario Ministry of Agriculture and Food; Queen's Printer of Ontario, © 2014. Reproduced with permission.)

Several *Candida* species have been cited as capable of causing sour rot. The *Candida* spp. associated with sour rot have ovoid to elongated cells and undergo multilateral budding. Some species produce simple to elaborate pseudohyphae. They assimilate glucose, glycerol, and ethanol. *C. krusei* (Castellani) Berkhout (teleomorph *Issatchenkia orientalis*) produces white to cream butryous colonies. It assimilates glucose, N-acetyl-D-glucosamine, ethanol, glycerol, DL-lactate, and succinate. *Candida zemplinina* (Kroemer & Krumbholz) Meyer & Yarrow (previously *C. stellata)* is acidogenic, highly osmotolerant, and grows well in the presence of ethanol and at low temperatures.

Pichia membranaefaciens Hansen is a film-forming yeast that causes off aroma and flavor in table wine. Its cells are ovoidal to elongate, (1.8–4.5) × (2.5–17) µm, and occur singly or in pairs, chains or clusters. Colonies are yellowish tan, dull, and smooth or wrinkled. Asci may be unconjugated or show conjugation between independent cells or between a parent cell and a bud. One to four spheroidal or hemispheroidal ascospores with or without a ledge are produced per ascus in culture. It produces acetaldehyde, ethyl acetate, and isoamyl acetate.

Pichia kluyveri (Bedford) forms white to cream butryous colonies, undergoes vegetative reproduction by budding, produces simple pseudohyphae, and produces evanescent asci containing up to four hat-shaped ascospores.

Metschnikowia pulcherrima Miller and van Uden produces a red diffusing pigment, pulcherin. Its colonies are smooth and cream colored with an entire margin. It produces pseudohyphae and buds multilaterally. One to two needlelike ascospores per ascus are produced in culture, sometimes only after mixing sexually compatible strains.

Saccharomycopsis crataegensis Kurtzman and Wickerham (syn. *S. vini*, *Endomycopsis vini*) does not cause sour rot on its own; however, when co-inoculated with *M. pulcherrima*, it can cause sour rot symptoms. In nutrient broth, cells are ovoid; elongate to cylindrical; (2.4–3.2) × (4.0–8.0) µm; and occur singly, in pairs, in short chains, or as multilateral buddings. Aerobic growth on agar is white to cream, butyrous, rough, glistening, and flat with filamentous margin. After 7 days at 25°C, growth on corn meal agar under a coverglass shows abundant true hyphae with conidia. Pseudohyphae are formed and the species appears heterothallic. Chains of two to four ellipsoidal asci, containing two ascospores, are borne directly on hyphae or supported by short stalks.

Disease Cycle and Epidemiology

Inoculations with bacteria or yeasts isolated from diseased berries have replicated the symptoms of berry decomposition accompanied by odor of acetic acid or ethyl acetate, respectively, although some investigators report that acetic acid production requires co-inoculation of ethanol-producing yeasts with the acetic acid bacteria. These microorganisms occur on the surface of healthy grape berries, although in significantly lower numbers than on diseased ones. They are not aggressive pathogens and appear to require mechanical injury or predisposing infections by other microorganisms to gain entry into the berries. It is likely that the additional organisms involved vary among different regions and habitats.

The volatile compounds in the juice that leaks from injured berries, especially those with sour rot, attracts vinegar flies (*Drosophila* spp.). Adult flies can vector yeasts and bacteria to uninfected clusters and into neighboring, subsequently maturing cultivars. Adults typically lay eggs near the juncture of the berry and the pedicel. It is hypothesized that when berries swell and pull away from the pedicel in the final stages of ripening, as is particularly likely in tightly compressed clusters, hatching larvae can enter the berries and carry yeasts and bacteria with them to the infection court.

In wetter climates, severe sour rot development is usually associated with high daytime temperatures (optimum being 25–28°C), high humidity, and heavy precipitation during the final 2–3 weeks before harvest. In Ontario, Canada, sour rot does not occur in the field until berries reach a minimum of 15°Brix, nor do inoculated berries become diseased before this stage of maturity. Heavy rains provide numerous infection courts by causing berries to swell rapidly and split, and by facilitating pathogen development and dispersal to infection sites. Similarly, sour rot is exacerbated in arid regions by use of over-the-canopy sprinklers as berries mature. The disease frequently develops from the interior of the cluster, where berries become crushed as they and their neighbors swell, then progresses outward. Wounding from hail, birds, insects, and fungal diseases may also allow entry of sour rot causal organisms. For example, sour rot is sometimes found as a secondary infection following botrytis bunch rot, and diffuse powdery mildew infections also can cause minute wounds in the skin that allow the proliferation of contaminating yeasts and bacteria, some of which can cause sour rot under favorable conditions. In the vineyard, there is likely a succession of organisms causing sour rot. For example, apiculate yeasts (*Hanseniaspora*, anamorph *Kloeckera*)

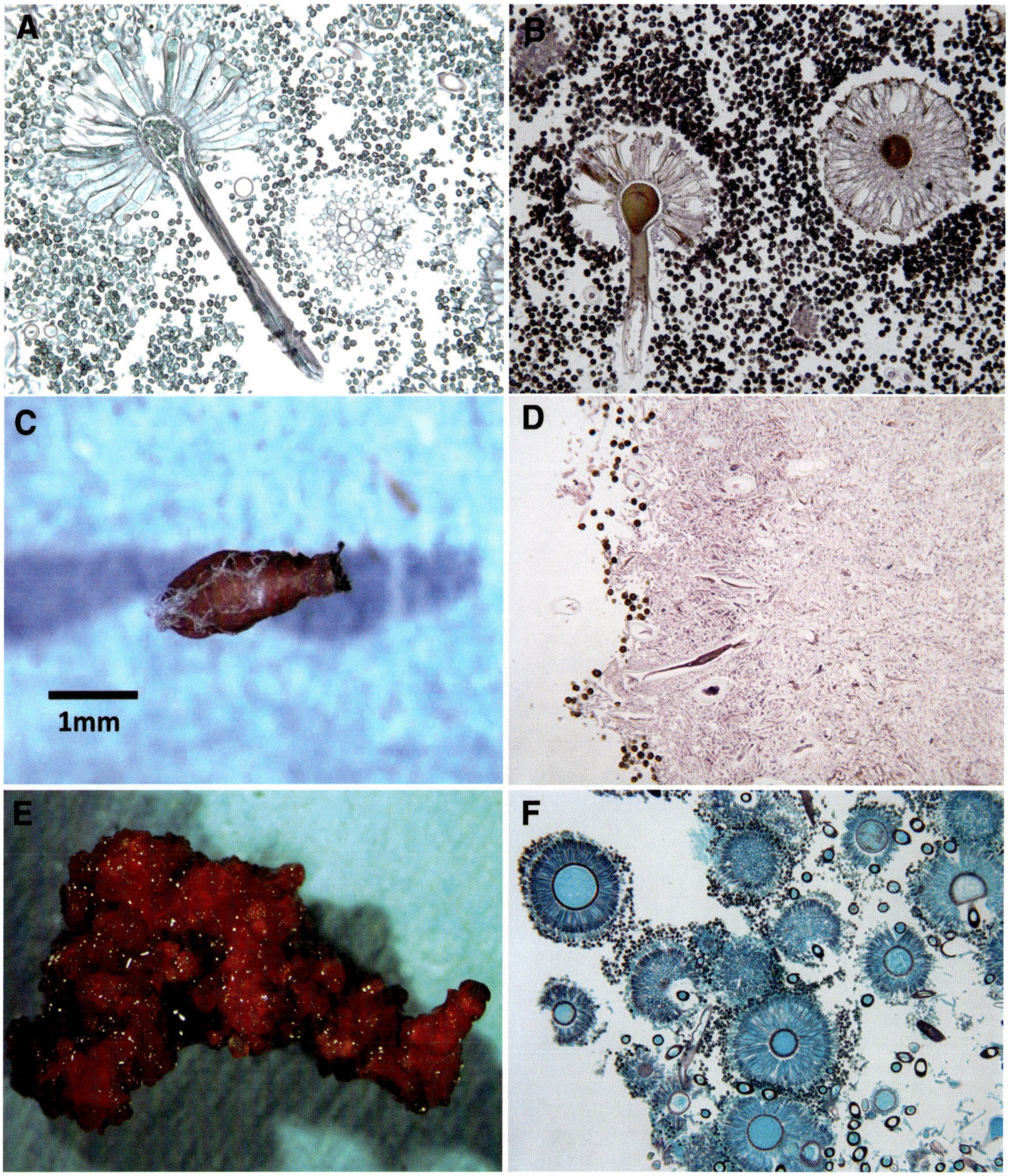

Fig. 178. Inverted stroma of *Aspergillus niger*, **C,** after drying in soil or, **E,** freshly recovered from within a decaying grape berry. **A, B, D,** and **F,** cross-sectional views of conidiophores and conidia produced from such a stroma. (Courtesy C. Pisani)

initially colonize many fruit surfaces and are dominant at early stages of ripeness. As the fruit deteriorates, yeasts capable of utilizing a broader spectrum of carbon sources become apparent on the surface. *Pichia* spp. and *Candida* spp. are in this latter category.

In Mediterranean climates, *Aspergillus niger* Tiegh. and *Aspergillus carbonarius* (Bainier) Thom are usually the first fungi to colonize berries that have been injured after veraison, often by insects and birds. *Aspergillus* spp. colonize berry pulp rapidly, causing leakage onto berries below the rot. Although the *Aspergillus* fungi do not cause true sour rot, they are soon followed by other fungi and the acetic acid bacteria that produce the vinegar smell. Both species of *Aspergillus* have been shown to produce a previously undescribed fruiting body in rotted berries, characterized as a polymorphic, inverted stroma from which conidia of the pathogen are produced (Fig. 178). These stromata fall to the ground in late summer and become leathery. It is hypothesized that these serve as overwintering structures that survive in the soil and provide spores for the following year's infections.

Non-Botrytis slip skin is a related postharvest disease of table grapes and occurs mostly on thin-skinned varieties such as Red Globe. The disease occurs only on clusters from which other berries with sour rot have been trimmed before packing. It is caused in storage (0°C, 94% relative humidity) by an SO_2-resistant *Hanseniospora uvarum* biotype able to ferment the sugar residues on healthy fruit that resulted from leakage by neighboring fruit with sour rot, eventually resulting in a berry decay similar to Botrytis box rot. This storage rot can be avoided by not harvesting clusters with sour rot symptoms.

Management

The primary strategy for managing sour rot is to minimize various forms of berry injury to the greatest extent practical. Standard programs for managing fungal diseases, especially powdery mildew and botrytis, should be implemented and adjusted as necessary for the particular location and year. Insects that wound berries, such as wasps and grape berry moth, should be controlled; the use of bird netting, where appropriate, should also reduce injuries subject to sour rot initiation. Canopy management and removal of leaves to improve cluster exposure and promote drying have been shown to reduce sour rot development, sometimes dramatically. Berry damage caused by cluster compaction may be reduced by loosening clusters through viticultural practices and/or the use of chemical growth regulators, as appropriate for the cultivar and region. Broad-spectrum fungicides applied during the postveraison period sometimes provide modest control of the secondary decay fungi often associated with sour rot, but have little effect on sour rot *per se* beyond their control of botrytis. Over-the-canopy irrigation should be limited when sour rot or predisposing injuries are present.

Selected References

Barata, A., Santos, S. C., Malfaeto-Ferreira, M., and Loureiro, V. 2012. New insights into the ecological interaction between grape berry microorganisms and *Drosophila* flies during the development of sour rot. Microb. Ecol. 64:416-430.

Bisiach, M., Minervini, G., and Zerbetto, F. 1986. Possible integrated control of grapevine sour rot. Vitis 25:118-128.

Duncan, R. A., Stapleton, J. J., and Leavitt, G. M. 1995. Population dynamics of epiphytic mycoflora and occurrence of bunch rots of wine grapes as influenced by leaf removal. Plant Pathol. 44:956-965.

Gadoury, D. M., Seem, R. C., Wilcox, W. F., Henick-Kling, T., Conterno, L., Day, A., and Ficke, A. 2007. Effects of diffuse colonization of grape berries by *Uncinula necator* on bunch rots, berry microflora, and juice and wine quality. Phytopathology 97:1356-1364.

Gravot, E., Blancard, D., Fermaud, M., Lonvaud, A., Joyeux, A., 2001. La pourriture acide. I. Étiologie: Recherché de causes de cette pourriture dans le vignoble bordelaise. Phytoma 543:36-39.

Guerzoni, M. E., and Marchetti, R. 1987. Analysis of yeast flora associated with grape sour rot and of the chemical disease markers. Appl. Environ. Microbiol. 53:571-576.

Marchetti, R., Guerzoni, M. E., and Gentile, M. 1984. Recherche sur l'étiologie d'une nouvelle maladie de la grappe: la pourriture acide. Vitis 23:55-65.

Nisiotou, A. A., and Nychas, G.-J. E. 2007. Yeast populations residing on healthy or *Botrytis*-infected grapes from a vineyard in Attica, Greece. Appl. Environ. Microbiol. 73:2765-2768

Pisani, C. 2011. Identification of a novel fungal fruiting structure formed by *Aspergillus niger* and *Aspergillus carbonarius* in sour rot infected grape berries under artificial and field conditions. M.S. thesis, University of California, Davis.

Sholberg, P. L., Lowery, T., Bowen, P., Haag, P., and Walker, M. 2008. Effect of early leaf stripping on bunch rot, powdery mildew and sour rot of wine grapes. Pages 199-211 in: Crop Protection Research Advances. E. N. Burton and P. V. Williams, eds. Nova Science Publishers, New York.

Zoecklein, B. W., Williams, J. M., and Duncan, S. E. 2000. Effect of sour rot on the composition of White Riesling (*Vitis vinifera* L.) grapes. Small Fruits Rev. 1:63-77.

Zoecklein, B. W., Wolf, T. K., Duncan, N. W., Judge, J. M., and Cook, M. K. 1992. Effects of fruit zone leaf removal on yield, fruit composition, and fruit rot incidence of Chardonnay and White Riesling (*Vitis vinifera* L.) grapes. Am. J. Enol. Vitic. 43:139-148.

(Prepared by W. McFadden-Smith and W. D. Gubler)

Verticillium Wilt

Verticillium wilt is a serious vascular disease of hundreds of woody and herbaceous plant species throughout the temperate areas of world, including some common dicot weeds and fruit and vegetable crops such as strawberries, raspberries, potatoes, and tomatoes. It occurs sporadically on grapes in several countries in North and South America, Europe, Asia, and Oceania. Because affected vines exhibit symptoms similar to those caused by other stresses, it is likely that the disease is present but has been misdiagnosed or unrecognized in some additional locations. Although first described on grapes in the 1950s in Germany, the disease is most destructive in warmer, irrigated regions. During the rapid expansion of the grape industry in California in the 1970s, the disease developed in many new vineyards planted at sites where crop plants susceptible to the causal pathogen had been grown previously. Following another major round of new planting in California in the 1990s, there was another resurgence of the disease. Typically, Verticillium wilt only affects grapevines less than 6–7 years old.

Symptoms

Symptom expression is the result of vascular occlusion and root death. Infected vines do not show symptoms in the early part of the growing season, but as transpirational demand increases as a result of rising temperatures (and declining soil moisture in some cases), a few shoots start to die and the vascular elements in the wood of these shoots become discolored (Fig. 179). New shoot growth from the base of such vines is often vigorous.

By early summer, the leaves on declining shoots wilt and show marginal burning. By midsummer, many shoots that had developed normally during the early season collapse completely. The leaves on these shoots become desiccated, and some may drop from the vine (Fig. 180). Fruit clusters at the base of these affected shoots dry up, and the individual berries remain attached to the pedicels as shriveled mummies, a result of their rapid desiccation before formation of an abscission zone. Symptoms typically are confined to one side of the vine, with leaves and shoots on a single cordon showing wilting. Occasionally, all shoots on a vine may collapse, whereas only a few shoots may show symptoms on others.

Fig. 179. Discolored xylem in the trunk of a young cv. Chenin Blanc vine infected with *Verticillium dahliae.* (Courtesy R. C. Pearson)

Fig. 180. Wilted leaves and dying shoots on one side of a cv. Semillon vine infected with *Verticillium dahliae.* (Courtesy A. C. Goheen)

Causal Organism

Verticillium dahliae Kleb., the causal fungus, grows rapidly on potato-dextrose agar at 24°C. Mycelium is hyaline to whitish at first but later turns black with the formation of microsclerotia. Microsclerotia arise from individual hyphae by repeated budding. They are dark brown to black and torulose or botryoidal. They consist of swollen, almost globular cells that vary in shape and size (generally 15–50 µm in diameter, and 100 µm maximum). Chlamydospores are absent, and dark-brown resting mycelium is formed only in association with microsclerotia.

Conidiophores are erect, hyaline, and verticillately branched, with three to four phialides arising at each node. Phialides (16–35 × 1–2.5 µm) are secondarily branched on occasion. Conidia (2.5–8 × 1.4–3.2 µm) arise singly at phialide apices and are ellipsoidal to subcylindrical and single-celled, rarely uniseptate.

V. dahliae is distinguished from the closely related species *V. albo-atrum* Reinke & Berth. primarily by the presence of its true microsclerotia and ability to grow at 30°C. In late spring, it is difficult to isolate the fungus from shoots that are starting to show symptoms, but isolation from leaves, petioles, and discolored wood in the canes becomes easier as the season progresses.

Disease Cycle and Epidemiology

Infection apparently takes place through the roots of young vines. This leads to vascular plugging by gums and tyloses that are produced by the plant in response to occlusion of the vessels by spores and mycelium of the fungus. Disease foci usually occur at locations where susceptible crops or weeds grew previously, causing inoculum to build up in the soil. The distribution of affected vines at such sites is erratic.

Vines that are planted in infested soil usually do not show disease symptoms during the first year. Some vines show symptoms during the second year, and new infections continue to appear for several subsequent seasons. By the fifth or sixth season, symptomatic vines that have not died typically recover, and the disease is no longer noticeable in the vineyards. This apparently spontaneous recovery is one reason it took many years of investigation before it was demonstrated that *Verticillium* infects grapes.

Full production is delayed in vineyards that show early wilting, but yields do not seem to be adversely affected once vines reach an age when they become asymptomatic. Unless young vines are killed by the initial attack, after 8–10 years it is difficult to find the areas in the vineyard where disease occurred.

Verticillium wilt of grapes is uncommon in the southern San Joaquin Valley of the California, even though it appears to be caused by the mild (SS-4) strain of the fungus that affects cotton in that area. For unknown reasons, it is more common in the northern San Joaquin Valley and in the Salinas Valley, where previous hosts, such as strawberries, apricots, and tomatoes, have favored the increase of the mildly virulent strain of the fungus.

Management

Sites in the few locations where the fungus has been known to kill grapevines, or where the immediate cropping history suggests high levels of soilborne inoculum, should be avoided. More specific control measures for *Verticillium* wilt usually are unnecessary, although fumigation may be warranted if microsclerotia populations are very high (e.g., as a result of the previous cropping history).

Selected References

Braun, A. J. 1953. Ills of the American bunch grapes. Pages 754-760 in: Plant Diseases: The Yearbook of Agriculture, 1953. U.S. Department of Agriculture, Washington, DC.

Canter-Visscher, T. W. 1970. Verticillium wilt of grapevine, a new record in New Zealand. N. Z. J. Agric. Res. 13:359-361.

Gubler, W. D., Baumgartner, K., Browne, G. T., Eskalen, A., Latham, S. R., Petit, E., and Bayramian, L. A. 2004. Root diseases of grapevines in California and their control. Austral. Plant Pathol. 33:157-165.

Hawksworth, D. L., and Talboys, P. W. 1970. *Verticillium dahliae.* Descriptions of Pathogenic Fungi and Bacteria, no. 256. C.A.B. International Mycological Institute, Kew, England.

Mace, E. M., Bell, A. A., and Beckman, C. H. 1981. Fungal Wilt Diseases of Plants. Academic Press, New York.

Schnathorst, W. C., and Goheen, A. C. 1977. A wilt disease of grapevines (*Vitis vinifera*) in California caused by *Verticillium dahliae.* Plant Dis. Rep. 61:909-913.

Thate, R. 1961. Die Apoplexie der Rebe: Eine Verticilliose. Mitt. Biol. Bundesanst. Land Forstwirtsch. Berlin-Dahlem 104:100-103.

(Prepared by A. Eskalen)

White Rot

White rot was first reported in Italy and the pathogen was described by Spegazzini in 1878. The disease can be found in most of the world's viticultural regions that have warm, humid summers. The white rot fungus can infect all green tissues of the grapevine, although clusters are the most susceptible. Uninjured berries are resistant to infection, whereas injured berries, pedicels, or rachises are susceptible. Severe epidemics and the greatest damage to clusters occur in areas particularly prone to hailstorms, where losses may reach 20–80% of the yield. For this reason, white rot is sometimes known as "hail disease," although hailstorms are not essential for its development. *Vitis* species differ greatly in their susceptibility to white rot. The Asiatic species are resistant to highly resistant (*V. amurensis*, *V. davidii*, and *V. qinlingensis* belong to the latter category), whereas *V. vinifera* and American species are susceptible to highly susceptible. For example, the rootstock Kober 5 BB is considered highly susceptible, as are the *V. vinifera* cultivars Chardonnay, Chasselas, and Rhine Riesling.

Symptoms

Symptoms of white rot vary among different plant organs. Rachis infections appear initially as brown pin-point spots, which later become elongated and depressed. Persistently high relative humidity favors girdling of the rachis, resulting in a sudden desiccation and death of the distal portion of the cluster (Fig. 181). Immature infected berries become pale green and shrivelled, later turning pinkish and then brown (Fig. 182). Early desiccation of infected berries prevents complete fungal colonization, whereupon pycnidia can be found only around the infection point. These symptoms can mistakenly be diagnosed as physiological drying, also known as stem necrosis, attributed to nutritional imbalances (improper Ca and Mg/K ratios) or water deficit problems.

Before veraison, berries on hailstorm-injured clusters turn beige within a few days of infection, then become pinkish blue and shrivelled. Erumpent pycnidia develop densely from the infected tissue, appearing brownish violet in hue while immature (Fig. 183). A result of the infection and colonization is the raising of the cuticle from the epidermal layer, thus conferring a whitish appearance to the berries. At maturity, pycnidia turn grayish white, hence the name white rot. Warm temperatures and persistently high relative humidity promote internal growth of mycelia from infected berries into the pedicel and then the rachis, thereby girdling the cluster and causing it to desiccate and die (Fig. 184).

Directly infected pedicels show a small, pale-brown, elongated depression. Under high relative humidity, the fungus expands throughout the entire pedicel and into uninjured berries. Pycnidia cover the wrinkled, withering surface of the berry within a few days, whereas under dry conditions they develop inside the berry on the surface of the seeds.

Fungal infection of nonlignified shoots can cause cankers. This symptom rarely occurs on *V. vinifera* cultivars, but it is frequently observed in nurseries producing American or interspecific hybrid rootstock selections, where shoots develop freely on the soil surface. Infections on green shoots are often localized around the nodes. Colonized tissue becomes necrotic, appearing brown, elongated, and depressed. The bark covering necrotic tissue splits, with the fibrous material conferring a characteristic frayed appearance to the canker. Cankers on canes in a nursery cause a significant loss of grafting wood and reduce the percentage of successful grafts. The fungus rarely infects leaves.

Fig. 181. Infection on the rachis of cv. Rhine Riesling by *Pilidiella diplodiella*, causing desiccation and death of the cluster. (Courtesy M. Bisiach)

Fig. 182. Immature berries of cv. Rhine Riesling infected through the pedicels by *Pilidiella diplodiella*. (Courtesy M. Bisiach)

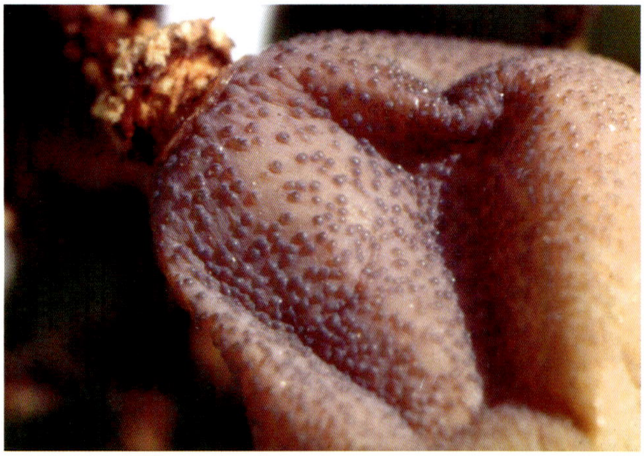

Fig. 183. Brownish-violet, immature pycnidia of *Pilidiella diplodiella* erupting across the surface of an infected berry. (Courtesy M. Bisiach)

Causal Organism

White rot is caused by the fungus *Pilidiella diplodiella* (Speg.) Crous & Van Niekerk [syns. *Coniella diplodiella* (Speg.) Petrak & Sidow, *Coniothyrium diplodiella* (Speg.) Sacc., *Phoma diplodiella* Speg.]. Intercellular, dichotomously branched, hyaline to yellowish septate hyphae (12–16 µm in diameter) give rise to appressoria and haustoria. Hyphal anastomoses are abundant and chlamydospores form. Subglobose pycnidia, 100–200 µm wide and with a central ostiole, arise from subcuticular stromata, turning dark brown at maturity (Fig. 185A). Conidiophores (10–20 × 3–4 µm) are simple or branched at the base, hyaline, with 0–3 septa. Smooth, single-celled, multiguttulate conidia (8–16 × 5–7 µm) are hyaline when immature but become pale to medium brown with time, and are elliptical or ovate with an obtuse apex and truncate base (Fig. 185B).

Disease Cycle and Epidemiology

The fungus overwinters as mycelium and pycnidia in infected canes and rachises, and in mummified berries that fall to the soil or are retained on the vines. Inoculum remains viable for months and accumulates in vineyards affected by the disease, where soil may contain 300–2000 conidia per gram. Under humid or wet conditions, thousands of conidia are extruded in a mucilaginous matrix through the ostiole of a pycnidium, and individual conidia are then separated and dispersed primarily by rainfall. Soil-borne inoculum can also be distributed on contaminated soil particles dispersed by rain, hailstones, or farm equipment. The pathogen infects berries through wounds, most often caused by hail; wounds caused by insects or other fungal diseases such as powdery mildew can also afford the pathogen entry but are epidemiologically less important. In contrast, rachises and pedicels can be infected via direct penetration by the fungus. Berries may also be colonized by mycelial expansion of the fungus from infected pedicels. Cluster susceptibility is highest during flowering, decreasing after.

Conidia germinate in free water at temperatures of 11–30°C. Infection of injured berries occurs within about 6 h of wetness at the optimal temperatures of 22–27°C, and is triggered by the presence of juice or berry exudates. Infection of unwounded rachises and pedicels requires longer durations of wetness. Severe disease development can follow hailstorms when temperatures are near the optimum, whereas negligible incidence has been recorded if the temperature drops below 15°C for 24–48 h after a hailstorm. Mycelium grows rapidly and the fungus produces abundant pycnidia at optimal temperatures, whereas growth is slower and no pycnidia are formed at temperatures <15°C or >34°C. The incubation period typically varies from 3–8 days in the field, depending on the type of tissue infected, cultivar susceptibility, and weather conditions. Conidia released from newly formed pycnidia may cause secondary infections, thereby creating the potential for epidemic disease development should favorable conditions persist or recur.

Management

Control of white rot in commercial vineyards is generally based on the use of fungicides, although some degree of control can also be provided by cultural practices, such as minimizing wounds, planting tolerant varieties, utilizing a high trellis system, and employing sanitation to reduce primary inoculum sources within the canopy and on the soil. Numerous traditional protective fungicides, including pthalimides, dichlofluanid, and chlorothalonil, are effective against white rot; dicarboximides and the DMI fungicide difenoconazole also are efficacious. Captan or folpet applied within 12–18 h after a hailstorm provides about 75% control of new infections, but efficacy falls to 50% if treatment is delayed for 21 h and is generally ineffective after 24 h, especially at temperatures >20°C. The dicarboximides also provide good control if applied within 12–18 h after a hailstorm at temperatures <20°C. Efficacious systemic materials such as difenoconazole should provide a longer period of postinfection control than the traditional protective materials.

Fig. 184. White rot on a cv. Barbera cluster, girdled by expansion of *P. diplodiella* from infected berries into the rachis; air introduced between the epidermis and raised cuticle has caused some berries to appear white. (Courtesy M. Bisiach)

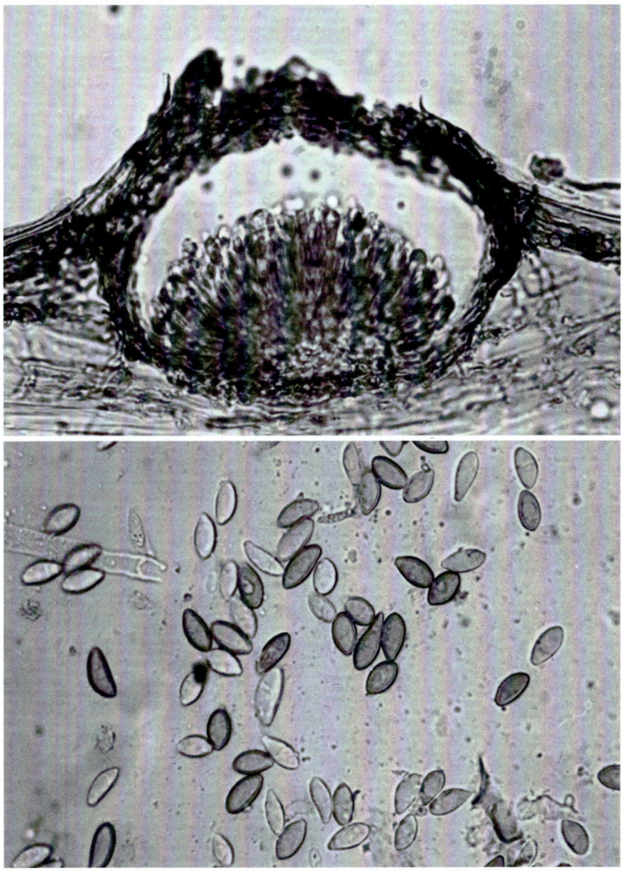

Fig. 185. Reproductive structures of *Pilidiella diplodiella*: (top) Cross section of a pycnidium; (bottom) conidia. (Courtesy P. Cortesi)

Selected References

Bisiach, M., and Battino-Viterbo, A. 1973. Attività *in vitro* di alcuni composti chimici contro *Coniothyrium diploidella*. Not. Mal. Piante 88-89(III S., N. 14-15):73-79.

Bisiach, M., and Battino-Viterbo, A. 1973. Further researches on grapevine cluster drying-off caused by *Coniothyrium diploidella*. Meded. Fac. Landbouwwet. Rijksuniv. Gent 38:1561-1571.

Bolay, A. 1977. Le point actuelle sur le traitement des vignes par les fongicides après la grele. Prog. Agric. Vitic. 94:233-234.

Li, D., Wan, Y., Wang, Y., and He, P. 2008. Relatedness of resistance to anthracnose and to white rot in Chinese wild grapes. Vitis 47:213-215.

Locci, R., and Quaroni, S. 1972. Studies on *Coniothyrium diploidella*. I. Isolation, cultivation and identification of the fungus. Riv. Patol. Veg. 8:59-82.

Van Niekerk, J. M., Groenewald, J. Z., Verkley, G. J. M., Fourie, P. H., Wingfield, M. J., and Crous, P. W. 2004. Systematic reappraisal of *Coniella* and *Pilidiella*, with specific reference to species occurring on *Eucalyptus* and *Vitis* in South Africa. Mycol. Res. 108:283-303.

(Prepared by P. Cortesi, in memory of M. Bisiach)

Diseases Caused By Bacteria

Bacterial Blight

Bacterial blight of grapevine is a destructive disease that was first reported in the late nineteenth century in Italy and France. After almost a century of confusion regarding the nature of grapevine decline, the causal agent was isolated and identified in Greece as the bacterium *Xanthomonas ampelina* and later renamed *Xylophilus ampelinus* on the basis of DNA and RNA studies.

Bacterial blight is a chronic, systemic disease that causes significant economic loss due to decreases in yield (up to 80% reduction from infected vines) and the shortened life spans of affected vineyards. Serious damage has been reported on highly susceptible cultivars such as Thompson Seedless, Ugni Blanc, Grenache, and Macabeu. To date, the disease has been observed across Southern Europe, where it is also known as *mal nero* (Italy), *maladie d'Oléron* (France), *tsilik marasi* (Greece), *necrosis bacteriana* (Spain), *mal negro* (Portugal); bacterial blight has also been observed in South Africa, where it is also known as *vlamsiekte*. *X. ampelinus* is a quarantine pest for the EU and many other countries, including the United States.

Symptoms

Symptoms can appear on all aerial parts of the plant. In early spring, budbreak on affected spurs either fails or is delayed, and growth of these developing shoots is stunted. Buds and shoots from other spurs on the same vine may develop normally. The first symptoms on young shoots become visible when they are 2–3 weeks old. Initially, linear reddish-brown streaks appear, extending from the base to the shoot tip; then, longitudinal cracks develop surrounded by brown to black necrotic tissue, and cankers form as these deepen and extend to the pith (Fig. 186). Shoots subsequently wilt, droop, and may eventually die. Longitudinal- and cross-sections through stems show a one-sided reddish or brown discoloration of the xylem vessels. Leaves on infected shoots may exhibit sectorial or marginal necrosis of the blades (Fig. 187) and one-sided cracking of their petioles (Fig. 188). When leaves are penetrated directly via the stomata, angular, reddish-brown spots about 1–2 mm in diameter can be observed, usually surrounded by a yellow halo (Fig. 187); a pale yellow bacterial ooze may be seen emanating from them when humidity is high. Cracks and cankers may also appear on the peduncle and rachis of the clusters (Fig. 188).

Symptoms may vary considerably depending on the cultivar and possibly environmental conditions. In several cultivars, cracks and cankers do not develop on the shoots and canes or do so only occasionally. It is not unusual for chronically infected vines that show little sign of disease to suddenly develop

Fig. 186. Symptoms of bacterial blight at the base of a young shoot, including a deep canker. (Courtesy W. Gärtel)

Fig. 187. Spots and marginal necrosis on a leaf infected by *Xylophilus ampelinus*. (Courtesy W. Gärtel)

Fig. 188. Shoot canker and necrosis and cracking of petioles and rachis due to bacterial blight. (Courtesy W. Gärtel)

acute symptoms and become debilitated. Affected vines may be scattered throughout the vineyard but more often occur in pockets.

Although the symptoms described above are typical of bacterial blight, they are not specific to it and can be confused with those of several other diseases. For example, bacterial blight's shoot cankers and leaf spots are somewhat similar to those of anthracnose (*Elsinöe ampelina*), although the latter does not cause brown discoloration of the xylem vessels, and of Phompsis cane and leaf spot (*P. viticola*), although this disease produces leaf lesions that are more discrete and shoot cankers less sunken than found with bacterial blight. Spur failure and the development of dead branches could also be caused by various wood canker and vascular pathogens. Laboratory identification of the causal agent by means of techniques such as direct isolation on nonselective media, serological tests (immunofluorescence, ELISA), or molecular tests (e.g., PCR) is therefore necessary for definitive diagnosis.

Causal organism

X. ampelinus is a gram-negative, aerobic, nonspore-forming, rod-shaped bacterium, motile by a single polar flagellum. In vitro growth is generally very slow on most media. On nutrient agar, colonies are round with entire margins, semitranslucent, slightly raised, glistening and pale yellow; colony diameters are 0.2–0.3 mm after 6 days incubation and 0.6–0.8 mm after 15 days. Best growth is reported on yeast extract-galactose-calcium carbonate (YGC) medium. Minimum, maximum, and optimal growth temperatures are 6, 30, and 24°C, respectively.

X. ampelinus strains are oxidase negative and catalase positive and use only a limited number of carbohydrates, organic acids and amino acids for growth. All strains of *Xylophilus* are sensitive to phage P *X. a.*15.

Disease Cycle and Epidemiology

Grapevine is the only known host of *X. ampelinus*. The pathogen survives and multiplies in the vascular tissues of infected plants and cuttings. It moves through the xylem vessels in late winter, thereby spreading to healthy canes and spurs and the shoots and clusters that later arise from them. It also can be moved from external sources to healthy plants by rain or overhead irrigation, infecting them through multiple types of wounds or leaf stomata. Wind and rain not only facilitate the spread of bacteria, but heavy winds and storms may cause wounds that provide additional entry points for the pathogen. It also can be transmitted through human activity, both locally via the use of contaminated pruning tools and over long distances through the distribution of infected vines or grafting material, which may be asymptomatic. Up to 50% of apparently healthy canes obtained from diseased vineyards have been found to be infected.

Sap bleeding from infected vines in the spring provides an important source of inoculum. Young green shoots that develop from overwintering buds are easily injured and very susceptible to infection, and can be contaminated and subsequently infected by bacteria contained in sap dripping from nearby pruning wounds. Cankers developing on new growth provide additional inoculum, although contaminated sap is considered a more important source. Bacteria oozing from infected leaves during humid periods constitute another source of inoculum.

Disease severity in the vineyard may vary considerably from year to year. Some cases of complete recovery were observed in Greece, probably the result of environmental conditions unfavorable to the pathogen and modified agricultural practices.

Management

Control efforts should be directed toward preventing the spread of the pathogen into unaffected grape-growing regions and into newly established vineyards in regions where it is already present. All planting and grafting material should be obtained from disease-free areas. New vines should be inspected before planting and quarantine restrictions strictly followed.

In contaminated vineyards, disease spread can be minimized by removing and burning all symptomatic vine parts, or, in cases of severe infections, removing and burning the entire vine. Pruning tools should be disinfested before moving from one plant to the next. Sprays of copper fungicides are recommended immediately after pruning and periodically up to the stage of half-expanded leaves and appear to be especially helpful in rainy areas. Overhead sprinkler irrigation should be avoided.

Selected References

Botha, W. J., Serfontein, S., Greyling, M. M., and Berger, D. K. 2001. Detection of *Xylophilus ampelinus* in grapevine cuttings using a nested polymerase chain reaction. Plant Pathol. 50:515-526.

Dreo, T., Gruden, K., Manceau, C., Janse, J. D., and Ravnikar, M. 2007. Development of a real-time PCR-based method for detection of *Xylophilus ampelinus*. Plant Pathol. 56:9-16.

EPPO/OEPP. 2009. PM 7/96(1): *Xylophilus ampelinus*. Bulletin 39:403-412.

Grall, S., Roulland, C., Guillaumes, J., and Manceau, C. 2005. Bleeding sap and old wood are the two main sources of contamination of emerging organs of vine plants by *Xylophilus ampelinus*, the causal agent of Bacterial Necrosis. Appl. Environ. Microbiol. 71:8292-8300.

Panagopoulos, G. C. 1969. The disease "Tsilik marasi" of grapevine: Its description and identification of the causal agent (*Xanthomonas ampelina* sp. nov.) Ann. Inst. Phytopathol. Benaki 9:59-81.

Willems, A., and Gillis, M. 2005. Genus Incertae Sedis XIX. *Xylophilus* Willems, Gillis, Kersters, Van Den Broecke and De Ley 1987, 428^{VP}. Pages 759-763 in: Bergey's Manual of Systematic Bacteriology, 2nd ed. D. J. Brenner, N. R. Krieg, J. T. Staley, and G. M. Garrity, eds. Williams & Wilkins, Baltimore, MD.

Willems, A., Gillis, M., Kersters, K., Van Den Broecke, L., and De Ley, J. 1987. Transfer of *Xanthomonas ampelina* Panagopoulos 1969 to a new genus, *Xylophilus* gen. nov., as *Xylophilus ampelinus* (Panagopoulos 1969) comb. nov. Int. J. Syst. Bacteriol. 37:422-430.

(Prepared by F. Peduto Hand)

Crown Gall

Crown gall is a disease of numerous plant species, characterized by the development of fleshy tumors that can affect plant vigor. Grapevines are affected by crown gall wherever they are cultivated, but the problem is most serious in areas with low winter temperatures. The primary crown gall disease of grapes is caused by the bacterium *Agrobacterium vitis*. This disease is similar in many ways to crown gall caused by the related species *Agrobacterium tumefaciens*, which affects over 600 dicotyledonous plants worldwide (including grapes) and is particularly important on fruit, nut, and ornamental crops such as almonds, cherries, peaches, raspberries, and roses. In 1990, the predominant grapevine crown gall pathogen, previously considered "biovar 3" of *A. tumefaciens*, was reclassified as a separate species, *A. vitis*.

Infected vines may grow poorly or die, depending on the extent of infection. Crop yield may be reduced, new trunks may need to be established, or entire vines may have to be replaced. The appearance of galls can be particularly significant in grapevine nursery operations, where diseased plants are usually discarded, resulting in significant economic loss.

Symptoms

The major symptom of crown gall is the fleshy galls that are produced in response to infection. Galls are composed of disorganized primary and secondary phloem tissues. Irregularly shaped parenchyma cells and disorganized vascular bundles can also be found in the gall tissue. Because galls are initiated in the cambium, normal nutrient flow is disrupted, compromising grapevine health. Galls on grapevines are mostly found on the lower trunk (Fig. 189), often at graft unions and near the soil line. Small, localized galls or masses of small, pimply galls extending up the trunk also are common (Fig. 190), with galls sometimes forming on woody tissues extending more than 2 m up into the trellis. Galled vines frequently produce inferior shoot growth, and portions of the vine above the galls may decline and die.

A high incidence of galling may occur in nurseries at the graft unions and above (Fig. 191), and at the below-ground disbudded nodes of some cultivars and rootstocks. Galls on lateral roots, however, are uncommon. Large galls may develop rapidly and completely girdle young vines in one season.

Gall expression is determined by the extent of the wound that initiated infection, the grape cultivar, and the strain of the pathogen. Current-season galls are first apparent in early to mid-summer as white, fleshy callus growths developing near injury sites, often near the periphery of old galls. The new galls turn brown by late summer and in the fall become dry and corky. After one or two years, the dead galls may flake off the vine. In some cases, galling that occurs at graft unions may not be possible to distinguish from abundant callus formation, and the pathogen must be isolated and identified for diagnosis. Galls at graft unions have been attributed to inoculum spread on grafting tools, but inoculum may also come from systemically contaminated propagation material.

Both tumorigenic and nontumorigenic strains of *A. vitis* have been shown to incite necrotic lesions in grape roots, a symptom not seen with *A. tumefaciens*. In the field, these lesions appear on current-season roots as dark, sunken areas about 3–4 mm in diameter (Fig. 192).

Fig. 189. Crown gall, which is most apparent on the basal portions of trunks, developing upward. (Courtesy T. J. Burr)

Fig. 190. Small, pimplelike galls resulting from infection by *Agrobacterium vitis*, extending up a trunk. (Courtesy T. J. Burr)

Fig. 191. Crown gall at the graft unions of young grapevines in a commercial nursery. (Courtesy T. J. Burr)

Fig. 192. Dark, sunken necrosis in a current-season grape root caused by *Agrobacterium vitis*. (Courtesy T. J. Burr)

Causal Organisms

Agrobacterium vitis (Ophel and Kerr 1990), the primary cause of crown gall of grapevines, belongs to the family Rhizobiaceae. This family includes bacteria such as *A. tumefaciens* (Smith and Townsend 1907) Conn 1942 and *Rhizobium* spp., the symbionts that cause nodules on leguminous plants associated with atmospheric nitrogen fixation. Colonies of *A. vitis* and *A. tumefaciens* growing on nonspecific culture media are white, convex, circular, glistening, and translucent; the bacteria are gram negative, motile, and rod shaped. The two species are morphologically indistinguishable

The pathogenicity of both *A. vitis* and *A. tumefaciens* is associated with the presence of a tumor-inducing (Ti) plasmid, an extrachromosomal fragment of DNA. For *A. vitis*, a second plasmid encoding tartrate utilization is also involved in the host-pathogen interaction. The genetic makeup of *A. vitis* strains is diverse; they are generally categorized by the presence and makeup of their Ti plasmid. *A. vitis* and *A. tumefaciens* have been shown to form a biofilm on tomato roots and on abiotic surfaces, abilities that may be associated with both survival and host infection. Nonpathogenic strains of *A. vitis* and *A. tumefaciens* are also found in nature. They may be detected as cohabitants with pathogenic strains in soil, in apparently healthy plant tissue, and in galls.

Selective culture media make it possible to isolate *Agrobacterium* spp. from soil and plant tissues. Some of the media are selective for *A. vitis*, and others allow growth of both *A. tumefaciens* and *A. vitis*. Because colonies of tumorigenic and nontumorigenic isolates of both species look identical and grow equally well on selective media, inoculation tests are necessary to confirm their pathogenicity following recovery from grapevine tissue or the environment. Indicator plants commonly used in greenhouse pathogenicity tests for *A. tumefaciens* include tomato, sunflower, and tobacco (particularly *Nicotiana glauca*). For *A. vitis*, grape plants must be used to test virulence for some host-specific strains, although most strains produce galls on common *A. tumefaciens* hosts. Recently, a method employing magnetic capture hybridization (MCH) technology in conjunction with real-time PCR was developed, which allows sensitive detection of tumorigenic strains of *A. vitis* from grapevine tissues. This method will further the understanding of the pathogen in grapevines as well as in other potential sources in the environment.

Disease Cycle and Epidemiology

Tumorigenic *A. vitis* is associated exclusively with cultivated grapevines and vineyard soils. It has not been found to cause crown gall on other plants under natural conditions, although there is one report of infection of chrysanthemum. Nontumorigenic strains have also been isolated from roots of wild grapevines growing in locations far removed from vineyards. The bacterium is routinely disseminated in symptomless grapevine propagation material in which it is endophytically established, and it is probably present in most grape plantings as an endophyte. However, *A. vitis* does not cause crown gall until it infects plant cells.

The disease cycle is shown in Figure 193. Infection is initiated in the cambium. Like other *Agrobacterium* spp., *A. vitis* infects by transferring into the plant cell a group of genes (the T-DNA) that reside on its Ti plasmid. These genes are incorporated into the DNA of one of the plant's chromosomes, and when transcribed by the plant their expression causes it to produce increased levels of hormones (cytokinin and auxin). This results in rapid and uncontrolled growth of the cambial cells, leading to the development of the fleshy, unorganized gall tissue characteristic of the disease. The process of how *Agrobacterium*

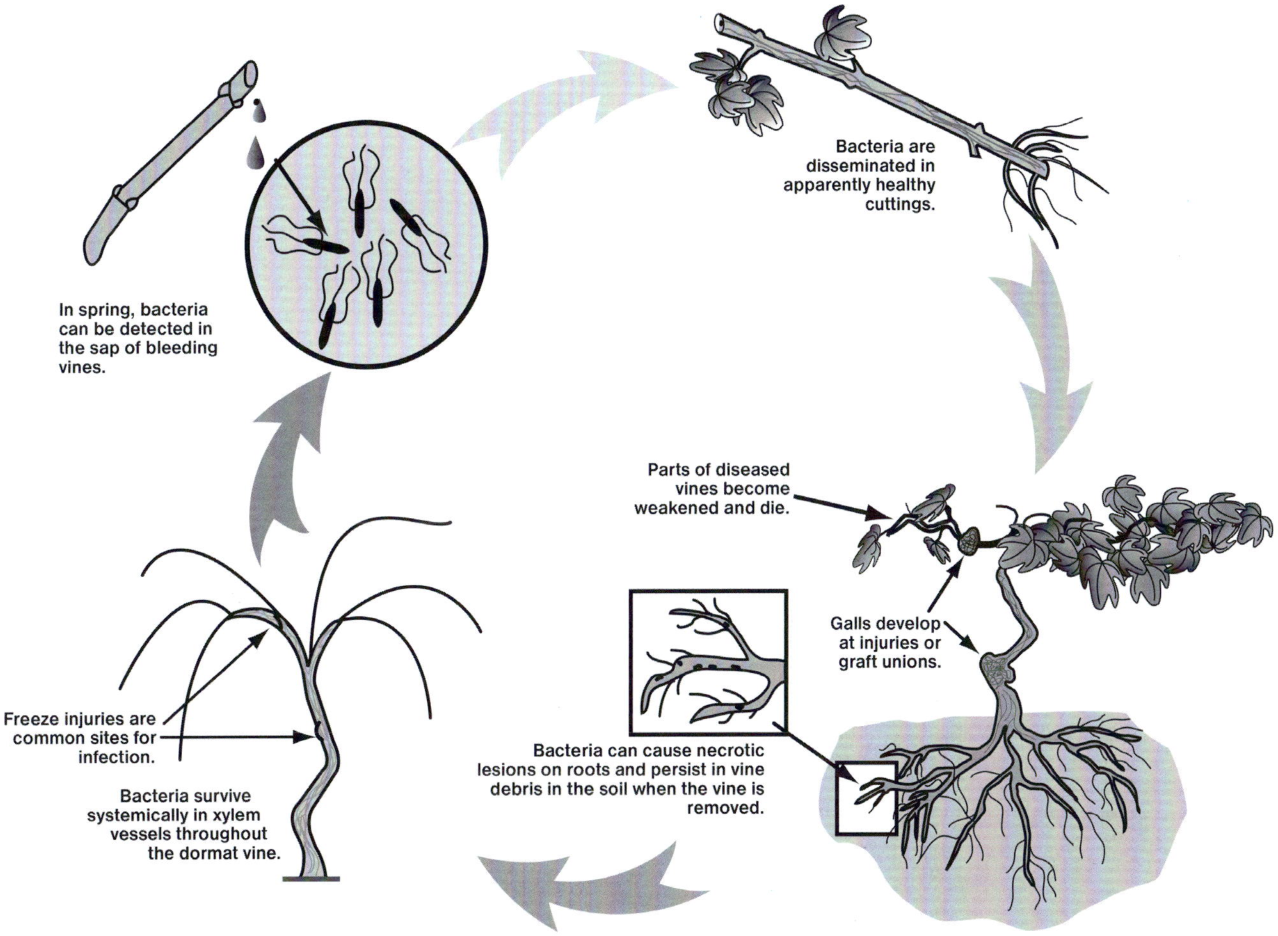

Fig. 193. Crown gall disease cycle. (Drawing by D. Boyce)

species infect plant cells has been studied extensively for many years, and excellent reviews on how they transfer their T-DNA to plants have been written.

Management

Grape crown gall generally develops following attachment of the bacteria to plant cells at sites where wounding has occurred. Wounds may play several functions, including the release of chemical signals that attract the bacterium and promote infection. Wounds also stimulate cell division in the plant cambium, thereby generating cells that are susceptible to infection. Internationally, the most common form of wounds leading to the development of crown gall are those initiated by freeze injuries on trunks, although mechanical injuries such as those associated with tillage equipment or grafting are sometimes involved. Therefore, all practices that help to prevent trunk injury are important in the management of crown gall. These include selecting planting sites with good air and water drainage to reduce the chance of freeze injuries, selecting an appropriate cultivar for planting in sites where winter injury is a potential threat, and using cultural practices that are less likely to result in trunk injuries.

When possible, vineyards should be established in soil not previously planted to grapevines, using planting stock derived from pathogen-free sources if available. Training vines with multiple trunks (3–5 per vine) allows the removal of individual galled trunks while maintaining productivity from the retained healthy trunks and training suckers as new replacement trunks for those removed. Hilling above the union on grafted vines in late autumn protects basal latent buds from freezing and ensures the development of new scion shoots that may be needed for trunk renewal the following season.

Grape varieties differ in their susceptibility to crown gall development, with winter hardiness being a key trait governing this response. In climates with cold winters, crown gall is a significantly greater problem on *V. vinifera* cultivars than on cultivars derived primarily from native American species such as *V. labrusca* or interspecific hybrid cultivars. Although *Agrobacterium* is sensitive to antibiotics such as streptomycin, treating plants with bactericides has not provided effective control, possibly because the bacteria reside within plant tissues and do not come into direct contact with topically applied compounds. Similarly, soil fumigation has not been effective for eradicating *Agrobacterium* from soils, probably because bacteria within infected root fragments are protected from exposure to the fumigants. Treatment of dormant cuttings by submersion in water at 50°C for 30 minutes reduces internal levels of *A. vitis* but cannot be relied on for eradication.

The K-84 strain of *A. radiobacter* has been used successfully for control of crown gall caused by *A. tumefaciens* in other crop systems, but this biocontrol is not effective against *A. vitis* on grapevines. Some nonpathogenic strains of *A. vitis* are now being tested for their utility as biocontrol agents. Transgenic grapevine rootstocks that show reduced susceptibility to crown gall disease also have been produced.

The treatment of established galls with eradicant chemicals has produced varying results. In regions where freeze injury to vines occurs only occasionally, such treatments may be useful in the very early stage of the disease to minimize further gall development, although these vines do remain infected. Some eradicants are effective in killing existing gall tissues, but new galls frequently develop at treated sites the following year. Effective treatments for preventing infection within existing, infested vineyards have yet to be developed, so every effort should be made to obtain pathogen-free planting material for establishing new vineyard sites.

Selected References

Bini, F., Kuczmog, A., Putnoky, P., Otten, L., Bazzi, C., Burr, T. J., and Szegedi, E. 2008. Novel pathogen-specific primers for the detection of *Agrobacterium vitis* and *Agrobacterium tumefaciens*. Vitis 47:181-189.

Burr, T. J., and Reid, C. L. 1993. Biological Control of grape crown gall with nontumorigenic *Agrobacterium vitis* strain F2/5. Am. J. Enol. Vitic. 45:213-219.

Johnson, K. L., Zheng, D., Kaewnum, S., Reid, C., and Burr, T. J. 2013. Development of a magnetic capture hybridization real-time PCR assay for detection of tumorigenic *Agrobacterium vitis* in grapevines. Phytopathology 103:633-640.

Kerr, A. 1972. Biological control of crown gall: Seed inoculation. J. Appl. Bacteriol. 35:493-497.

Lowe, N., Gan, H. M., Chakravartty, V., Scott, R., Szegedi, E., Burr, T. J., and Savka, M. A. 2011. Quorum-sensing signal production by *Agrobacterium vitis* strains and their tumor-inducing and tartrate-catabolic plasmids. FEMS Microbiol Lett. 296:102-109.

Ophel, K., and Kerr, A. 1990. *Agrobacterium vitis* sp. nov. for strains of *Agrobacterium* biovar 3 from grapevines. Int. J. Syst. Bacteriol. 40:236-241.

Zhu, J., Oger, P. M., Shrammeijer, B., Hooykaas, P. J. J., Farrand, S. K., and Winans, S. C. 2000. The basis of crown gall tumorigenesis. J. Bact. 182:3885-3895.

Zupan, J., Muth, T. R., Draper, O., and Zambryski, P. 2000. The transfer of DNA from *Agrobacterium tumefaciens* into plants: A feast of fundamental insights. Plant J. 23:11-28.

(Prepared by T. J. Burr)

Pierce's Disease

Pierce's disease, caused by the xylem-limited bacterium *Xylella fastidiosa*, is a lethal disease of European (*Vitis vinifera*), some American (*V. labrusca*), and French-American interspecific hybrid grapevines. The disease is found throughout areas of North and Central American with mild winter temperatures; it has also been reported in northwestern South America. It is not found in other viticultural regions of the world, where *X. fastidiosa* is usually rated as a pathogen of quarantine significance. Pierce's disease is not found in North American viticultural regions that experience low winter temperatures, such as inland portions of the Pacific Northwest, the Upper Midwest, and the Northeast. Research has shown that grapevines experimentally infected with *X. fastidiosa* and displaying symptoms of Pierce's disease emerge pathogen- and disease free following exposure to low winter temperatures that typically include many hours below 0°C. Although the "cold curing phenomenon" has been clearly documented, the physiological/biochemical basis for it has not been elucidated.

Pierce's disease, named after USDA scientist Newton B. Pierce who demonstrated the disease to be graft transmissible, was first described in 1897 in the Anaheim region of Orange County, California. Within a matter of years, the disease had killed approximately 50,000 acres (20,000 ha) of vineyards, which led to the closure of more than 50 wineries in this region. Decades later, the disease was reported in Florida and other southeastern states. However, this was likely a result of the expanded planting of susceptible grapevine genotypes into the region, as many scientists believe that *X. fastidiosa* is endemic to the southeastern United States and was introduced into California on rootstocks or cultivars imported from there. Today, Pierce's disease is a limiting factor in the cultivation of susceptible grapevine species and hybrids throughout the southeastern United States, and it is believed to have prevented the culture of *V. vinifera* in Florida beginning with the early Spanish settlement in the sixteenth century. The disease continues to cause significant economic losses throughout California's coastal and northern viticultural regions, including Napa and Sonoma valleys. It caused epidemic losses in California's central San Joaquin valley in the 1940s, but losses have remained low since then in this large viticultural region.

Pierce's disease is spread by sucking insects that specialize in feeding on xylem sap. Leafhoppers (Hemiptera: Cicadellidae) commonly referred to as sharpshooters (subfamily Cicadellinae) are the principal vectors, but other families in the order Hemiptera, such as spittle bugs (Cercopidae) and cicadas (Cicadidae), also can transmit the causal bacterium. In the southeastern United States, there are at least 40 species of xylem feeding insects thought to be capable of infecting grapevines. The most important vectors vary with region and immediate habitat near the vineyard. The introduction of the glassy wing sharpshooter (*Homalodisca vitripennis*) into southern California in the late 1980s resulted by the early 2000s in epidemic losses from Pierce's disease in such regions as Temecula, and this insect vector of *X. fastidiosa* continues to cause significant disease losses in several southern California counties. Major investments have funded research to find a cure for Pierce's disease. Some of the most promising methods include genetically engineering grapevines for resistance to the pathogen, although consumer acceptance of any resulting products remains a concern.

Symptoms

Symptoms of Pierce's disease are most severe in the European grapevine species, *V. vinifera*. All varieties of *V. vinifera* are susceptible to the disease and most infected vines eventually die, although there are some differences among cultivars in the severity of symptom expression and frequency of mortality. Chardonnay and Pinot Noir wine grapes and Red Globe table grapes are examples of cultivars that express severe symptoms and die more rapidly than Cabernet Sauvignon, Chenin Blanc, and Merlot, cultivars that typically take up to 5 years to die. In contrast, muscadine (*Vitis rotundifolia* Michx. cultivars and some hybrids between *V. vinifera* and other bunch-grape *Vitis* species native to the southeastern United States become systemically infected with *X. fastidiosa* but bear quality fruit and show few symptoms beyond mild leaf scorching.

Following inoculation with *X. fastidiosa* by infectious insect vectors, symptoms of Pierce's disease typically appear during the same year in the inoculated shoot and may appear in other shoots, depending on varietal susceptibility, date of infection, and temperatures favorable for pathogen growth. The bacterium then moves systemically into older wood and throughout the entire vine within 2–5 years, depending upon varietal susceptibility. Foliar symptoms typically occur in mid- to late summer in most North American viticulture regions, although they may appear earlier near the Gulf of Mexico. In California, foliar symptoms begin to develop in the warmer southern counties in early to mid-July but not until late July or early August in northern production counties. Overall, symptoms of Pierce's disease are associated with the plugging of xylem elements by bacterial cells and biofilms as well as by host reactions to the pathogen such as the formation of tyloses and gums (Fig. 194), and are more severe in hotter climates and in vines with moisture stress. Small spots or sections of the leaf margin initially develop chlorosis; as the season progresses, this chlorosis extends further into the leaf blade and the margin of the leaf turns brown and necrotic (Fig. 195). Eventually, the entire leaf may become chlorotic and necrotic, and late in the season, the leaf blade detaches from the end of the petiole and the petiole remains attached to the cane (Fig. 196). This late-season "match stick" symptom is very diagnostic for Pierce's disease when observed in conjunction with other foliar symptoms. In red varieties, there is often a dark-red band of pigmentation near the margin of the chlorotic zone (Fig. 197), whereas white varieties typically produce only the chlorotic band. Infections initiated in early summer often produce only yellowed or reddened leaves without distinctive marginal discoloration. Fruit on diseased portions of the vine typically shrivel and can become "raisined" at any time following veraison (Fig. 198); such fruit is economically worthless. Infected canes may fail to uniformly lignify late in the summer, resulting in "green islands" of nonlignified tissue that often form at the bases of leaf petitoles or fruit rachises attached to the canes (Fig. 199). There is a direct correlation between the severity of symptoms and the concentration of *X. fastidiosa* bacteria in the tissue. The presence of the bacteria in symptomatic tissues is easily confirmed by culture, ELISA, and PCR. However, bacteria are very hard to detect in the new growth of infected but asymptomatic vines early in the season, before overwintering bacteria can move to them from overwintering sites in woody tissues and begin multiplying.

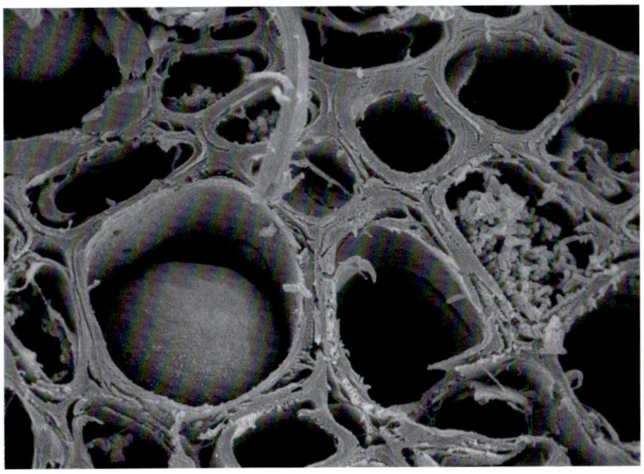

Fig. 194. Electron micrograph of a cross-sectional view through the vascular system of a grapevine infected with *Xylella fastidiosa*, with two xylem vessels plugged by tyloses (upper left, left) and another plugged by a mass of bacterial cells (right). (Courtesy H. C. Hoch)

Fig. 195. Marginal leaf chlorosis developing into necrosis on a grapevine with Pierce's disease. (Courtesy J. S. Kamas)

Fig. 196. Retained petioles following abscission of affected leaf blades late in the season, resulting in the "matchstick" symptom characteristic of Pierce's disease. (Courtesy J. S. Kamas)

In the years following initial infection, the vine continues to produce symptomatic leaves, fruit, and canes but permanent woody cordons begin to die back as the disease progresses. Canes that develop on infected cordons often push out later in the spring and their growth is very stunted early in the season compared with healthy vines. Spring growth on vines that were severely infected the year before may be stunted and distorted, with interveinal chlorosis resembling zinc deficiency. New canes can have shortened internodes with a zigzag pattern of growth. However, retarded cane growth can be caused by abiotic factors as well as other grapevine pathogens, and this particular symptom occurring by itself should not be considered diagnostic for Pierce's disease. The dieback of permanent wood continues until the entire vine fails to leaf out in the spring, although most growers remove vines affected long before they die since they fail to set viable fruit once the entire vine shows these symptoms.

Symptoms of Pierce's disease can be confused with only a few other conditions or diseases. Similar chlorotic leaves are produced by esca-affected vines, but there are important differences. Unlike that of Pierce's disease, foliar chlorosis due to ecsa usually occurs throughout the leaf blade and in its initial stages is not necessarily confined to the leaf margins. In addition, esca foliar symptoms tend to occur earlier in the growing season than is typical for Pierce's disease. Thus, in California the observation of significant foliar chlorosis in May or June is unlikely to be caused by Pierce's disease. The stunted, chlorotic shoots produced as a result of Eutypa dieback also can be confused with the stunted canes associated with Pierce's disease. However, distinctive pruning wound-associated cankers can usually be found on the cordon basal to shoots stunted by *Eutypa* infection, whereas cankers are not a symptom of Pierce's disease (although unrelated co-infection by canker fungi and *X. fastidiosa* is possible). Vines with Armillaria root rot (*Armillaria* spp.) or cotton root rot (*Phymatotrichopsis omnivora*) may also exhibit foliar symptoms and dieback that can resemble Pierce's disease, although such vines often show diagnostic signs of those causal pathogens as described within their respective sections of this book. Severe drought stress can also mimic the foliar symptoms of Pierce's disease, although an examination of climatic data and/or any irrigation regime can usually confirm or eliminate this as a possibility. Ultimately, however, *X. fastidiosa* is readily detected in symptomatic leaves by a number of conventional laboratory tests, thereby conclusively confirming the presence of the pathogen. Suspensions of cultured *X. fastidiosa* can be efficiently inoculated into plants by means of a pin-prick method to confirm their pathogenicity.

Causal Organism

Studies in the late nineteenth century established that Pierce's disease was caused by a graft-transmissible pathogen. Bacteria were later found in the xylem of affected vines by electron microscopy, but not until 1978 was a bacterium successfully cultured and shown conclusively to be the causal agent of Pierce's disease. This organism, *Xylella fastidiosa*, is a gram negative, rod-shaped bacterium with a rippled cell wall. In addition to the strain that causes Pierce's disease, several additional genetically and pathologically distinct strains of *X. fastidiosa* also cause diseases of citrus (citrus variegated chlorosis), almond (almond leaf scorch), and oleander (oleander leaf scorch).

In plants, the bacterium colonizes xylem vessels exclusively; in its insect vectors, it also colonizes the lining of the foregut. *X. fastidiosa* strains are nutritionally fastidious and typically are grown on complex, undefined media such as Periwinkle wilt (PW) or PD3, although some strains have been grown recently

Fig. 197. Band of dark-red pigmentation near the margin of the chlorotic and necrotic zone on a leaf of a red-fruited grapevine cultivar with Pierce's disease. (Courtesy H. C. Hoch)

Fig. 198. Shriveled fruit on a vine infected with *Xylella fastidiosa*. (Courtesy J. S. Kamas)

Fig. 199. "Green islands" of stem tissue failing to form periderm on a vine infected with *Xylella fastidiosa*. (Courtesy J. S. Kamas)

on chemically defined media. *X. fastidiosa* grows very slowly on all types of culture media, with Pierce's disease strains typically producing small, opalescent colonies within 7–10 days.

The complete genomes of *X. fastidiosa* strains that cause disease in citrus, grapevine, almond, and oleander have been sequenced, allowing researchers to study numerous specific interactions between the bacterium, its host, and its vector. For example, studies found that *X. fastidiosa* polygalacturonase is necessary for the pathogen to systemically colonize grapevines. Its Type I secretion system was shown to be necessary for the bacterium to survive in grapevine xylem, and mutants unable to synthesize a "diffusible signal factor" mediating the expression of a number of *X. fastidiosa* genes were hypervirulent in grapevines and no longer transmissible by sharpshooter vectors. Similarly, mutations in the bacterium's hemagglutinin genes, which play a role in cell-cell attachment and biofilm formation, were likewise hypervirulent and somewhat impaired in insect transmissibility. The question of how the bacterium is able to systemically move throughout plants against the xylem transpirational stream without a flagellum was answered by the discovery that the bacterium exhibits a "twitching" motility that is mediated by Type IV pili. A recent discovery found that a breakdown product of pectin triggers gene expression needed for the bacterium to attach to vector mouthparts, which is the crucial step of the vector transmission process.

These discoveries and others have greatly increased our understanding of how *X. fastidiosa* causes disease and how it is transmitted by insect vectors. This basic information has suggested strategies for developing Pierce's disease-resistant transgenic grapevines as well as other nontransgenic approaches for managing the disease, which are being evaluated.

Disease Cycle and Epidemiology

In nature, *Xylella fastidiosa* is transmitted from infected host plants to healthy plants by xylem sap-feeding insects and root grafts. After acquiring *X. fastidiosa*, a vector can immediately transmit the bacterium to a healthy plant, without a transmission latent period. However, the longer an insect feeds on an infected host plant, the greater the probability of it acquiring the bacterium. (Similarly, the more often a grapevine is challenged by vectors that have acquired the bacterium, the greater the probability of it becoming infected). Once acquired, the vector inoculates a healthy plant by egesting *X. fastidiosa* into its xylem. Nearly all xylem-feeding insects that have been tested can acquire and transmit *X. fastidiosa* with some efficiency, although a vector-specific strain of the bacterium was reported recently. Some vectors, such as the blue-green sharpshooter (*Graphocephala atropunctata*) are extremely efficient at acquiring and transmitting *X. fastidiosa* to grape, whereas others, such as the glassy-winged sharpshooter, are considerably less efficient. However, even an inefficient vector such as glassy-winged sharpshooter can cause extensive spread of Pierce's disease if it is abundant and moves extensively within vineyards. Nymphal stages of the insect vectors can transmit the bacterium, but the pathogen is shed when the insect molts to its next nymphal or adult stage. *X. fastidiosa* is not transmitted transovarially from an infested female to its offspring, nor is it seedborne.

Grapevine strains of *X. fastidiosa* can infect a large number of plant species. Most of these are not systemic hosts of *X. fastidiosa* but rather nonsystemic, "microsite" hosts where the bacteria are restricted to a few xylem vessels at the feeding site of the vector. Vectors can acquire *X. fastidiosa* from such plants, but they are less important as pathogen reservoirs than systemically infected hosts. In California, some of the most epidemiologically significant *X. fastidiosa* reservoirs are wild grapevines (*Vitis californica* and feral *V. californica*/*V. vinifera* hybrids), ground periwinkle (*Vinca minor*), and Himalayan blackberry (*Rubus armeniacus*) growing in riparian habitats adjacent to vineyards. If a systemic host of *X. fastidiosa* is also a preferred feeding and reproductive host of an *X. fastidiosa* vector, the nymphs that emerge on an infected plant have a much greater probability of acquiring and transmitting the pathogen to healthy plants.

Infected *V. vinifera* and tolerant native *Vitis* spp. are considered to be important pathogen reservoirs in the southeastern United States and Southern California. This is largely because glassy-winged sharpshooter is the most important Pierce's disease vector in these areas. In California, it has been shown to feed on and acquire *X. fastidiosa* from woody canes as well to transmit the pathogen into new woody canes, although feeding on woody tissues is rarely observed in the southeastern United States. Because *X. fastidiosa* overwinters in woody grapevine tissues, glassy-winged sharpshooter likely can acquire bacterium from an infected vine much earlier in the season than other vectors. Experiments have shown that glassy-winged sharpshooter can transmit the pathogen to dormant vines in the southern San Joaquin Valley of California, although the same experiments failed to show acquisition from diseased dormant vines. If an infectious glassy-winged sharpshooter feeds on a pruned woody spur, the bacterium has only a short distance to travel in order to become systemically established within the permanent framework of that vine. In contrast, many other sharpshooter vectors can only acquire and transmit *X. fastidiosa* from green leaves and shoots. If an infectious vector such as the blue-green sharpshooter feeds on the succulent ends of a 2-m-long shoot, the bacterium has a much greater distance to travel before it can cause systemic infection in woody tissues. Indeed, research has shown that *X. fastidiosa* only causes systemic infection if vines are inoculated very early in the season by vectors such as blue-green sharpshooter or green- and red-headed sharpshooters, *Draeculacephala minerva* and *Carneocephala fulgida*, respectively, which feed exclusively on succulent tissues. The latter three sharpshooters are the most important vectors of Pierce's disease in the North Coast and Central Valley regions of California.

The epidemiology of Pierce's disease in California is driven by the feeding behavior, preferences, and prevalence of its vectors. The glassy-winged sharpshooter is a very dispersive and polyphagous insect. As such, it can acquire *X. fastidiosa* from a variety of monocot and dicot hosts. In mixed plantings and natural vegetation, a glassy-winged sharpshooter shifts plant hosts during the course of the day. In contrast, other Pierce's disease vectors often remain on single plant hosts for days at a time. Because the glassy-winged sharpshooter can acquire and transmit the pathogen to woody grapevine tissues, it can inoculate healthy vines during most of the growing season. Its ability to acquire *X. fastidiosa* from an infected vine and then migrate to and infect adjacent healthy vines results in efficient spread within a vineyard. If glassy-winged sharpshooter populations are high, this can result in epidemic Pierce's disease losses such as those documented in Temecula, California in the early 2000s. Vineyards inoculated by the glassy-winged sharpshooter ultimately suffer widespread vine loss throughout the planting, although it is common to see a clumped distribution characteristic of localized vine-to-vine spread early in the course of an epidemic.

In contrast to the glassy-winged sharpshooter, the blue-green sharpshooter, which is the primary vector of *X. fastidiosa* in the coastal and North Coast regions of California, prefers to feed on select dicot hosts that live in riparian areas near vineyards. Adults of the blue-green sharpshooter overwinter in the riparian areas where they may acquire *X. fastidiosa* from various plant hosts. Adult sharpshooters that acquire the bacterium from infected *V. vinifera* vines in the fall remain infective through the winter. As temperatures increase in the spring and vines begin new growth, infectious blue-green sharpshooter migrate out of the riparian zone and into adjacent vineyards. The early-season distribution pattern of blue-green sharpshooters resembles that of Pierce's disease in these vineyards: the highest incidence near the riparian source area of overwintering vectors and diminishing with distance, with most Pierce's disease-affected vines found within 100 m of the riparian areas. This spatial disease pattern occurs very stably over time (80- to 100-year-old vineyards show this same pattern), regardless of whether blue-green sharpshooter populations are found in the vineyard during the summer or if the planting is treated with insecticides to eliminate detectable populations of the insect. Thus, vector populations in vineyards after late spring seem unimportant for systemic Pierce's disease spread.

When, during the season, an infection is initiated determines the likelihood of its persistence through the following winter. In California, vines that were experimentally exposed to infested vectors every month from April through August became infected with *X. fastidiosa* at similar rates throughout that period. However, almost all vines inoculated in April were still diseased the next year, whereas Pierce's disease persisted in far fewer vines inoculated in May and only a few that were inoculated in July. Thus, infections that occur during the initial period of growth in the spring appear to be those most likely to persist as chronic Pierce's disease. Populations of *X. fastidiosa* become sufficiently high in the new growth of chronically diseased vines to be detected by late June or July, when insect vectors most readily acquire the bacterium. However, the nonchronic nature of new infections established during summer months explains why vector control within vineyards after spring has not slowed Pierce's disease spread in coastal and Northern California vineyards and why removing diseased vines from vineyards has had no detectable effect in slowing the spread of Pierce's disease in these regions.

Climatic variables are important factors in both the rate of Pierce's disease development and overwinter recovery from the disease. Summer temperatures of 22–32°C are ideal for the continuous growth of *X. fastidiosa*, which may explain the faster appearance of symptoms of Pierce's disease in Florida and Texas than in regions of California with cooler night temperatures. Within the United States, the geographic areas in which Pierce's disease transitions from a regular to a sporadic problem and then to a rarity generally reflect the geographic transitions in measures of winter severity. In general, Pierce's disease severity decreases in the Appalachian Mountains with increases in altitude and northern latitude as well as distance from the moderating influence of the Atlantic Ocean. Both altitude and latitude correlate broadly with winter cold, but there are no quantitative measures of the precise climatic conditions most likely to deter Pierce's disease development or favor its spread. Disease severity has increased following mild winters in the Texas Hill Country west of Austin and in mountain vineyards in Georgia.

Both the red-headed and green sharpshooters that are the main vectors of Pierce's disease in California's Central Valley prefer to feed and reproduce on monocot plant hosts. They then move into adjacent vineyards when their annual weed hosts dry up in the summer or are physically removed and can cause noticeable Pierce's disease problems if populations are sufficiently high. Alfalfa fields are good breeding habitats for both red-headed and green sharpshooters if their grass host plants occur as weeds there. Alfalfa also serves as a systemic, propagative host of *X. fastidiosa*, as do some grass and sedge species. The typical vineyard infected by these vectors is adjacent to weedy alfalfa fields or irrigation ditches, and the distribution of infected vines typically reveals a gradient from the weed habitat. Like the blue-green sharpshooter, red-headed and green sharpshooters prefer to feed only on succulent green tissues of the grapevine.

Management

To date, there is no therapeutic cure for Pierce's disease-affected *V. vinifera* grapevines, but much can be done to prevent disease spread if vectors can be excluded or controlled before they transmit *X. fastidiosa* to healthy vines. In areas with reasonably cold winters, there is the possibility that vines showing limited, first-year infections may emerge pathogen and disease free the following spring owing to the cold-curing phenomenon discussed above in Disease Cycle and Epidemiology. Vines showing a few symptomatic canes should be identified and marked in mid- to late August and pruned as normal the following fall or winter. Growers should inspect these vines for poor shoot growth in the spring and for typical foliar symptoms in August. If it is apparent that the vine did not recover over the winter and additional symptomatic canes are observed the following season, the vine should be considered systemically infected and removed to eliminate it as a pathogen reservoir. Severe pruning of affected cordons, and even removing the entire trunk close to the graft unions, has not eliminated the pathogen in research trials. Although severely pruned vines may produce disease-free, vigorous growth the year after pruning, nearly all develop Pierce's disease symptoms throughout the vine in the second year.

No potential bactericides (e.g., antibiotics, copper, and zinc) or compounds that induce systemic acquired resistance in other plants have successfully cured or prevented new infections in evaluation trials, regardless of the mode of delivery employed. Likewise, attempts to find bacterial endophytes antagonistic to *X. fastidiosa* for deployment as biological control agents have been unsuccessful. However, research in Florida identified a strain of *X. fastidiosa* from elderberry that can systemically infect *V. vinifera* vines but not cause disease in them. Furthermore, vines that were prophylactically inoculated with this nonpathogenic strain did not develop Pierce's disease when grown in Florida, where *X. fastidiosa* is endemic and Pierce's disease pressure is high, whereas unprotected control vines died from the disease within a few years. Additional evaluations of this *X. fastidiosa*-biocontrol strain are currently underway in the southeastern United States and California.

In Southern California, where glassy-winged sharpshooter is an important vector, applications of systemic and topical insecticides have greatly reduced populations relative to those that first developed after it was introduced into the state. Vineyards are typically treated with systemic neonicotenoids through irrigation water once a year, whereas contact insecticides are applied on an as-needed basis. Insecticides are also applied to citrus, a preferred feeding and reproductive host for glassy-winged sharpshooter, as well as to other perennial vector hosts near vineyards. Because the glassy-winged sharpshooter congregates and breeds in habitats outside vineyards, an area-wide program is necessary to manage Pierce's disease where this insect is the primary vector. Removal of chronically infected vines may also be important to reduce inoculum where glassy-winged sharpshooter is the primary vector, since citrus and other crops traditionally grown near grapes are not alternative host plants of *X. fastidiosa*. Although some vine losses still occur in treated vineyards, these are minimal compared to the heavy losses sustained in the early 2000s. In California, state and federal governments in cooperation with ornamental nursery and grape-grower organizations fund an extensive nursery plant inspection system, the goal of which is to prevent the spread of glassy-winged sharpshooter into noninfested counties. This effort has been very successful to date, with only a few localized detections of glassy-winged sharpshooter outside the infested southern counties. In such cases, local infestations have been eradicated by aggressive insecticide applications.

Control of Pierce's disease in coastal and Northern California, where the blue-green sharpshooter is the primary vector, has focused on managing riparian plant composition and timely springtime insecticide applications in riparian areas. Work is ongoing with various federal and state regulatory agencies to develop a riparian plant management program that selectively removes the breeding hosts of the vector while simultaneously providing environmental benefits to the riparian habitats. Growers can apply for permits to remove vector plant hosts while replanting the area with native vegetation that is not attractive to the blue-green sharpshooter. This program has been very effective at reducing blue-green sharpshooter populations and lowering the incidence of Pierce's disease in adjacent vineyards; unfortunately, it also is very labor intensive and expensive. Additionally, implementing the program is not feasible unless the vegetation can be managed on both sides of the waterway, which sometimes requires the cooperation of a neighboring landowner.

Management of weeds in adjacent fields, especially those of alfalfa, and along irrigation ditches is the primary means of reducing populations of the red-headed and green sharpshooter. If undertaken, weed control significantly reduces the incidence of Pierce's disease in California's Central Valley and other viticultural regions where these insects are its primary vectors.

In the southeastern United States where the disease is clearly polycyclic, rouging of infected vines inhibits within-vineyard spread of the pathogen and is an integral part of disease management.

Currently, the only satisfactory control of Pierce's disease in areas where the disease is endemic and chronic is to plant grapevines that are genetically resistant to *X. fastidiosa*. Currently available resistant genotypes do not produce fruit that is comparable to that from *V. vinifera* cultivars, and the market for it is limited to affected regions. Whereas all *V. vinifera* varieties are susceptible and eventually die from Pierce's disease once infected, some are much more susceptible and die sooner than others. Where commercially feasible, growers should plant less-susceptible cultivars near known sources of the pathogen and insect vectors, such as riparian areas. The discovery that genes for Pierce's disease resistance reside in a single locus in *V. arizonica* has facilitated the utilization of traditional breeding methodologies to develop resistant hybrids with *V. vinifera* that maintain fruit qualities of the latter. Field testing of these selections is currently underway. A number of strategies to develop Pierce's disease-resistant transgenic *V. vinifera* cultivars or rootstocks are being evaluated nationwide. Development and use of transgenic rootstocks that could deliver anti-*X. fastidiosa* compounds into nontransgenic fruiting varieties are seen by some as the most desirable and likely approach to be accepted by growers, wineries, and consumers.

Selected References

Almeida, R. P. P., and Purcell, A. H. 2003. Transmission of *Xylella fastidiosa* to grapevines by *Homalodisca coagulata* (Hemiptera: Cicadellidae). J. Econ. Entomol. 96:264-271.

Almeida, R. P. P., Wistrom, C., Hill, B. L., Hashim, J., and Purcell, A. H. 2005. Vector transmission of *Xylella fastidiosa* to dormant grape. Plant Dis. 89:419-424.

Chatterjee, S., Almeida, R. P. P., and Lindow, S. E. 2008. Living in two worlds: The plant and insect lifestyles of *Xylella fastidiosa*. Ann. Rev. Phytopathol. 46:243-271.

Davis, M. J., Purcell, A. H., and Thomson, S. V. 1978. Pierce's disease of grapevines: Isolation of the causal bacterium. Science 199:75-77.

Feil, H., and Purcell, A. H. 2001. Temperature-dependent growth and survival of *Xylella fastidiosa* in vitro and in potted grapevines. Plant Dis. 85:1230-1234.

Feil, H., Feil, W. S., and Purcell, A. H. 2003. Effects of date of inoculation on the within-plant movement of *Xylella fastidiosa* and persistence of Pierce's disease within field grapevines. Phytopathology 93:244-251.

Guilhabert, M. R., and Kirkpatrick, B. C. 2005 Identification of *Xylella fastidiosa* avirulence genes: Hemagglutinin adhesions contribute to *Xylella fastidiosa* biofilm maturation and colonization and attenuate virulence. Mol. Plant-Microbe Interact. 18:856-868.

Hopkins, D. L. 2005. Biological control of Pierce's disease in the vineyard with strains of *Xylella fastidiosa* benign to grapevine. Plant Dis. 89:1348-1352.

Hopkins, D. L., and Purcell, A. H. 2002. *Xylella fastidiosa*: Cause of Pierce's disease of grapevine and other emergent diseases. Plant Dis. 86:1056-1066.

Kamas, J., Appel, D., Black, M., Labay, A., Lewis, J., and Morano, L. 2012. Pierce's Disease Overview and Management Guide: A Resource for Grape Growers in Texas and other Eastern U. S. Growing Regions. Texas A&M AgriLife Ext. Serv., College Station.

Killiny, N., and Almeida, R. P. P. 2009. Host structural carbohydrate induces vector transmission of a bacterial plant pathogen. Proc. Natl. Acad. Sci. U. S. A. 109:22416-22420.

Meng, Y., Li, Y., Galvani, C. D., Hao, G., Turner, J. N., Burr, T. J., and Hoch, H. C. 2005. Upstream migration of *Xylella fastidiosa* via pilus-driven twitching motility. J. Bacteriol. 187:5560-5570.

Newman, K. L., Almeida, R. P. P., Purcell, A. H., and Lindow, S. E. 2004. Cell-cell signaling controls *Xylella fastidiosa* interactions with both insects and plants. Proc. Natl. Acad. Sci. U. S. A. 101:1737-1742.

Purcell, A. H., and Sanders, S. R. 1999. Fate of Pierce's disease strains of *Xylella fastidiosa* in common riparian plants in California. Plant Dis. 83:825-830.

Redak, R. A., Purcell, A. H., Lopes, J. R. S., Blua, M. J., Mizell, R. F., and Andersen, P. C. 2004. The biology of xylem fluid–feeding insect vectors of *Xylella fastidiosa* and their relation to disease epidemiology. Annu. Rev. Entomol. 49:243-270.

Reddy, J. D., Reddy, S. L., Hopkins, D. L., and Gabriel, D. W. 2007. TolC is Required for Pathogenicity of *Xylella fastidiosa* in *Vitis vinifera* Grapevines. Mol. Plant-Microbe Interact. 20:403-410.

Roper, M. C., Greve, L. C., Warren, J. G., Labavitch, J. M., and Kirkpatrick, B. C. 2007. *Xyella fastidiosa* requires polygalacturonase for colonization and pathogenicity in *Vitis vinifera* grapevines. Mol. Plant-Microbe Interact. 20:411-419.

Walker, A. 2008. Breeding Pierce's disease resistant winegrapes. Pages 248-253 in: Proc. Pierce's Disease Research Symposium. California Dep. of Food Agric., Sacramento.

(Prepared by B. C. Kirkpatrick)

Diseases Caused by Phytoplasmas

Grapevine Yellows Diseases

Grapevine yellows are a group of diseases reported in several countries and continents. All varieties of grapevines are susceptible to grapevine yellows, but some may be more affected than others. Symptoms of all grapevine yellows diseases are similar: canes, shoots, leaves, and berries are damaged, resulting in a decline of the affected vines and significant loss in crop yield and quality. All grapevine yellows diseases are associated with phytoplasmas, obligate intracellular parasites similar to bacteria but lacking a true cell wall. They are restricted to the phloem of plants and are transmitted only by specific, phloem-feeding insect vectors (leafhoppers, planthoppers, and psyllids) or through vegetative propagation techniques. Because phytoplasmas can be transmitted within infected cuttings and budwood, grapevine yellows diseases may become epidemic when infected mother plants are used as sources of propagating material and the resulting progeny are established in vineyard sites inhabited by vector species.

The history of grapevine yellows diseases dates back to the 1950s, when *flavescence dorée* was first reported in southwest France. At that time *flavescence dorée* was regarded as a physiological disorder, with symptoms of bright golden leaves on the interspecific hybrid cultivar Baco 22A. Later, after aerial spread and graft transmission were demonstrated, the causal agent was assumed to be a virus. In 1967, however, a different class of pathogens was ascribed to the so-called yellows group of plant diseases. First called mycoplasmalike organisms and now renamed phytoplasmas, these microbes were subsequently identified in affected grapevines and in vector insects, allowing *flavescence dorée* to be compared with *Bois noir* and *Vergilbungskrankheit* diseases occurring in northern France, Italy, and Germany, and to yellows diseases of grapevines reported in other countries around the world. The generic name of grapevine yellows was recommended for application to these individual diseases in 1994 by the Steering Committee of ICVG (International Council for the Study of Virus and Virus-like Diseases of the Grapevine).

With the development of molecular tools to detect and characterize phytoplasmas in plants and insects, the causal organisms of each disease can now be characterized and vector species identified. Because phytoplasmas are presently unculturable, they cannot be classified according to formally recognized species with Latin binomials, which requires a description of the organism grown in pure culture. They therefore have been classified mainly on the basis of the sequence homology of their ribosomal DNA (16S rDNA and 16S-23S intergenic spacer region), and the convention of '*Candidatus* Phytoplasma' species has been adopted as one means of referring to distinct phytoplasma lineages and putative species. Biological traits, when available, are also taken into account to differentiate closely related organisms grouped in the same cluster. To date, nine different phytoplasmas belonging to five different groups have been associated with the various grapevine yellows diseases. Table 3 lists some characteristics of these phytoplasmas. The most outstanding feature of the classification of grapevine yellows phytoplasmas is their great diversity. The association of phytoplasmas within the same ribosomal group with different vectors, for example, those causing *flavescence dorée* and Palatinate grapevine yellows, further accentuates grapevine phytoplasmas diversity.

Grapevine yellows diseases have been identified in all grapevine-growing countries of Europe, in Israel and Lebanon, in Asia Minor, in several states along the eastern seaboard of the United States, and in Australia. The reason for the apparent absence of grapevine yellows diseases in California is unknown. In the eastern United States, North American grapevine yellows is most destructive in Virginia, where nearly 30% of the vines have been lost in some Chardonnay plantings over a 5- to 10-year period, but becomes less severe moving both northward and southward; e.g., the disease is observed only sporadically and with little general consequence in New York.

Phytoplasmas are acquired by vector insects while feeding in the phloem sieve tubes of infected host plants. After acquisition, the phytoplasmas move, multiply, and persist in the body of infected vectors, which transmit them to other susceptible plants while feeding. Phytoplasmas occupy only the phloem sieve tubes of the vascular system but are distributed throughout the grapevine in this tissue, including roots, trunk, canes, shoots, leaf petioles and veins, and bunch rachises. However, their distribution within infected plants is uneven and their movement and multiplication are not fully understood. Affected current-year shoots do not lignify (become woody) and they therefore die in the winter. Phytoplasmas, nevertheless, persist over the winter in cordons, trunks, roots, and in (asymptomatic) dormant, lignified canes, and therefore may be propagated by grafting or rooting dormant wood, providing a mechanism for long-distance spread of the pathogens. Although phytoplasmas are persistent in vector insects once acquired, they generally are not passed on to their progeny transovarially. Hence, the role of reservoir plants in maintaining inoculum is considered to be critical.

Natural spread of each grapevine yellows diseases depends on the presence and abundance of its vector species in addition to specific aspects of vector biology, behavior, and feeding activity. Disease transmission is also dependent on the distribution of infected vines within the vineyard, and the presence outside the vineyard of reservoir plants from which insect vec-

TABLE 3. Grapevine Yellows Diseases and Grouping of Associated Phytoplasmas

GY Disease	Phytoplasma Name	Ribosomal Group (and Subgroup)	Known Insect Vector to Grapevine	Preferred Host Plants of Vector	Geographical Occurrence
Flavescence dorée	Grapevine flavescence doree ('*Candidatus* Phytoplasma vitis'-related strains)	EY or 16SrV (-C, -D)	*Scaphoideus titanus* Ball	*Vitis* spp.	France, Italy, Spain
Palatinate grapevine yellows (PGY)	PGY	EY or 16SrV	*Oncopsis alni* Schrank	*Alnus glutinosa* L.	Germany
Bois noir/Vergilbungskrankheit	Bois noir ('*Candidatus* Phytoplama solani')	stolbur or 16SrXII-A	*Hyalesthes obsoletus* Sign	*Convolvulus arvensis* L.	European Community, Israel, Lebanon
North American grapevine yellows	Western X ('*Candidatus* Phytoplasma pruni'-related strain)	W-X or 16SrIII-I	ND	ND	Eastern United States
	'*Candidatus* Phytoplasma asteris'-related strain	16SrI-A	ND	ND	
Australian grapevine yellows	AGY ('*Candidatus* Phytoplasma australiense')	stolbur or 16SrXII-B	ND	ND	Australia
	Tomato big bud ('*Candidatus* Phytoplasma australasia')	FBP or 16SrII	ND	ND	
	Buckland Valley grapevine yellows (BVGY)	16SrXXIII-A	ND	ND	
Grapevine yellows	Aster Yellows ('*Candidatus* Phytoplasma asteris')	AY or 16SrI	ND	ND	Italy

tors may acquire inoculum and subsequently become infective. Hence, epidemiological features of the various grapevine yellows diseases do differ. In general, two modes of spread have been recognized, i.e., clustered and random, depending on the preferred- or occasional-feeding status of specific vectors on grapevines. Individual diseases have exhibited contrasting epidemiological traits among different vineyards subject to variable environments and viticultural practices.

Management

Although spontaneous remissions have occurred, in general grapevine yellows-affected grapevines cannot be purposely cured. Therefore, disease management is largely predicated on preventative measures based upon knowledge of the vector identity, biology, and behavior, and of phytoplasma host reservoirs. For example, in some portions of western Europe, *flavescence dorée* has been managed with a well-timed application of an insecticide targeting a single leafhopper vector, which has only one generation per year and lives only on grapevines. In contrast, North American grapevine yellows is vectored by multiple leafhopper species with most having multiple generations per year; thus, effective insecticidal control would require repeated treatments throughout the year, which is both economically and environmentally impractical. For this disease, management recommendations include the avoidance of highly susceptible cultivars, removal of wild reservoir hosts, and early roguing of symptomatic vines

New vineyards should be established from propagation sources that are free of grapevine yellows phytoplasmas, and movement of phytoplasma-infected and/or vector-infested materials to areas not already harboring them must be avoided. The *flavescence dorée* phytoplasma is a quarantine organism in the European Community and importation of infected planting material is prohibited. Soaking dormant planting material in hot water (50°C for 45 min) can be used to eradicate phytoplasmas and destroy insect eggs deposited in the bark.

Selected References

Batlle, A., Martínez, M. A., and Laviña, A. 2000. Occurrence, distribution and epidemiology of Grapevine Yellows in Spain. Eur. J. Plant Pathol. 106:811-816.

Bianco, P. A., Davis, R. E., Casati, P., and Fortusini, A. 1996. Prevalence of aster yellows (AY) and elm yellows (EY) group phytoplasmas in symptomatic grapevines in three areas of northern Italy. Vitis 35:195-199.

Boudon-Padieu, E., and Grenan, S. 2002. Hot water treatment. Methods page of website of International Council for the Study of Virus and Virus-like Diseases of the Grapevine (ICVG). http://web.pppmb.cals.cornell.edu/fuchs/icvg/data/icvghotw.pdf

Bovey, R., and Martelli, G. P. 1992. Directory of Major Virus and Virus-like Diseases of Grapevines: Description, Historical Review and Bibliography. Mediterr. Fruit Crop Improv. Counc. and Int. Coun. for the Study of Viruses and Virus-like Diseases of the Grapevine. Imprimerie Finzi, Tunis, Tunisia.

Carraro, L., Loi, N., Kuszala, C., Clair, D., Boudon-Padieu, E., and Refatti, E. 1994. On the ability-inability of *Scaphoideus titanus* Ball to transmit different grapevine yellows agents. Vitis 33:231-234.

Daire, X., Clair, D., Larrue, J., and Boudon-Padieu, E. 1997. Survey for grapevine yellows phytoplasmas in diverse European countries and Israel. Vitis 36:53-54.

Davis, R. E., Dally, E. L., Tanne, E., and Rumbos, I. C. 1997. Phytoplasma associated with grapevine yellows in Israel and Greece belong to the stolbur phytoplasma subgroup, 16SrXII-A. J. Plant Pathol. 79:181-187.

Gibb, K. S., Constable, F. E., Moran, J. R., and Padovan, A. C. 1999. Phytoplasmas in Australian grapevines—detection, differentiation and associated diseases. Vitis 38:107-114.

Lee, I.-M., Gundersen-Rindal, D. E., Davis, R. E., and Bartoszyk, I. M. 1998. Revised classification scheme of Phytoplasmas based on RFLP analyses of 16S rRNA and ribosomal protein gene sequences. Int. J. Syst. Bacteriol. 48:1153-1169.

Magarey, P. A., and Wachtel, M. F. 1986. Australian grapevine yellows. Int. J. Trop. Plant Dis. 4:1-14.

Maixner, M., Daire, X., Boudon-Padieu, E., Laviña, A., Batlle, A., and Reinert, W. 1997. Phytoplasmas. Pages 183-195 in: Sanitary Selection of the Grapevine: Protocols for Detection of Viruses and Virus-like Diseases. Les Colloques 86, INRA Editions, Paris, France.

Seemüller, E., Marcone, C., Lauer, U., Ragozzino, A., and Göschl, M. 1998. Current status of molecular classification of the phytoplasmas. J. Plant. Pathol. 80:3-26.

Uyemoto, J. K., Cummins, J. R., and Abawi, G. S. 1977. Virus and virus-like diseases affecting grapevines in New York vineyards. American J. Enol. Vitic. 28:131-136.

Wolf, T. K., Prince, J. P., and Davis, R. E. 1994. Occurrence of a grapevine yellows in Virginia vineyards. Plant Dis. 78:208.

(Prepared by E. Boudon-Padieu)

Australian Grapevine Yellows

Symptoms of grapevine yellows were first reported in Australia in 1976. The disease was originally named the "Rhine Riesling problem" in South Australia and "Chardonnay decline" in Victoria, but was later renamed Australian grapevine yellows because of its similarity to other grapevine yellows diseases. The disease is now found in most viticultural regions of Australia. A high incidence occurs in the hotter inland grape-growing districts of Sunraysia in New South Wales and Victoria, the Riverina in New South Wales, and the Riverland in South Australia, with cvs. Chardonnay and Riesling most commonly affected. However, disease symptoms have been observed and phytoplasmas detected in other cultivars as well.

Symptoms

During spring and summer, disease symptoms on white-fruited cultivars include one or more shoots with an irregular chlorosis of the leaves (Fig. 200). Later, the chlorotic areas of the leaves often turn necrotic. In red-fruit cultivars, affected leaves can show an irregular pattern of reddening, but they can also be solid red in color (Fig. 201). In both red- and white-fruited cultivars, the leaf margins tend to curl downward and overlay one another. Leaf blade abscission occurs with the petioles initially remaining attached to the shoots, although they eventually abscise as well. Shoots with Australian grapevine yellows symptoms may produce aborted fruit clusters early in the season or berry shrivel later in the season. A dieback begins at the shoot tips and progresses downward node by node. Affected canes develop a blue, waxy appearance, lignify unevenly or not at all, and become rubbery in strength. Often only a few

Fig. 200. Irregular chlorosis and downward curling of leaves on a white-fruited cultivar affected with Australian grapevine yellows. (Courtesy F. E. Constable)

canes are affected on a grapevine, but occasionally all may be symptomatic. Significant yield reductions have been reported in some Australian grapevine yellows-affected vineyards. Symptoms mimicking those of the disease may be confused with certain abiotic causes of vascular disruption, such as strangulation by tendrils, mechanical breakage, or insect feeding. However, symptomatic leaves are always distal to the point of injury in these examples.

Causal Organism

Three distinct phytoplasmas have been associated with Australian grapevine yellows on the basis of electron microscopy studies, antibiotic treatments, and polymerase chain reaction (PCR) assays: the Australian grapevine yellows phytoplasma of the stolbur (16SrXII) group; the Australian tomato big bud phytoplasma of the faba bean phyllody (16SrII) group; and the Buckland Valley grapevine yellows phytoplasma, which is closely related to the stolbur and the aster yellows (16SrI) groups. The Australian grapevine yellows phytoplasma was further assigned the *Candidatus* (*Ca.*) species name '*Ca.* Phytoplasma australiense' because analysis of its 16S rRNA gene sequence showed it to be distinct from other members of the stolbur group, which includes the grapevine phytoplasma associated with the *Bois noir/Vergilbungskrankheit* disease. In addition, '*Ca.* Phytoplasma australasia' is the designated *Candidatus* species name for the tomato big bud phytoplasma. A *Candidatus* species name has not been proposed for the Buckland Valley grapevine yellows phytoplasma. The Australian grapevine yellows phytoplasma occurs more frequently than the tomato big bud or Buckland Valley grapevine yellows phytoplasmas; the latter is currently limited to two vineyards in the Buckland Valley of Victoria. None of the other phytoplasmas known to infect grapevines in other parts of the world have been found in Australia.

Disease Cycle and Epidemiology

The incidence of Australian grapevine yellows disease in individual vineyards can fluctuate from year to year. Chronic symptoms can persist over several years in some grapevines while the disease goes into remission in others. The reason for the apparent recovery of some symptomatic grapevines is not known. Both Australian grapevine yellows and tomato big bud phytoplasmas have been detected throughout the year in shoot, cordon, trunk, and root tissues by means of PCR assays, indicating that they can persistently infect grapevines. Phytoplasmas also have been detected in symptomless Australian grapevines, albeit less frequently than in symptomatic vines.

Alternative hosts for the Australian grapevine yellows phytoplasma have not been found consistently in or around diseased

Fig. 201. Solid red discoloration and downward curling of leaves on a red-fruited cultivar affected with Australian grapevine yellows. (Courtesy F. E. Constable)

vineyards. However, strains of '*Ca.* Phytoplasma australiense' (genetically similar to the Australian grapevine yellows phytoplasma) were detected in papaya and strawberry in northern Australia and in flax (*Phormium tenax*) in New Zealand. Insect vectors have not been identified for any strains of '*Ca.* Phytoplasma australiense' in Australia; however, in New Zealand, the vector of the Phormium yellow leaf strain of '*Ca.* Phytoplasma australiense' is the planthopper *Oliarus atkinsonii* (Myers) of the family Ciixidae.

In contrast, the tomato big bud phytoplasma has a broad plant host range and is vectored by *Orosius argentatus* (Evans). Studies have demonstrated acquisition of tomato big bud phytoplasma by *O. argentatus* when the insects are exposed to diseased grapevines, and successful retransmission to faba bean test plants. Vineyard surveys using PCR assays for detection of the tomato big bud phytoplasma have detected low numbers of potentially inoculative *O. argentatus*. No insect vector of the Buckland Valley grapevine yellows phytoplasma has been identified. The propagation of Australian grapevine yellows and tomato big bud phytoplasmas through green cuttings has been demonstrated, albeit at a frequency of less than 1%.

Management

Because the existence of aerial vectors for any Australian grapevine phytoplasma has not been proven, insecticide sprays are not recommended as a general management strategy to prevent the spread of Australian grapevine yellows disease. However, there is a risk of spreading the phytoplasmas through conventional propagation practices, so only clean sources should be used. On the basis of research done elsewhere with other grapevine yellows diseases, hot water treatment of 50°C for 30 min may be tried.

Selected References

Bonfiglioli, R. G., Guerrini, S., and Symons, R. H. 1996. Cooperative Research Centre for Viticulture: Sampling program for grapevine yellows diseases. Aust. Grapegrower and Winemaker 394:22-24.

Constable, F. E. 2002. The biology and epidemiology of Australian grapevine phytoplasmas. Ph.D. thesis. University of Adelaide, Australia.

Constable, F. E., Jones, J., Gibb, K. S., Chalmers, Y. M., and Symons, R. H. 2004. The incidence, distribution and expression of Australian grapevine yellows, restricted growth and late season leaf curl diseases in selected Australian vineyards. Ann. Appl. Biol. 144:205-218.

Constable, F. E., Whiting, J. R., Gibb, K. S., and Symon, R. H. 2002. A new grapevine yellows phytoplasma from the Buckland Valley of Victoria, Australia. Vitis 41:147-154.

Constable, F. E., Whiting, J. R., Jones, J., Gibb, K. S., and Symons, R. H. 2003. The distribution of grapevine yellows disease associated with the Buckland Valley grapevine yellows phytoplasma. J. Phytopathol. 151:65-73.

Constable, F. E., Gibb, K. S., and Symons, R. H. 2003. Seasonal distribution of phytoplasmas in Australian grapevines. Plant Pathol. 52:267-276.

Davis, R. E., Dally, E. L., Gundersen, D. E., Lee, I-M., and Habili, N. 1997. "*Candidatus* Phytoplasma australiense," a new phytoplasma taxon associated with Australian grapevine yellows. Int. J. Syst. Bacteriol. 47:262-269.

Gibb, K. S., Constable, F. E., Moran, J. R., and Padovan, A. C. 1999. Phytoplasmas in Australian grapevines—detection, differentiation and associated diseases. Vitis 38:107-114.

Liefting, L. W., Beever, R. E., Winks, C. J., Pearson, M. N., and Forster, R. L. S. 1997. Planthopper transmission of Phormium yellow leaf phytoplasma. Aust. Plant Pathol. 26:148-154.

Magarey, P. A., and Wachtel, M. F. 1986. Australian grapevine yellows. Int. J. Trop. Plant Dis. 4:1-14.

Osmelak, J. A., Emmett, R. W., and Pywell, M. 1989. Monitoring for potential leafhopper vectors (Hemiptera: Cicadelloidea and Fulgoroidea) of the causal agent of Australian grapevine yellows. Plant Prot. Quart. 4:8-10.

Schneider, B., Padovan, A., De la Rue, S., Eichner, R., Davis, R., Bernuetz, A., and Gibb, K. 1999. Detection and differentiation of phytoplasmas in Australia: An update. Aust. J. Agric. Res. 50:333-342.
White, D. T., Blackall, L. L., Scott, P. T., and Walsh, K. B. 1998. Phylogenetic positions of phytoplasmas associated with dieback, yellow crinkle and mosaic diseases of papaya, and their proposed inclusion in '*Candidatus* Phytoplasma australiense' and a new taxon, '*Candidatus* Phytoplasma australasia'. Int. J. Syst. Bacteriol. 48:941-951.

(Prepared by F. E. Constable)

Bois noir or *Vergilbungskrankheit*

Bois noir was first described in northeastern France and later in the Moselle and Rhine river valleys of Germany as '*Vergilbungskrankheit*'. The two terms are synonyms, describing the same disease. *Bois noir/Vergilbungskrankheit* is the most widespread grapevine yellows disease in Europe and southern Mediterranean countries and continues to spread in Germany. It has occurred in conjunction with *flavescence dorée* in vineyards in southern France and Italy and with other grapevine yellows diseases in Italy, Germany, Greece and Israel.

Symptoms

The overall symptoms of *Bois noir/Vergilbungskrankheit* diseases are essentially similar to those of other grapevine yellows diseases. All involve leaf discoloration, even if restricted to secondary shoots (Fig. 202), accompanied by flower abortion, cluster collapse (Fig. 203), or berry shrivel (Fig. 204), and delayed lignification of shoots and canes. However, symptoms of *Bois noir/Vergilbungskrankheit* may differ slightly among grape cultivars. It causes black pustules arranged in longitudinal rows along the stems of *V. vinifera* cvs. White Riesling and Scheurebe (Fig. 205), a symptom that is absent on cvs. Pinot Noir, Pinot Blanc, and Pinot Gris. Leaf symptoms are pronounced in Pinot Noir (Fig. 206), White Riesling (Figs. 207 and 208), Chardonnay, Scheurebe, and Kerner, but are less distinctive in cv. Müller-Thurgau.

Generally, symptoms on infected grapevines are either restricted to a few shoots on an otherwise normal cane or involve entire canes. Symptoms usually develop throughout the vine if it is infected while young. As with *flavescence dorée*, infected grapevines may be in symptom remission one year and develop symptoms the next. Whether grapevines recover completely,

Fig. 202. Leaf symptoms of *Bois noir/Vergilbungskrankheit* on cv. White Riesling, confined to random shoots developing on otherwise normal appearing canes. (Courtesy M. Maixner)

Fig. 203. Cluster symptoms of *Bois noir/Vergilbungskrankheit*, including partial abortion of set berries on cv. White Riesling. (Courtesy M. Maixner)

Fig. 204. Cluster symptoms of *Bois noir/Vergilbungskrankheit*, including berry shrivel in cv. Pinot Noir. (Courtesy M. Maixner)

Fig. 205. An unlignified shoot of cv. White Riesling affected by *Bois noir/Vergilbungskrankheit*, with rows of black pustules arranged longitudinally and a bluish layer of wax (largely removed from the middle internode). (Courtesy M. Maixner)

remain in a state of latency for one or more years, or are reinoculated via infective leafhoppers is uncertain. As a rule, *Bois noir/Vergilbungskrankheit* is not lethal.

Symptom severity is influenced by ambient weather conditions. Cool, wet weather leads to delayed symptom development whereas hot and dry conditions favor an early, rapid development of pronounced symptoms. Symptom intensity is reduced when grapevines are deficient in nitrogen or coinfected by nepoviruses.

Fig. 206. Early leaf symptoms on cv. Pinot Noir affected by *Bois noir/Vergilbungskrankheit*; dark sectors will eventually expand across the entire leaf surface. (Courtesy M. Maixner)

Fig. 207. Early leaf symptoms of *Bois noir/Vergilbungskrankheit* on cv. White Riesling. Yellow discoloration and necrosis begins along the main veins and extends across the leaf blade. (Courtesy M. Maixner)

Fig. 208. Advanced leaf symptoms of *Bois noir/Vergilbungskrankheit* on cv. White Riesling, with tissue chlorosis and necrosis extending over substantial portions of the leaf blade and leaf margins rolled downwards. (Courtesy M. Maixner)

Causal Organism

Bois noir and *Vergilbungskrankheit* are associated with the same phytoplasma of the stolbur (16SrXII-A) group, which is found in Austria, Croatia, France, Germany, Greece, Hungary, Italy, Slovenia, Spain, Switzerland, Israel, and Lebanon. Although stolbur phytoplasmas are widespread and infect a wide variety of plants, the group is quite homogenous with little genomic variation between isolates; however, a molecularly distinct isolate of *Vergilbungskrankheit* was detected in one viticulture area in Germany.

Disease Cycle and Epidemiology

The cixiid planthopper, *Hyalesthes obsoletus* Signoret, reportedly vectors *Bois noir/Vergilbungskrankheit*, although it is possible that alternative vectors might be involved in at least some locations. *H. obsoletus* is a circum-Mediterranean species whose distribution extends from North Africa and the Middle East to Central Europe and is a known vector of the stolbur phytoplasma to solanaceous plants. Among the additional herbaceous host species susceptible to this phytoplasma is the bindweed (*Convolvulus arvensis* L.), upon which *H. obsoletus* completes its entire life cycle as this insect's preferred breeding host. Eggs are deposited on the root collar; after hatching, all larval instars remain on the roots and acquire the phytoplasma from infected plants. Consequently, emerging adults may already be infective. *H. obsoletus* has a single generation in central Europe and Turkey, two in Israel.

Although *H. obsoletus* feed on a wide range of plant species, feedings on grapevines are regarded as accidental events. Nevertheless, incidental probes are sufficient to transmit the pathogen and cause infections. Overall, infected grapevines are regarded as a "dead-end" host for this phytoplasma from which no additional transmission occurs, and these vines serve no significant function in the disease epidemiology. Regardless, disease incidence can be high in some areas and years, albeit with considerable annual fluctuation. The extent of disease development is strongly influenced by abiotic and biotic factors that affect the vector species and breeding host. For example, vineyards in Germany on steep slopes are characterized by stony soils with high average soil temperatures, supporting high densities of bindweed plants and thereby contributing to large populations of potentially infectious vectors. In some regions, large vector populations attributed to above-normal summer temperatures have also been associated with *Bois noir/Vergilbungskrankheit* outbreaks.

Management

Currently, no satisfactory disease management strategy has been identified. Applications of insecticides directed at the vector *H. obsoletus* are not recommended owing to its subterranean behavior and the low affinity of the adults to grapevines. However, the use of sticky traps to monitor adults emerging from the soil may provide some estimate of potential infection pressure.

As bindweed plants are considered the principal host for the pathogen and vector species, control of this weed may provide benefit. A combination of different strategies, including but not limited to herbicide treatments and planting of competitive cover crops, should be considered in vineyards. Management of this weed should be extended to neighboring fallowed fields to whatever extent possible, especially where disease spread is chronic. In northern Europe, infected grapevines play an insignificant role in the disease epidemiology and removal of diseased vines from affected vineyards is not warranted. Thorough inspection of propagation sources and/or hot water treatment to eradicate phytoplasmas from propagating wood should be practiced to limit disease spread to new vineyards and disease-free regions.

Selected References

Batlle, A., Martínez, M. A., and Laviña, A. 2000. Occurrence, distribution and epidemiology of Grapevine Yellows in Spain. Eur. J. Plant Pathol. 106:811-816.

Carraro, L., Loi, N., Kuszala, C., Clair, D., Boudon-Padieu, E., and Refatti, E. 1994. On the ability-inability of *Scaphoideus titanus* Ball to transmit different grapevine yellows agents. Vitis 33:231-234.

Caudwell, A., Larrue, J., Kuszala, C., and Bachelier, J. C. 1971. Pluralité des jaunisses de la vigne. Ann. Phytopathol. 3:95-105.

Daire, X., Clair, D., Larrue, J., and Boudon-Padieu, E. 1997. Survey for grapevine yellows phytoplasmas in diverse European countries and Israel. Vitis 36:53-54.

Daire, X., Clair, D., Reinert, W., and Boudon-Padieu, E. 1997. Detection and differentiation of grapevine yellows phytoplasmas belonging to the elm yellow group and to the stolbur subgroup by PCR amplification of non-ribosomal DNA. Eur. J. Plant Pathol. 103:507-514.

Klein, M., Weintraub, P. G., Davidovich, M., Kuznetsova, L., Zahavi, T., Ashanova, A., Orenstein, S., and Tanne, E. 2001. Monitoring phytoplasma-bearing leafhoppers/planthoppers in vineyards in the Golan Heights, Israel. J. Appl. Entomol. 125:19-23.

Maixner, M., Ahrens, U., and Seemüller, E. 1995. Detection of the German grapevine yellows (Vergilbungskrankheit) MLO in grapevine, alternative hosts and a vector by a specific PCR procedure. Eur. J. Plant Pathol. 101:241-250.

Reinert, W., and Maixner, M. 2000. Distribution and differentiation of grapevine phytoplasmas in Germany. 13th ICVG Conference, International Council for the Study of Virus and Virus-like Diseases of the Grapevine (ICVG). http://web.pppmb.cals.cornell.edu/fuchs/icvg/data/abstrb.pdf

Sforza, R., Clair, D., Daire, X., Larrue, J., and Boudon-Padieu, E. 1998. The role of *Hyalesthes obsoletus* (Hemiptera: Cixiidae) in the occurrence of Bois noir of grapevines in France. J. Phytopathol. 146:549-556.

Weber, A., and Maixner, M. 1998. Survey of populations of the planthopper *Hyalesthes obsoletus* Sign. (Auchenorrhyncha, Cixiidae) for infection with the phytoplasma causing grapevine yellows in Germany. J. Appl. Entomol. 122:375-381.

(Prepared by M. Maixner)

Flavescence dorée

Flavescence dorée was the first grapevine yellows disease to be recognized, in 1955 in southwestern France. Its leafhopper vector, *Scaphoideus titanus* Ball (= *S. littoralis* Ball), a nearctic species that in Europe lives only on grapevine, was identified in 1961. A native of eastern North America, it was introduced to Europe after World War II. The associated *flavescence dorée* phytoplasma was visualized in 1971. Although some other grapevine yellows diseases were originally described as "*flavescence dorée*-like," molecular characterization of phytoplasmas now permits their differentiation.

In Europe, the vector, *S. titanus*, occurs in the northern parts of Portugal, Spain, and Italy; in the southern half of France; and in a portion of Switzerland and western Slovenia. Distribution of the disease in France is confined to all southern and southwestern viticulture regions and a few localities in the east. Although large populations of *S. titanus* are found in the viticultural regions to the north (i.e., the Rhone Valley, Beaujolais, Bourgogne, Franche-Comté, and Pays de Loire), these currently are not affected by *flavescence dorée*. In Northern Italy, *flavescence dorée* extends to the provinces of Piemonte, Lombardia, Trentino-Alto Adige, Veneto, Friuli-Venezia Giulia, Liguria, and some sites in Emilia-Romagna. In Spain, *flavescence dorée* has been identified in a limited number of sites in northeastern Catalonia.

Flavescence dorée is a quarantine disease in the European Community and is considered to be among the most dangerous grapevine diseases in Europe. All *V. vinifera* scions are susceptible and losses from the disease have caused vineyards to be abandoned in some regions. Even in regions where *flavescence dorée* is absent, the presence of *S. titanus* should be regarded as posing a threat of a disease outbreak.

Symptoms

All *Vitis vinifera* cultivars are susceptible to *flavescence dorée*, albeit to different degrees. Highly susceptible cultivars may die after 2–3 years or may display a crisis-recovery-relapse cycle, whereas more resistant cultivars can recover completely the year after symptoms appear. Symptom expression is probably dependent also on the response of the rootstock partner. Bud burst in the spring is typically delayed or even absent on infected vines and entire inflorescences may dry up before or during bloom. The most characteristic symptoms occur in the summer and include discoloration and downward rolling of the leaves, shriveling of berries on developing clusters, and poor lignification of shoots and canes that results in flexuous wood and a "weeping" posture of conspicuously symptomatic vines in autumn. On black-fruited cultivars, all parts of the leaf (both lamina and veins) most exposed to the sun turn red (Fig. 209), whereas affected leaves become golden yellow on white-fruited cultivars (Fig. 210). Because the vector leafhopper *S. titanus* completes its life cycle on grapevine, symptomatic vines commonly occur in clusters that progressively expand in area over time.

Leaf symptoms of *flavescence dorée* may be confused with those caused by certain virus diseases (leafroll disease, corky

Fig. 209. Red, downward-curled leaves and "weeping" growth habit symptomatic of *flavescence dorée*, on red-fruited cv. Grenache Noir in September. (Courtesy E. Boudon-Padieu)

Fig. 210. Golden-yellow, downward-curled leaves symptomatic of *flavescence dorée*, on white-fruited cv. Sémillon in September. (Courtesy E. Boudon-Padieu)

bark, and the yellow mosaic syndrome of fanleaf degeneration); magnesium deficiency; or mechanical injury. However, with leafroll disease, veins remain green and lignification of the stems is not affected; with leafroll or corky bark, berries do not shrivel; and with the yellow mosaic strain of *Grapevine fanleaf virus* or magnesium deficiency, leaves do not roll downward. Also, *flavescence dorée*-like symptoms associated with mechanical injury are limited to the part of the cane or shoot beyond the point of injury. Infected mother plants used in rootstock production either develop weak symptoms (delayed bud burst and/or irregular lignification) or remain asymptomatic. However, rootstocks derived from such asymptomatic vines can serve as a source of inoculum to infect and cause typical disease symptoms in otherwise-healthy scion varieties that are budded or grafted onto them.

Causal Organism

The *flavescence dorée* phytoplasma belongs to the elm yellows or 16SrV group, which also includes phytoplasmas of forest trees and other woody plants in North America, Europe, and China. The genome size is about 700 Kb. The *flavescence dorée* phytoplasma is specifically acquired from and transmitted to grapevines by the vector leafhopper *S. titanus*. It has been transmitted experimentally to the herbaceous broadbean plant (*Vicia faba* L.) by means of *S. titanus* collected in *flavescence dorée*-affected vineyards, and from infected broadbean to grape seedlings by means of *S. titanus* reared in an insectary. The *flavescence dorée* phytoplasma also has been transmitted from broadbean to broadbean with another leafhopper, *Euscelidius variegatus* Kirschbaum.

Although belonging to the same group of phytoplasmas as those associated with Palatinate grapevine yellows, the *flavescence dorée* phytoplasma and these latter organisms are clearly distinguished in terms of their vectors, molecular assays, and specific antibodies formed against them. A single isolate of the *flavescence dorée* phytoplasma appears to be widespread and commonplace in all affected provinces in France, Italy, and Spain.

Disease Cycle and Epidemiology

Cultivated and wild grapevines are the sole source of *flavescence dorée* inoculum. In Europe, the vector species *S. titanus* lives exclusively on *Vitis* spp. Eggs are laid in summer, overwinter beneath bark and between bud scales, and begin hatching around the second week of May in southern France. The length of the hatching period depends on the degree of winter chilling, which is necessary for terminating egg diapause, typically ranging from 5 weeks in southern France to 12 weeks in Corsica. After five larval instar stages at moulting intervals of approximately 10 days, the adults emerge in mid-July, mate, lay eggs, and perish in September, providing just one generation per year.

The *flavescence dorée* phytoplasma overwinters within infected canes in an incubation phase. It is acquired through feeding of *S. titanus* nymphs in the spring or by adults once they have developed in the summer, a process that usually takes 7–8 (but as few as 4) days, after which the pathogen multiplies and circulates inside the insect, which becomes infective after approximately 4 additional weeks of incubation. The *S. titanus* adults are very mobile and, since they transmit the *flavescence dorée* phytoplasma from vine to vine and feed only on grapes, can spread the disease rapidly within a viticultural district. Symptoms generally appear the year after infection.

Flavescence dorée can also be transmitted through the use of dormant wood for grafting, although only from canes that were inoculated the previous growing season and, thus, still in a state of incubation. Because such canes are asymptomatic, they pose a danger of inadvertent spread of the disease if used for propagation.

Management

As long-distance spread of *flavescence dorée* is associated with infected planting materials, care must be taken in selecting sources for vine propagation. Only clean sources should be used and propagations should be avoided in affected viticultural regions. Alternatively, a hot water treatment (50°C for 45 min) may be employed to destroy the *flavescence dorée* phytoplasma in dormant canes.

Consideration must also be given to the long distance transport of the vector species on infested canes and plants, where up to 1,000 viable eggs per kilogram of cane have been reported. Once the disease is established in a region, insecticides may be applied during the egg hatching period to control the single generation of the vector. Because the insect is not infectious during its incubation period, insecticide treatments do not need to begin until 3 weeks after the first hatch and can discontinue once the hatch is completed.

Selected References

Angelini, E., Bianchi, G. L., Filippin, L., Morassutti, C., and Borgo, M. 2007. A new TaqMan method for the identification of phytoplasmas associated with grapevine yellows by real-time PCR assay. J. Microbiol. Meth. 68:613-622.

Angelini, E., Clair, D., Borgo, M., Bertaccini, A., and Boudon-Padieu, E. 2001. *Flavescence doree* in France and Italy - Occurrence of closely related phytoplasma isolates and their near relationships to Palatinate grapevine yellows and an alder yellows phytoplasma. Vitis 40:79-86.

Batlle, A., Laviña, A., Kuszala, C., Clair, D., Larrue, J., and Boudon-Padieu, E. 1997. Detection of *Flavescence dorée* phytoplasma in grapevine in northern Spain. Vitis 36:211-212.

Borgo, M., and Angelini, E. 2002. Diffuzione della Flavescenza dorata della vite in Italia e relazioni con vitigni, pratiche agronomiche et materiali di propagazione. ATTI Giornate Fitopatologiche 2002. 1:35-50.

Boudon-Padieu, E., Larrue, J., and Caudwell, A. 1989. ELISA and dot-blot detection of *Flavescence dorée* MLO in individual leafhopper vectors during latency and inoculative state. Curr. Microbiol. 19:357-364.

Boudon-Padieu, E. 2002. *Flavescence dorée* of the grapevine: Knowledge and new developments in epidemiology, etiology and diagnosis. ATTI Giornate Fitopatologiche 2002 1:15-34.

Caudwell, A., Larrue, J., Boudon-Padieu, E., and McLean, G. D. 1997. *Flavescence dorée* elimination from dormant wood of grapevines by hot-water treatment. Aust. J. Grape Wine Res. 3:21-25.

Daire, X., Boudon-Padieu, E., Bervillé, A., Schneider, B., and Caudwell, A. 1992. Cloned DNA probes for detection of grapevine *Flavescence dorée* mycoplasma-like organism (MLO). Ann. Appl. Biol. 121:95-103.

Lefol, C., Lherminier, J., Boudon-Padieu, E., Larrue, J., Louis, C., and Caudwell, A. 1994. Propagation of the *Flavescence dorée* MLO (Mycoplama-like organism) in the leafhopper vector *Euscelidius variegatus* Kbm. J. Invertebr. Pathol. 63:285-293.

Meignoz, R., Boudon-Padieu, E., Larrue, J., and Caudwell, A. 1992. *Flavescence dorée* de la vigne. Présence de MLO et effets cytopathogènes associés, dans le liber de la vigne. J. Phytopathol. 134:1-9.

Quartau, J. A., Guimarães, J. M., and André, G. 2001. On the occurrence in Portugal of the nearctic *Scaphoideus titanus* Ball (Homoptera, Cicadellidae), the natural vector of the grapevine "*Flavescence dorée*" (FD). IOBC/wprs Bulletin 24:273-276.

Schvester, D., Carle, P., and Moutous, G. 1961. Sur la transmission de la Flavescence dorée des vignes par une cicadelle. C. R. Acad. Agric. Fr. 47:1021-1024.

Steffek, R., Reisenzein, H., and Zeisner, N. 2007. Analysis of the pest risk from grapevine *Flavescence dorée* phytoplasmas to Austrian viticulture. EPPO Bull. 37:191-203.

Vidano, C. 1964. Scoperta in Italia dello *Scaphoideus littoralis* Ball Cicalina americana collegata all "*Flavescence dorée*" della Vite. Ital. Agric. 101:1031-1049.

(Prepared by E. Boudon-Padieu)

North American Grapevine Yellows

The presence of a North American grapevine yellows was first suggested in 1974 on the basis of similarities between *flavescence dorée* and expression of a previously undescribed disease of the *Vitis* interspecific hybrid cv. DeChaunac in the Finger Lakes region of New York State. Later termed leaf curl and berry shrivel, the DeChaunac disease affected a relatively low percentage of grapevines in several vineyards. However, the causal pathogen of leaf curl and berry shrivel was not determined. *Flavescence dorée*-like symptoms were also observed on *V. vinifera* cv. Riesling in the same region, where the indigenous leafhopper, *Scaphoideus titanus* Ball, was proposed as a possible vector. North American grapevine yellows was first reported in 1993 in Virginia, where *flavescence dorée*-like symptoms had been observed since 1987, initially on the highly susceptible *V. vinifera* cultivars Chardonnay and Riesling and, in time, on other vinifera cultivars and rarely on interspecific hybrids. Symptomatic vines have more recently been observed and found positive for the causal pathogen in Maryland, southeast Pennsylvania, the Finger Lakes region of New York, and the midwestern states of Ohio and Missouri. In Virginia, the highest incidence of North American grapevine yellows has been observed in the western piedmont and Blue Ridge Mountain region of the state.

Symptoms

The symptoms originally observed with North American grapevine yellows in DeChaunac grapevines included yellowing, hardening, and downward rolling of leaves, and a late-summer shriveling of berries that easily dislodged from the cluster rachis. Shoots remained green or exhibited delayed periderm development in late summer, when healthy shoots matured into woody canes. Diseased grapevines in the vineyards had expressed chronic infections. However, when individual grapevines were transplanted and established in large containers, or when diseased canes were rooted and grown in a greenhouse, all recovered and developed into apparently healthy plants that failed to reexpress leaf curl and berry shrivel symptoms within the two years they were observed.

Symptoms of North American grapevine yellows on *V. vinifera* cultivars are essentially similar to those of other grapevine yellows diseases, and bear many similarities to the symptoms described for leaf curl and berry shrivel on DeChaunac. Symptoms initially involve one to several shoots of an otherwise apparently healthy grapevine, although all shoots of an affected cordon or grapevine may simultaneously express symptoms, particularly in highly susceptible varieties such as Chardonnay. In cases where disease symptoms appear early in the season, they often advance to affect other shoots on the grapevine as the season progresses. In the first season of expression, overt symptoms can appear anytime after bloom, and are often first apparent as a destruction of fruit clusters. Cluster abortion is noticed initially as a loss of berry turgor, followed by rapid desiccation of the entire cluster, including the rachis. Shriveled berries adhere to the rachis, but the entire cluster may eventually abscise from the shoot. Fruit clusters and tendrils on affected shoots are subject to abortion any time from fruit set (Fig. 211) through berry maturity. Additional symptoms on Chardonnay can include significant shoot dieback and the acquisition of a blue-gray cast to the shoot stems, rather than the shades of green with red highlights that stems normally express in mid-summer. Shoot stems fail to develop brown periderm, or develop periderm in irregular islands or patches (Fig. 212). Dark pustules less than 2 mm in diameter may appear on the base of the stems, particularly with cv. Riesling. North American grapevine yellows-affected shoots are pliable and often droop, a condition thought to be related to the failure of affected shoots to form lignified secondary phloem fibers. Internodes are often shortened, leading to an overlapped pattern of leaves on a shoot.

Leaf symptoms are variable, but commonly include abaxial leaf rolling; the acquisition of a crisp, hardened, or brittle texture; and discoloration. Discoloration is variable and a function of cultivar, leaf age, and stage of symptom development. Foliar symptoms on Chardonnay initiate on the older, more basal leaves of the shoot, and the discoloration is first apparent as an expanding chlorosis, starting at the leaf margins. As the season progresses, symptoms extend to younger primary leaves, as well as to the leaves of lateral shoots. The chlorosis may appear as a diffuse yellowing of the normal dark-green interveinal region of the leaf (Fig. 213), or it may involve the development of discrete chlorotic zones either confined to interveinal regions or including veinal tissue. Chlorotic tissue can become necrotic before leaves abscise. The appearance of

Fig. 211. Aborted clusters on Chardonnay grapevine affected with North American grapevine yellows (NAGY), approximately 2 weeks post bloom. (Courtesy T. K. Wolf)

Fig. 212. Lack of late-season periderm development and aborted fruit cluster on the shoot of a NAGY-affected cv. Chardonnay vine (right) compared with healthy shoot (left) that has well-developed periderm and a fully developed fruit cluster. (Courtesy T. K. Wolf)

Fig. 213. Chlorotic cv. Chardonnay leaf symptomatic of NAGY (left) compared with healthy leaf (right). (Courtesy T. K. Wolf)

discrete zones of chlorotic tissue is more commonly observed on cv. Riesling, where leaves may also develop a coppery or metallic sheen. Symptoms in red-fruited cultivars are similar, except the leaf discoloration causes leaves to turn red rather than yellow. Leaf abscission occurs as the symptoms intensify and the season progresses.

Causal Organism

Early studies of North American grapevine yellows-like diseases in New York State implicated phytoplasmas as the causal agent, first on the basis of the similarity of the disease's symptoms to those described for *flavescence dorée* in Europe, and later owing to enzyme-linked immunosorbent assay (ELISA) tests of *S. titanus* leafhoppers collected from symptomatic vines and wild *V. riparia* grapevines growing in the vicinity, in which up to 20% reacted positively to antibodies produced from *flavescence dorée*-infected plants.

Additional evidence of phytoplasma involvement was obtained in the 1990s when phytoplasmas genetically similar to those that cause X-disease of stone-fruit trees (including peach) were detected in North American grapevine yellows-symptomatic Chardonnay grapevines in Virginia by means of a polymerase chain reaction (PCR) test. Shortly thereafter, a second phytoplasma was also detected and described from affected Virginia vineyards and other hosts in the vicinity of the affected vineyards. The second phytoplasma was closely associated with the rRNA group 16SrI phytoplasmas, or aster yellows group ('*Candidatus* Phytoplasma asteris'-related strains) and was related to subgroup A (tomato big bud). Both this 16SrI-group phytoplasma and the previously detected 16SrIII phytoplasma (a '*Candidatus* Phytoplasma pruni'-related strain) occurred alone or in mixed infections in Chardonnay grapevines in Virginia. Both phytoplasmas also were detected frequently in wild *V. vulpina* (syn. *V. cordifolia*) grapevines outside the affected vineyards, while the 16SrI phytoplasma also was detected in black cherry (*Prunus serotina* Ehrh.) trees. Neither *V. vulpina* nor *P. serotina* expressed disease symptoms. The 16SrI North American grapevine yellows phytoplasma is known to have a wide host range, including wild and cultivated *Vitis* species, whereas the 16SrIII phytoplasma has been detected only in *V. vinifera*. To date, neither of the specific North American grapevine yellows phytoplasmas has been detected outside North America.

Disease Cycle and Epidemiology

The North American grapevine yellows diseases observed in eastern North America share common symptoms and may share one or more causal agents. Regardless of cultivar, the appearance of initial symptoms in the first year is followed by a more severe expression the next growing season, where shoot growth is stunted and systemic spread is evident throughout the canopy. Symptomatic grapevines often fail to survive into a third season, and rarely survive into a fourth. Symptom expression sometimes varies among locations, which may be the result of environmental factors that affect the titer or spatial distribution of phytoplasmas within the host vine. There is no evidence to date that the two phytoplasmas present in Virginia vineyards elicit different symptoms.

The epidemiological aspects of North American grapevine yellows diseases are incompletely understood, but some components are known and some can be inferred from knowledge of other grapevine yellows diseases. Wild grapevines have tested positive for the presence of phytoplasmas in New York State and Virginia, and, hence, may serve as phytoplasma reservoirs. The highest incidence of North American grapevine yellows diseases in Virginia has been observed in vineyards closely bounded by deciduous woods, especially those harboring *Prunus* and *Vitis* species. Specific vectors of North American grapevine yellows phytoplasmas have not been conclusively identified; however, there is mounting evidence of one or more insect vectors.

Scaphoideus titanus, vector of the *flavescence dorée* phytoplasma in Europe, is abundant in the vicinity of North American grapevine yellows-affected vineyards in the Finger Lakes region of New York State. This leafhopper acquires phytoplasmas from cultivated and wild grapevines in New York State and transmits them to *Vicia faba* plants, which develope grapevine yellows-like symptoms. *S. titanus* and a closely related species, *S. minor*, are also found in and around vineyards in Virginia. In addition, several other leafhopper species known to vector other phytoplasmas to different hosts occur abundantly in Virginia vineyards, and currently these are being investigated as potential vectors of North American grapevine yellows phytoplasmas. Preliminary laboratory transmission studies in Virginia have implicated several leafhopper species as potential vectors.

As with other grapevine yellows diseases, Chardonnay, Malbec, and Pinot Noir are particularly sensitive to North American grapevine yellows symptom expression. Multiyear survey data in Virginia found an incidence of recently-infected Chardonnay grapevines as high as 6% and cumulatively exceeding 25% over a 6-year period. In Virginia vineyards that exhibit up to 3% annual loss of Chardonnay grapevines to North American grapevine yellows, infections and plant losses have also occurred in *V. vinifera* cvs. Cabernet Franc, Cabernet Sauvignon, Sauvignon Blanc, Petit Manseng, and Viognier, but at a far lower (< 0.5%) annual incidence.

Management

The most effective means of North American grapevine yellows management is to avoid planting highly susceptible cultivars such as Chardonnay, Malbec, and Riesling in or near the vicinity of diseased vineyards. Efforts to prune out affected cordons or entire trunks in the first season of symptom expression has failed thus far to limit the systemic movement of the causal organism in grapevines. Prophylactic insecticide treatments might be warranted if specific insect vectors and their seasonal biology are resolved.

Selected References

Beanland, L., Noble, R., and Wolf, T. K. 2006. Spatial and temporal distribution of North American grapevine yellows disease and of potential vectors of the causal phytoplasmas in Virginia. Environ. Entomol. 35:332-344.

Davis, R., Dally, E. L., Zhao, Y., Lee, I., Wei, W., Wolf, T. K., Beanland, L., LeDoux, D. G., Johnson, D. A., Fiola, J. A., Walter-Peterson, H., Dami, I., and Chien, M. L. 2015. Unraveling the etiology of North American Grapevine Yellows (NAGY): Novel NAGY phytoplasma sequevars related to '*Candidatus* Phytoplasma pruni'. Plant Dis. doi: http://dx.doi.org/10.1094/PDIS-11-14-1185-RE

Davis, R. E., Jomantiene, R., Dally, E. L., and Wolf, T. K. 1998. Phytoplasmas associated with grapevine yellows in Virginia belong to group 16SrI, subgroup A (tomato big bud phytoplasma subgroup), and group 16SrIII, new subgroup I. Vitis 37:131-137.

Maixner, M., Pearson, R. C., Boudon-Padieu, E., and Caudwell, A. 1993. *Scaphoideus titanus*, a possible vector of grapevine yellows in New York. Plant Dis. 77:408-413.

Pearson, R. C., Pool, R. M., Gonsalves, D., and Goffinet, M. C. 1985. Occurrence of *flavescence dorée*-like symptoms on 'White Riesling' grapevines in New York, USA. Phytopathol. Mediterr. 24:82-87.

Prince, J. P., Davis, R. E., Wolf, T. K., Lee, I. M., and Dally, E. L. 1994. Genomic diversity and possible wild plant sources of mycoplasmalike organisms (MLOs) infecting grapevines: Implications for epidemiology. IOM Letters 3:288 (abstract).

Prince, J. P., Davis, R. E., Wolf, T. K., Lee, I.-M., Mogen, B. D., Dally, E. L., Bertaccini, A., Credi, R, and Barba, M. 1993. Molecular detection of diverse mycoplasmalike organisms (MLOs) associated with grapevine yellows and their classification with aster yellows, X-disease, and elm yellows MLOs. Phytopathology 83:1130-1137.

Stoepler, T. M., and Wolf, T. K. 2013. North American grapevine yellows disease: Current knowledge and management recommendations for wine growers. Virginia Coop. Ext. Publ. AREC-48P. http://pubs.ext.vt.edu/AREC/AREC-48/AREC-48.html

Uyemoto, J. K. 1975. A new disease affecting the grapevine variety de Chaunac. Proc. Amer. Phytopathol. Soc. 1:146 (abstract).

Uyemoto, J. K., Cummins, J. R., and Abawi, G. S. 1977. Virus and virus-like diseases affecting grapevines in New York vineyards. Am. J. Enol. Vitic. 28:131-136.

Wolf, T. K., Prince, J. P. and Davis, R. E. 1994. Occurrence of grapevine yellows in Virginia vineyards. Plant Dis. 78:208.

(Prepared by T. K. Wolf)

Palatinate Grapevine Yellows

Currently, Palatinate grapevine yellows occurs only in four viticultural areas of Germany, where it was first detected in the Palatinate (Pfalz) region. This disease is closely related to other European grapevine yellows of the elm yellows group such as *flavescence dorée*, but its epidemiology differs significantly from that of the latter.

Symptoms

Symptoms of Palatinate grapevine yellows resemble those typical of all grapevine yellows diseases, such as leaf discoloration and rolling, lack of shoot lignification, and cluster abortion or shriveling of berries. Therefore, when Palatinate grapevine yellows and *Bois noir*/*Vergilbungskrankheit* occur in the same vineyard, their symptoms are indistinguishable. However, Palatinate grapevine yellows is almost completely restricted in the field to cv. Scheurebe, although other cultivars have produced typical symptoms when inoculated experimentally.

Most grapevines infected by Palatinate grapevine yellows develop systemic symptoms with severe downward rolling of uniformly discolored leaves (Fig. 214). Symptoms develop late in the season, which might explain the nearly complete lignification of primary shoots (Fig. 215) and attendant difficulty in identifying infected wood during dormancy. Although the vigor of infected grapevines is reduced considerably, they usually survive dormancy.

Causal Organism

An elm yellows (16SrV) group phytoplasma has been associated with Palatinate grapevine yellows, representing four different types designated as PGY-A, -B, -C, and -D. Black alder (*Alnus glutinosa* (L.) Gaertn.) trees appear to be the natural host of the PGY phytoplasma, as they are commonly infected, albeit in latent fashion. PGY phytoplasmas may be distinguished by molecular traits from other elm yellows group phytoplasmas infecting grapevine and other plants in Europe. These molecular differences also are important to differentiate Palatinate grapevine yellows from *flavescence dorée*, which have yet to be found together in the same region. Furthermore, Palatinate grapevine yellows is vectored by the alder leafhopper *Oncopsis alni* Schrank, and the vector of *flavescence dorée*, *S. titanus*, does not occur in the Palatinate region.

Disease Cycle and Epidemiology

Palatinate grapevine yellows is typically restricted to old vineyards of cv. Scheurebe planted on the valley floor alongside creeks and rivers populated with black elder trees. The incidence of infected vines usually remains below 1%. In contrast, *Vergilbungskrankheit* naturally infects several cultivars and grapevines of all ages and is more commonly found in vineyards situated on steep slopes. Occasionally, both disease agents may occur in the same vineyard.

Fig. 214. Grapevine cv. Scheurebe affected by Palatinate Grapevine Yellows (PGY), exhibiting a stunted, non-lignified secondary shoot with downward rolled yellow leaves on a primary shoot in an advanced stage of lignification. (Courtesy M. Maixner)

Fig. 215. A grapevine of cv. Scheurebe with systemic symptoms of PGY, adjacent to healthy vines. (Courtesy M. Maixner)

The phytoplasma associated with alder yellows in Germany resembles the one associated with Palatinate grapevine yellows from grapevine. It is transmitted by the leafhopper *Oncopsis alni* Schrank, a strictly monophagous species that preferentially feeds on different species of *Alnus*. Exposing healthy grapevines to *O. alni* collected from symptomatic alder trees produced characteristic Palatinate grapevine yellows symptoms, and field surveys have confirmed the occasional presence of *O. alni* in vineyards adjacent to alder trees despite the strict host preference of the leafhopper. Nevertheless, transmission efficiency of the phytoplasma to grapevines is very low because the vector ceases to feed on them after just a few probes, perhaps explaining in part the low incidence of Palatinate grapevine yellows in vineyards.

Management

No specific control measures are required because the incidence and economic impact of this disease have been insignificant. Infected grapevines are typically retained in vineyards because further spread between them does not occur.

Selected References

Angelini, E., Clair, D., Borgo, M., Bertaccini, A., and Boudon-Padieu, E. 2001. *Flavescence dorée* in France and Italy - Occurrence of closely related phytoplasma isolates and their near relationships to Palatinate grapevine yellows and an alder yellows phytoplasma. Vitis 40:79-86.

Daire, X., Clair, D., Reinert, W., and Boudon-Padieu, E. 1997. Detection and differentiation of grapevine yellows phytoplasmas belonging to the elm yellow group and to the stolbur subgroup by PCR amplification of non-ribosomal DNA. Eur. J. Plant Pathol. 103:507-514.

Maixner, M., and Reinert, W. 1999. *Oncopsis alni* (Schrank) (Auchenorrhyncha: Cicadellidae) as a vector of the alder yellows phytoplasma of *Alnus glutinosa* (L.) Gaertn. Eur. J. Plant Pathol. 105:87-94.

Maixner, M., Reinert, W., and Darimont, H. 2000. Transmission of grapevine yellows by *Oncopsis alni* (Schrank) (Auchenorrhyncha: Macropsinae). Vitis 39:83-84.

Maixner, M., Rüdel, M., Daire, X., and Boudon-Padieu, E. 1995. Diversity of grapevine yellows in Germany. Vitis 34:235-236.

Reinert, W., and Maixner, M. 2000. Distribution and differentiation of grapevine phytoplasmas in Germany. 13th ICVG Conference, International Council for the Study of Virus and Virus-like Diseases of the Grapevine (ICVG). http://web.pppmb.cals.cornell.edu/fuchs/icvg/data/abstrb.pdf

(Prepared by M. Maxiner)

Diseases Caused by Viruses and Viruslike Agents

Grapevine Virus Diseases

Until the mid-1800s, the traditional viticultural regions throughout Europe enjoyed relative freedom from widespread problems with virus diseases. This envious scenario from viticulture's premodern era was likely due to limited trade of grapevine cultivars among countries and to plantings predominately of own-rooted scions; both served to minimize wholesale spread of viruses and vectors. Later, however, a sundry array of virus and viruslike diseases began to plague the well-established viticultural regions. This change was brought about by the introduction of the phylloxera root aphid, *Daktulosphaira vitifoliae* (Fitch) (Homoptera: Phylloxeridae), from North America to Europe in the 1860s, and the destruction of essentially all vineyards in France in the ensuing 20 years. Thus began an intensive effort to combat this exotic pest. Breeders made interspecific crosses with the aim of producing resistant fruiting scions and rootstocks. The subsequent practice of establishing (or re-establishing) vineyards in Europe and elsewhere with grafted vines utilizing phylloxera-resistant rootstocks, many of them symptomlessly or latently infected with different viruses, proved to be a significant factor in the eventual global dissemination of virus diseases.

In 1953, the International Council for the Study of Viruses and Virus Diseases of Grapevine (ICVG) was organized to promote communication and collaboration among scientists studying grapevine virus diseases. The first formal meeting was held in Switzerland in 1964 and the ICVG met regularly thereafter. Over the years, council members helped to establish standards and procedures for implementing virus disease control measures and assist in promulgating rules for certification programs. Adoption of a clean stock program was critical because all viral agents may be increased and disseminated long distances in propagating materials and planting stocks.

In the 1980s and 1990s, a new biotype of phylloxera arose in California and destroyed numerous vineyards planted to the previously resistant rootstock AXR#1 (*V. vinifera* × *V. rupestris*), leading to the replanting of vines on alternative, phylloxera-resistant rootstocks typically derived from virus-indexed mother blocks. However, unlike AXR#1, which tolerated latent viruses, bud-grafts of diseased scion sources onto several of these rootstocks declined rapidly, sometimes even when elite foundation material was used. Subsequent investigations led to the discovery and characterization of *Grapevine leafroll-associated virus 2* (GLRaV-2) isolates that are involved in graft-incompatibility disorders and decline of young grapevines grafted onto susceptible rootstocks. For more information, see "*Grapevine leafroll-associated virus 2* and Graft Incompatibility." Furthermore, based on the differential responses in two rootstock trials, other sources were determined to contain additional, yet unidentified, graft-transmissible agents (GTAs) associated with decline/death of bud-inoculated test plants grown on specific rootstocks.

Currently, more than 70 viruses, viroids, and graft-transmissible agents are known to infect grapevines. Some have restricted distribution, such as *Peach rosette mosaic virus*, which is restricted to Michigan and southwestern Ontario in the Great Lakes region of North America, whereas others such as grapevine leafroll associated viruses occur worldwide.

(Prepared by J. K Uyemoto and G. P. Martelli)

Enation Disease

First identified in California in the 1950s as a disorder possibly caused by a viruslike agent, enation disease was known to occur in Germany as early as the turn of the nineteenth century. It was later reported from most European countries, Venezuela, South Africa, Tunisia, New Zealand, and Australia.

Symptoms

Affected vines of sensitive cultivars (e.g. Italia, Panse precoce, Primus) show delayed bud burst, slow growth of the shoots, and produce a bushy appearance. As the season progresses, the shoots resume a more normal growth pattern. Enations, the foliar alterations for which the disease is named, are cuplike in outline and develop along main veins as linear outgrowths of variable length (up to 4–5 cm) and height (2–3 mm). Enations occur primarily on the undersides of basal leaves (Fig. 216). However, in certain *V. vinifera* cultivars and American rootstocks, enations have developed on the upper leaf surfaces (Fig. 217). Symptomatic leaves are misshapen, fanlike in outline with deep lacination, prominent veins, and a thickened lamina. Severely affected leaves drop prematurely, whereas leaves developed late in the season are asymptomatic and remain attached. The basal internodes of affected shoots are short, irregular and misshapen with longitudinal bark cracks. Expression of leaf enations can vary greatly from year to year, and chronically infected grapevines may be in remission 2 or 3 years following the appearance of severe symptoms. Fruit from infected vines is poor in quality and yields may be reduced by up to 50%.

Causal Agent

The causal agent is unknown but successful graft transmission suggests that it may be a virus. Transmission by grafting is erratic, presumably owing to uneven distribution of the causal agent within infected vines. The hybrid LN33 is the best indicator currently available. Symptoms appear between 2 and 6 years after grafting, with the greatest frequency in the third and fourth years. Transmission has not been obtained using budwood from infected asymptomatic plants.

Fig. 216. Severe deformation and enations (outgrowths) on undersides of basal leaves, symptomatic of grapevine enation disease. (Courtesy U. Prota)

Fig. 217. Symptoms of enation disease appearing on the upper side of affected leaves. (Courtesy U. Prota)

Disease Cycle and Epidemiology

Enation disease is not known to spread naturally in the field. Its dissemination apparently depends entirely on the use of infected propagative material.

Management

Disease-free propagative material obtained through selection and indexing should be used. There is no information on whether the disease can be eliminated by heat treatment or meristem tip culture.

Selected References

Credi, R. 1995. Epidemiological observations on grapevine enation disease in Emilia-Romagna (Italy). Phytopathol. Mediterr. 34:88-92.

Hewitt, W. B. 1954. Some virus and virus-like diseases of grapevines. Bull. Calif. Dep. Agric. 43:47-64.

Prota, U., Garau, R., and Cugusi, M. 1982. Quantitative aspects of yield of grapevine affected by enation disease in Sardinia. Pages 41-48 in: Proc. 7th Meeting ICVG, Niagara Falls 1980, Canada Agriculture.

(Prepared by U. Prota and R. Garau)

The Fleck Complex

The fleck complex is comprised of the diseases fleck proper, asteroid mosaic, Rupestris necrosis, Rupestris vein feathering, Grapevine redglobe virus (GRGV), and viruses that cause latent or semilatent infections in *Vitis vinifera* and most American *Vitis* species and rootstock interspecific hybrids. Fleck and asteroid mosaic diseases were first described in California, but whereas fleck occurs worldwide, asteroid mosaic is known only in California. Rupestris necrosis has been reported from Japan and Rupestris vein feathering from Greece and Italy. GRGV has been reported from Italy and California.

Symptoms

All diseases in the fleck complex except GRGV are detected in the indicator host, *Vitis rupestris* 'St. George'. Symptoms of fleck disease include localized translucent spots along the tertiary leaf veins (Fig. 218). Leaves with intense flecking are wrinkled, twisted, and may curl upward. Severe forms of the disease also cause stunting, reduced rooting ability, and poor graft take. The disease is latent in *V. vinifera* cultivars and in most American rootstocks.

Fig. 218. Translucent spots along tertiary veins of a *V. rupestris* leaf with fleck disease. (Courtesy G. P. Martelli)

Symptoms of asteroid mosaic in *V. vinifera* consist of star-shaped chlorotic spots with occasional necrotic centers, irregularly distributed on leaf blades. Affected grapevines may be stunted and produce little or no fruit. Symptoms in *V. rupestris* appear as creamy-yellow bands along the main leaf veins, and affected leaves are twisted and asymmetric.

Rupestris necrosis is characterized by localized necrosis of leaf petioles and veinlets (Fig. 219). In contrast, symptoms of Rupestris vein feathering consist of transient, mildly chlorotic discolorations of the primary and secondary leaf veins (Fig. 220).

Causal Agents

The agent of fleck is *Grapevine fleck virus* (GFkV), the type species of genus *Maculavirus* in the family *Tymoviridae*. GFkV has isometric particles approximately 30 nm in diameter with rounded contour and a prominent surface structure, provided by clusters of coat protein subunits (molecular mass of approximately 25 kDa) arranged as pentamers and hexamers. Virions contain a single molecule of capped, positive-sense, single-stranded RNA of 7564 nt, four open reading frames, and high cytosine content (about 50%). In density gradient centrifugation, virus preparations sediment as two components: T, empty protein shells; and B, virions containing 35% RNA.

GRGV, Grapevine asteroid mosaic-associated virus (GAMaV), and Grapevine rupestris vein feathering virus (GRVFV) have morphological, chemico-physical, and molecular properties similar to GFkV, and all are phylogenetically related. However, whereas GRGV is in the genus *Maculavirus*, GAMaV and GRVFV are in the genus *Marafivirus*. Both genera are in the family *Tymoviridae*.

GFkV-infected grapevine cells contain multivesiculated bodies, i.e. cytopathic structures derived from deranged mitochondria that undergo peripheral vesiculation following invagination of both lamellae of the organelle's limiting membrane. Identical structures and particles of an unidentified isometric virus were observed in the phloem of *V. rupestris* displaying leaf vein necrosis. Also, vesiculated chloroplasts and mitochondria are observed in the phloem elements of grapevines infected by GRVFV and GAMaV, respectively. Vesicles in mitochondria and chloroplasts are likely sites of viral RNA replication.

Polyclonal antisera and monoclonal antibodies raised to GFkV are widely used for diagnosis, and virus-specific primers have been designed for the molecular detection of GFkV, GAMaV, GRGV. and GRVFV.

Fig. 219. Localized necrosis of the veins in *V. rupestris*, symptomatic of Rupestris necrosis. (Courtesy S. T. Ohki)

Fig. 220. *Vitis rupestris* leaf with symptoms of Rupestris vein feathering, consisting of mild chlorotic discolorations of the primary and secondary leaf veins. (Courtesy G. P. Martelli)

Disease Cycle and Epidemiology

All recognized or putative disease agents of the fleck complex are phloem-limited, nonmechanically transmitted viruses that persist in infected propagative material and are disseminated with it. GFkV is not seedborne. Vectors are unknown, but natural spread of fleck has been reported in South Africa and Italy.

Management

GFkV can be eliminated by heat therapy and meristem tip culture. Control depends on the production and distribution of healthy nursery material. Symptomless infections make sanitary selection based on symptoms of *V. vinifera* cultivars and most American rootstock hybrids unreliable. Hence, all sources must be indexed onto *V. rupestris* St. George and/or assayed by means of antisera or primer sets. Only pathogen-free materials should be used for propagations.

Selected References

Abou Ghanem-Sabanadzovic, N., Sabanadzovic, S., and Martelli, G. P. 2003. Sequence analysis of the 3′ end of three grapevine fleck virus-like viruses from grapevine. Virus Genes 27:11-16.

Boscia, D., Martelli, G. P., Savino, V., and Castellano, M. A. 1991. Identification of the agent of grapevine fleck disease. Vitis 30:97-105.

Boscia, D., Sabanadzovic, S., Savino, V., Kyriakopoulou, P. E., Martelli, G. P., and Lafortezza, R. 1994. A non-mechanically transmissible isometric virus associated with asteroid mosaic of the grapevine. Vitis 33:101-102.

El Beaino, T., Sabanadzovic, S., Digiaro, M., Abou-Ghanem-Sabanadzovic, N., Rowhani, A., Kyriakopoulou, P. E., and Martelli, G. P. 2001. Molecular detection of Grapevine fleck virus-like viruses. Vitis 40:65-68.

Martelli, G. P., Sabanadzovic, S., Abou Ghanem-Sabanadzovic, N., Edwards, M. C., and Dreher, T. 2002. The family *Tymoviridae*. Arch. Virol. 147:1837-1846.

Martelli, G. P., Sabanadzovic, S., Abou Ghanem-Sabanadzovic, N., and Saldarelli, P. 2002. Maculavirus, a new genus of plant viruses. Arch. Virol. 147:1847-1853.

Matsumoto, T., and Ohki, S. T. 1998. A possible new necrotic diseases of grapevine associated with small isometric particles and novel membrane-bound large particles. Ann. Phytopath. Soc. Japan 64:560-564.

Sabanadzovic, S., Abou Ghanem, N., Castellano, M. A., Digiaro, M., and Martelli, G. P. 2000. Grapevine fleck virus-like viruses in *Vitis*. Arch. Virol. 145:553-565.

Sabanadzovic, S., Abou Ghanem-Sabanadzovic, N., Saldarelli, P., and Martelli, G. P. 1998. Complete nucleotide sequence and genome organization of grapevine fleck virus. J. Gen. Virol. 82:2009-2015.

(Prepared by G. P. Martelli, A. Rowhani, and S. Sabanadzovic)

Fig. 221. Chlorotic, deformed leaf blades of a Kyoho grapevine infected with *Grapevine berry inner necrosis virus*. (Courtesy Y. Terai)

Grapevine Berry Inner Necrosis

Originally named grapevine mosaic, the disease was first recognized in the table grape cultivar Kyoho (*V. vinifera* L.) in 1984. In 1992, concomitant with identification of the causal agent, the disease was renamed grapevine berry inner necrosis (GBIN) disease and the causal agent *Grapevine berry inner necrosis virus* (GBINV). The disease is economically important in the table grape cultivars Kyoho and Pione but is unknown outside of Japan.

Symptoms

Diseased grapevines exhibit delayed bud break and reduced vegetative vigor. With chronic infection, the canopy volume is limited in part as a result of shortening of internodes and vascular necrosis in young shoots. Also, affected leaves develop chlorotic rings and line patterns beginning in early spring that remain visible thereafter, and leaves become malformed (Fig. 221). Grape clusters and berries are small with dark-green spots developing on the skin surface (Fig. 222) and an internal necrosis of the flesh tissue (Fig. 223). Berries mature late and the flesh is hard.

Graft inoculations to other grape cultivars produce strong symptoms in cultivar Campbell Early; the related cultivars Takao, Kyoho, and Pione; and the rootstock *V. riparia* Gloire. All are useful indicators for GBINV. Latent infections occurre in cultivars Delaware (*V.* × *labrusca* L.), Koshu, Kaiji, Semillon, Merlot, Cabernet Sauvignon, Cabernet Franc, and Pinot Noir (all *V. vinifera*).

Causal Agent

GBINV (genus *Trichovirus*) is a flexuous filamentous particle, ca. 740 nm in length by 12 nm in width, containing a polyadenylated, single-stranded RNA, mol. wt. 2.5×10^6 Da; a genome organization with three open reading frames (encoding for replication-associated protein, movement protein, and coat protein); and a coat protein size of 22 kDa. The virus can be sap-transmitted to *Chenopodium quinoa* Willd. and *Nicotiana occidentalis* L. and is readily detected in extracts of young grape leaves by ELISA or RT-PCR.

Disease Cycle and Epidemiology

Natural virus spread was demonstrated when healthy plants of cultivars Kyoho and Pione were transplanted into a diseased vineyard site and exposed for a year. In Yamanashi Prefecture, disease incidences of 9.1 and 73.1% were correlated with low and high vector mite populations in their respective vineyard sites. In transmission experiments, the vector, grape erineum mite (*Colomerus vitis*), acquired GBINV from diseased grapevines and transmitted it to healthy ones. The virus-vector relationship and transmission mode are not known. GBINV is the first miteborne virus reported in grapes.

Fig. 222. Dark-green spots developing on the surface of fruit from a Kyoho grapevine infected with *Grapevine berry inner necrosis virus*. (Courtesy Y. Terai)

Fig. 223. Internal view of Kyoho grapes expressing symptoms of grapevine berry inner necrosis disease. (Courtesy Y. Terai)

Management

Only grapevines propagated from certified sources should be planted. Applications of lime sulfur during the dormant season may be used to reduce populations of the vector species.

Selected References

Kunugi, Y., Asari, S., Terai, Y., and Shinkai, A. 2000. Studies on the grapevine berry inner necrosis virus disease: 2. Transmission of grapevine berry inner necrosis virus by the grape erineum mite, *Colomerus vitis* in Yamanashi. Bull. Yamanashi Fruit Tree Exp. Stn. 10:57-64.

Nishijima, T., Terai, Y., and Kunugi, Y. 2000. Studies on the grapevine berry inner necrosis virus disease: 1. Symptoms on vines, varietal susceptibility and natural spread. Bull. Yamanashi Fruit Tree Exp. Stn. 10:47-56.

Yoshikawa, N., Iida, H., Goto, S., Magome, H., Takahashi, T., and Terai, Y. 1997. Grapevine berry inner necrosis, a new trichovirus: Comparative studies with several known trichoviruses. Arch. Virol. 142:1351-1363.

(Prepared by Y. Terai)

Grapevine Leafroll Disease

Grapevine leafroll disease was first recognized as rougeau in France (ca. 1853). In Germany, typical disease symptoms were observed as early as 1910, and the viral nature of the disease was demonstrated in 1936. Today, grapevine leafroll is a major virus disease occurring in every grape-growing region worldwide. This status was reached by disseminating, albeit unknowingly, infected grape cuttings and propagating scions on infected but symptomless rootstocks. In California, the disorder was originally known as White Emperor disease.

Grapevine leafroll may have originated in the Near East and co-evolved with *Vitis vinifera*. Disease severity and yield losses are dependent on a combination of factors, including but not limited to virus strain, scion cutltivar and rootstock, climate, soil, and cultural practices. Under certain conditions, including the use of rootstocks susceptible to graft-incompatibility disorders and young vine decline, certain isolates of *Grapevine leafroll-associated virus 2* (GLRaV-2) can cause plant death. Although long recognized as a graft-transmissible disease, the etiology of grapevine leafroll disease has only recently been elucidated after several Grapevine leafroll-associated viruses (GLRaVs) were purified and antisera produced. Further advances in virus strain characterization and detection have been accomplished utilizing molecular techniques. Once considered to be a disease transmitted only through propagative materials, mealybugs (Hemiptera: Pseudococcidae) and soft-scale insects (Hemiptera: Coccidae) have now been shown to vector some GLRaVs.

Fig. 224. Blotchy interveinal reddening on a Cabernet Sauvignon vine testing positive for *Grapevine leafroll associated virus 3* but negative for Grapevine red blotch associated virus. (Courtesy M. Fuchs)

Symptoms

Grapevine leafroll disease involves a degeneration of primary phloem tissue in young shoots, leaves, and fruit pedicels. Infected vines often show delayed bud break and shoot development, slightly smaller leaves, and reduced shoot length with an overall attendant loss of vigor and yield of fruit. Fruit clusters borne on diseased grapevines are smaller than normal, and both sugar accumulation and pigmentation (in dark-fruited cultivars) are often greatly reduced. Leaf symptoms are most striking on dark-fruited cultivars, and consist of reddening of the interveinal regions contrasting with green veins. The intensity of this symptom expression can range from scattered blotches of red virtually indistinguishable from symptoms of red blotch disease (Fig. 224) to progressively greater areas of the lamina being discolored (Figs. 225 and 226) to the entire leaf turning bright red with the exception of the green veins (Figs. 227 and 228). Leaf symptoms on light-fruited cultivars are much less pronounced, consisting of chlorotic interveinal tissue and normal green veins (Fig. 229). Foliar symptoms typically begin to develop midseason, initially on the more basal-positioned leaves. However, their onset and intensity are dependent on virus strain, scion cultivar, and growing conditions. Mixed virus infections are common and may act additively or synergistically, thereby increasing disease severity. In addition to showing discoloration, affected leaves may roll downward, giving rise to the disease name; the extent of this reaction is dependent on the virus strain and scion/rootstock combination.

The damage to infected grapevines is chronic. Even so, the severity of leafroll disease may vary from one growing season to the next. Most American rootstocks with GLRaV infections

Fig. 225. Incomplete reddening of the interveinal regions of a leaf from a cv. Pinot Noir vine with grapevine leafroll disease. (Courtesy M. Fuchs)

Fig. 226. Pronounced interveinal reddening, contrasting green veins, and downward rolling on a cv. Cabernet Franc leaf exhibiting symptoms of grapevine leafroll disease. (Courtesy M. Fuchs)

Fig. 227. Severe foliar symptoms of grapevine leafroll disease on a cv. Pinot Noir vine. (Courtesy M. Fuchs)

Fig. 228. Severe foliar symptoms of grapevine leafroll disease on a cv. Cabernet Franc vine. (Courtesy M. Fuchs)

Fig. 229. Interveinal chlorosis and downward rolling of leaves on a cv. Chardonnay vine exhibiting symptoms of grapevine leafroll disease characteristic for white-fruited cultivars. (Courtesy M. Fuchs)

are symptomless. Limited surveys have shown leafroll infection to be common in juice grapes (e.g., *Vitis* × *labruscana* 'Concord' and *V. labrusca* 'Niagara'), although typical symptoms generally are not present. The consequence of such infections in these cultivars remains unclear.

Causal Agents

Although conclusive scientific proof of their individual etiologies is still lacking, there are five recognized, serologically distinct viruses associated with grapevine leafroll disease. The association of disease and virus strain has been corroborated for GLRaV-1 and GLRaV-3 through large-scale serological testing of diseased grapevines worldwide. This strict association has been further affirmed by transmission experiments using graft inoculations, vector transmissions, and heat therapy sanitation.

Taxonomically, all five GLRaVs are classified in the family *Closteroviridae*, characterized by large, flexuous rod-shaped particles with lengths ranging from 1250 to 2200 nm; a genome size ranging from 13,696 nucleotides (nt) encoding seven open reading frames to one of 18,498 nt encoding 13 open reading frames; and coat protein size classes of either 26 or 35 to 39 kDa. Phylogenetically, GLRaVs-1, 3, and, 4 are classified in the genus *Ampelovirus* (vectored by mealybug); GLRaV-2 in the genus *Closterovirus* (vectored by aphids); and GLRaV-7 in the genus *Velarivirus* (no known vector). GLRaV-8 is now considered an erroneously established or invalid virus species. On the basis of sequence information available to date, the formerly recognized GLRaV-5, - 6, -9, -Pr, -De, and -Car are now regarded as more or less widely divergent variants of GLRaV-4.

Disease Cycle and Epidemiology

GLRaVs have been widely disseminated through the propagation of infected plant material. However, in some viticulture regions, natural spread also is evident locally. Experimentally, transmissions have been demonstrated with scale insects (GLRaV-1 by *Pulvinaria innumnerabilis*, *P. vitis*, and *Parthenolecanium corni*; and GLRaV-3 by *P. innumnerabilis* and *P. vitis*) and mealybugs (GLRaV-1 by *Heliococcus adenostomae* and *Phenacoccus aceris*; GLRaV-3 by *H. adenostomae*, *P. aceris*, *Planococcus citri*, *Pl. ficus*, *Pseudococcus calceolariae*, *Ps. comstocki*, *Ps. longispinus*, *Ps. maritimus*, and *Ps. viburni*; and GLRaV-4 by *Pl. ficus* and *Ps. longispinus*).

Management

The best control option is to propagate planting materials using only sources testing negative for GLRaVs. All GLRaVs, with the exception of some genetic variants of GLRaV-2, can be detected in a 2-year field and/or a 2- to 3-month greenhouse indexing of candidate sources through the use of *V. vinifera* cvs. Cabernet Franc or Pinot Noir as indicators. Alternatively, laboratory-based tests such as enzyme-linked immunoassay and reverse-transcription polymerase chain reaction assay are available with results in 1–2 days.

If spread of the viruses transmitted by vectors within or between plantings is problematic, roguing, mealybug control, and sanitation measures can reduce new infections. Removal of whole blocks should also be considered to control the disease when disease incidence exceeds a certain threshold, followed by a 2-year fallow period to manage remnant roots on which some vector species can be present. Economic thresholds can be developed to help identify the best control options for individual vineyard blocks and devise optimal profit-maximizing management strategies.

Selected References

Almeida, R. P. P., Daane, K. M., Bell, V. A., Blaisdell, G. K., Cooper, M. L., Herrbach, E., and Pietersen, G. 2013. Ecology and management of grapevine leafroll disease. Front. Microbiol. 4:94.

Al Rwahnih, M., Dolja, W. V., Daubert, S., Koonin, E. V., and Rowhani, A. 2012. Genomic and biological analysis of *Grapevine leafroll-associated virus 7* reveals a possible new genus within the family *Closteroviridae*. Virus Res. 163:302-309.

Atallah, S. S., Gomez, M. I., Fuchs, M. F., and Martinson, T. E. 2012. Economic impact of grapevine leafroll disease on *Vitis vinifera* cv. Cabernet franc in Finger Lakes vineyards of New York. Am. J. Enol. Vitic. 63:73-79.

Belli, G., Fortusini, A., Casati, P., Belli, L., Bianco, P. A., and Prati, S. 1994. Transmission of a grapevine leafroll associated closterovirus by the scale insect *Pulvinaria vitis* L. Riv. Patol. Veg. 4:105-108.
Bertazzon, N., Borgo, M., Vanin, S., and Angelini, E. 2010. Genetic variability and pathological properties of *Grapevine leafroll-associated virus 2* isolates. Eur. J. Plant Pathol. 127:185-197.
Daane, K. M., Almeida, R. P. P., Bell, V. A., Walker, J. T. S., Botton, M., Fallahzadeh, M., Mani, M., Miano, J. L., Sforza, R., Walton, V. M., and Zaviezo, T. 2012. Biology and management of mealybugs in vineyards. Pages 271-307 in: Arthropod Management in Vineyards: Pests, Approaches, and Future Directions. N. J. Bostanian, C. Vincent, and R. Isaacs, eds. Springer Science, Berlin, Germany.
Fazeli, C. F., and Rezanian, M. A. 2000. Nucleotide sequence and organization of ten open reading frames in the genome of grapevine leafroll-associated virus 1 and identification of three subgenomic RNAs. J. Gen. Virol. 81:605-615.
Gambino, G., and Gribaudo, I. 2006. Simultaneous detection of nine grapevine viruses by multiplex reverse transcription-polymerase chain reaction with coamplification of a plant RNA as internal control. Phytopathology 96:1223-1229.
Ghanem-Sabanadzovic, N. A., Sabanadzovic, S., Gugerli, P., and Rowhani, A. 2012. Genome organization, serology and phylogeny of Grapevine leafroll-associated viruses 4 and 6: Taxonomic implications. Virus Res. 163:120-128.
Goheen, A. C., Harmon, F. N., and Weinberger, J. H. 1958. Leafroll (white Emperor disease) of grapes in California. Phytopathology 48:51-54.
Golino, G. A., Sim, S., Gill, R., and Rowhani, A. 2002. California mealybugs can spread grapevine leafroll disease. Calif. Agric. 56:196-201.
Gugerli, P., Brugger, J.-J., and Bovey, R. 1984. L'enroulement de la vigne: Mise en évidence de particules virales et développement d'une méthode immuno-enzymatique pour le diagnostic rapide. Rev. Suisse Vitic. Arboric. Hortic. 16:299-304.
Harmon, F. N., and Snyder, E. 1946. Investigations on the occurrence, transmission, spread, and effect of "white" fruit color in the Emperor grape. Proc. Am. Soc. Hortic. Sci. 47:190-194.
Jarugula, S., Alabi, O. J., Martin, R. R., and Naidu, R. A. 2010. Genetic variability of natural populations of *Grapevine leafroll-associated virus 2* in Pacific Northwest vineyards. Phytopathology 100:698-707.
Jelkmann, W., Mikona, C., Turturo, C., Navarro, B., Rott, M. E., Menzel, W., Saldarelli, P., Minafra, A. and Martelli, G. P. 2012. Molecular characterization and taxonomy of grapevine leafroll-associated virus 7. Arch. Virol. 157:359-362.
Maree, H. J., Almeida, R. P. P., Bester, R., Chooi, K-M., Cohen, D., Dolja, V. V., Fuchs, M. F., Golino, D. A., Jooste, A. E. C., Martelli, G. P., Rayapati, N., Rohawni, A. K., Saldarelli, P., and Burger, J. T. 2013. *Grapevine leafroll-associated virus 3*. Front. Microbiol. 4:82.
Martelli, G. P., Abou Ghanem-Sabanadzovic, N., Agranovsky, A. A., Al Rwahnih, M., Dolja, V. V., Dovas, C. I., Fuchs, M., Gugerli, P., Hu, J. S., Jelkmann, W., Katis, N. I., Maliogka, V. I., Melzer, M. J., Menzel, W., Minafra, A., Rott, M. E., Rowhani, A., Sabanadzovic, S., and Sladarelli, P. 2012. Taxonomic revision of the family *Closteroviridae* with special reference to the Grapevine leafroll-associated members of the genus *Ameplovirus* and the putative species unassigned in the family. J. Plant Pathol. 94:7-19.
Martelli, G. P., Agranovsky, A. A., Bar-Joseph, M., Boscia, D., Candresse, T., Coutts, R. H. A., Dolja, V. V., Falk, B. W., Gonsalves, D., Jelkmann, W., Karasev, A. V., Minafra, A., Namba, S., Vetten, H. J., Wisler, G. C., and Yoshikawa, N. 2002. The family *Closteroviridae* revised. Arch. Virol. 147:2039-2044.
Sforza, R., Boudon-Padieu, E., and Greif, C. 2003. New mealybug species vectoring Grapevine leafroll-associated viruses-1 and -3 (GLRaV-1 and -3). Eur. J. Plant Pathol. 109:975-981.
Thompson, J. R., Fuchs, M., and Perry, K. 2012. Genomic analysis of *Grapevine leafroll-associated virus-5* and related viruses. Virus Res. 163:19-27.
Tsai, C-W., Rowhani, A., Golino, D. A., Daane, K. M., and Almeida, R. P. P. 2010. Mealybug transmission of grapevine leafroll viruses: An analysis of virus–vector specificity. Phytopathology 100:830-834.

(Prepared by A. Rowhani and P. Gugerli)

Grapevine leafroll-associated virus 2 and Graft Incompatibility

The involvement of a viral agent or agents, including *Grapevine leafroll-associated virus 2* (GLRaV-2), in graft incompatibility and young vine decline was suspected in France and Italy in the early 1990s. In California, failure of the AXR#1 rootstock to control a new strain of the grape phylloxera (*Daktulosphaira vitifoliae*), first recognized in the mid-1980s, led to wide-scale replantings in winegrape vineyards where it had formerly predominated, employing alternative rootstocks with higher levels of phylloxera resistance. However, whereas AXR#1 tolerated latent viruses, infected scions grafted onto certain phylloxera-resistant rootstocks declined soon after transplanting. One graft-incompatible-associated agent was later characterized. Initially called Grapevine rootstock stem lesion-associated virus, it is now referred to as GLRaV-2 strain Red Globe (GLRaV-2 RG). In parallel, additional distinct, genetically related GLRaV-2 variants associated with graft incompatibility also were described.

Symptoms

In bench grafts of GLRaV2 RG-infected scions onto sensitive rootstocks, disease symptoms consist of weak plant growth, stunted shoots, and leaves with early fall color. The unions are enlarged and the majority of affected plants succumb during the growing season. On virus-insensitive rootstocks such as Freedom and 101-14 Mtg, where the infections are latent, infected plants develop normal unions (Fig. 230) and canopies.

In 2-year-old test plants of Cabernet Sauvignon propagated on different phylloxera-resistant rootstocks and graft-inoculated with GLRaV-2 RG, the canopies developed a similar reddish color a year later (Fig. 231), and in dormancy several symptomatic plants died. A few surviving plants produced weak shoot growth and reddish, small leaves in the next growing season (Fig. 232). Survivor plants on rootstock Kober 5BB developed small, discrete necrotic lesions on the rootstock portion, whereas survivors on rootstock 3309C developed more extensive lesions. Lesion development required two growing seasons.

The mechanistic interaction between GLRaV-2 RG and the sensitive rootstocks resembles a girdling effect at the scion/

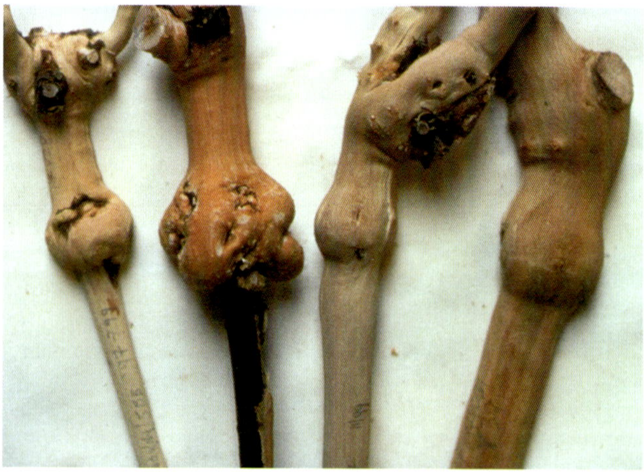

Fig. 230. Specimens of Redglobe scions infected with GLRaV-2 RG bench-grafted onto two acutely sensitive and two insensitive rootstocks. Rootstocks 5BB (far left) and 3309C (second from left) with enlarged unions and necrotic fissures (rootstock 3309C also with severe stem necrosis); rootstocks Freedom (second from right) and 101-14 Mgt (far right) with normal woody cylinders. (Courtesy J. K. Uyemoto)

rootstock junction, owing in part to completely reddened leaves throughout the canopy. Cabernet Sauvignon scions die when bench-grafted on the acutely sensitive rootstocks 5C, 1103P, 1616C, 3309C, and Kober 5BB. In contrast, GLRaV-2 RG causes latent infections in the rootstocks O39-16, 101-14Mtg, 110R, 140R, 420A, Boerner, Freedom, Harmony, Ramsey, Schwarzmann, *Vitis riparia* Gloire, and *Vitis rupestris* St. George. Some of these virus-insensitive test plants, e.g., rootstocks O39-16, 101-14Mtg, Freedom, Harmony, and Ramsey, are devoid of stem lesions.

In two trials, a common strain of GLRaV-2 induced typical leafroll symptoms on the various test plants. All symptomatic test plants in these trials survived in an apparently robust fashion throughout the 5–8 year duration of the trials.

Causal Agent

GLRaV-2 RG was first isolated and characterized from an infected Redglobe grapevine. Molecularly, it shares 74% sequence homology with other GLRaV-2 strains classified in the genus *Closterovirus*, family *Closteroviridae*. Members in this virus family have a genome organization consisting of 9 open reading frames and a coat protein subunit size of 22 to 25 kDa. Furthermore, the coat protein of GLRaV-2 RG cross-reacts with GLRaV-2 antiserum in western blot assays, confirming the virus relationship and demonstrating it as a molecular and biological variant.

Four GTAs in addition to GLRaV-2 RG have been found to cause stem lesions on selective grapevine rootstocks. Source GTA-1 induced red foliage 1 year post-inoculation in test plants of Cabernet Sauvignon propagated on the rootstocks 101-14 Mtg, 3309C, and Freedom. Only test plants on 101-14 Mtg survived dormancy; when these were later sacrificed and bark removed, the rootstock stems contained scattered, discrete small necrotic lesions (Fig. 233). Bud grafts on other test plants developed typical leafroll symptoms and did not decline. Test plants developed stem lesions or were killed when on rootstock 5C inoculated with source GTA-2; on Harmony, 5BB and 1616C rootstocks inoculated with source GTA-3; and on rootstock 5BB inoculated with GTA-4. Attempts are underway to characterize the unknown GTAs.

Disease Cycle and Epidemiology

Although first discovered in the table grape cv. Redglobe, about 20% of samples collected from diseased wine grape cultivars in various California viticulture regions also tested positive for GLRaV-2 RG. Propagation with infected source material is likely the principal method of virus dissemination.

Fig. 231. Test plants of cv. Cabernet Sauvignon scions on 3309C rootstocks 1 year after chip-bud inoculation with GLRaV-2 RG; two of these plants did not survive through the following dormant season. (Courtesy J. K. Uyemoto)

GLRaV-2 RG causes latent infections in the standard virus disease indicators used in grapevine certification programs, e.g., it infected but failed to induce observable symptoms in own-rooted indicator plants *V. vinifera* 'Cabernet Franc', LN33, and *V. rupestris* St. George, and on test plants of Cabernet Sauvignon grafted to GLRaV-2 RG-insensitive rootstocks.

Management

Propagation material should be derived only from pretested, clean sources. GLRaV-2 RG likely causes latent infections in

Fig. 232. Reddish mottled leaves and weak spring growth of cv. Cabernet Sauvignon scions on 3309C rootstocks 2 years after chip-bud inoculation with GLRaV-2 RG. (Courtesy J. K. Uyemoto)

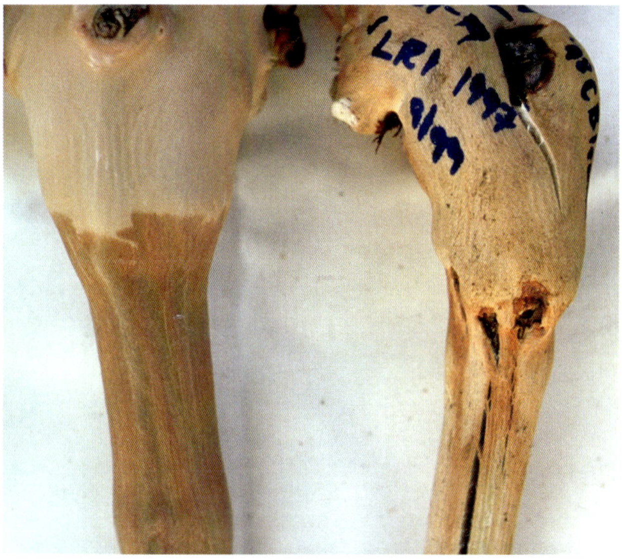

Fig. 233. Discrete stem lesions produced on rootstock 101-14 Mtg (right) graft inoculated with an unknown transmissible agent capable of inducing graft-incompatibility symptoms similar to those on sensitive rootstocks inoculated with GLRaV-2 RG; healthy plant is on the left. (Courtesy J. K. Uyemoto)

most scions and rootstocks, and therefore virus contamination is difficult to assess visually. Source plants may be assayed by RT-PCR or graft-indexed using indicator plants of Cabernet Sauvignon propagated on virus-sensitive rootstocks 5BB, 5C, 1103P, 1616C, or 3309C.

Selected References

Bertazzon, N., Borgo, M., Vanin, S., and Angelini, E. 2010. Genetic variability and pathological properties of *Grapevine leafroll-associated virus 2* isolates. Eur. J. Plant Pathol. 127:185-197.

Bertazzon, N., Borgo, M. and Angelini, E. 2010. The complete genome sequence of the BD variant of grapevine leafroll-associated virus 2. Arch. Virol. 155:1717-1719.

Greif, C., Garau, R., Boscia, D., Prota, V. A., Fiori, M., Bass, P., Walter, B., and Prota, U. 1995. The relationship of grapevine leafroll-associated closterovirus 2 with a graft incompatible condition of grapevines. Phytopathol. Mediterr. 34:167-173.

Martelli, G. P., Agranovsky, A. A., Bar-Joseph, M., Boscia, D., Candresse, T., Coutts, R. H. A., Dolja, V. V., Falk, B. W., Gonsalves, D., Jelkmann, W., Karasev, A. V., Minafra, A., Namba, S., Vetten, H. J., Wisler, G. C., and Yoshikawa, N. 2002. The family *Closteroviridae* revised. Arch. Virol. 147:2039-2044.

Meng, B., Li, C., Goszczynski, D., and Gonsalves, D. 2005. Genome sequence and structure of two biologically distinct strains of Grapevine leafroll-associated 2 and sequence analysis. Virus Genes 31:31-41.

Uyemoto, J. K., Rowhani, A., Luvisi, D., and Krag, C. R. 2001. New closterovius in 'Redglobe' grape causes decline of grafted plants. Calif. Agric. 55(4):28-31.

Uyemoto, J. K., and Rowhani, A. 2003. Discovery of different grapevine sources with graft-transmissible agents causing union-incompatibility on sensitive rootstocks. Pages 139-140 in: Proc. 14 Meeting, International Council for the Study of Virus and Virus-like Diseases of the Grapevine (ICVG). http://web.pppmb.cals.cornell.edu/fuchs/icvg/data/2003Session45.pdf

(Prepared by J. K. Uyemoto and A. Rowhani)

Grapevine Red Blotch

Grapevine red blotch disease was first recognized in 2008 in vineyards in Napa County, California. The name "red blotch" was given to distinguish disease symptoms from those caused by leafroll viruses and other graft-transmissible agents involved in graft union disorders on red-fruited *V. vinifera* cultivars, as well as those caused by various injuries to trunks and shoots. In addition to north coast vineyards in Napa and Sonoma counties, diseased vineyards have also been found in the central coast and San Joaquin Valley regions of California; in the states of Arizona, Arkansas, Georgia, Maryland, New York, North Carolina, Oregon, Pennsylvania, Texas, Virginia, and Washington; and in the provinces of British Columbia and Ontario, Canada. This widespread distribution of the virus was determined through surveys in North America during the 2012 and 2013 growing seasons, and it seems likely that the disease will be confirmed in many additional locations as further investigations continue. The occurrence of red blotch outside of North America has not yet been reported, although plant material introduced to the United States from abroad was found to be infected.

Symptoms

Grapevines of various ages (2–25 years old) have been found exhibiting red blotch symptoms. Disease symptoms generally start appearing in late August through September as irregular red blotches on leaf blades located on basal portions of shoots, either between secondary or tertiary veins (Fig. 234) or extending from the leaf margin (Fig. 235). The veins of affected leaves can turn partly or fully red (Fig. 236). Occasionally, reddening of the leaf blade in the interveinal zones between secondary

Fig. 234. Symptoms of red blotch on a leaf of cv. Cabernet Franc. (Courtesy M. Fuchs)

Fig. 235. Symptoms of red blotch extending inward from the margins on a leaf of cv. Carignane. (Courtesy M. Fuchs)

Fig. 236. Symptoms of red blotch on the underside of a leaf of cv. Cabernet Sauvignon leaf, including reddening of primary and secondary veins and the petiole. (Courtesy M. Fuchs)

Fig. 237. Interveinal reddening on the leaf blade of a cv. Pinot Noir vine infected with Grapevine red blotch-associated virus, similar to certain symptoms caused by leafroll-associated viruses. (Courtesy M. Fuchs)

Fig. 238. Symptoms on the leaf blade of a cv. Cabernet Franc vine infected with a leafroll-associated virus, easily confused with red blotch symptoms. (Courtesy M. Fuchs)

veins (Fig. 237) may resemble certain symptoms of grapevine leafroll disease (Fig. 238), with which red blotch has sometimes been confused. However, in contrast with leafroll disease, the margins of red blotch-affected leaves do not roll downward. To date, disease symptoms have been observed in vineyards of *V. vinifera* cultivars Cabernet Franc, Cabernet Sauvignon, Malbec, Merlot, Mourvèdre, Petite Sirah, Petit Verdot, Pinot Noir, and Zinfandel. Infections of white *V. vinifera* cultivars such as Chardonnay, Riesling, Semillon, and Viognier also have been detected, but disease symptoms on them are less conspicuous. Red blotch symptoms may appear earlier (mid- to late June) in the season in some cultivars (e.g., Cabernet Franc and Merlot) than others (e.g., Cabernet Sauvignon). Severity of red blotch symptoms may also vary from year to year.

The most significant impact of red blotch is on the accumulation of total soluble solids (°Brix), which can be as much as 4–5 units lower in fruit from diseased vines than in fruit from healthy vines. Also, fruit from affected grapevines exhibit increased acidity. It is not yet established whether red blotch disease has any effect on vine vigor or fruit yield.

Causal Agent

Although red blotch symptoms have sometimes been confused with those caused by nutritional deficiencies, stress, or other infectious agents such as viruses associated with leafroll disease, independent studies using rolling circle amplification and deep sequencing have identified a previously unknown virus, now named Grapevine red blotch-associated virus (GRBaV), in vines exhibiting red blotch symptoms. (This virus was initially reported as Grapevine cabernet franc-associated virus). GRBaV has a circular, single-stranded DNA genome and is proposed to be a member of the family *Geminiviridae*. Based on the GRBaV genome sequence information, a PCR test was developed for accurate diagnosis, and by means of this test, the virus has been detected consistently in grapevines exhibiting red blotch symptoms. Recently, GRBaV was proven to be the cause of this disease through inoculation studies utilizing *Agrobacterium tumefaciens*.

Disease Cycle and Epidemiology

The primary source of inoculum appears to be infected planting material. Disease incidence has exceeded 90% in several vineyards surveyed and secondary spread is suspected in some situations. The Virginia creeper leafhopper (*Erythroneura ziczac* Walsh) is reported to transmit GRBaV from vine to vine in the greenhouse. The role of this and other potential insect vectors in transmitting GRBaV in vineyards has yet to be determined. GRBaV is graft transmissible but is not likely to be seedborne, as there is no evidence for seed transmission of any other members of the family *Geminiviridae*.

Management

The occurrence of red blotch disease may be difficult to assess visually owing to symptomatic similarities with grapevine leafroll disease and deficiencies of certain nutrients, such as potassium. Therefore, vines should be assayed by PCR. Diseased grapevines should be replaced with healthy vines derived from clean, certified virus-tested stocks.

Selected References

Al Rwahnih, M., Dave, A., Anderson, M. M., Rowhani, A., Uyemoto, J. K., and Sudarshana, M. R. 2013. Association of a DNA virus with grapevines affected by red blotch disease in California. Phytopathology 103:1069-1076.

Krenz, B., Thompson, J. R., Fuchs, M., and Perry, K. L. 2012. Complete genome sequence of a new circular DNA virus from grapevine. J. Virol. 86:7715.

Poojari, S., Alabi, O. J., Fofanov, V. Y., and Naidu, R. A. 2013. A leafhopper-transmissible DNA virus with novel evolutionary lineage in the family *Geminiviridae* implicated in grapevine redleaf disease by next-generation sequencing. PLoS ONE 8:e64194.

Sudarshana, M. R., and Wolpert, J. A. 2012. Grapevine red blotch disease. UC Cooperative Ext. Fact Sheet. http://ucanr. edu/sites/viticulture-fresno/files/157139.pdf

(Prepared by M. Sudarshana and M. Fuchs)

Nematode-transmitted Virus Diseases

European Nepoviruses and *Strawberry latent ringspot virus*

Several nepoviruses and *Strawberry latent ringspot virus* (SLRSV) infect grapevines in Europe, causing degenerative diseases (collectively termed infectious degeneration) whose symptoms are similar to or indistinguishable from those of fanleaf degeneration induced by *Grapevine fanleaf virus* (GFLV). Fanleaf is the oldest known virus disease of *V. vinifera*. Its name comes from the peculiar malformation of infected leaves, which exhibit widely open petiolar sinuses and abnormally gathered primary veins that produce an appearance resembling that of an open fan (Fig. 239). In European literature, records of this disease date back some 200 years, and grapevine leaves with typical symptoms were found in herbaria established before the introduction of American rootstock hybrids. The consensus is that fanleaf degeneration may have existed in the Mediterranean Basin and the Near East since the earliest time of grape cultivation. Now the disease is known to occur worldwide.

GFLV and several of the other grapevine-infecting European nepoviruses have distorting and chromogenic strains and may occur in mixed infections. Their economic impact varies with the tolerance of the cultivar to the viruses. Tolerant cultivars produce moderately good crops. In contrast, sensitive cultivars are severely affected, showing progressive vine decline, low yields (up to 80% losses) with poor fruit quality, shortened productive life of the vineyard, a low proportion of graft take, reduced rooting ability of propagation material, and decreased tolerance of adverse climatic factors.

Symptoms

Infectious degeneration is characterized by two distinct syndromes, infectious malformation and yellow mosaic, caused by different strains of the causal agents. Infectious malformation is a syndrome induced by distortion-causing strains of the virus. Affected leaves are variously and severely distorted, and may be asymmetrically shaped, puckered, and/or exhibit open petiolar sinuses, deep lobes, and acute denticulations (Figs. 239 and 240). Chlorotic mottling may accompany these deformations (Figs. 241, 242, and 243). Foliar symptoms develop early in the spring and persist throughout the vegetative season, becoming less distinct in summer. Shoots also are malformed, showing abnormal branching, double nodes, short internodes, fasciations, and zigzag growth. Fruit clusters are smaller and fewer in number, and berries ripen irregularly and are small and set poorly (Fig. 244).

Yellow mosaic is induced by chromogenic virus strains. Leaves of infected grapevines develop bright chrome-yellow

Fig. 239. Fan-shaped and denticulated leaves of a cv. Montepulciano vine infected with *Grapevine fanleaf virus*. (Courtesy G. P. Martelli)

Fig. 241. Mottled leaf infected by *Grapevine fanleaf virus*. (Courtesy G. P. Martelli)

Fig. 240. Fan-shaped leaf of a cv. Sylvaner vine infected by *Strawberry latent ringspot virus*. (Courtesy M. Fuchs)

Fig. 242. Mottling and malformations induced by *Arabis mosaic virus* in a leaf of cv. Zirock. (Courtesy M. Fuchs)

discolorations early in the spring that may also affect stems, tendrils, and inflorescences. Chromatic alterations of the leaves vary from a few scattered, yellow spots that sometimes appear as rings or lines, to extensive mottling of the veins and/or interveinal areas, to a typical yellowing of most or all of the leaf blade (Fig. 245). Often, infected grapevines occur in patches (Fig. 246) that are easily seen from a distance. The foliage and shoots show little if any malformation, but fruit clusters and berries are small and few. With increased ambient temperatures during summer, the chlorosis fades rapidly and the canopy develops a normal green color.

"Vein banding" is a disease characterized by chrome-yellow flecks first localized along the main veins of mature leaves and progressing into the interveinal areas (Fig. 247). The discolorations appear in mid- to late summer in a limited number of leaves, which show little malformation. Fruit set is poor, clusters are straggly, and the yield may be virtually zero. This disorder was first reported from California as a syndrome ascribed to a specific strain of GFLV. More recently, however, vein banding symptoms were demonstrated to be induced by a co-infection by *Grapevine yellow speckle viroid* and GFLV.

Trabeculae, or endocellular cordons, are diagnostic of grapevines infected by GFLV. These are radial bars crossing the lumen of epidermal, parenchyma, phloem, and xylem cells, and they are made up of pectic cores surrounded by cellulose sheaths encrusted with lignin, suberin, or cutin, depending on the tissue in which they form. These structures are readily

Fig. 245. Typical yellow mosaic symptoms induced by a chromogenic strain of *Grapevine fanleaf virus*. (Courtesy G. P. Martelli)

Fig. 243. Mottled leaf of cv. Chardonnay infected by the grapevine strain of *Raspberry ringspot virus*. (Courtesy M. Fuchs)

Fig. 246. Patchy distribution of grapevines infected with a chromogenic strain of *Grapevine fanleaf virus*. (Courtesy G. P. Martelli)

Fig. 244. Healthy fruit of Cabernet Sauvignon (left) compared with fruit from a vine infected with *Grapevine fanleaf virus* (right). (Reprinted, by permission, from Raski et al., 1983)

Fig. 247. Vein-banding symptoms in a leaf of cv. Sultanina doubly infected with *Grapevine fanleaf virus* and *Grapevine yellow speckle viroid 1*. (Courtesy G. P. Martelli)

visible by light microscopy in lignified shoots, especially in the basal internodes.

Causal Agents

With one proposed exception, all recognized causal agents of degenerative diseases are members of the genus *Nepovirus* belonging to one of three subgroups, discussed below. They are subgroup A (GFLV, *Arabis mosaic virus* [ArMV], *Raspberry ringspot virus* [RpRSV], and the newly described *Grapevine deformation virus* [GDefV]); subgroup B (*Tomato black ring virus* [TBRV], *Grapevine chrome mosaic virus* [GCMV], and a newly described virus from Turkey denoted *Grapevine Anatolian ringspot virus* [GARSV]); and subgroup C [*Cherry leafroll virus* (CLRV)]. *Strawberry latent ringspot virus* (SLRSV), also discussed below, is an unassigned species in the family *Secoviridae*. These viruses have polyhedral particles about 30 nm in diameter with a positive-sense, single-stranded RNA genome existing as two functional species (RNA-1 and RNA-2), both being required for infectivity and separately encapsidated. Virus preparations sediment as three centrifugal components: component T, made up of empty protein shells; component M, containing 27–31% RNA-2; and component B, containing 38–43% RNA-1. Except for SLRSV, which has capsid protein subunits of two sizes (27 and 43 kDa, respectively), all other viruses have subunits of a single size, 52–57 kDa.

Individual viruses can be identified by biological assays (differential reactions on an herbaceous host range) and laboratory tests. ELISA using polyclonal antisera and/or monoclonal antibodies, when available, is rapid, inexpensive, and reliable. The best antigen sources for serological diagnosis are leaves collected in spring or cortical shavings from mature dormant canes. Molecular assays using radioactive or digoxigenin-labelled probes, reverse transcription-polymerase chain reaction (RT-PCR), and immunocapture RT-PCR are becoming increasingly popular. RT-PCR is estimated to be four- to sixfold more sensitive than ELISA.

Nepoviruses of subgroup A. GFLV was the first grapevine virus to be recovered by mechanical inoculation and to be thoroughly characterized physico-chemically and molecularly. RNA-1 has 7342 nt and RNA-2 has 3774 nt. Virus populations are antigenically uniform and occur as families with minor molecular variants. A satellite RNA with 1114 nt is associated with some isolates. GFLV has been disseminated extensively via infected grapevines and now occurs worldwide.

ArMV is serologically related to GFLV. Its RNA-1 is 7334 nt in size and there are two types of RNA-2 molecules differing slightly in size (3711 and 3852 nt). ArMV supports the replication of two satellite RNAs, a linear form 1104 nt and a circular model about 350 nt in size. Although cross-protection between ArMV and GFLV has been reported, these viruses often occur in mixed infections in certain areas of central Europe and with other nepoviruses in western Germany. ArMV was also recorded from grapevines in Switzerland, Bulgaria, former Yugoslavia, Hungary, Romania, Turkey, Israel, Canada, USA (California and Missouri), France, and Japan.

Serologically, the grapevine strain of RpRSV is very distantly related to the two main serotypes, Scottish and English, and it differs from the type strain because it sediments as a single centrifugal component. These differences suggest that grapevine-infecting RpRSV may be a different virus species. RNA-1 has 7935 nt and RNA-2 has 3928 nt. RpRSV occurs as two different strains of different virulence in the Palatinate (western Germany), where harvests of infected grapevines can be reduced by more than 30%.

GDefV is distantly related serologically to ArMV but not to GFLV, the two viral species from whose recombination it originated. RNA-1 contains 7386 nt and RNA-2 contains 3753 nt. GDefV has no recognized vector, is not seedborne, and has been reported only from southeastern Turkey.

Nepoviruses of subgroup B. TBRV is distantly related serologically to GCMV. RNA-1 has 7356 nt and RNA-2 4662 nt. TBRV supports the replication of a satellite RNA with 1327 nt. Joannes Seyve virus, known to cause a severe disease of cv. Joannes Seyve in Ontario, Canada, is a strain of this virus. TBRV was first recorded from grapevines in Germany, and then occasionally in former Yugoslavia, Greece, Israel, Turkey, and Ontario.

GCMV is serologically distantly related to TBRV. RNA-1 has 7212 nt and RNA-2 4441 nt. GCMV was first found in Hungary, near Lake Balaton, and was originally called Hungarian chrome mosaic virus. It occurs also in former Czechoslovakia, Croatia, and Austria.

GARSV is not serologically related to any of the known grapevine nepoviruses but is phylogenetically related to TBRV and GCMV. RNA-1 has 7286 nt and RNA-2 has 4607 nt. GARSV has no recognized vector, is not seedborne, and has been reported from southeastern Turkey and Iran.

Nepovirus of subgroup C. CLRV is not serologically related to any of the other known nepoviruses. RNA-1 has 7918 nt and RNA-2 has 6360 nt. The virus is transmitted through seeds and pollen in walnut, but whether these types of transmission operate in grapevines is unknown. Grapevine infection by CLRV has been reported in Chile, Poland, and Germany.

Strawberry latent ringspot virus. SLRSV, a virus serologically unrelated to all known nepoviruses, was recently classified as an unassigned species in the family *Secoviridae*. Its RNA-1 has 7496 nt and RNA-2 has 3824 nt. SLRSV supports the replication of a satellite RNA with 1118 nt. SLRSV has been recovered from grapevines in the Palatinate (Germany) and in northern Italy and was also detected occasionally in other European countries and Turkey.

Disease Cycle and Epidemiology

Most of the grapevine-infecting European nepoviruses are transmitted from grapevine to grapevine by nematodes belonging to the *Longidoridae* family. GFLV is transmitted by *Xiphinema index*. A single brief feeding on an infected plant is sufficient for virus acquisition. In nature, *X. index* acquires GFLV only from roots of infected grapevines (Figs. 248), and retains it for up to four years in the absence of host plants or up to three months when the nematode feeds on virus-immune host plants. The virus is retained in the cuticular lining of the esophageal lumen of the vector. Adults and juvenile stages of *X. index* can transmit the virus experimentally. Specific transmission by *X. index* is determined by the viral coat protein. The sequence determining viral transmission consists of a stretch of 11 conserved amino acids located in an exposed region of the coat protein.

Fig. 248. *Xiphinema index* nematodes feeding on a grapevine root. (Courtesy G. P. Martelli)

ArMV and SLRSV are transmitted by *Xiphinema diversicaudatum*. As with GFLV, both viruses are retained in the esophageal lumen of the vectors. Their particles apparently attach to a discontinuous layer of carbohydrates that lines the nematode's odonthophore (the posterior portion of the stylet) and esophagus, from which they dissociate when secretions of the esophageal glands are regurgitated during the first phases of feeding on plant roots.

The grapevine strain of RpRSV is transmitted by *Paralongidorus maximus*, whereas the vectors of TBRV to grapevine are *Longidorus attenuatus* in Germany and *L. elongatus* in Canada. Although field association between GCMV and *Xiphinema vuittenezi* has been reported and virus particles were detected by immunosorbent electron microscopy in *X. index* that had fed on infected hosts, experimental evidence for transmissions of GCMV by either nematode species has not been obtained.

Short-distance spread of all grapevine-infecting nepoviruses is dependent on their vectors. For instance, the yellow mosaic strain of GFLV may be spread in the field to a distance of 1.3 to 1.5 meters per year by nematodes. Although pollenborne, GFLV is rarely transmitted through grape seeds. It has been recorded from naturally infected weeds (*Aristolochia clematis, Lagenaria siceraria* var. *turbinata*, and more recently, *Cynodon dactylon* and *Polygonum aviculare*), but the importance of such infections in the epidemiology of disease on grapevines is not known. GCMV is seedborne, though at a very low rate, and ArMV, SLRSV, TBRV, and RpRSV have alternative hosts, which may facilitate virus survival, representing a factor to consider in disease epidemiology. However, the primary source of GFLV and nematode acquisition is root tissue of infected grapevines. Infected grapevine roots can remain undegraded in the soil for many years after removal of the mother plants, supporting the reproduction and survival of viruliferous *X. index* populations during that period.

Although none of the grapevine-infecting nepoviruses is disseminated over long distances by natural means because of the limited range of vector movement, long-distance spread is achieved through transport of infected propagation material. Anecdotal evidence suggests that GFLV had long been localized along with its vector in scattered enclaves in the traditional European grape-growing regions during the era when all vineyards were established with own-rooted plants. The disastrous dissemination of GFLV in Europe and throughout the world began in the late 1800s, a period during which the practice of grafting scions onto rootstocks to combat the grape phylloxera (*Daktulosphaira vitifoliae*) became commonplace. Many phylloxera-resistant rootstocks were largely asymptomatic when infected with GFLV and other deleterious viruses, and were unwittingly used to propagate finished plants.

Management

It is virtually impossible to eradicate an established population of nematode vectors in vineyard sites. Therefore, several steps must be taken in the preplanting stages to break the ecological cycle of the nematode-virus-grapevine complex. In replant situations, the thorough removal of grapevine roots should be attempted in conjunction with prolonged fallow and weed control. To reduce nematode vector populations, high rates of soil fumigants have been used. However, fumigants only slow the virus transmission rate in shallow soils, and have largely been ineffective at deep soil sites, especially in clay soils. Thus, in some contaminated sites, soil fumigants offer at best temporary disease control. Furthermore, the future availability of fumigants is questionable in many locations as a result of increased regulatory control of their use.

In France, use of GFLV-infected grapevines to offer protection against subsequent infections by ArMV via nematode vectors and vice versa has shown some promise. However, cross protection has many limitations and is not applicable against other unrelated nepoviruses.

Host resistance to GFLV and to *X. index* has been identified in some eastern *V. vinifera* cultivars and in *Muscadinia* or *Vitis* species other than *V. vinifera*. Work is under way in different laboratories to create resistant rootstocks or cultivars through traditional breeding methods or genetic transformation technology. Various rootstocks have been successfully transformed with the coat protein gene of GFLV, ArMV, and GCMV for induction of resistance and are currently under evaluation.

Combining a careful selection of planting materials with vineyard sanitation practices provides a powerful tool for reducing or altogether avoiding the introduction of degenerative diseases into new vineyards. Healthy stocks planted in nematode-free soil or soil with populations of nonviruliferous vectors remain healthy for the productive life of the vineyard. Relative to those planted with infected stock, vineyards established with GFLV-free stocks are very homogeneous in morphology and productivity, grape yields are improved by 40–70%, berries contain more sugar, and the wine is of better quality. Virus elimination is readily achieved from vegetative shoot tips by heat treatment (38–40°C for 4 weeks), in vitro meristem tip culture, micrografting, or somatic embryogenesis, providing healthy source material for subsequent multiplication and propagation.

Selected References

Andret-Link, P., Schmitt-Keichinger, C., Demangeat, G., Komar, G., and Fuchs, M. 2004. The specific transmission of *Grapevine fanleaf virus* by its nematode vector *Xiphinema index* is solely determined by the viral coat protein. Virology 320:12-22.

Demangeat, G., Voisin, R., Minot, J. C., Bosselut, N., Fuchs, M., and Esmenjaud, D. 2005. Survival of *Xiphinema index* in vineyard soil and retention of *Grapevine fanleaf virus* over extended time in the absence of host plants. Phytopathology 95:1151-1156.

Hewitt, W. B., Goheen, A. C., Raski, D. J., and Gooding, G. V. 1962. Studies on virus diseases of grapevine in California. Vitis 3:57-83.

Komar, V., Vigne, E., Demangeat, G., Lemaire, O., and Fuchs, M. 2008. Cross-protection as control strategy against *Grapevine fanleaf virus* in naturally infected vineyards. Plant Dis. 92:1689-1694.

Martelli, G. P. 1978. Nematode-borne viruses of grapevine, their epidemiology and control. Nematol. Mediterr. 6:1-27.

Martelli, G. P., ed. 1993. Graft-transmissible Diseases of Grapevines. Handbook for Detection and Diagnosis. FAO Publ. Div., Rome, Italy.

Mayo, M. A., Fritsch, C., Leibowitz, M. J., Palukaitis, P., Scholthof, K. B., Simons, G., and Taliansky, M. 2000. Satellite nucleic acids. Pages 1028-1032 in: Virus Taxonomy: 7th Report of the International Committee on Taxonomy of Viruses. M. H. V. van Regenmortel, C. M. Fauquet, D. H. L. Bishop, E. B. Carstens, M. K. Estes, S. M. Lemon, J. Maniloff, M. A. Mayo, D. J. McGeoch, C. R. Pringle and R. B. Wichner, eds. Academic Press, San Diego, CA.

Raski, D. J., Goheen, A. C., Lider, L. A., and Meredith, C. P. 1983. Strategies against grapevine fanleaf virus and its nematode vector. Plant Dis. 67:335-339.

Sanfaçon, H., Iwanami, T., Karasev, A. V., van der Vlugt, R., Wellink, J., Wetzel, T., and Yoshikawa, N. 2012. Secoviridae. Page 881 in: Virus Taxonomy: Classification and Nomenclature of Viruses. A. M. Q. King, M. J. Adams, E. B. Carstens, and E. J. Lefkowitz, eds. Academic Press, San Diego, CA.

Sanfaçon, H., Wellink, J., LeGall, O., Karasev, A., van der Vlugt, R., and Wetzel, T. 2009. *Secoviridae*: A proposed family of plant viruses within the order Picornavirales that combines the families *Sequiviridae* and *Comoviridae*, the unassigned genera *Cheravirus* and *Sadwavirus*, and the proposed genus *Torradovirus*. Arch. Virol. 154:889-907.

Szychowski, J. A., McKenry, M. V., Walker, M. A., Wolpert, J. A., Credi, R., and Semancik, J. S. 1995. The vein-banding disease syndrome: A synergistic reaction between grapevine viroids and fanleaf virus. Vitis 34:229-232.

Taylor, C. A., and Brown, D. J. F. 1997. Nematode Vectors of Plant Viruses. CAB International, Wallingford, UK.

(Prepared by G. P. Martelli, M. Fuchs, and V. Savino)

North American Grapevine Nepoviruses

To date, seven nepoviruses have been described in North American vineyards: *Arabis mosaic virus* (ArMV); *Blueberry leaf mottle virus* (BLMoV) [syn. *Grapevine Bulgarian latent virus* (GBLV)]; *Grapevine fanleaf virus* (GFLV); *Peach rosette mosaic virus* (PRMV); *Tobacco ringspot virus* (TRSV); *Tomato black ring virus* (TBRV); and *Tomato ringspot virus* (ToRSV). However, only ToRSV, TRSV, and PRMV are regarded as North American in origin, whereas the others were reported previously in Europe and presumably introduced into North America via infected imports. These viruses cause similar symptoms, i.e., delayed bud burst in the spring accompanied by slow development of shoots and leaves, which may be mottled and malformed. With chronic infections, grapevines degenerate, decline, and suffer severe yield loss. In North American vineyards, GFLV is the predominant nepovirus, followed by ToRSV and to a lesser extent TRSV. The occurrence of BLMoV and TBRV has not been described in commercial vineyards in the past 25 years, PRMV only rarely, and ArMV, although present occasionally in California, Missouri, and Canada, causes limited economic damage.

Fig. 249. Foliar symptoms of ToRSV infection in the interspecific hybrid cv. Cascade. (Courtesy J. K. Uyemoto)

Fig. 250. Symptoms of infection by ToRSV in grapevine cv. Cascade, consisting of a stunted, weak cane and lack of fruit production (right). Compared with cane from a healthy plant (left). (Courtesy J. K. Uyemoto)

Symptoms

In northeastern North America, ToRSV and TRSV cause a disease known as virus-induced grapevine decline, which involves extensive damage in own-rooted vines of *V. vinifera* cultivars (Chardonnay, Mission, and Riesling) and interspecific hybrids (Baco Noir, Cascade, Chelois, DeChaunac, Dutchess, Ventura, Siegfriedrebe, Vidal Blanc, and Vincent). Although cultivars vary in susceptibility, both viruses induce similar disease symptoms. The first symptom of infection on an otherwise healthy appearing grapevine consists of leaves on isolated shoots that have chlorotic mottling, oak leaf pattern, and/or ring spots (Fig. 249). Thereafter, in chronically infected grapevines, buds can be killed or injured by low winter temperatures and surviving buds produce characteristically stunted, weak shoot growth (Fig. 250) and small, distorted leaves. Yields are nil or greatly reduced as fruit clusters are few and berries set poorly and ripen unevenly (Fig. 251).

ToRSV also incites diseases known as grapevine yellow vein in California and grapevine little berry in Maryland, where the overall symptoms differ markedly from those produced in the colder viticultural regions of New York State, Pennsylvania, and Ontario, Canada. With grapevine yellow vein, symptoms appear as a general chlorosis on portions of the leaf blade and yellow flecks concentrated along leaf veins. In Maryland, infected grapevines do not develop leaf symptoms, although both diseases reduce fruit set. Furthermore, in conjunction with this fruit thinning effect and in the absence of severe low temperatures damaging buds during dormancy (e.g., in California), diseased grapevines show increased vegetative vigor, producing "bull canes" and appearing more robust than healthy ones.

Peach rosette mosaic virus is restricted to Michigan and to southwestern Ontario, Canada. Other names for disease incited by this virus include grapevine degeneration, grapevine decline, berry shelling, and delayed budding. PRMV was first described infecting peach trees, where it causes leaf mosaic and rosetting of shoots. In infected Concord grapevines (*V.* × *labruscana*), bud burst is delayed, leaves are malformed and mottled, shoots are stunted, fruit set is poor, and ripening uneven (Fig. 252). Affected grapevines generally decline slowly over a period of several years.

Blueberry leaf mottle virus typically remains latent in cultivars of *V. vinifera*. When present, symptoms are subtle, consisting of canes with reduced growth and leaves with widened sinuses. In *V.* × *labruscana* cv. Concord, infected vines exhibit delayed bud burst, erratic shoot growth, and fan-like leaves (Fig. 253), although these symptoms are largely absent by late summer and diseased vines do not decline appreciably. Nevertheless, fruit clusters are few, small and poorly set. Seed trans-

Fig. 251. Berry cluster of the interspecific hybrid Le Commandant (Bertille Seyve 2862) infected with TRSV. (Courtesy M. Fuchs)

mission of BLMoV occurs in cv. Concord, producing seedlings that show leaves with flattened sinuses.

An isolate of *Tomato black ring virus* was reported from Ontario, Canada, in the interspecific hybrid cv. Chambourcin (Joannes-Seyve 26-205). Infected grapevines exhibited leaf deformity, severe stunting and had few, if any, fruit clusters. In an extensive graft-inoculation trial, chronic symptoms developed in *V. labrusca* cvs. Dutchess, Elvira, Fredonia, and Vinered; in the interspecific hybrids Baco 22A, Chambourcin, and LN 33; *V. vinifera* cv. Mission; and rootstocks 4453, 110R, and *V. rupestris* St. George. Other cultivars and rootstocks developed latent infections or were resistant to the virus. Overall, eight of 10 rootstocks, 11 of 16 *V. labrusca* cultivars, eight of 10 interspecific hybrids, and one of five *V. vinifera* cultivars were successfully infected.

Grapevine fanleaf virus is widespread in California and has been found in Washington State and Missouri. *Arabis mosaic virus* is documented in California, Missouri, and the Canadian provinces of British Columbia, Ontario, and Quebec. For more information on GFLV and ArMV, see "European Nepoviruses and *Strawberry latent ringspot virus*."

Fig. 252. Uneven ripening on a vine of cv. Concord infected with peach rosette mosaic virus. (Courtesy A. Schilder)

Fig. 253. Two diseased leaves with flat sinuses (leaf lobes near petiole) from a BLMoV-infected cv. Concord grapevine, compared with a healthy leaf (right). (Courtesy J. K. Uyemoto).

Causal Agents

All abovementioned viruses belong to the genus *Nepovirus* in the family *Secoviridae* and most have nematode vectors. In addition to those already discussed under European nepoviruses, ToRSV is vectored by the dagger nematodes *Xiphinema americanum*, *X. californicum*, and *X. rivesi*; TRSV by *X. americanum*; and PRMV by *X. americanum* and *Longidorus diadecturus*. No nematode vector is known for BLMoV.

These nepoviruses are transmissible by mechanical inoculation of grapevine leaf extracts to herbaceous host plants, e.g., *Chenopodium quinoa*. They can also be seedborne in a variety of weed hosts. With the exception of GFLV and ArMV, which are serologically related, the others are antigenically distinct, readily detected and identified by serological assays of extracts prepared from succulent leaf tissues of grapevines and herbaceous host plants.

Disease Cycle and Epidemiology

These viruses are easily disseminated through propagation of infected source material, a particular concern with BLMoV, TBRV, GFLV, and ArMV, for which latent infections are common and visual inspections unreliable for detecting contaminated candidates. Also, with the possible exception of BLMoV, various weed hosts can serve as reservoirs for each of the individual nepoviruses, providing a reservoir source from which the nematode vector species can acquire and later transmit them to grapevines. These viruses are commonly seed transmitted and are easily perpetuated in their weed hosts. Weed hosts of primary concern with ToRSV and PRMV are dandelion (*Taraxacum officinale*); with TRSV, broadleaf plantain (*Plantago major*); with TBRV, 24 species in 15 plant families; and with ArMV, 15 species in 12 plant families.

BLMoV is seedborne in grapevine and *C. quinoa*. However, lacking a nematode vector, the epidemiological importance of such infections in grapevine seedlings is uncertain. The seemingly extensive occurrence of the serologically related GBLV in Bulgarian vineyards is assumed to be due to vine propagation from infected, asymptomatic sources.

Although GFLV is the causal agent of a production-limiting disease in various vineyard sites in California, it has not yet posed a worrisome problem in other regions of the United States or in Canada. Although the virus can be distributed occasionally through infected propagation material, it does not spread in areas where the nematode vector species, *X. index*, does not occur or fails to survive extreme winter temperatures.

Management

Only certified, clean planting stocks should be used when establishing new vineyards or replacing missing vines. Some resistant scion cultivars and/or rootstocks also are available. For example, all cultivars of *V. labrusca* are resistant to ToRSV and TRSV and certain ones are also resistant to the grapevine isolate of TBRV. Although several interspecific hybrids are highly susceptible to ToRSV and TRSV, Aurore (Siebel 5279), Seyval Blanc, and Vidal 256 appear to be resistant to ToRSV. Furthermore, susceptible scions may be propagated on the ToRSV-resistant rootstocks 5A, 5C, 18-815, 41B Millardet et de Grasset, 44-53 Malegue, 110 Richter, 1616C, Clinton, Sanona, Shakoka, *V. riparia* Gloire, and *V. rupestris* St. George. Host resistance to GFLV has been identified, albeit not in commercially available cultivars and rootstocks. However, the use of *X. index*-tolerant rootstocks can substantially delay GFLV infection of scions

Assuming the use of clean or resistant planting stock, the most effective control tactic for these diseases is careful preplant preparation of known infested sites. Large grape roots should be thoroughly destroyed when replanting previous vineyard ground, as they are capable of surviving for several years after vine removal and can sustain residual populations of viruliferous vector nematodes. Nematicides are generally not efficient for disease control due to their inability to reach nematodes present in deeper layers of the soil. Fallowing for up to a 10-year period is recommended for eradication of *X. index* and, therefore, control of GFLV. However, this practice is difficult to implement for economic reasons. Thorough control of potential virus-reservoir weed hosts is a necessity.

Selected References

Allen, W. R., Van Schagen, J. G., and Eveleigh, E. S. 1982. Transmission of peach rosette mosaic virus to peach, grape and cucumber by *Longidorus diadecturus* obtained from diseased orchards in Ontario. Can. J. Plant Pathol. 4:16-18.

Dias, H. F. 1975. Peach rosette mosaic virus. Descriptions of Plant Viruses, no. 150. Commonwealth Mycological Institute and Association of Applied Biologist, Kew, England.

Gonsalves, D. 1982. Reaction of grape varieties to Tomato ring spot virus. Dev. Ind. Microbiol. 23:91-97.

Lunden, S., Meng, B., Avery, J., and Qiu, W. 2010. Association of *Grapevine fanleaf virus, Tomato ringspot virus* and *Grapevine rupestris stem pitting-associated virus* with a grapevine vein-clearing complex on var. Chardonnay. Eur. J. Plant Pathol. 126:135-144.

McKenzie, D. J., Johnson, R. C., and Warner, C. 1996. Incidence of four important viral pathogens in Canadian vineyards. Plant Dis. 80:955-958.

Martelli, G. P., Gallitelli, D., Abracheva, P., Savino, V., and Quacquarelli, A. 1977. Some properties of Grapevine Bulgarian latent virus. Ann. Appl. Biol. 85:51-58.

Martelli, G. P., Quacquarelli, A., and Gallitelli, D. 1978. Grapevine Bulgarian latent virus. Descriptions of Plant Viruses, no. 186. Commonwealth Mycological Institute and Association of Applied Biologist, Kew, England.

Mekuria, T. A., Gutha, L. R., Martin, R. R., and Naidu, R. A. 2009. Genome diversity and intra- and interspecies recombination events in *Grapevine fanleaf virus*. Phytopathology 99:1394-1402.

Milkus, B. N. 2001. Incidence of four NEPO viruses in Missouri vineyards. Am. J. Enol. Vitic. 52:56-57.

Mountain, W. L., Powell, C. A., Forer, L. B., and Stouffer, R. F. 1983. Transmission of Tomato ring spot virus from dandelion via seed and dagger nematodes. Plant Dis. 67:867-868.

Murant, A. F. 1970. Arabis mosaic virus. Descriptions of Plant Viruses, no. 16. Commonwealth Mycological Institute and Association of Applied Biologist, Kew, England.

Murant, A. F. 1970. Tomato black ring virus. Descriptions of Plant Viruses, no. 38. Commonwealth Mycological Institute and Association of Applied Biologist, Kew, England.

Ramsdell, D. C., and Stace-Smith, R. 1983. Blueberry leaf mottle virus. Descriptions of Plant Viruses, no. 267. Commonwealth Mycological Institute and Association of Applied Biologist, Kew, England.

Stobbs, L. W., and Van Schagen, J. G. 1985. Relationship between Grapevine Joannes-Seyve virus and Tomato blackring virus. Can J. Plant Pathol. 7:37-40.

Uyemoto, J. K., Cummins, J. R., and Abawi, G. S. 1977. Virus and viruslike diseases affecting grapevines in New York vineyards. Am. J. Enol. Vitic. 28:131-136.

Uyemoto, J. K., Taschenberg, E. F., and Hummer, D. K. 1977. Isolation and identification of a strain of Grapevine Bulgarian latent virus in Concord grapevines in New York State. Plant Dis. Rep. 61:949-953.

(Prepared by J. K. Uyemoto)

The Rugose Wood Complex

Descriptions of woody cylinder alterations on grapevines resembling rugose wood symptoms are found in the French literature of the early 1900s, with no indication of their possible infectious nature. As a graft-transmissible disease, rugose wood was first identified and described from southern Italy in the early 1960s. It is now known to occur worldwide, with records existing from Europe, the Mediterranean basin, Near and Far East, Australia, South Africa, and North and South America.

Symptoms

Affected grapevines appear less vigorous than normal and show delayed bud opening in spring; some decline and die within a few years from planting. Grafted grapevines often show swollen scions at the bud union and a marked difference in the relative diameters of their scion and rootstock (Fig. 254). With certain cultivars, the bark above the graft union is exceedingly thick and corky, spongy in texture, and has a rough appearance, a condition aptly named "corky rugose wood." The woody cylinder is typically marked by pits and grooves, which correspond to peg- and ridgelike protrusions originating from the cambial surface of the bark (Fig. 255). These alterations may occur on scions, rootstocks, or both. In most cases, no specific foliage symptom develops, although certain cultivars show leaf rolling, yellowing, or reddening. Bunches may be few and small, and the crop reduced by 20–30%. Highly susceptible cultivars can be unfruitful. Most, if not all, *Vitis vinifera* cultivars and American rootstocks are susceptible. Infections are latent in ungrafted scions and rootstocks; symptoms develop after grafts are complete, although some infected grafted plants remain asymptomatic. The severity of pitting and grooving in symptomatic wood varies according to climate (e.g., symptoms are much reduced or not expressed under cool and wet conditions) and possibly in relation to the scion/rootstock combination as well. For example, grapevine shoot necrosis reported to affect cv. Corniola in the Apulia region of Italy may represent a cultivar reaction to a rugose wood component.

Rugose wood is a complex consisting of four different diseases recognized by the differential reactions induced in the indicators *Vitis rupestris* St. George, LN33, and Kober 5BB.

Rupestris stem pitting disease. Symptoms on the woody cylinder appear as a distinct band of basipetal pitting extending downward from the point of inoculation in *V. rupestris* St. George (Fig. 256), whereas LN33 and Kober 5BB remain symptomless.

Fig. 254. Rugose wood symptoms at the graft union of an infected grapevine with a swollen scion portion with thick, rough bark. (Courtesy G. P. Martelli)

Corky bark disease. Symptoms occur as grooves and pits over the entire stem surface of *V. rupestris* and LN33. Moreover, LN33 is severely stunted with rolled, reddish leaves and internodal swelling of the canes (Fig. 257). No disease symptoms develop in Kober 5BB.

Kober stem grooving disease. Symptoms of marked grooving develop on Kober 5BB (Fig. 258), whereas wood markings are absent on *V. rupestris* and LN33.

LN33 stem grooving disease. The grooves formed on LN33 are indistinguishable from those of corky bark, but are not accompanied by the foliar discolorations or internodal swellings associated with corky bark. No diagnostic symptoms are induced in *V. rupestris* or Kober 5BB.

Causal agents

Viruses in the genera *Vitivirus* or *Foveavirus* are recognized as putative agents of the individual diseases of the rugose wood complex. All have flexuous filamentous particles of about 800 × 12 nm, with distinct transverse cross banding and containing a single species of linear positive-sense, single-stranded RNA that accounts for 5% of the total particle weight. Coat protein subunits have a molecular weight of 22–28 kDa. Vitiviruses, but not foveaviruses, are mechanically transmissible to a restricted range of herbaceous hosts, albeit with difficulty. Individual viruses can be identified through ELISA tests with polyclonal antisera and/or monoclonal antibodies. The best antigen sources for serological diagnosis are cortical shavings from mature dormant canes. Alternative assays include single-step or nested reverse transcription-polymerase chain reaction (RT-PCR), immunocapture RT-PCR, and spot-RT-PCR using degenerate or virus-specific primers. Immunocapture RT-PCR is 1000-fold more sensitive than ELISA for virus detection in grapevines.

Grapevine rupestris stem pitting-associated virus (GRSPaV), a definitive member of the genus *Foveavirus*, is the associated

Fig. 255. Prominent peglike protrusions originating from the cambial surface of the bark of a rugose wood-affected grapevine, with corresponding pits and grooves in the underlying wood cylinder. (Courtesy G. P. Martelli)

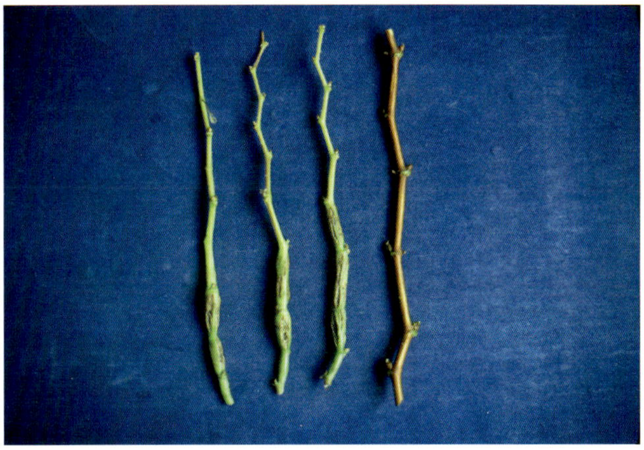

Fig. 257. Symptoms of corky bark disease on LN33, including swelling and cracking of shoot basal internodes on plants inoculated with *Grapevine virus B* (healthy cane on right). (Courtesy G. P. Martelli)

Fig. 256. Symptoms of rupestris stem pitting disease on *V. rupestris* 'St. George', including a line of small pits developing basipetally from the point of inoculation with *Grapevine rupestris stem pitting-associated virus*. (Courtesy G. P. Martelli)

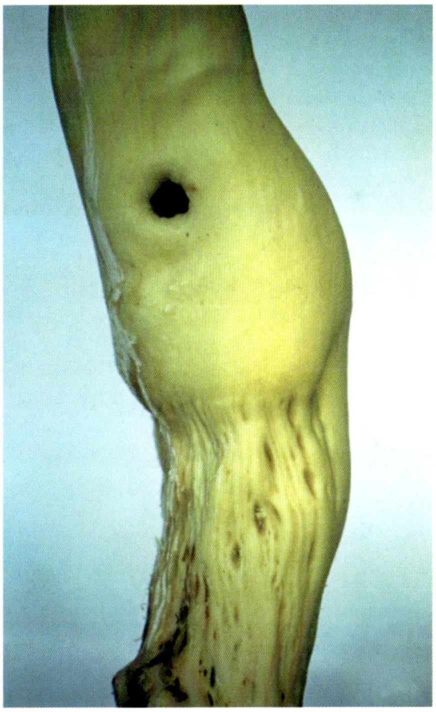

Fig. 258. Symptoms of Kober stem grooving disease in Kober 5BB rootstock, associated with infections by *Grapevine virus A*. (Courtesy G. P. Martelli)

agent of rupestris stem pitting disease. Virus particles are not readily observed with the electron microscope. GRSPaV occurs in nature as a family of molecular variants, with a viral RNA genome of 8726 nt.

Grapevine virus A (GVA), the type species of the genus *Vitivirus*, has been associated with Kober stem grooving. Its viral RNA has a genome of 7349 nt. Biological, serological, and molecular variants of the virus are known. Some of these variants are thought to be implicated in the etiology of Shiraz disease in South Africa and Australia.

Grapevine virus B (GVB) is a vitivirus distantly related serologically to GVA and one of the etiological agents associated with corky bark. GVB is also involved in young grapevine decline, a graft incompatibility condition recorded from California. GVB has a genome of 7599 nt and occurs in nature as a family of molecular variants.

Grapevine virus D (GVD) is a vitivirus distantly related serologically to GVA and GVB. An association with corky rugose wood has been reported, but there is no experimental evidence of its implication in the genesis of this disease.

Grapevine virus E (GVE) is a newly described vitivirus. Although identified from a table grape cultivar with stem pitting symptoms, its implication in the rugose wood complex is uncertain. GVE has also been detected in Alaska in domesticated *Ribes* spp. exhibiting a vein-clearing symptom, but not in native *Ribes* spp. GVE has a genome of 7568 nt. The Ribes isolate of GVE contains a 322 intergenic region that is absent in the grapevine isolates.

Grapevine virus F (GVF) is a novel member of the genus *Vitivirus* found in California. Its genome is 7551 nt in size. It induces graft incompatibility in cv. Cabernet Sauvignon grafted on different rootstocks.

Disease Cycle and Epidemiology

GVA and GVB are transmitted among grapevines by pseudococcid mealybugs and/or scale insects in a semipersistent manner. Specific GVA vectors are the mealybugs *Planococcus citri*, *Pl. ficus*, *Pseudococcus longispinus*, *Ps. viburni*, *Heliococcus bohemicus*, *Phenacoccus aceris*, and the scale insect *Neopulvinaria innumerabilis*. The virus was consistently detected in over 70% of mealybug populations collected in rugose wood-infected vineyards from different European and Mediterranean countries. GVB is transmitted by *Ps. longispinus*, *Ps. viburni*, *Ps affinis*, *Ph. aceris*, and *Pl. ficus*. GVE is transmitted by *Pseudococcus comstocki*. GRSPaV has no known vectors, but is suspected to be borne by pollen.

All putative disease agents of the rugose wood complex are phloem-limited viruses that persist in infected stocks and generally are mechanically transmittable with difficulty. No alternative hosts in nature are known, and natural dissemination is limited owing to the relatively limited range of vector movement. Long-distance spread is, therefore, achieved through transport of infected propagative material. The presence of rugose wood and its causal agents in phylloxera-free countries with a millennial history of own-rooted grapevine cultivation suggests that the disease complex originated in the Old World and was distributed worldwide in infected plants.

Management

Sanitary selection of propagating material can efficiently reduce or minimize the introduction of rugose wood into vineyards. However, symptomless infections limit the reliability of sanitary selection; hence, all sources must be indexed on three indicators (*V. rupestris*, Kober 5BB, and LN33) and/or assayed using antisera or PCR primer sets. Rugose wood agents can be eliminated with high efficiency (up to 85%) by heat therapy, meristem tip culture, or a combination of the two. Other efficient methods are somatic embryogenesis (GRSaV, GVA), cryotherapy, and electrotherapy (GVA). Only clean material should be used for propagation.

No information is available regarding the utility of controlling mealybug vectors. However, these insects are not easy to control by pesticides as they overwinter under the bark of grapevines and possess an unwettable waxy covering. Scale insects are also difficult to control with pesticides. No natural sources of resistance to any rugose wood agent are known but the possibility of using pathogen-derived resistance in transformed *Vitis* constructs has shown potential and is being explored.

Selected References

Abou Ghanem, N., Saldarelli, P., Minafra, A., Buzkan, N., Castellano M. A., and Martelli, G. P. 1997. Properties of grapevine virus D, a novel putative trichovirus. J. Plant Pathol. 79:15-25.

Alabi, O. J., Martin, R. R., and Naidu, R. A. 2010. Sequence diversity, population genetics and potential recombination events in *Grapevine rupestris stem pitting-associated virus* in Pacific North-West vineyards. J. Gen. Virol. 91:265-276.

Boscia, D., Savino, V., Minafra, A., Namba, S., Elicio, V., Castellano, M. A., Gonsalves, D., and Martelli G. P. 1993. Properties of a filamentous virus isolated from grapevines affected by corky bark. Arch. Virol. 130:109-120.

Boscia, D., Minafra, A., and Martelli, G. P. 1997. Filamentous viruses of the grapevine: Putative trichoviruses and capilloviruses. Pages 19-28 in: Filamentous Viruses of Woody Plants. P. L. Monette, ed. Research Signpost, Trivandrum, India.

Chevalier, S., Greif, C., Clauzel, J. M., Walter, B., and Fritsch, C. 1995. Use of an immunocapture-polymerase chain reaction procedure for the detection of grapevine virus A in Kober stem grooving-infected grapevines. J. Phytopathol. 143:369-373.

Coetzee, B., Maree, H. J., Stephan, D., Freeborough, M-J., and Burger, J. T. 2010. The first complete nucleotide sequence of a grapevine virus E variant. Arch. Virol. 155:1357-1360.

Conti, M., Milne, R. G., Luisoni, E., and Boccardo, G. 1980. A closterovirus from a stem-pitting-diseased grapevine. Phytopathology 70:394-399.

Du Preez, J., Stephan, D., Mawassi, M., and Burger, J. T. 2011. The grapevine infecting vitiviruses with particular reference to grapevine virus A. Arch. Virol. 156:1495-1503.

Goszczynski, D. E., du Preez, J., and Burger J. T., 2008. Molecular divergence of Grapevine virus A (GVA) variants associated with Shiraz disease in South Africa. Virus Res. 138:105-110.

Hewitt, W. B., Goheen, A. C., Raski, D. J., and Gooding, G. V. Jr. 1962. Studies on virus diseases of the grapevine in California. Vitis 3:57-83.

La Notte, P., Buzkan, N., Choueiri, E., Minafra, A., and Martelli, G. P. 1997. Acquisition and transmission of Grapevine virus A by the mealybug *Pseudococcus longispinus*. J. Plant. Pathol. 79:79-85.

Martelli, G. P., ed. 1993. Graft-transmissible Diseases of Grapevine. Handbook for Detection and Diagnosis. FAO Publication Division, Rome, Italy.

Martelli, G. P., Minafra, A., and Saldarelli, P. 1997. *Vitivirus*, a new genus of plant viruses. Arch. Virol. 42:1929-1932.

Martelli, G. P., and Jelkmann, W. 1998. *Foveavirus*, a new plant virus genus. Arch. Virol. 142:1245-1249.

Martelli, G. P., Adams, M. J., Kreuze, J. F., and Dolja, V. V. 2007. Family *Flexiviridae*: A case study in virion and genome plasticity. Annu. Rev. Phytopathol. 45:73-100.

Meng, B., Johnson, R., Peressini, S., Forsline, P. L., and Gonsalves D. 1999. Rupestris stem pitting associated virus-1 is consistently detected in grapevines that are infected with rupestris stem pitting. Eur. J. Plant Pathol. 105:191-199.

Meng, B., Pang, S. Z., Forsline, P. L., McFerson, J. R., and Gonsalves, D. 1998. Nucleotide sequence and genome structure of grapevine rupestris stem pitting-associated virus-1 reveal similarities to apple stem pitting virus. J. Gen. Virol. 79:2059-2069.

Minafra, A., Saldarelli, P., and Martelli, G. P. 1997. Grapevine virus A: Nucleotide sequence, genome organization, and relationships in the *Trichovirus* genus. Arch. Virol. 142:417-423.

Nakaune, R., Toda, S., Mochizuki, M., and Nakano, M. 2008. Identification and characterization of a new vitivirus from grapevine. Arch Virol. 153:1827-1832.

Saldarelli, P., Minafra, A., and Martelli, G. P. 1996. The nucleotide sequence and genomic organization of Grapevine virus B. J. Gen. Virol. 77:2645-2652.

Zhang, Y.-P., Uyemoto, J. K., Golino, D. A., and Rowhani, A. 1998. Nucleotide sequence and RT-PCR detection of a virus associated with grapevine rupestris stem pitting disease. Phytopathology 88:1231-1237.

(Prepared by G. P. Martelli, D. Boscia, A. Minafra, and P. Saldarelli)

Vein Mosaic

Vein mosaic is a semi-latent disease first described from France, and reported to occur in other European countries with incidence of up to 50–60%. It also has been reported from California, Brazil, and Australia.

Symptoms

In sensitive *V. vinifera* cultivars, disease symptoms include pale green discoloration of the leaf tissues, primarily along the main veins (Fig. 259). The veins themselves produce a feathering or banding, which is not necessarily restricted to these zones. Grapevine size and vigor may be adversely affected. Symptom severity can vary with cultivar and year, although specific climatic influences have not been determined. Infected *V. rupestris* St. George vines are reduced in vegetative growth and basal leaves develop chlorosis. However, infections are commonly symptomless in most cultivars. Ultrastructural studies of symptomatic leaf tissues have depicted modifications in the shape and outward appearance of chloroplasts.

Fig. 259. Symptoms of vein mosaic disease with pale green discoloration along the veins of a *V. vinifera* leaf. (Courtesy U. Prota)

Fig. 260. Typical vein mosaic symptoms on the indicator *Vitis riparia* 'Gloire de Montpellier'. (Courtesy U. Prota)

Causal agent

Vein mosaic is caused by an unidentified graft-transmissible agent. Transmission to *Vitis riparia* Gloire de Montpellier, the specific indicator, is readily achieved by chip budding in the field and by green grafting. Under greenhouse conditions at 26°C, symptoms may appear within 3 weeks after inoculation (Fig. 260). In field indexing experiments, symptoms develop much more slowly, requiring incubation of up to one year.

Disease cycle and epidemiology

Vein mosaic is spread through propagation of infected materials. Although seed transmission is suspected, there is no evidence of natural spread.

Management

Propagation should be done using only clean sources. The disease agent can be eliminated from contaminated sources by heat therapy.

Selected References

Kassemeyer, H.-H., Grenan, S., and Greif, C. 1997. Use of green grafting for the biological indexing of grapevine virus and virus-like diseases. Pages 119-127 in: Sanitary Selection of the Grapevine. Protocols for Detection of Viruses and Virus-like Diseases. B. Walter, ed. Les Colloques 86. INRA-INITA, Bordeaux, France.

Krake, L. R., Steele Scott, N., Rezaian, M. A., and Taylor, R. H. 1999. Graft-transmitted diseases of grapevines. CSIRO Publishing, Collingwood, Victoria, Australia.

Legin, R., and Vuittenez, A. 1973. Comparaison des symptomes et transmission par greffage d'une mosaique nervaire de *Vitis vinifera*, de la marbrure de *V. rupestris* et d'une affection necrotique des nervures de l'hybride Rup.-Berl. 110 R. Riv. Patol. Veg. Ser. IV, 9 (Suppl.):57-63.

Legin, R., and Walter, B. 1986. Regenerierung von Virosen befallener Reben durch Thermotherapie oder Wärmebehandlung und Schnellvermehrung del Rebe. Wein-Wiss. 41:14-25.

(Prepared by U. Prota and R. Garau)

Vein Necrosis

Vein necrosis was first found in France. It also occurs in other European countries, California, Brazil, and Australia. Disease incidence exceeding 70% has been observed in commercial vineyards.

Symptoms

The disease is latent in all *V. vinifera* cultivars and most American rootstocks. On the indicator host for vein necrosis, *V. rupestris* × *V. berlandieri* 110R, symptoms include veinlet necrosis (Fig. 261) appearing first on the basal leaves and progressing to other leaves toward the shoot tip. Necrotic symptoms are best seen on the lower leaf surfaces but are visible on the upper leaf surfaces also. Severe forms of the disease may induce necrosis of tendrils, dieback of green shoots, stunting, and occasionally plant death. Symptoms persist and are readily observed throughout the vegetative season.

Causal agent

Vein necrosis is caused by a graft-transmissible agent whose identity has not been determined, although some genetic variants of *Grapevine rupestris stem pitting-associated virus* are associated with the disease. Transmissions to the 110R indicator host is accomplished by chip bud grafts and by green grafts. Symptoms appear 4 weeks after inoculation under greenhouse conditions at 26°C or 8–10 weeks in the field. Mild disease

Fig. 261. Necrotic veins and veinlets symptomatic of vein necrosis disease on the indicator host *V. rupestris* x *V. berlandieri* 110R. (Courtesy U. Prota)

forms may require incubation of up to one year before symptoms develop.

Disease cycle and epidemiology

Vein necrosis spreads only through propagation of infected materials.

Management

Propagation only from healthy sources is recommended. Detection is accomplished by grafting on 110R; RT-PCR with virus-specific primers and western blot with an antiserum to recombinant coat protein of GRSPaV allow sensitive and reliable detection in symptomatic 110R plants. The disease agent can be eliminated from contaminated sources by heat therapy.

Selected References

Alliaume. A., Spilmont. A. S., Beuve. M., and Lemaire. O. 2012. Grapevine vein necrosis is not exclusively associated to GRSPaV group 2. Pages 84-85 in: Proc. 17th Congress, International Council for the Study of Virus and Virus-like Diseases of the Grapevine (ICVG). http://web.pppmb.cals.cornell.edu/fuchs/icvg/data/ICVG-2012-Proceedings.pdf

Bouyahia, H., Boscia, D., Savino, V., La Notte, P., Pirolo, C., Castellano, M. A., Minafra, A., and Martelli, G. P. 2005. Grapevine rupestris stem pitting-associated virus is linked with grapevine vein necrosis. Vitis 44:133-137.

Kassemeyer, H. H., Grenan, S., and Greif, C. 1997. Use of green grafting for the biological indexing of grapevine virus and virus-like diseases. Pages 119-127 in: Sanitary Selection of the Grapevine. Protocols for Detection of Viruses and Virus-like Diseases. B. Walter, ed. Les Colloques 86. INRA-INITA, Bordeaux, France.

Legin, R., and Vuittenez, A. 1973. Comparaison des symptomes et transmission par greffage d'une mosaique nervaire de *Vitis vinifera*, de la marbrure de *V. rupestris* et d'une affection necrotique des nervures de l'hybride Rup.-Berl. 110 R. Riv. Patol. Veg. Ser. IV, 9 (Suppl.):57-63.

Savino, V., Boscia, D., and Martelli, G. P. 1985. Incidence of some graft-transmissible viruslike diseases of grapevine in visually selected and heat-treated stocks from southern Italy. Phytopathol. Mediterr. 24:204-207.

(Prepared by U. Prota and R. Garau)

Minor Virus Diseases

Several graft-transmissible diseases of minor economic importance have been recorded from Europe.

Fig. 262. Bright yellow foliar discolorations symptomatic of grapevine yellow mottle disease. (Courtesy G. P. Martelli)

Grapevine yellow mottle. This disease is characterized by different patterns of chrome-yellow discolorations of the foliage. The spring growth shows more or less extensive yellowing of leaf blades that usually does not extend to the veins (Fig. 262), whereas typical summer responses are faint yellow speckling and ring and line patterns. The putative causal agent is *Alfalfa mosaic virus* (AMV), a mechanically-transmissible virus with differently shaped particles (from quasi-isometric to bacilliform, 30–57 nm in size) and a tripartite RNA genome consisting of 3644, 2593, and 2037 nt for RNA-1, -2, and 3, respectively. AMV is aphidborne but new infections in grapevines are very rare, suggesting that spread is primarily through the use of infected propagating materials. Yellow mottle has been reported from Bulgaria, former Czechoslovakia, France, Germany, Hungary, Switzerland, and Turkey.

Grapevine line pattern. Symptoms are bright yellow discolorations of the leaves forming marginal rings, scattered spots, blotches, or maple leaf patterns that typically are confined to the petiolar areas or upper part of the leaf blade, roughly following its contour (Fig. 263). Grapevine line pattern virus (GLPV), the presumed causal agent of the disease, is a poorly characterized putative *Ilarvirus* with differently shaped particles, quasi-spherical 25–30 nm in diameter and bacilliform 40–75 nm in length. GLPV has no known vector, is seed-transmitted and spread by propagating diseased materials. This disease has been reported only from Hungary.

Grapevine Roditis leaf discoloration. Symptoms of this disease are prominent in late summer and consist of yellow and/or reddish discolorations of the tissues along the veins, in the interveinal areas, or across variously extended and deformed sectors of the leaf blade, especially near the petiole (Fig. 264). Bunches are reduced in number and size and berries have low sugar content. Symptomatic grapevines were reported to be doubly infected with GFLV and *Carnation mottle virus* (CarMV), an isometric virus 30 nm in diameter with a monopartite RNA genome of 4003 nt. More recent findings indicate, however, that GFLV is not involved in the etiology of the disease. Instead, *Grapevine virus B* (GVB), one of the putative agents of corky bark (rugose wood complex) has a very high association with diseased grapevines. Roditis leaf discoloration

Fig. 263. Marginal yellow line pattern presumed to be elicited by Grapevine line pattern virus infection. (Courtesy G. P. Martelli)

Fig. 264. Symptoms of grapevine Roditis leaf discoloration in cv. Roditis. (Courtesy of A.D. Avgelis)

is graft-transmissible and spreads with infected propagative material. It has been recorded only from Greece.

Grapevine angular mosaic. The symptoms of this disease include chlorotic angular spots on leaf blades, discoloration of tissues bordering the veins, and crinkling and deformation of the leaves (Fig. 265). Infected grapevines are stunted, decline gradually, and some die. Fruit bunches are poorly developed because of floral abortion. Berries when present remain small, wrinkled, and bear nonviable seeds. The disease agent is Grapevine angular mosaic virus (GAMV), a virus with a tripartite RNA genome molecularly related to a number of ilarviruses, the closest being *Tobacco streak virus*. GAMV is pollenborne in herbaceous hosts, but whether this mechanism operates in grapevines is not known. The disease has been reported only from Greece.

Raspberry bushy dwarf virus (RBDV), the type species of the genus *Idaeovirus*, is a pollen- and seedborne virus with quasi-spherical particles about 30 nm in diameter and a bipartite single-stranded RNA genome consisting of two functional RNA species with 5449 nt (RNA-1) and 2214 nt (RNA-2).

Fig. 265. Leaf symptoms of grapevine angular mosaic in a self-cross of *V. vinifera* cv. Baresana. (Courtesy S.M. Girgis)

RBDV is widespread in *Rubus* spp. It was recently isolated from grapevines cv. Laski Rizling in Slovenia that showed yellow line pattern symptoms resembling those reported for GLPV.

Selected References

Bovey, R., and Brugger, J. J. 1975. Le virus de la mosaïque de la luzerne sur la vigne. Rev. Suisse Vitic., Arboric., Hortic. 7:63-65.

Girgis, S. M., Bem, F. P., Kyriakopoulou, P. E., Dovas, C. I., Avgelis, A., and Katis, N. I. 2003. The etiology of a new virus disease: Grapevine angular mosaic. Page 19 in: Extended Abstracts 14th Meeting, International Council for the Study of Virus and Virus-like Diseases of the Grapevine (ICVG). http://web.pppmb.cals.cornell.edu/fuchs/icvg/data/2003Session12.pdf

Mavrič Pleško, I., Viršček Marn, M., Širca, S., and Urek, G. 2009. Biological, serological and molecular characterisation of *Raspberry bushy dwarf virus* from grapevine and its detection in the nematode *Longidorus juvenilis*. Eur. J. Plant Pathol. 123:261-268.

Rumbos, I. C., and Avgelis, A. D. 1989. Roditis leaf discoloration: A new virus diseases of grapevine: Symptomatology and transmission to indicator plants. J. Phytopathol. 125:274-278.

(Prepared by G. P. Martelli and A. D. Avgelis)

Other Viruses and Viruslike Agents

The most significant viruses and viruslike diseases of grapevines have been described elsewhere in this compendium. However, other pathogens and diseases that are less known or limited in distribution have been reported in grapevines over the years (Table 4). One disease found in South America, Elqui yellows, might be caused by a phytoplasma. The distribution and horticultural significance of these viruses and diseases vary considerably. Whereas a few are mentioned in isolated reports, others have occurred in one or more cultivars in a specific region or were found in several viticultural areas. Some agents were latent and produced no obvious or measurable effect, whereas others caused significant symptoms and yield losses in specific geographic areas. The causal agents for 11 of these diseases are unknown; their detection was based on symptoms displayed or transmissions to healthy grape plants or herbaceous indicators.

Grapevine viruses have being characterized in significant numbers in recent years, and new diseases caused by unknown pathogens have been reported. Modern diagnostic and molecular technologies have assisted in detecting both known viruses newly associated with grapes and newly described viruses associated with symptoms of formerly unknown etiology.

TABLE 4. Other Viruses and Viruslike Agents Reported in Grapevines[a]

Pathogen	Natural Symptoms	Transmission[b]	Distribution
Artichoke Italian latent virus	Fanleaf-like, stunted	(N), Sa	Bulgaria
Broad bean wilt virus	Latent or mild mosaic	?G, (I), Sa	Bulgaria, South Africa
Cucumber mosaic virus	Latent	Sa	Denmark, Turkey
Elqui yellows agent	Leaf yellowing, sudden leaf drop, uneven lignification, poor fruit quality	?	?Argentina, Chile, ?South Africa
Grapevine ajinashika virus	No leaf symptoms, fruit unpalatable due to low sugar content	G	Japan
Grapevine Algerian latent virus	Latent	Sa	Algeria, *Germany, *Italy (*in river water)
Grapevine asteroid mosaic-associated virus	Star-shaped chlorosis, leaf asymmetry, twisting, puckering, stunted with little or no fruit	G	Greece, Italy, South Africa, U.S.A. (California)
Grapevine Bratislava mosaic agent	Leaf blotches, yellowing, low yield	Sa	Former Czechoslovakia
Grapevine bushy stunt agent	Bushy stunted plants, reduced yield	G	Italy
Grapevine Chasselas latent agent	Deformed leaves, stunting, delayed growth	G	France
Grapevine flat trunk agent	Elliptical or flattish trunk on opposite sides	G	Hungary, Italy, Israel, USA (California)
Grapevine little leaf agent	Small, cupped, discolored leaves, short internodes, no flowers or fruit	G	India
Grapevine shoot necrosis agent	Shoot necrosis & death, pale leaves, less yield	G	Italy
Grapevine stunt virus	Delayed growth, leaf rolling and stunting, undersized inflorescences	G, I	Japan
Grapevine Tunisian ringspot virus	Mild leaf mottling, leaf deformation	Sa	Tunisia
Grapevine yellow dwarf agent	Yellow leaf mottling, deformed leaves, necrotic spots, stunting	?	Taiwan
Infectious chlorosis and leaf reddening of Pinot noir agent	Leaf chlorosis becoming red-purple, premature leaf drop, strong growth depression	G	France
Potato virus X	Latent	(I), Sa	Italy, Tunisia
Sowbane mosaic virus	Latent	Sa	Slovak Republic, Germany
Summer mottle agent	Pale green to yellowish discolorations adjacent to main or secondary veins	G	Australia
Tobacco mosaic virus	Latent	Sa, Se	Bulgaria, Germany, Italy, Russia, USA (New York), former Yugoslavia
Tobacco necrosis virus	Latent	(Fu), Sa	Germany, South Africa, USA
Tomato bushy stunt virus	Yellow vein banding, yellow spots	(Fu), Sa	Bulgaria, Czech Republic, Germany, Italy
Unidentified isometric grapevine virus	Latent	G	Italy, Switzerland

[a] A question mark before an entry indicates an unproven association only.
[b] Transmission: Fu = soil fungi; G = grafting; I = insects; N = nematodes; Sa = sap transmissible (mechanically) to herbaceous hosts; Se = seeds of grapevine. Brackets are placed around means of transmission that have been demonstrated for other hosts but not proven for grapevine.

Innovations in viticultural practices have contributed in the recognition of some new diseases. Changes in cultivars and rootstocks have led to the emergence of latent infections into newly recognized diseases, often with serious economic consequences for specific scion/rootstock combinations. New transmission opportunities and, sometimes, economic damage may result when a pathogen is introduced into an environment with different populations of potential insect or nematode vectors. Plant damage and yield losses may also result when grapevines or formerly exotic *Vitis* species are introduced into a region and exposed to endemic pathogens and vectors for the first time.

The primary means of virus spread over long distances is by movement of infected propagative material. In today's market, international transport of plants for commercial and research purposes is significant and increasing. Viticulturists and growers alike are interested in importing and evaluating new cultivars and rootstocks. Viruses and viruslike graft-transmissible agents are difficult to detect in dormant plants or propagative materials without diagnostic testing. Once established, these agents are often difficult to control. Infected plants cannot be therapeutically cured.

Selected References

Bovey, R., Gärtel, W., Hewitt, W. B., Martelli, G. P., and Vuittenez, A. 1980. Virus and Virus-like Diseases of Grapevines. Payot Lausanne, Switzerland.

Bovey, R., and Martelli, G. P. 1992. Directory of Major Virus and Viruslike Diseases of Grapevines. Imprimerie Finzi, Tunis.

Frison, E. A., and Ikin, R., eds. 1991. Technical Guidelines for the Safe Movement of Grapevine Germplasm. Food and Agriculture Organization of the United Nations (FAO) and International Board for Plant Genetic Resources (IBPGR), Rome, Italy.

Koklu, G., Digiaro, M., Sabanadzovic, S., and Savino, V. 1999. Natural infections by cucumber mosaic virus in Turkish grapevines. Phytopathol. Medit. 38:33-36

Smith, I. M., Dunez, J., Lelliot, R. A., Phillips, D. H., and Archer, S. A., eds. 1988. European Handbook of Plant Diseases. Blackwell Scientific, Oxford, UK.

(Prepared by R. Johnson and J. A. Foster)

Grapevine Viroids

Viroids are small, circular RNA molecules that are a fraction of the size of viruses. The first viroid reported in grapevines was *Hop stunt viroid* (HSVd-g) in Japan. Subsequent research revealed multiple viroid infections in grapevines worldwide. Although viroids can cause severe diseases in other plant hosts, e.g., potato spindle tuber and citrus exocortis, there is no consistent evidence that they cause adverse effects in grapevines in and of themselves. However, two related viroids serve as components of the deleterious grapevine leaf banding disease.

Grapevine yellow speckle disease is the name given to the only syndrome attributed solely to infection by viroids.

Symptoms

Foliar symptoms of Grapevine yellow speckle disease are often absent or difficult to diagnose when only a few yellowish spots or flecks develop along major or minor leaf veins. In some years in Australia, the symptoms can be spectacular, consisting of bright yellow blotches (Fig. 266, A and D); vein-banding patterns (Fig. 266, E) are a variant resulting from co-infections with *Grapevine fanleaf virus* (GFLV). Occasionally, the symptoms may develop during mid-spring, but more often during late summer to autumn. When symptoms occur early, the speckles appear bleached and by late summer turn white. Speckle symptoms that develop mid-season are pale yellow and those developed during late summer light green. Yellow speckle symptoms are most likely to develop on exposed basal leaves.

Causal agents

To date, five different viroids have been identified in grapevines (Table 5). Two of these, *Grapevine yellow speckle viroids 1* and *2* (GYSVd-1 and -2), are the causal agents of yellow speckle disease and are only known to infect grapevines. Grapevine vein banding disease, common in Europe and California and previously attributed to GFLV as the sole incitant, is now known to require the presence of GFLV plus GYSVd-1 or -2, with its symptoms the product of a synergistic interaction between GYSVds and GFLV.

Australian grapevine viroid (AGVd) also is unique to grapevines, in which it is latent. It was originally recovered from grapevines in Australia co-infected with the other four viroids from this host but was distinguished from them by differences in electrophoretic properties, ability to replicate in cucumber and tomato, and in failed hybridization assays using previously characterized grapevine viroid probes. AGVd also is found in California.

Fig. 266. Variations in grapevine yellow speckle symptoms in inoculated cv. Cabernet Franc leaves (A-D). **A,** Severe symptoms resulting from seedling inoculation with plus-sense dimeric *Grapevine yellow speckle-1* transcript; **B,** scattered speckles from seedling inoculation with GYSVd-2; **C,** moderate symptoms from co-inoculation of a seedling with GYSVd-1 and -2; and **D,** severe leaf symptoms from a field-grown grapevine. **E,** Leaf of cv. Sultana with vein-banding symptoms, resulting from co-infection by GYSVd and GFLV. (Courtesy L. R. Krake)

The remaining two viroids also are presumed to be latent in grapevines. As early as 1952, stunted hop plants were observed in Japan. A biological assay using cucumber was developed and the causal agent was identified and named *Hop stunt viroid* (HSVd). In grapevine, the same infectious low-molecular-weight RNA (designated HSVd-g) was detected by molecular hybridization assays with HSVd probe, and on cucumber plants it caused symptoms indistinguishable from those incited by the hop isolate. HSVd-g has been detected in nearly all grapevines tested in Japan and Australia, and is reported from the United States and Europe. Its significance in grapevines may be as a reservoir to infect hop plants.

Symptoms of citrus exocortis were first described on that host in 1948. The causal agent, *Citrus exocortis viroid* (CEVd), was transmitted to herbaceous plants and *Gynura aurantiaca* as an indicator host. In Spain, a viroid with 89–92% sequence homology with CEVd was isolated from a symptomless grapevine and designated CEVd-g. It was also detected in symptomless grapevines in Australia.

Disease Cycle and Epidemiology

Long-distance spread of viroids is largely attributed to the propagation and distribution of infected plant material. The apparent ubiquitous presence of viroids within individual vineyards is likely attributable to mechanical transmissions with contaminated pruning tools.

Management

Propagation should be done using only clean source material. Although grape seedlings were once regarded as viroid-free germplasm, the grapevine viroids GYSVd-1, -2, AGVd, and GEVd have been detected in seedlings by means of molecular diagnostic techniques, which also have provided additional evidence of seed to seedling transmissions of GYSVd-1 and HSVd-g in multiple grape cultivars.

Selected References

García-Arenal, F., Pallas, V., and Flores, R. 1987. The sequence of a viroid from grapevine closely related to severe isolates of citrus exocortis viroid. Nucleic Acids Res. 15:4203-4210.

Koltunow, A. M., and Ali Rezaian, M. 1988. Grapevine yellow speckle viroid: Structural features of a new viroid group. Nucleic Acids Res. 16:849-864.

Koltunow, A. M., and Rezaian, M. 1989. Grapevine viroid 1B, a new member of the apple scar skin viroid group contains the left terminal region of tomato planto macho viroid. Virology 170:575-578.

Krake, L. R., and Woodham, R. C. 1983. Grapevine yellow speckle agent implicated in the aetiology of vein banding disease. Vitis 22:40-50.

Little, A., and Rezaian, M. A. 2003. Grapevine viroids. Pages 195-206 in: Viroids: Properties, Detection, Diseases and their Control. A. Hadidi, R. Flores, J. Randles, and J. Semancik, eds. CSIRO Publishing, Australia.

Rezaian, M. A. 1990. Australian grapevine viroid—evidence for extensive recombination between viroids. Nucleic Acids Res. 18:1813-1818.

Sano, T., Uyeda, I., Shikata, E., Meshi, T., Ohno, T., and Okada, Y. 1985. A viroid-like RNA isolated from grapevine has high sequence homology with hop stunt viroid. J. Gen. Virol. 66:333-338.

TABLE 5. Identities and Properties of Viroids Infecting Grapevine

Viroid	Size (nt)	Genus	Symptoms in Grapevine	Herbaceous Host	Reference
GYSVd-1	366	*Apscaviroid*	Yellow speckle	None known	Koltunow & Rezaian, 1988
GYSVd-2	363	*Apscaviroid*	Yellow speckle	None known	Koltunow & Rezaian, 1989
AGVd	369	*Apscaviroid*	None	Cucumber	Rezaian, 1990
HSVd-g	297	*Hostuviroid*	No report	Cucumber	Sano et al., 1985
CEVd-g	369	*Pospiviroid*	No report	Tomato	Garcia-Arenal et al., 1987

Szychowski, J. A., McKenry, M. V., Walker, M. A., Wolpert, J. A., Credi, R., and Semancik, J. S. 1995. The vein-banding disease syndrome: A synergistic reaction between grapevine viroids and fanleaf virus. Vitis 34:229-232.

Wolpert, J. A., Szychowski, J. A., and Semancik, J. S. 1996. Effect of viroids on growth, yield, and maturity indices of Cabernet Sauvignon grapevines. Am. J. Enol. Vitic. 47:21-24.

(Prepared by M. A. Rezaian and A. Little)

Nematode Parasites of Grapevines

Nematode diseases of grape were first reported in the mid-1880s. These initial reports were concerned exclusively with root-knot nematodes, primarily owing to the visible galls they produced on affected roots. Scarcely more than a handful of additional reports were published until the mid-1950s. The impetus for the study of nematodes was the discovery of chemicals in the 1940s and 1950s that reduced nematode numbers while increasing plant growth. Since 1954, research has shown that grapes are afflicted by a wide array of root-parasitic nematode species. Many of the species occur worldwide wherever grapes are grown, as they have been transported with rooted planting material. Although some nematode species cause characteristic galls on the root system, aboveground symptoms are generally nonspecific and may be manifested as unthrifty vines, poor growth, reduced yields, nutrient deficiency, and greater sensitivity to stress. These effects of damage to the root system are often misdiagnosed as water stress or nutrient deficiency. Indeed, in some cases the effect of the root damage may be partially compensated by increased water or nutrient supply, but that is usually a short-term solution. The types of plant-feeding nematodes that become established in a vineyard are determined by the nematode assemblage present in the soil at planting, sanitation and cleanliness of nursery stock, susceptibility of the selected rootstock, the nematode host status of rotation and cover crops, water source and irrigation method, and movement of soil by vehicles and tillage. The damage caused by nematodes may be exacerbated by climatic conditions, geographic region, moisture and nutrient stress, and viticultural practices.

Citrus Nematode

The citrus nematode was first found on citrus roots in 1913 and first reported on grape in 1956, in both California and Australia; since that time, it has been reported from all continents where grapes are grown. This nematode has a relatively narrow host range primarily consisting of woody perennial species: citrus, olive, persimmon, and grape. The citrus nematode is highly pathogenic to grapevines, particularly in finer-textured soils where root-knot nematodes are less competitive. Vine vigor is markedly reduced, and affected plants do not have the resilience to withstand stressful conditions. Yields gradually and inevitably decline, and vineyards become uneconomical.

Fig. 267. Roots damaged by citrus nematodes (right) versus healthy roots (left). (Courtesy D. J. Raski)

Symptoms

No specific symptoms diagnostic for the citrus nematode are apparent on aboveground parts of the vine. The main effect belowground is the death of feeder roots, although some rootlets may remain alive and support large populations of nematodes (Fig. 267). The bodies of adult females protrude from the root, often with many individuals in close proximity to each other and probably feeding from the same nurse cell system. Females produce a gelatinous matrix in which they deposit the eggs, and this covers the mass of female nematodes when they are located close together. Soil particles adhere to the matrix and are not easily dislodged when soil is shaken from the roots, giving infested rootlets a dirty appearance.

Causal Organism

The citrus nematode is *Tylenchulus semipenetrans* Cobb, the specific epithet reflecting that in the adult female stage the anterior of the body is embedded in the root and the posterior remains exposed on the root surface.

Life Cycle and Epidemiology

Juvenile citrus nematodes hatch from the egg in the second developmental stage. Those individuals destined to become males develop quite differently from those that will become females. Male juveniles have a degenerate stylet and esophagus and do not feed, and therefore do not cause damage to plants. They pass through three molts, with the intestine developing a progressively starved appearance with some residual bands of food reserves and clear areas where reserves have been depleted. Males are not necessary for reproduction; females produce viable eggs without being fertilized.

Juveniles destined to become adult females do feed, as ectoparasites on root epidermal or outer cortical cells and in areas where lateral rootlets are emerging, as they pass through three more stages after hatching before becoming adult; therefore, they do not have the starved appearance of males. When they reach the young female stage, the nematodes penetrate the root so that the head is in the inner cortical region, near the endodermis. There, the nematode stimulates formation of a system of somewhat enlarged, metabolically active nurse cells. As the females feed, their bodies enlarge; this occurs primarily on that portion of the body outside the root, as enlargement of the anterior portion embedded in the root tissue is restricted by the walls of root cells. Since the enlarged nematodes are partially inside the root and no longer able to move, these nematodes in the adult female stage are described as sedentary semi-endoparasites. The females, either in groups or as individuals, deposit several hundred eggs into the gelatinous matrix surrounding them. The life cycle from egg to egg requires 4–8 weeks to complete.

Selected References

Anwar, S. A., and Van Gundy, S. D. 1989. Influence of four nematodes on root and shoot growth parameters in grape. J. Nematol. 21:276-283.

Rahman, L., Creecy, H., and Orchard, B. 2008. Impact of citrus nematode (*Tylenchulus semipenetrans*) densities in soil on yield of grapevines (*Vitis vinifera* 'Shiraz') in south-eastern New South Wales. Vitis 47:175-180.

Siddiqi, M. R. 1974. *Tylenchulus semipenetrans*. Descriptions of Plant-Parasitic Nematodes. Set 3. No. 34. Commonwealth Institute of Helminthology, St. Albans, Hertfordshire, England.

Van Gundy, S. D. 1958. The life history of the citrus nematode *Tylenchulus semipenetrans* Cobb. Nematologica 3:283-294.

(Preparted by I. A. Zasada, H. Ferris, and M. V. McKenry)

Dagger and Needle Nematodes

Dagger nematodes, *Xiphinema* spp., are present in all major grape-growing areas of the world. This is a diverse group of nematodes with over 50 described species. Whereas many different species of *Xiphinema* have been reported from vineyards during surveys, the pathogenicity of many of these species to grapevines has not been determined. *Xiphinema index* Thorne & Allen, the most thoroughly studied of the dagger nematodes, is worldwide in distribution and can be a devastating pathogen of grapevines. Affected vines within highly porous soils decline markedly, producing fewer shoots until ultimately vines are totally unproductive. Besides its direct damage to plant roots, this nematode is also a vector of *Grapevine fanleaf virus* (GFLV), and infestations of *X. index* that are viruliferous have even more devastating effects on vineyards. *Xiphinema vuittenezi* Luc, Lima, Weischer & Flegg is most common in Europe, although it also is reported from other locations; direct damage associated with the species plus its ability to vector Nepoviruses make it a serious pest. *Xiphinema diversicaudatum* (Micoletzky) Thorne also is most common in Europe but found elsewhere as well. *Xiphinema americanum* Cobb is considered a complex of morphologically similar species and relationships within the group will no doubt become clearer with application of molecular technologies. *Xiphinema americanum* sensu lato is widespread in the United States but less so in European vineyards. Although not a strong pathogen of grapevines unless present at very high population densities, it also can cause damage as a vector of *Tomato ringspot virus* (ToRSV, a serious problem on some interspecific hybrid cultivars in eastern North America) and *Tobacco ringspot virus* (TRSV).

At least 15 species of needle nematodes, *Longidorus* spp., are reported from vineyard soils worldwide. Of these, *L. africanus* has been described as causing necrosis and malformations of grape roots. The remaining species have not been studied and are known only as part of recorded collections.

Both dagger and needle nematodes cause slow, gradual decline, seldom killing the vines outright. Once decline caused by these nematodes begins, little can be done to restore the affected vines' vigor and productivity. When these species also transmit plant viruses, their effects are significantly more devastating.

Symptoms

The root system of a vine attacked by *X. index*, *X. vuittenzi*, or *X. diversicaudatum* exhibits terminal root galls and many dead feeder roots if population levels are high (Fig. 268). Most feeding takes place near root tips, where proliferation and enlargement of affected cells often causes a bending of the root accompanied by slight swelling. Multiple prolonged attacks can result in darkened, necrotic spots that spread over the entire root tip. Parasitism by *L. africanus* causes root growth to stop soon after feeding begins.

Causal Organisms

The species of dagger nematodes that have been associated with grape are: *Xiphinema index* Thorne & Allen; *X. italiae* Meyl; *X. diversicaudatum* (Micoletzky) Thorne; *X. mediterraneum* Martelli & Lamberti; *X. pachtaicum* (Tulagonov) Kirjanova; *X. algeriense* Luc & Kostadinov; *X. vuittenezi* Luc, Lima, Weischer & Flegg; *X. turcicum* Luc & Dalmasso; and *X. americanum* Cobb. Within the *X. americanum* species complex the following species have been reported: *X. rivesi* Dalmasso; *X. californicum* Lamberti & Bleve-Zacheo; and *X. brevicolle* Lordello & da Costa.

The species of needle nematodes that have been associated with grape are: *Longidorus attenuatus* Hooper; *L. africanus* Merny; *L. carniolensis* Sirca; *L. elongatus* (de Man) Thorne & Swanger; *L. juvenilis* Micoletzky; *L. leptocephalus* Hooper; *L. sylphus* Thorne; *L. diadecturus* Eveleigh & Allen; *L. iranicus* Sturhan & Barooti; *L. macrosoma* Hooper; *L. poessneckensis* Altherr; *L. protae* Lamberti & Bleve-Zacheo; and *L. raskii* Lamberti & Agostinelli. In addition, *Paralongidorus maximus* has been reported from Europe.

There are several Nepoviruses that are vectored by dagger and/or needle nematodes. Specific virus/vector associations include: *Grapevine fanleaf virus* (*X. index*, *X. italiae*, and *X. vuittenzi*); *Tomato ringspot virus* (*X. californicum*, *X. americanum*, and *X. revesi*); *Tobacco ringspot virus* (*X. americanum* and *X. rivesi*); *Peach rosette mosaic virus* (*X. americanum*, *X. revesi*, and *L. diadecturus*); *Arabis mosaic virus* (*X.diversicaudatum*); *Raspberry ringspot virus* (*L. elongatus*, *L. macrosoma*, and *P. maximus*); *Tomato black ring virus* (*L. attenuates* and *L. elongatus*), and *Artichoke Italian latent virus* (*L. attenuatus*). In addition, the sadwavirus *Strawberry latent ringspot virus* is vectored by *X. diversicaudatum*.

Life Cycle and Epidemiology

Dagger and needle nematodes hatch from eggs as first-stage juveniles and most molt four times in the soil to become adults (Fig. 269). Those within the *X. americanum* species complex are reported to have only three juvenile stages. Adults resemble the young, remaining vermiform without swellings or modifications such as cysts. The duration of the cycle from egg to adult female in *X. index* varies by region. It has been reported as 22–27 days in California, a very short time for a nematode with such a large body size, and as 7–9 months in Israel; reasons for such a difference are not known. *Xiphinema index* has been reported to survive in soil for up to 4 years; its survival in vineyard soils following vine removal is extended when

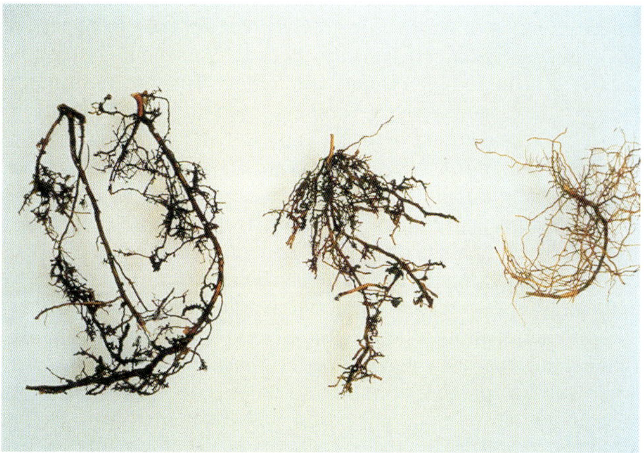

Fig. 268. Roots damaged by dagger nematodes (left and center) compared with healthy roots (right). (Courtesy D. J. Raski)

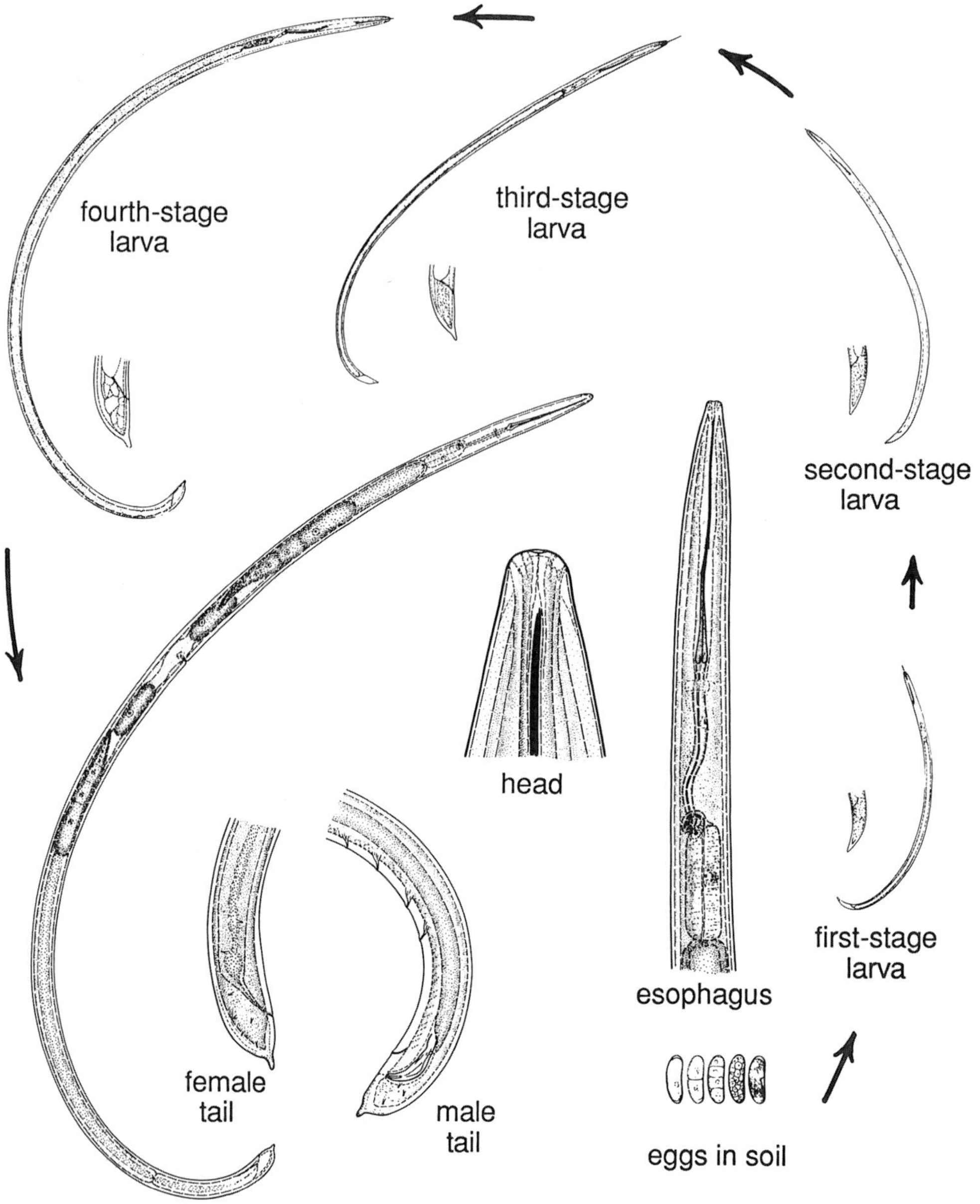

Fig. 269. Life cycle of the dagger nematode. (Courtesy D. J. Raski; drawing revised by R. Sticht)

residual grape roots remain alive. The longevity of *X. americanum* individuals is 3–5 years, with only one generation produced per year.

Dagger and needle nematodes are strictly ectoparasitic. They feed by means of a very long stylet that they use to penetrate the root tip cells. Their feeding alters cells in the meristematic region, leading to hypertrophy and hyperplasia, and commonly the nematode feeds from several enlarged nurse cells. Each juvenile stage must feed before it can molt and continue growth. Reproduction in most species is parthenogenic and males often are rare or unknown, although they are present in about the same numbers as females in a few species. These nematodes spread to new fields primarily by transport of infected planting material or as a result of cultural practices that move infested soil, sometimes in contaminated irrigation water.

Upon acquisition of appropriate Nepoviruses, needle nematodes retain the virus particle on the inner surface of the stylet, whereas virus particles bind to the esophageal lining of dagger nematodes. Those cuticular portions of the nematode digestive system are shed during the molts between developmental stages so that the adhering virus is also lost and must be reacquired by the nematode during subsequent feeding events. Since the nematodes probably do not move far from the root-tip feeding site during or between molts, virus particles are readily reacquired when feeding commences again. *Xiphenema index* is able to retain GFLV for at least 4 years in the absence of host plants, revealing an extremely long survival and viral retention period within that nematode.

Vectoring of plant viruses by dagger nematodes has been most thoroughly studied in the *X. index*/GFLV system. Acquisition of the virus by the nematode occurs passively as the nematode withdraws plant cell contents through the stylet. The virus particles selectively adhere to the cuticular lining of the esophageal lumen and are lost when that cuticle is molted. Particles are thought to be released from the esophageal cuticle when nematode enzymes are secreted into the plant cell. The virus does not seem to directly harm the nematode beyond its effects on debilitating or killing the host plant. The virus particles do not replicate in the nematode and are not passed through the egg stage to the juveniles.

Selected References

Cohn, E., and Mordechai, M. 1969. Investigations on the life cycles and host preference of some species of Xiphinema and Longidorus under controlled conditions. Nematologica 15:295-302.

Decraemer, W., and Robbins, R. T. 2007. The who, what and where of Longidoridae and Trichodoridae. J. Nematol. 39:295-297.
Demangeat, G., Voisin, R., Minot, J. C., Bosselut, N., Fuchs, M., and Esmenjaud, D. 2005. Survival of *Xiphinema index* in vineyard soil and retention of *Grapevine fanleaf virus* over extended time in the absence of host plants. Phytopathology 95:1151-1156.
Ferris, H., and McKenry, M. V. 1974. Seasonal fluctuations in the spatial distribution of nematode populations in a California vineyard. J. Nematol. 6:203-210.
Ferris, H., and McKenry, M. V. 1975. Relationship of grapevine yield and growth to nematode densities. J. Nematol. 7:295-304.
Halbrendt, J. M., and Brown, D. J. F. 1992. Morphometric evidence for three juvenile stages in some species of *Xiphinema americanum* sensu lato. J. Nematol. 24:305-309.
Martelli, G. P. 1978. Nematode-borne viruses of grapevine, their epidemiology and control. Nematol. Mediterr. 6:1-27.
Pinochet, J., Raski, D. J., and Goheen, A. C. 1976. Effects of *Pratylenchus vulnus* and *Xiphinema index* singly and combined invine growth of *Vitis vinifera*. J. Nematol. 8:330- 335.
Robbins, R. T. 1993. Distribution of *Xiphinema americanum* and related species in North America. J. Nematol. 25:344-348.
Taylor, C. E., and Brown, D. J. F. 1997. Nematode vectors of plant viruses. CAB International, Oxford, UK.

(Prepared by I. A. Zasada, H. Ferris, and M. V. McKenry)

Ring Nematode

The ring nematode, *Mesocriconema xenoplax*, was first reported from California in 1952, and has been subject to several taxonomic reclassifications since then. This status remains unsettled, and both the names *Mesocriconema xenoplax* and *Criconemoides xenoplax* appear in current literature. Herein, we use *Mesocriconema* for the generic reference.

The ring nematode is a migratory ectoparasite of plant roots. It does not enter the root but feeds deep into root tissues using a long, stout stylet. Since the nematodes are relatively inactive, they do not readily pass through the tissue barrier of the Baermann funnel commonly used for extracting most other nematode species. Consequently, they are not readily detected by that method and must be extracted by alternative methods that float the nematodes from soil rather than rely on their active movement. Because of this inefficiency of detection by means of common methodology, it is highly likely that the importance of the ring nematode in grape vineyards has been underestimated.

Symptoms

Porous soils such as sands, coarse sandy loams, or well-aggregated clay loams appear necessary for the development of the ring nematode at the high population levels associated with symptom expression on vines. Damage by ring nematodes often results in reduction of small feeder roots, and there may be a "witches'-broom" appearance of small roots emanating from larger roots below the tillage zone. Almost all grapevine cultivars and rootstocks are susceptible to the ring nematode; vines exhibiting the greatest aboveground damage appear to be those with the least inherent vigor, including such cultivars as Chardonnay and Pinot Noir. In severe cases, more than half the buds along a cordon or spur may fail to develop. Current season growth that develops from an apparently healthy spur can appear as a mass of worthless twigs. In a few vineyards, leaves on affected plants have been observed to exhibit interveinal necrosis by early summer. Ring nematode may also negatively impact vine associations with arbuscular mycorrhizal fungi, suppressing arbuscule frequency and leading to symptoms of phosphorus deficiency.

Causal Organisms

Within the *Criconematidae* family, *Mesocriconema xenoplax* (Raski) Loof & de Grisse (= *Criconemoides xenoplax* Raski) is the primary species associated with damage to grapevines. Other species from this family also may occur in the vineyard feeding on roots of weeds or ground cover. The common name of the ring nematode is derived from its thick, deeply striated cuticle, which gives the nematode the appearance of being surrounded by a sequence of collocated rings. In some areas additional nematodes in the same family, including the sheathoid nematode *Hemicriconemoides californianus* Pinochet & Raski and other *Hemicriconemoides* species have been reported from vineyards. These nematodes have the protection of a double cuticle and appear to have similar parasitic habits as *M. xenoplax*.

Life Cycle and Epidemiology

Adult stages of *M. xenoplax* are sedentary ectoparasites with a life cycle of 30–60 days. It is likely that roots encounter ring nematode adults as they grow, rather than the nematodes actively migrating toward roots. However, juvenile stages are more active and are likely involved in moving toward host roots. Once the stylet has been inserted into plant tissue, the nematodes remain at that location, somewhat protected from other rhizosphere organisms by their thick cuticle. In a time frame similar to that of root-knot nematodes, ring nematodes can develop to very high population levels in porous soils. Ring nematodes population densities are highest in areas where grape roots are most abundant, usually within the top 45 cm of soil.

Selected References

Brzeski, M., Loof, P. A. A., and Choi, Y. E. 2002. Compendium of the genus *Criconemoides* Taylor, 1936 (Nematoda: Criconematidae). Nematology 4:325-339.
Brzeski, M., Loof, P. A. A., and Choi, Y. E. 2002. Compendium of the genus *Mesocriconema* Andrássy, 1965 (Nematoda: Criconematidae). Nematology 4:341-360.
Decraemer, W., and Hunt, D. 2006. Structure and classification. Pages 3-32 in: Plant Nematology. R. N. Perry and M. Moens, eds. CAB International, Oxford, UK.
Ferris, H., Zheng, L., and Walker, M. A. 2013. Soil temperature effects on the interaction of grape rootstocks and plant-parasitic nematodes. J. Nematol. 45:49-57.
Klingler, J. 1975. Observations on the parasitic activity of the nematode *Macroposthonia xenoplax* on grape vine roots. Z. Pflanzenkrankh. Pflanzenschutz 82:722-728.
Pinkerton, J. N., Schreiner, R. P., Ivors, K. L., and Vasconcelos, M. C. 2004. Effects of *Mesocriconema xenoplax* on *Vitis vinifera* and associated mycorrhizal fungi. J. Nematol. 36:193-201.
Pinochet, J., and Raski, D. J. 1975. Four new species of the genus *Hemicriconemoides* (Nematoda: Criconematidae). J. Nematol. 7:263-270.
Santo, G. S., and Bolander, W. J. 1977. Effects of *Macroposthonia xenoplax* on the growth of Concord grape. J. Nematol. 9:215-217.

(Prepared by I. A. Zasada, H. Ferris, and M. V. McKenry)

Root-knot Nematodes

Root-knot nematodes were first found and described on species of *Vitis* in Florida in 1911. Since then, at least seven species of *Meloidogyne* have been reported from vineyards worldwide and this number is likely to increase as molecular identification methods are employed to distinguish between morphologically similar forms currently assigned to a single taxon. Currently, the species reported most frequently and considered the most damaging are *M. hapla*, which is dominant in cooler regions of the world, and *M. incognita*, *M. arenaria*, and *M. javanica*, which are common in warmer regions. Specific rootstocks once considered as resistant to *Meloidogyne* spp. can become highly susceptible to certain *Meloidogyne* populations or pathotypes over time.

Symptoms

The distinctive response of grape roots to root-knot nematodes is the production of small swellings or galls on young feeder roots or secondary rootlets (Fig. 270). Larger galls may result from multiple infections. The galling response varies across cultivars and rootstocks. For example, galls are barely visible on Teleki 5C, Schwarzmann, or RS-3 rootstocks and nematode feeding sites and galls are restricted to new roots. Other rootstocks, including 99R and 110R do not exhibit galls but washing of the roots can reveal the presence of numerous glistening, reproductive adult females along root surfaces. Root-knot resistant rootstocks, like the UCD-GRN series and others, do not gall and do not support nematode development and reproduction. The galling response also differs with nematode species; for example, penetration of susceptible roots by *M. hapla* or *M. chitwoodi* may result in very small galls or indiscernible swellings. Among other root-knot species, gall size can be larger or smaller depending on the specific host genotype attacked. When the galls are broken apart, the tiny, glistening, white bodies of mature females can be detected with the aid of a hand lens. Gelatinous egg masses are often attached to the female bodies. Second-stage juveniles and males can be found only by microscopic examination after extracting nematodes from soil. Usually, the size of root systems of grapevines damaged by root-knot nematodes is severely limited, with few feeder roots.

Root-knot nematodes cause their greatest damage in newly replanted vineyards with high populations of nematodes associated with the previous vines. Young vines, which may already be weakened owing to other replant factors, fail to establish an effective root system under the nematode pressure. Vines that do become established may be weak and fail to grow sufficiently to permit training onto trellises. Root-knot nematodes seldom kill vines; more often, plants decline in vigor and become more susceptible to other stress factors. On occasion, scorching of leaves, cordons, and even the berries may occur on devigorated, nematode-infected vines. Yields of infested blocks may be less than half those of healthy vineyards in the same region. Greatest damage by root-knot nematodes usually occurs in vineyards planted in sandy soils in areas where higher soil temperatures are experienced.

Causal Organisms

Root-knot nematodes are members of the genus *Meloidogyne* Goeldi. The four species currently considered to have greatest impact on grape production are *M. incognita* (Kofoid & White) Chitwood, *M. javanica* (Treub) Chitwood, *M. arenaria* (Neal) Chitwood, and *M. hapla* Chitwood. Other root-knot nematode species reported on grapevines include: *M. nataliei* Golden, Rose & Bird from Michigan; *M. ethiopica* Whitehead from Chile; and *M. chitwoodi* Golden, O'Bannon, Santo & Finley from California. Biotypes within the nematode species differ in their virulence to grapevine cultivars and rootstocks. As resistant rootstocks are deployed, the diversity of virulence among root-knot nematode species gradually becomes apparent through the selection of virulent pathotypes.

Fig. 270. Roots of cv. Cabernet Sauvignon vine damaged by root-knot nematodes, with galls and swellings on secondary rootlets. (Courtesy D. J. Raski)

Life Cycle and Epidemiology

Root-knot nematode females do not retain their eggs; they deposit them outside their bodies, typically in a mass of 400–500 (occasionally up to 1,500) eggs within a gelatinous matrix that is secreted by glands in the posterior region of the body. The matrix is usually located on the outside of the root but may be inside and completely surrounded by root tissue. The embryo within the egg develops into a juvenile nematode, molts once, and emerges from the egg as a second-stage juvenile, the migratory and infective stage. Their root penetration capabilities are restricted to a zone just behind the growing root tip; however, some also appear to successfully reinvade maternal feeding sites, thus perpetuating a survival of galls on the roots that may last a decade. Following penetration, the juveniles quickly migrate toward the root cap, then reverse direction and migrate toward developing vascular tissues where they establish a feeding site and complete their life cycle as sedentary endoparasites. The plant response to their feeding is the production of undifferentiated, multinucleate, thick-walled "giant cells" within the vascular region. Typically, a gall becomes visible within 15 days after nematode entry, although there are exceptions with different rootstocks. Galls are barely visible on Teleki 5C, Schwarzmann, or RS-3 rootstocks and their feeding sites and galls are restricted to new roots. Rootstocks such as 99R and 110R do not exhibit galls but washing of the roots can reveal the presence of numerous glistening, reproductive adult females along root surfaces. Penetration by *M. hapla* or *M. chitwoodi* does not yield readily visible galls. Among other root-knot species, gall size can be larger or smaller depending on the specific host genotype attacked.

Second-stage infective juveniles emerging from eggs have sufficient energy reserves at 25°C to persist for about 2 weeks, during which time they must penetrate the roots and establish feeding sites. Once established, the juveniles commence to feed, become enlarged, pass through two more developmental stages without further feeding, and become adult, pear-shaped females. The life cycle from egg to egg takes about 25 days at 25°C, and at this temperature adult females may survive approximately 60 days. *Meloidogyne incognita*, *M. javanica*, and *M. arenaria* can complete five to six generations while temperature and other conditions are favorable for nematode and plant physiological activity. With approximately 500 eggs produced per female within each generation, the potential for population increase during a single growing season is enormous. When vines enter dormancy, root-knot nematodes overwinter principally as developing juveniles inside eggs within the gelatinous matrix of the egg mass.

Root-knot nematodes are mostly parthenogenic, i.e., females produce viable eggs without fertilization by a male. Males are rare or absent from the soil but may become apparent when roots become overcrowded or in response to other conditions unfavorable to the nematodes. The life cycle of a male is similar to that of a female up to the third juvenile stage. At that point, a metamorphosis takes place and the enlarged body is reorganized into a coiled-worm shape, which emerges as an adult male from the cuticle of the fourth stage juvenile.

In deep vineyard soils, root-knot nematodes can be found wherever roots occur, but they are most prevalent at a depth of 15–90 cm. Even after the removal of a vineyard, old grapevine roots can remain alive and support root-knot nematode populations for 8 years or more. Additionally, root-knot nematode species have wide host ranges and are sustained and increased on both weed and crop hosts that follow the vineyard.

Selected References

Anwar, S. A., McKenry, M. V., and Faddoul, J. 2001. Reproductive variability of field populations of *Meloidogyne* spp. on grape rootstock. J. Nematol. 32:265-270.

Anwar, S. A., and McKenry, M. V. 2002. Development response of a resistance-breaking population of *Meloidogyne arenaria* on *Vitis* spp. J. Nematol. 34:28-33.

Cain, D. W., McKenry, M. V., and Taralio, R. E. 1984. A new pathotype of root-knot nematode on grape rootstocks. J. Nematol. 16:207-208.

Carneiro, R. M. D. G., Almeida, M. R. A., Cofcewicz, E. T., Magunacelaya, J. C., and Aballay, E. 2007. *Meloidogyne ethiopica*, a major root-knot nematode parasitizing *Vitis vinifera* and other crops in Chile. Nematology 9:633-639.

Ferris, H., and McKenry, M. V. 1974. Seasonal fluctuations in the spatial distribution of nematode populations in a California vineyard. J. Nematol. 6:203-210.

Ferris, H. 1976. Observations on the distribution of *Meloidogyne* spp. in a vineyard soil. J. Nematol. 8:183-184.

Ferris, H., Schneider, S. M., and Stuth, M. C. 1982. Probability of penetration and infection by *Meloidogyne arenaria* in grape cultivars. Am. J. Enol. Vitic. 33:31-55.

Ferris, H., Zheng, L., and Walker, M. A. 2012. Resistance of grape rootstocks to plant-parasitic nematodes. J. Nematol. 44:377-386.

McKenry, M. V. 1984. Grape root phenology relative to control of parasitic nematodes. Am. J. Enol. Vitic. 35:206-211.

(Prepared by I. A. Zasada, H. Ferris, and M. V. McKenry)

Root-lesion Nematodes

Root lesion nematodes (*Pratylenchus* spp.) are commonly encountered in vineyard soils, although in many cases there has been no attempt to identify these nematodes to species. Ten species have been identified but only a few have been tested for their pathogenicity to grapevines. *Pratylenchus vulnus* is reported from California, Spain, and Australia and has received the most research attention. The presence of *P. zeae* in Australia and Brazil and *P. brachyurus* in Brazil may also warrant concern. The remaining species have limited distribution or are associated primarily with ground covers present in the vineyard.

Damage caused by root-lesion nematodes can be severe, but almost exclusively in settings where vines are planted following removal of an orchard crop upon which they develop far more readily and have achieved high population levels. An early attack by root-lesion nematodes particularly inhibits growth of the main root system. Vineyard plantings that have not grown well for their first decade for this reason often can be invigorated by a series of nematicide treatments. Upon receiving adequate protection, the vine can develop a new root system that outgrows subsequent increases in the nematode population.

Symptoms

There are no specific symptoms on grapevines that are diagnostic for root-lesion nematode damage. Root systems exhibit poor development, including an absence of major roots and many dead feeder roots. Occasionally, rootlets are killed shortly after they are produced. In somc hosts, certain phenolic compounds are oxidized, causing necrosis and lesion formation at feeding sites, but these symptoms are rare on grapevines. The two best indications of the presence of these nematodes are the cropping history of the soil and appropriate soil sampling.

Causal Organisms

Root-lesion nematodes are members of the genus *Pralylenchus* Filipjev. Species that have been reported from vineyards worldwide are: *P. vulnus* Allen & Jensen; *P. brachyurus* (Godfrey) Filipjev & Schuurmans Stekhoven; *P. scribneri* Steiner; *P. neglectus* (Rensch) Filipjev & Schuurmans Stekhoven; *P. pratensis* (de Man) Filipjev; *P. zeae* Graham; *P. hexincisus* Taylor & Jenkins; *P. coffeae* Zimmermann; *P. crenatus* Loof; and *P. thornei* Sher & Allen.

The genus *Zygotylenchus* Siddiqi belongs to the same nematode family as *Pratylenchus*. A few records of *Zygotylenchus* spp. have been published in surveys of vineyard soils, but nothing more is known about their abundance, general incidence, or significance in grape production.

Life Cycle and Epidemiology

Root-lesion nematodes are migratory and endoparasitic in life habit, spending a portion of their lives in the soil and a portion in root tissues. The female deposits eggs singly in the soil or in root tissues. The young develop and elongate in the egg, molt once, and emerge as second-stage juveniles. The nematode invades host roots and moves through the cortex, penetrating, feeding on, and killing cells. All juvenile stages and adults are capable of invading roots and feeding. Most *Pratylenchus* spp. are parthenogenic; however, in some species, such as *P. vulnus*, males are common and sexual reproduction occurs.

Selected References

Allen, M. W., and Jensen, H. J. 1951. *Pratylenchus vulnus*, new species (Nematoda: Pratylenchinae), a parasite of trees and vines in California. Proc. Helminthol. Soc. Wash. 18:47-50.

Pinochet, J., and Raski, D. J. 1977. Observations on the host-parasite relationship of *Pratylenchus vulnus* on grapevine, *Vitis vinifera*. J. Nematol. 9:87-88.

Pinochet, J., Raski, D. J., and Goheen, A. C. 1976. Effects of *Pratylenchus vulnus* and *Xiphinema index* singly and combined invine growth of *Vitis vinifera*. J. Nematol. 8:330- 335.

Puerari, H. H., Dias-Arieira, C. R., Moura, M. F., Biela, F., Chiamolera, F. M., and da Cunha, T. P. L. 2012. Reaction of grapevine rootstocks to *Pratylenchus brachyurus* and *Pratylenchus zeae*. Trop. Plant Pathol. 37:220-222.

Sauer, M. R. 1962. Distribution of plant-parasitic nematodes in irrigated vineyards at Merbein and Robinvale. Aust. J. Exp. Agric. Anim. Husb. 2:8-11.

Walker, G., and Morey, B. 2000. Effects of lesion nematodes associated with cereals on grapevine growth. Aust. Grapegr. Winemaker 438:130-132.

(Prepared by I. A. Zasada, H. Ferris, and M. V. McKenry)

Miscellaneous Ectoparasitic Nematodes

Surveys in several countries have revealed a wide array of ectoparasitic species in vineyard soils, including: pin nematodes, *Paratylenchus* spp., especially *P. hamatus*; reniform nematodes, *Rotylenchulus* spp.; spiral nematodes, *Helicotylenchus* spp.; lance nematodes, *Hoplolaimus* spp. and *Rotylenchus* spp.; stubby-root nematode, *Paratrichodorus minor* (Allen) Siddiqi; and stunt nematodes, *Tylenchorhynchus* (syn. *Telotylenchus)* spp. The importance of these nematodes to grape production is poorly understood.

Selected References

Ferris, H., and McKenry, M. V. 1975. Relationship of grapevine yield and growth to nematode densities. J. Nematol. 7:295-304.

Ferris, H., and McKenry, M. V. 1976. Nematode community structure in a vineyard soil. J. Nematol. 8:131-137.

Ferris, H., and McKenry, M. V. 1976. A survey of nematode distribution in California vineyard soils. J. Am. Soc. Hortic. Sci. 101:332-336.

Mohamed, R. M. S. 2009. Plant parasitic nematodes associated with grapevine in Sana'a and Sadah Governorates of Yemen. Mansoura J. Agric. Sci. 34:1339-1345.

Raski, D. J. 1955. Additional observations on the nematodes attacking grapevines and their control. Am. J. Enol. Vitic. 6:29-31.
Raski, D. J., and Radewald, J. D. 1958. Reproduction and symptomatology of certain ectoparasitic nematodes on roots of Thompson seedless grape. Plant Dis. Rep. 42:941-943.
Sulan, L., Puchao, H., and Dongting, Z. 2003. Investigation on parasitic nematodes associated with the wild grapes native to China. Acta Hortic. Sin. 30:127-130.

(Prepared by I. A. Zasada, H. Ferris, and M. V. McKenry)

Nematode Management

Effective nematode management programs for grape vineyards rely on a suite of practices. If an area is free of damaging species of plant-parasitic nematodes, every precaution should be taken to exclude them, as nematode infestations are permanent once established. Since the 1950s, control programs in vineyards have been centered on nematicidal chemicals. However, as the availability of such chemicals becomes progressively more regulated, strategies based on nematode ecology and hostplant resistance or tolerance become increasingly important alternatives. In essence, current and evolving nematode management strategies are increasingly information-based and require consideration of the host ranges, life cycles, survival strategies, and longevity of the plant-parasitic nematode species present in a vineyard. Knowledge of the ability of resident nematode species to feed upon the grape rootstock and cover crops present, and an appreciation for the importance and stewardship of natural regulators of pest species populations, are required. Important decisions must be made, and practices implemented, prior to vineyard establishment. Continued monitoring and management of soil health are necessary throughout the development and productive cycle of the vineyard. Damage from plant-parasitic nematodes is much less common in soils with physical, chemical, and biological characteristics that sustain plant growth while maintaining biological homeostasis and providing beneficial ecosystem services such as organic matter decomposition, nutrient cycling, and water purification.

Preplant Management

Field assessment. Fields should be sampled to determine the diversity and abundance of plant-parasitic nematodes prior to planting a vineyard. Nematode population densities are highest in the fall, and assessments at this time are most indicative of the damage potential. Both soil and root samples must be collected; root samples provide a better diagnostic measure of population densities of root lesion nematode. The distribution of plant-parasitic nematodes in a field tends to be patchy and is determined by soil conditions and previous management history. Therefore, samples for nematode analysis should consist of composites of soil cores or subsamples representing each area of the field that appears reasonably uniform in terms of soil texture and previous crop vigor. One such composite sample of about 20 cores or subsamples should be collected per 2–5 hectares of the field. Sample depth may vary with the root profile of the previous crop, but generally samples taken to a depth of 30 cm are appropriate. If the previous crop is still in the field, samples should be taken from the root zone of these plants in areas where active root growth has been favored, e.g., within zones of irrigation when used. Samples should be clearly labeled and kept cool until processed by a diagnostic laboratory.

Exclusion. Plant-parasitic nematodes have become established in vineyards throughout the world because they were introduced with rooted planting material. Where nursery certification programs are available, only certified nematode-free stock should be planted. There is a high risk of introducing nematodes in rooted cuttings that have been propagated in soil that is not known to be free of pest nematode species. Nematodes are easily moved and introduced on soil adhering to vehicles, implements, and animals; ensure that such potential sources are eliminated by appropriate sanitation practices. Nematodes are also moved and introduced by water, either through flooding or runoff or through irrigation with water drained from neighboring fields or drawn from rivers that have passed through infested areas. Several tactics of varied effectiveness have been tested and implemented for reducing infestation levels of irrigation water, including redesign of the supply system, settling ponds, use of water from deep wells, and ozone treatment.

Vineyard establishment. *Soil preparation.* Several preparatory measures and decisions must be made prior to planting. Deep tillage may be necessary to disrupt restrictive layers that may be naturally occurring or that have resulted from previous cultural practices. Some nematodes are susceptible to soil disturbance. For example, population levels of large-bodied dagger nematodes, *Xiphinema* spp., can be reduced by repeated tillage of fallow soil. However, this practice can be detrimental to soil structure and does not affect nematodes below the tilled profile.

Population levels of potentially harmful nematode species may be reduced by rotation with appropriate nonhost or resistant crop species. Cover crops and organic amendments may be used to build soil carbon levels and to provide resources for the beneficial organisms of the soil food web.

Soil fumigation. Application of fumigant nematicides can be an effective way to reduce plant-parasitic nematode populations prior to planting; the availability of soil fumigants varies depending upon location. Soil fumigants that have been widely used in vineyards include 1,3-dichloropropene, metam sodium, and chloropicrin. Proper planning to ensure adequate soil preparation and conditions at the time of treatment are the keys to successful fumigation.

Rootstocks. One of the most important decisions prior to planting is the selection of a cultivar or rootstock that is resistant to, or tolerant of, the plant-parasitic nematodes present in the specific site or that are known to exist in the general area. Rootstocks available for grapevines differ in their susceptibility to nematodes, and various selections have been developed and released in recent years that have resistance to several species of plant-parasitic nematodes. Besides their nematode-resistance characteristics, it is important to select rootstocks that have appropriate vigor and horticultural characteristics for the local conditions. A summary of available data on the susceptibility of various rootstocks to individual nematode species is provided in Table 6.

Vineyard renovation. Replanting a vineyard that has declined in productivity provides special challenges because the biological, chemical, and physical causes of the decline may still be present in the soil. Old vines must be extracted with equipment that removes the greatest possible root mass from the soil profile. There may be advantages in applying an appropriate herbicide to vine stumps before their removal so that root fragments remaining in the soil are killed and do not provide a resource for pest nematode survival. Following vine removal, cultural practices such as fallow, crop rotation, or cover cropping may be implemented to reduce survival and reproduction of resident nematode populations.

When a vineyard is removed, a crop-rotation period should be allowed before replanting in order to limit nematode pressure on the new planting. Rotations may include nonhosts, resistant hosts, allelopathic plants, fallow, trap crops, or green manures. The length of rotation necessary for good nematode management depends upon the survival capabilities of the target species, with a shorter time period for the citrus nematode (1 year) and longer for dagger nematodes (≥4 years). If a fallow period is implemented, good weed management must be achieved since many weeds are hosts to nematodes. The selection of a rotation

TABLE 6. Susceptibility of Grapevine Rootstocks to Various Nematode Species (Data from Ferris et al., 2012)[a]

Rootstock genotype	*M. incognita* Race 3	*M. javanica*	*Meloidogyne* pathotypes Harmony A&C	*M. chitwoodi*	*X. index*	*M. xenoplax*	*P. vulnus*	*T. semi-penetrans*	*X. americanum*	*Para. hamatus*
101-14Mgt			R		S	S	MR			S
1103Paulsen			S		S	S	MS			S
110Richter			MR		S	S	S			S
140Ruggeri			MR		S	S	S			MS
1613Couderc	R	R	S	S	MR	S	MS	S	S	
1616Couderc			MR		S	S	MS			S
3309Couderc	S	S	S		MS	S	S	S	S	S
420A			R		S	S	MS			S
44-53Malegue			S		S	MR	MS			S
AxR1			S		S	S	S			S
Borner			R		R	S	MS			
Dog Ridge	R	R	S		S	S		MR	MR	MS
Freedom	R	R	S	S?	R	MS	MS	S	MS	MR
Harmony	R	R	S	S	MS	S	S	S	S	S
K51-32	MR				MS	S	R	S		S
Kober 5BB			R		S	S	MS			MR
Ramsey	R	R	S	S?	MR	S	MS	MS	S	S
Riparia Gloire			R		R	S	MR			S
RS-3	R	R	MR	MR	S	S	MR			S
RS-9	R	R	R	R	S	S	MS			S
Schwarzmann	S	MR	S		MR	MS	S	S	MS	S
St. George	S		S		S	S	MS			MS
Teleki 5C	MS	MR	S		MR	MS	S	S	S	MS
USDA 10-17A	R	R	R	R	R	MS	R	R		
USDA 10-23B	R	R	R	R	R	MR	R	R		
USDA 6-19B	R	R	MS	R	MR	MR	R	R	R	
VR O39-16	S	S	S		R	R	MR	S	MR	MR
UCD GRN1	R		R		R	R	MR	R		MR
UCD GRN2	R		R		R	MS	MR	MS		MR
UCD GRN 3	R		R		R	MR	MR	MR		MR
UCD GRN4	R		R		R	MR	MR	MR		MS
UCD GRN5	R		R		R	R	MR	MR		MR
UCD GRN5	R		R		R	R	MR	MR		MR

[a]R = resistant; S = susceptible; MR = moderately resistant; MS = moderately susceptible.

or cover crop should be made with knowledge of the host range of the target plant-parasitic nematode. Options may be limited for root knot nematodes, which have a very broad host range, and more abundant for the citrus nematode, which has only a few known hosts. Ring nematode also has a wide host range, including grasses and, particularly, legumes. Nemabase, a database on the Nemaplex website (http://plpnemweb.ucdavis.edu/Nemaplex) maintained by University of California, Davis has over 47,000 records on the nematode host status of plants, a valuable resource when selecting rotation and cover crops for vineyard renovation.

Virus vector considerations. For nematodes that are vectors of plant viruses, such as dagger nematodes, very low numbers of viruliferous individuals can cause serious damage by transmitting virus particles to new vines. *Xiphinema index* and the *X. americanum* species complex are two taxa that are particularly important in this regard, and it is important to attempt elimination of these species prior to establishing a new vineyard. In the past, this has been attempted through the use of high rates of soil fumigants. Where such practices are no longer available, a long nonhost crop rotation may be necessary. *X. americanum*, vector of *Tomato ringspot virus* (ToRSV) and *Tobacco ringspot virus*, has a wide host range. However, *Brassica* and grass cover crops, although supporting nematode feeding, are not hosts of the grapevine Nepoviruses, and new generations of the nematode are virus-free after a 2-year rotation with them. Management of broadleaf weeds is an important component of this strategy because many are hosts for ToRSV. *X. index*, vector of *Grapevine fanleaf virus*, has a narrow host range consisting primarily of woody perennial plants. Some rootstocks are available with resistance to *X. index*, although this does not necessarily confer resistance to the virus because the nematode may transmit particles during exploratory stylet thrusts. However, at least one rootstock, O39-16, which is resistant to *X. index*, also confers tolerance to the virus and vines remain productive for long periods even though they are infected.

Postplant Management

Vine health. The decline in vigor of vines damaged by nematodes can be slowed or offset in part by providing more frequent irrigation, improving soil water-holding capacity, avoiding over cropping, applying fertilizer levels to support a limited root system, and controlling other diseases and pests that stress the vines. Drip irrigation systems allow frequent irrigation and fertilizer application to alleviate nematode stress on vines.

Postplant nematicides. As with soil fumigants, the availability of registered postplant nematicides varies among grape-producing regions and countries. Postplant nematicides that have proven effective in vineyards are drip-applied sodium tetrathiocarbonate or 1,3-dichloropropene and foliar-applied systemic spirotetramat. The advantage of a systemic material is that relatively small applied quantities can move to the plant-nematode interaction zone at the root surface, avoiding treatment of the entire soil volume potentially occupied by the root system. Because of the complex carbon dynamics in perennial plants such as grapevines, repeated applications of nematicides over time may be necessary to realize yield benefits. Once a previously-damaged plant is protected from nematode damage, some time is required to rebuild its root system to the point that productive vine growth is supported.

Selected References

Ferris, H. 2010. Contribution of nematodes to the structure and function of the soil food web. J. Nematol. 42:63-67.

Ferris, H., Zheng, L., and Walker, M. A. 2012. Resistance of grape rootstocks to plant-parasitic nematodes. J. Nematol. 44:377-386.

Ferris, H., Zheng, L., and Walker, M. A. 2013. Soil temperature effects on the interaction of grape rootstocks and plant-parasitic nematodes. J. Nematol. 45:49-57.

Halbrendt, J. M., and LaMondia, J. A. 2004. Crop rotation and other cultural practices. Pages 909-930 in: Nematology, Advances and Perspectives, Vol. 2: Nematode Management and Utilization. Z. X. Chen, S. Y. Chen, D. W. Dickson, eds. CAB International, Wallingford, UK.

McKenry, M. V. 1999. The Replant Problem and Its Management. Catalina Publishing, Fresno, CA.

Nicol, J. M., Stirling, G. R., Rose, B. J., May, P., and Van Heeswijck, R. 1999. Impact of nematodes on grapevine growth and productivity: Current knowledge and future directions, with special reference to Australia. Aust. J. Grape Wine Res. 5:109-127.

Pinkerton, J. N., and Martin, R. R. 2008. Management of tomato ringspot virus in red raspberry with crop rotation. Int. J. Fruit Sci. 5:55-67.

Reisch, B. I., Owens, C. L., and Cousins, P. S. 2012. Grape. Pages 225-262 in: Fruit Breeding (Handbook of Plant Breeding). M. L. Badenes and D. H. Byrne, eds. Springer, Rueil-Malmaison, France.

Zasada, I. A., Halbrendt, J. M., Kokalis-Burelle, N., LaMondia, J., McKenry, M. V., and Noling, J. W. 2010. Managing nematodes without methyl bromide. Annu. Rev. Phytopathol. 48:311-328.

(Prepared by I. A. Zasada and H. Ferris)

Part II. Mites and Insects that Cause Diseaselike Symptoms in Grapes

Leafhoppers and Treehoppers

Leafhoppers and treehoppers are especially important insects to consider in a discussion of grape diseases because injuries they cause by feeding and oviposition on vines are often confused with symptoms produced by diseases. In addition, some species in this insect group serve as vectors of the Pierce's disease bacterium (*Xylella fastidiosa*) and the phytoplasmas that cause grapevine yellows diseases. Few insects have plagued grape growers during the last century more than leafhoppers. Leafhoppers (Cicadellidae, subfamily Cicadellinae) that attack grapes in North America include a complex of *Erythroneura* species and the potato leafhopper, *Empoasca fabae* (Harris). Important leafhoppers that attack grapes in Eurasia include *Empoasca vitis* (Gothe), *Jacobiasca libyca* (Bergevin & Zanon), *Zygina rhamni* Ferrari, *Arboridia dalmatina* (Novak & Wagner), and *Scaphoideus titanus* Ball (syn. *S. littoralis* Ball) (vector of the *flavescence dorée* yellows pathogen).

Erythroneura Leafhoppers

Important *Erythroneura* species included in the grape leafhopper complex are *E. calycula* McAtee, *E. coloradensis* Gillette, *E. comes* (Say) (eastern grape leafhopper), *E. elegantula* Osborn (western grape leafhopper), *E. maculator* Gillette, *E. tricincta* Fitch (threebanded leafhopper), *E. variabilis* Beamer (variegated leafhopper), *E. vitifex* Fitch, *E. vitis* (Harris), *E. vulnerata* Fitch, and *E. ziczac* Walsh (Virginia creeper leafhopper).

Vine injury from *Erythroneura* leafhoppers (and that from *Z. rhamni* and *A. dalmatina)* appears first as a white stippling or speckling on the upper leaf surface (Fig. 271) that can resemble virus infection. This damage is caused by the needlelike mouthparts of nymphs and adults piercing the mesophyll tissue and emptying the contents of groups of palisade cells; the feeding can occur from either the top or bottom side of the leaves. The speckling is at first limited to the areas along the main veins and later spreads over the leaf blade and turns a blotchy pale yellow or whitish yellow. For most leafhopper species the greatest injury occurs on the basal one-third of leaves on a shoot, simply because the leafhoppers tend to habitat this inner canopy section. When the leafhopper population is large enough, the damaged leaves may become necrotic and drop. In severe infestations, partial defoliation may occur, which leads to reduced fruit quality and vine vigor. However, research in California and Pennsylvania has indicated that mature, vigorous vines can tolerate high densities of leafhoppers.

The liquid excrement of *Erythroneura* leafhoppers settles onto leaves and berries and may trap dust particles or support the growth of a superficial sooty mold.

Empoasca Leafhoppers

The potato leafhopper, *Empoasca fabae*, breeds during the winter months in the Gulf Coast states. Adults increase in numbers during March and April and then migrate northward, becoming distributed over most of the United States. The leafhopper does not migrate south in autumn, and those in the colder regions perish with the occurrence of the first hard frost. Grape leaves injured by the potato leafhopper are mottled, with yellowed margins (Fig. 272) that often roll or curl downward and might be confused with symptoms of leafroll virus on white-fruited cultivars or grapevine yellows. Chlorotic areas may become necrotic late in the growing season.

Empoasca vitis is widely distributed in Europe. In Italy, it produces three generations per year and overwinters as adults

Fig. 271. Injury caused by eastern grape leafhoppers on cv. Concord leaves. (Courtesy G. L. Jubb, Jr.)

Fig. 272. Injury caused by potato leafhoppers. (Courtesy D. M. Gadoury)

on conifers. The feeding of *Empoasca vitis* cause leaf vein browning, downward rolling, thickening, bright coloration, angular reddening (Fig. 273) or yellowing, and occasionally a marginal leaf burn that is sometimes confused with leafroll or grapevine yellows. Severe foliar injury inhibits the normal hardening of shoots into canes. Injury by *J. libyca* is similar and occurs in the hotter areas of Eurasia.

Empoasca spp. can injure grapes by mechanically injuring and blocking vascular tissue, by injecting toxic enzymatic secretions, or by a combination of both mechanisms.

Sharpshooter Leafhoppers

Sharpshooters (Cicadellidae, subfamily Tettigellinae) differ physically from the leafhoppers discussed above in that the simple eyes are positioned differently on the head. Their chief importance in vineyards is the transmission of Pierce's disease. Some 20 species of sharpshooters are vectors of the Pierce's disease bacterium throughout the range of this disease. Feeding by sharpshooters on leaves and cluster stems results in wilting and eventual necrosis of the affected parts.

Fig. 273. Angular reddening produced by *Empoasca* leafhopper feeding on cv. Alden leaves. (Courtesy G. L. Jubb, Jr.)

Fig. 274. 'Pinot Noir' shoot girdled by treehopper feeding, with red leaves distal to the feeding site. (Courtesy R. C. Pearson)

Treehoppers

Common treehoppers (Membracidae) that feed on grapes include *Stictocephala bisonia* Kopp & Yonke (buffalo treehopper), *Enchenopa binotata* (Say) (twomarked treehopper), and *Spissistilus festinus* (Say) (threecornered alfalfa hopper). The treehoppers injure grapevines by feeding directly on the shoot and by depositing eggs therein. Shoots fed upon by the buffalo treehopper show a brownish girdling involving the epidermis and cortex and a thickening, reddening or yellowing, and downward rolling of the leaves distal to the girdle (Fig. 274). These symptoms resemble those of leafroll and grapevine yellows. Some treehoppers create a ring of feeding punctures around the shoot, resulting in an enlarged callus just above the wound.

Selected References

Beirne, B. P. 1956. Leafhoppers (Homoptera: Cicadellidae) of Canada and Alaska. Can. Entomol. 88:1-180.

Bournier, A. 1976. Grape insects. Annu. Rev. Entomol. 22:355-376.

DeLong, D. M. 1948. The leafhoppers, or Cicadellidae of Illinois (Eurymelinae-Balcluthinae). Ill. Bull.-Ill. Nat. Hist. Surv. 24:91-376.

Jubb, G. L., Jr., Danko, L., and Haeseler, C. W. 1983. Impact of *Erythroneura comes* Say (Homoptera: Cicadellidae) on caged 'Concord' grapevines. Environ. Entomol. 12:1576-1580.

McGiffen, K. C., and Neunzig, H. H. 1985. A guide to the identification and biology of insects feeding on muscadine and bunch grapes in North Carolina. N.C. Agric. Res. Serv. Bull. 470.

Smith, F. F., and Poos, F. W. 1931. The feeding habits of some leafhoppers of the genus Empoasca. J. Agric. Res. 43:267-285.

Vidano, C., and Arzone, A. 1983. Biotaxonomy and epidemiology of Typhlocybinae on vine. Pages 75-85 in: Proceedings of the 1st International Workshop on Biotaxonomy, Classification and Biology of Leafhoppers and Planthoppers (Auchenorrhyncha) of Economic Importance. W. J. Knight, N. C. Pant, T. S. Robertson, M. R. Wilson, eds. Commonwealth Institute of Entomology, London.

(Prepared by K. M. Daane)

Mealybugs

Mealybugs (Homoptera: Pseudococcidae) are important insect pests that can be found throughout the world's grape-growing regions. They feed primarily in phloem tissue and on most parts of the vine—trunk, canes, leaves, berries and, for some species, even roots—using piercing-sucking mouthparts. Once liquids are extracted and utilized, the remaining portion is excreted as a liquid called honeydew. This fluid contains a high concentration of carbohydrates and is often found sticking to surfaces below or around the mealybugs, as many species can "flick" honeydew droplets away from their feeding site.

Mealybugs damage the crop by directly infesting the grape clusters. In addition, the excreted honeydew, which initially has a shiny appearance, darkens and dries to directly foul the crop and leaves. The honeydew also becomes a substrate for the growth of dark, sooty molds, and leaf surfaces covered in such mold are rendered photosynthetically inactive. In addition to the presence of mealybugs and honeydew, and the attendant growth of sooty mold, other damage caused by mealybug infestations can include yellowing of leaves, stunting of new growth, and even defoliation when populations are very large. Feeding wounds, especially inside the clusters, allow for the introduction of bunch rot organisms such as *Botrytis* and members of the sour rot complex. Some species of mealybugs also vector viruses associated with grapevine leafroll disease.

Pseudococcus Mealybugs

Three species of mealybugs in the genus *Pseudococcus* have pest status in grapes. These are the grape mealybug, *P. maritimus* (Ehrhorn); obscure mealybug, *P. viburni* (Signoret); and longtailed mealybug, *P. longispinus* (Targioni-Tozzetti). All three species are vectors of leafroll associated viruses.

The grape mealybug is native to North America and can be found in vineyards throughout most viticultural regions, as well as on some fruit tree species such as apple and pear. On vines, the grape mealybug overwinters under the bark as eggs in a cottony mass (ovisac) or as young nymphs. As temperatures warm, the population develops and moves up the vine to feed on canes, leaves, and clusters. Those clusters that come in contact with the bark are especially vulnerable. There are typically two generations per year in most grape-growing regions. Grape mealybug populations are frequently controlled by resident parasitoids.

Obscure mealybug is closely related to the grape mealybug and has a similar appearance and life cycle. However, the obscure mealybug has less tolerance for high summer temperatures, more readily feeds on the grape leaves during the summer months, and can have three generations per year. It is commonly found in the cooler grape-growing regions in the United States, Europe, New Zealand, and Australia.

Longtailed mealybug has a wide host range but limited distribution on vines owing to its preference for regions with cooler temperatures. It is better known as an indoor and glasshouse pest and, in this venue, occurs throughout North America and in Africa, Europe, the Middle East, Asia, New Zealand, and Australia. When found on a vine, it produces damage similar to grape and obscure mealybugs.

Planococcus mealybugs

Vine mealybug, *Planococcus ficus* (Signoret), is a very serious pest of grapes. It is an economic pest in the Mediterranean regions of Europe; northern and southern Africa; the Middle East; Argentina; and most recently, in the southeastern United States, California, and Mexico. It produces more offspring and has more generations per year (three to six, depending on temperature) than the *Pseudococcus* mealybugs. Vine mealybug populations are often found feeding on roots underground, in vineyards with sandy soil or where ant populations aid their descent. This location provides buffering from temperature extremes and protection from many potential chemical and biological control agents. Vine mealybug feeds on all parts of the vine during the season, producing copious amounts of honeydew, and can cause leaf yellowing and defoliation. It has a great affinity for feeding within the cluster, where it causes contamination by cottony ovisacs, honeydew, and the attendant growth of sooty mold; it also causes feeding wounds that allow for the entry of bunch rot organisms. Additionally, vine mealybug is a known vector of leafroll-associated viruses.

An additional *Planococcus* species, the citrus mealybug, (*P. citri* [Risso]) has been found in grapes but is not known to cause economic damage or to vector viruses.

Pink hibiscus mealybug

Pink hibiscus mealybug, *Maconellicoccus hirsutus* (Green), feeds on over 200 species of plants, including grapes. It is found in Africa, Asia, Australia, South America, the Caribbean islands, and in parts of Mexico, Hawaii, Florida, and California. In grapes, it feeds near the tips of new growth, causing severe deformation and curling of leaves and stunting of new shoots. It is not documented as a vector of any grape disease. Because it has only recently been found in the grape-growing regions of southern California, there have been no reports of this mealybug species infesting any North American vineyards. It has been the successful target of a biological control program throughout the Caribbean.

Ferrisia mealybugs

Mealybugs in the genus *Ferrisia* also can feed on grapes. These include the striped mealybug, *F. virgata* (Cockerell), and *F. gilli* Gullan. Striped mealybug occurs throughout most of the tropical and subtropical parts of the world. *F. gilli* was identified as a new species in 2003 and is thought to be native to the southeast United States. Although there have been few reports of *Ferrisia* mealybugs infesting vineyards, the newly named *F. gilli* was found in isolated North America vineyards, feeding on all parts of the vine and aggregating in the cluster near harvest. *Ferrisia* mealybugs can be distinguished from other mealybugs by the long (5–10 mm) waxy filaments they excrete on the dorsal side of their bodies. During the winter, a white layer of these filaments produced by aggregations of mealybugs on the vine trunk and cordons gives the bark a bearded appearance, as if draped in cotton candy. *Ferrisia* mealybugs are not known to vector diseases in grapes.

Selected References

Flaherty, D. L., Phillips, P. A., Legner, E. F., Peacock, W. L., and Bentley, W. J. 1992. Mealybugs. Pages 159-165 in: Grape Pest Management. 2nd ed. D. L. Flaherty, L. P. Christensen, W. T. Lanini, J. J. Marois, P. A. Phillips, and L. T. Wilson, eds. Publ. 3343. Division of Agricultural Sciences, University of California, Berkeley.

Godfrey, K. E., Daane, K. M., Bentley, W. J., Gill, R. J., and Malakar-Kuenen, R. 2002. Mealybugs in California Vineyards. Publ. 21612. Division of Agriculture and Natural Resources, University of California, Berkeley.

(Prepared by D. R. Haviland and K. M. Daane)

Mites

Mites encompass a large, diverse group of small (0.1–0.5 mm and smaller) arthropods related to spiders and found in many habitats around the world. The feeding habits of different mite species vary from plant feeders to predators, parasites, and detritivores. There are two taxonomic groupings of economically important pest mites of grapes: (i) the superfamily Tetranychoidae, which includes the spider mites (Tetranychidae) and false spider mites (Tenuipalpidae); and (ii) the superfamily Eriophyoidea, which includes the bud mite and the grape rust mite. Several mite species discussed below are potentially serious pests of grapes in most of the world's viticultural regions. They damage plant tissue by using their modified mouthparts for piercing these cells and sucking their contents.

Spider Mites

Spider mites are common pests of grapevines in many regions. Two species are considered important pests in California: the Pacific spider mite, *Tetranychus pacificus* McGregor; and the Willamette spider mite, *Eotetranychus willamettei* (Ewing). In the Pacific Northwest, the twospotted spider mite, *T. urticae* Koch, and the McDaniel mite, *T. mcdanieli* McGregor, are the predominate pests, although in recent years Pacific spider mite and Willamette spider mite have become more problematic. In the eastern United States, the European red mite, *Panonychus ulmi* (Koch), is the principal spider mite pest, although in some years the twospotted spider mite can reach damaging levels. In Europe, *P. ulmi*, *E. carpini* Oudemans, *T. urticae*, and *T. mcdanieli* are grape pests. *Oligonychus vitis* Zaher & Shehata is reported as a serious pest of grapes in Egypt and Chile, and *O. punicae* (Hirst) is an occasional pest of grapes in tropical America. A number of spider mites, including *P. ulmi*, *E. pruni* (Oud.), *T. urticae*, *T. turkestani* Ugarov & Nikolski, and *Bryobia praetiosa* Koch, are reported to infest vineyards in central and Eastern Europe. Recently, *E. sexmaculatus* (Riley) has been reported feeding on grapevines in Western Australia, although its pest status is unclear.

Willamette and Pacific spider mites. In vineyards in the San Joaquin Valley in California, feeding by small colonies of Willamette and Pacific spider mites produces yellow spots on the upper leaf surfaces. Yellowing of foliage is characteristic of high densities of Willamette spider mites (Fig. 275). A bronze discoloration, heavy webbing, and cupping of the foliage are characteristic symptoms of high Pacific spider mite populations, particularly on foliage that is exposed to the sun on the tops of vines (Fig. 276). Later, heavily infested vines may become defoliated as leaves turn brown and dry.

Pacific spider mite injury in vineyards is spotty and tends to occur on weaker vines or in water-stressed areas. Both Pacific and Willamette spider mite populations are favored by dusty vineyard conditions, although the reasons are not clear. Pacific spider mites are more tolerant of high temperatures and are therefore more problematic in the hotter inland regions of California than Willamette mites, which tend to dominate in the cooler regions closer to the coast. Yellowing of vines caused by nitrogen deficiency is sometimes mistaken for Willamette spider mite injury. In the cooler coastal areas of California, Willamette spider mite feeding causes a reddening of the foliage on dark-fruited cultivars such as Cabernet Sauvignon. Such coloration is delimited by veinlets and occurs primarily on the north side of the vines in east-west rows or other shady areas. This injury may sometimes be mistaken for symptoms of certain virus diseases, i.e., grapevine leafroll or red blotch (Fig. 277).

Willamette spider mites may cause amber coloring of Thompson Seedless table grapes if clusters are exposed to direct sunlight. Severe injury by Willamette spider mites has been reported on Zinfandel in Northern California. Large numbers of overwintering Willamette or Pacific spider mites may cause necrotic areas on foliage in early spring, although serious injury more typically develops mid- to late season.

Twospotted spider mite. In California, twospotted spider mite injury is similar to that of the Pacific spider mite on young shoot growth. Infestations are often related to movement from nearby, heavily infested fruit orchards. The mites reproduce rapidly and cause severe injury to new growth on vines, but reproduction appears to cease as foliage matures. Twospotted spider mite is the most commonly reported mite pest of grapes in Washington. In New York, twospotted spider mite infestations occur sporadically, although they can reach damaging levels in some years. Twospotted spider mite overwinters as adult females on groundcover vegetation or possibly under loose bark on the trunk and older wood.

In Europe, the twospotted spider mite is an important pest of grapevines in regions with dry summers. It is particularly serious in Spain. The first symptoms of attack are chlorotic spots caused by feeding on lower leaf surfaces where dense colonies thrive. Later, defoliation is likely to affect berry maturation and quality. Twospotted mites also attack berries, producing dark spots on the skin.

European red mite. In the eastern United States, injury by European red mites is recognized as a fine speckling that begins at the base of leaves but soon develops into a bronze discoloration on dark-fruited cultivars or yellowing on white-fruited cultivars, covering the entire leaf (Fig. 278). Continued feeding by the pest may cause leaves to turn brown and fall. Oxidant stipple is sometimes confused with European red mite bronzing. However, leaves with oxidant stipple are usually darker brown than those bronzed by mites. European red mites do not produce significant amounts of webbing and are found mostly, but not exclusively, on the underside of leaves.

The European red mite overwinters as eggs on 1-year-old canes or older wood. Although populations tend not to reach damaging levels until mid- to late season, they may seriously injure new spring growth under some circumstances in which overwintering egg populations are high. This has been reported in France. The margins of young leaves appear brown as if damaged by a mild frost. The injured leaves remain small and deformed and often appear yellow-mottled. When the attack is heavy, injured leaves shrivel, desiccate, and fall. In gen-

Fig. 275. Typical yellowing caused by Willamette spider mite feeding on foliage of a white-fruited cultivar. (Reprinted, by permission, from Flaherty et al. 1981)

Fig. 276. Leaf burn caused by Pacific spider mite feeding on the top of a grapevine. (Reprinted, by permission, from Flaherty et al. 1981)

Fig. 277. Interveinal leaf reddening caused by Willamette spider mite feeding on a red-fruited cultivar. (Courtesy V. Walton)

eral, shoots emerging from buds at the base of canes or spurs are most severely injured because they are closest to the older wood where mites overwinter and also because they grow more slowly than shoots from terminal buds. During the period of rapid vine growth, injury becomes less visible, and mites may not be noticed again until the end of summer, when foliage is bronzed or yellowed.

E. carpini. Commonly called the yellow vine mite, *E. carpini* occurs on grapevines in Mediterranean viticultural areas, especially in France and Italy. The injury it causes is similar to that caused by the Willamette spider mite in California. Overwintering females attack the young shoots during bud break, causing small necrotic spots.

T. mcdanieli. Early-season injury caused by *T. mcdanieli* in France is similar to that caused by *T. pacificus* in California. Young tissues yellow while concomitantly exhibiting some diffused grayness and necrosis. Young leaves show cupping, with some drying of the raised margins. Summer injury resembles symptoms of leafroll disease. Pinot Noir is highly susceptible to attack by *T. mcdanieli*. *T. mcdanieli* is also reported as an economic pest of grapes in Washington.

O. vitis. In desert areas in northern Chile, *O. vitis*, which characteristically infests upper leaf surfaces, causes a heavy bronzing in late summer and may cause early defoliation of the cultivars Emperor, Thompson Seedless (Sultanina), and Muscat of Alexandria. In Egypt, the cultivar Romi-red is commonly infested.

Management. In vineyard systems, spider mite populations are often kept below economic injury levels by natural enemies, of which predatory mites (Phytoseiidae and to a lesser extent, Stigmatidae) are the most important. These tiny mite predators have the capacity to consume large numbers of spider mite prey and reproduce quickly. There are a great number of species of predatory mites found in vineyards, with the species of importance varying according to mite prey, geographic region, climate, and to some extent, cultivar characteristics. In California vineyards, the most important predatory mite species is *Galendromus occidentalis* (Nesbitt), the western predatory mite. In cool-climate growing regions of the United States and Europe, *Typhlodromus pyri* Scheuten and *Amblyseius andersoni* can play key roles in preventing serious spider mite outbreaks. In Australia, the introduced *G. occidentalis* and the native species *Euseius victoriensis* (Womersley), among others, have reportedly been effective in preventing outbreaks of spider mites and eriophyids. Broadly speaking, the most effective species of predatory mites in perennial cropping systems such as vineyards are characterized by a preference for spider mite prey with an ability to persist on alternative food sources such as nonpest mite species, pollen, and fungal spores.

Fig. 278. Yellowing of a leaf of cv. Riesling caused by feeding of the European red mite. (Courtesy S. Hessler)

Effective predation is also aided significantly by an ability to survive some exposure to commonly used pesticides. Although phytoseiid mites and other natural enemies such as predatory bugs and thrips can effectively maintain mite pests below damaging levels in vineyards if left undisturbed, they can be killed by certain pesticide inputs. These include several fungicides (e.g., mancozeb, wettable sulfur, and oils used for powdery mildew control), insecticides (e.g., fenpropathrin, methomyl), and acaracides (e.g., dicofol, pyridaben) often used in vineyards. Repeated use of such products may have a negative impact on beneficial organisms and consequently result in significant increases in pest mite populations. In many cases, it is possible to replace or significantly limit such materials by substituting alternatives that are less toxic to mite predators and thereby conserve natural enemy populations. This concept should be particularly considered in vineyards where spider mite management has become problematic.

There are a number of chemical control options, many registered since 2000, for managing spider mites in vineyards. These include: organochlorines (e.g., dicofol); organotins (e.g., fenbutatin-oxide); synthetic pyrethroids (e.g., bifenthrin); sulfide esters (e.g., propargite); mite growth inhibitors (e.g., hexythiazox); tetraonic and tetramic acid derivaties (e.g., spirodiclofen); mitochondrial electronic transport inhibitors (e.g., pyridaben); and microbials (e.g., abamectin). The mode of action varies among chemical options, as does the stage of mite development against which they are most effective. Some are contact poisons that impact all mobile stages whereas others only kill eggs and/or early immature stages. For these latter compounds, there often is a lag between the time of application and population decline. Some of these acaracides, especially older generation materials such as dicolfol and synthetic pyrethoids (e.g., bifenthrin), are harsh on beneficial predatory mites. Others are softer on these predators and can often be used to supplement their presence and provide an additive effect against the pest spider mites. Acaracides can vary in terms of their relative efficacy against individual species of mites, including spider mites. For example, bifenazate is more effective against twospotted spider mite than against European red mite, whereas the opposite is true for pyridaben; use rates should be adjusted accordingly.

Because spider mite problems are often more severe under dusty vineyard conditions, limiting vineyard dust also aids in their management.

False Spider Mites

False spider mites resemble very small (<0.5 mm long), flat, red spider mites. They are associated with feeding injury to grapes and also have been identified recently as competent vectors of virus and viruslike diseases in other crops, although not grapes. *Brevipalpus chilensis* Baker is a serious pest in Chile. *B. lewisi* McGregor is responsible for "bunch mite" symptoms of grapes in Australia, is also reported as a pest in Greece and in Eastern Europe, and in 1984, damaging populations were observed for the first time in California vineyards. Another false spider mite species, *Tenuipalpus granati* Sayed, is reported as a pest of grapevines in Egypt.

In Chile, *B. chilensis* injures shoots and foliage of dark cultivars. Cultivars grafted on American rootstocks are attacked to a lesser extent. During budbreak, overwintering *B. chilensis* females induce necrosis of leaves and shoots, which acquire an intense brown color similar to the damage produced by early frost. As populations increase and spread, leaves become discolored and their margins fold down. The leaves first appear greenish copper, then dark reddish, and finally grayish brown. Under heavy attack, newly developed leaves are smaller than normal and fruit production is reduced.

Injury may be especially severe in vineyards of Cardinal, Emperor, and Ribier table grapes. Severe desiccation and blackening of new shoot growth are not uncommon. Later, the mites distribute onto the rachis, branches, and pedicels of clusters, where desiccation results in darkened grooves that unite to form large lesions (Fig. 279) that resemble symptoms of several grape diseases such as Phomopsis cane and leaf spot and bunch stem necrosis. The stems and berries may completely dehydrate. On the foliage, *B. chilensis* causes loss of greenness and a concentration of anthocyanins on the lamina of dark cultivars, and affected parts appear reddish. Cultivars such as Moscatel de Austria tolerate high populations per leaf; the reddish mites concentrate on both sides of the main vein, especially in the vein angles.

Predatory mites (Phytoseiidae) have been reported to feed on *Brevipalpus* spp., although their role in regulating populations is not well understood. Some of the same chemical controls used for erineum mites (e.g., sulfur, see Management of Eriophyid Mites below) have been used for false spider mite control.

Bud Mite

The bud mite, *Colomerus vitis* (Pagenstecher), attacks several grape cultivars and is widely distributed geographically. Three morphologically identical "strains" of the mite have been recognized by the characteristic injury that each causes: the erineum strain, the bud mite strain, and the leaf curl strain. Recent molecular analyses strongly suggest the bud mite strain and the erineum strain are distinct species. Adults are whitish and wormlike, approximately 0.2 mm long and less than 0.05 mm wide.

Erineum strain. The erineum strain is found in many grape-growing regions of the world. It lives and feeds within dense hairs on the lower (abaxial) leaf surface, where it causes patches of concave, felty galls, called erinea, with corresponding blisterlike swellings on the opposite upper (adaxial) surface. The galls are whitish at first, then yellow, and finally reddish brown (Fig. 280). In the white stage, the underleaf galls can be confused with sporulating downy mildew lesions. A leaf may have 50 or more erinea. These leaves may drop somewhat earlier in the fall compared with uninfested leaves, but the overall economic impact of the erineum strain is thought to be negligible.

Bud strain. Bud mites live within grape buds and do not produce erinea on leaves. Bud mites are generally confined to feeding on the outer bud scales, although they may penetrate and feed on embryonic tissues of the shoot primordia. In severe cases, this feeding may result in total destruction of bud tissue. Common symptoms (Fig. 281) include short basal internodes, scarred epidermis of new shoots, flattened shoots, smaller wrinkled and cupped leaves, dead terminal buds on new shoots, "witches'-broom" proliferation of new shoots, zigzagged shoots, and dead overwintering buds. These symptoms are often similar to those caused by grape rust mite (see below). Flower clusters may also drop prematurely because of injury to immature leaf and flower buds, a symptom sometimes con-

Fig. 279. Darkened blotches caused by false spider mites (*Brevipalpus chilensis*) on the rachis of cv. Ribier grapes in Chile. (Courtesy R. H. Gonzalez)

Fig. 280. Erinea on the lower surface of a leaf, caused by the grape erineum mite. (Courtesy V. Walton)

Fig. 281. Stunted, zigzag growth and lateral shoot proliferation caused by the bud mite strain of the grape erineum mite feeding on *V. vinifera* in early spring. (Courtesy R. H. Gonzalez)

fused with boron deficiency in early spring. The economic impact of bud mite varies depending on their abundance and the grape cultivar, although under some conditions it can be significant. Bud mite appears to have a more restricted range than the erineum strain, being reported from the western United States, Chile, Spain, Crimea, Egypt, Israel, South Africa, and Australia.

Leaf curl strain. Symptoms of the leaf curl strain appear in summer and include downward curling or rolling of leaves and areas of necrosis on the underside of leaves (Fig. 282). These symptoms are sometimes confused with the leaf cupping symptoms of boron toxicity. Their intensity ranges from slight to severe, where the leaf tends to roll into a crude ball. Overall growth of the shoot may be stunted, and scarring is sometimes noted on it. Lateral shoot growth also may be induced. The leaf curl strain has been reported from California, Chile, Hungary, Romania, Egypt, and South Africa.

Grape Rust Mite

The grape rust mite, *Calepitrimerus vitis* (Nalepa), is amber, 0.1 mm long, and appears wormlike; the front end is broader than the rear, producing an overall triangular shape. It moves slowly on the leaf surface. The mite has been reported as a pest in most areas of the world where *V. vinifera* is grown. Grape rust mite overwinters as females within outer bud scales, under bark, and in crevices near 1-year and older wood close to the head of the vine. As soon as vines begin to break dormancy, these mites move from their overwintering sites and start to feed on young, susceptible tissue, as early as the wooly bud stage (Eichorn-Lorenz stage 3).

Symptoms are more pronounced when tissue development is slow because of cool conditions, and heavy infestations prevent vines from growing normally early in the season. In severe cases, feeding may result in death of primary buds and their clusters. In less severe cases, basal internodes are shortened and show zigzag growth with characteristic scarring in a pattern that resembles railroad tracks (Fig. 283). Leaves are cupped and small and fruit production is reduced. Shoot symptoms caused by the grape rust mite have been referred to as "Restricted Spring Growth' in Australia and as "Short Shoot Syndrome" in the Pacific Northwest. Mites colonize the underside of leaves as shoots elongate during the season. Clusters also can be infested. Later in the season, feeding on the surface of leaves of white grape cultivars causes a yellowing that closely resembles the appearance of leaves slightly injured by spider mites. On dark-fruited grape cultivars, injured leaves may become red, superficially resembling those associated with diseases such as leafroll. Bronzing of leaves late in the season is also a common symptom (Fig. 284).

Management of Eriophyid Mites

Eriophyid mites are fed upon by a number of mite and insect predators. Predatory mites, particularly in the families Phytoseiidae, Stigmatidae, and Tydeidae, are probably the most important for biological control, especially Phytoseiidae. Several of the phytoseiid mites that are important biological control agents for spider mites in different parts of the world also feed on eriophyid mites (e.g., *Amblyseius aberrans*, *A. andersoni*, *A. californicus*, *A. victoriensis*, *Galendromus occidentalis*, *Typhlodromus pyri*). Indeed, it has been observed that, where present, eriophyid mites serve as an alternative food source for phytoseiid mites, maintaining their populations and resulting in better control of spider mites. The impact of certain pesticides commonly used in vineyards needs to be considered within the context of their effect upon these predators and the attendant consequences as discussed above. For example, in the Pacific Northwest, there is a close inverse association between rust mite populations and those of the predator, *T. pyri*.

Fig. 282. Symptoms of the leaf curl strain of the grape erineum mite on cv. French Colombard. (Courtesy F. Jensen)

Fig. 283. Characteristic "railroad" scarring on basal internode tissue caused by feeding of the grape rust mite. (Courtesy V. Walton)

Fig. 284. Leaf bronzing caused by feeding of the grape rust mite. (Courtesy V. Walton)

Chemical control of bud mites and rust mites may be required under some circumstances. Wettable sulfur as well as some miticides have been shown to provide control of eriophyid mites, although timing is important to optimize contact with susceptible stages as they move to new feeding sites. For grape rust mite, treatment in early spring (wooly bud) followed by a second spray 10–14 days later provides a measure of control. For bud mite, treatment within one week after budbreak is recommended.

Selected References

Bernard, M. B., Horne, P. A., and Hoffmann, A. A. 2005. Eriphyoid mite damage in *Vitis vinifera* (grapevine) in austalia: *Calepitrimerus vitis* and *Colomerus vitis* (Acari: Eriophyidae) as the common cause of the widespread 'restricted spring growth' syndrome. Exp. Appl. Acarol. 35:83-109.

Buchanan, G. A., Bengston, M., Exley, E. M., 1980. Population growth of *Brevipalpus lewisi* McGregor (Acarina: Tenuipalpidae) on grapevines. Aust. J. Agric. Res. 31:957-965.

Carew, M. E., Goodisman, M. A. D., and Hoffmann, A. A. 2004. Species status and population genetic structure of grapevine eriophyoid mites. Entomol. Exp. Appl. 111:87-96.

Childers, C. C., French, J. V., and Rodrigues, J. C. V. 2003. *Brevipalus californicus*, *B. obovatus*, *B. phoenicis*, and *B. lewisi* (Acari: Tenuipalpidae): A review of their biology, feeding injury and economic importance. Exp. Appl. Acarol. 30:5-28.

Duso, C., and De Lillo, E. 1996. 3.2.5 Grape. Pages 571-582 in: Eriophyoid Mites: Their Biology, Natural Enemies and Control. E. E. Lindquist, M. W. Sabelis, and J. Bruin, eds. Elsevier Science Publishers, Amsterdam.

Flaherty, D. L., Wilson, L. T., Welter, S. C., Lynn, C. D., and Hanna, R. 1992. Spider mites. Pages 180-192 in: Grape Pest Management. D. L. Flaherty, L. P. Christensen, W. T. Lanini, J. J. Marois, P. A. Phillips, and L. T. Wilson, eds. Publ. 3343. Division of Agricultural Sciences, University of California, Berkeley.

Gadino, A. N., Walton, V. M., and Dreves, A. J. 2011. Impact of vineyard pesticides on a beneficial arthropod, *Typhlodromus pyri* (Acari: Phytoseiidae), in laboratory bioassays. J. Econ. Entomol. 104:970-977.

Gonzalez, R. H. 1983. Manejo de plagas de la vid. Cienc. Agric. 13. Departamento de Sanidad Vegetal, Universidad de Chile, Santiago.

Jeppson, L. T., Keifer, H. H., and Baker, E. W. 1975. Mites Injurious to Economic Plants. University of California Press, Berkeley.

Keifer, H. H., Baker, E. W., Kono, T., Delfinado, M., and Styer, W. E. 1982. An Illustrated Guide to Plant Abnormalities Caused by Eriophyid Mites in North America. U. S. Dep. Agric. Agric. Handb. 573.

Schruft, G. A. 1986. Grape. Pages 354-366 in: Spider Mites, Their Biology, Natural Enemies and Control. W. Helle and M. W. Sabelis, eds. Elsevier, Amsterdam.

Stavrinides, M. C., and Mills, N. J. 2009. Demographic effects of pesticides on biological control of Pacific spider mite (*Tetranychus pacificus*) by the western predatory mite (*Galendromus occidentalis*). Biol. Control 48:267-273.

Walton, V. M., Dreves, A. J., Gent, D. H., James, D. G., Martin, R. R., Chambers, U., and Skinkis, P. A. 2007. Relationship between rust mites *Calepitrimerus vitis* (Nalepa), bud mites *Colomerus vitis* (Pagenstecher) (Acari: Eriphyidae) and short shoot syndrome in Oregon vineyards. Internat. J. Acarol. 33:307-318.

Walton, V. M., Dreves, A. J., Coop, L. B., Jones, G. V., and Skinkis, P. A. 2010. Developmental parameters and seasonal phenology of *Calepitrimerus vitis* (Acari: Eriophyidae) in wine grapes of western Oregon. Environ. Entomol. 39:2006-2016.

(Prepared by G. Loeb, V. Walton, and F. Zalom)

Grape Phylloxera

Grape phylloxera, *Daktulosphaira vitifoliae* (Fitch) (Homoptera: Phylloxeridae), is an aphidlike insect that initiates the formation of galls (swellings) for feeding on grapevine roots and leaves. In its North American native range, this insect originally fed on leaves of indigenous grape species (*Vitis riparia*, *V. rupestris*, *V. champinii*, *V. berlandieri*, etc.) and to a limited extent on their immature roots. Phylloxera damage to these genotypes ranges from limited to nil. However, phylloxera's geographic range now extends to most grape-growing regions of the world, with a few specific exceptions (e.g., Chile, most viticultural districts of Australia, much of the Pacific Northwest). Worldwide, it is capable of feeding on immature and mature roots of the European grape (*V. vinifera*) wherever the pest is present, with the capacity to severely damage and kill these vines. Phylloxera is only rarely found galling *V. vinifera* leaves.

History

Prior to the 1850s, phylloxera were restricted to the Americas, where they were first described on wild American grapes by Fitch in 1855. When European settlers tried to establish *V. vinifera* grapes in North America, they did not succeed, though it is only in retrospect that we can blame grape phylloxera for a portion of that failure. Native North American fungal diseases of grapes (downy mildew, powdery mildew, and black rot) also appear to have played a significant role in these early failures.

In the 1850s, phylloxera were inadvertently moved to Europe on infested planting materials most likely destined for plant collections and for use in breeding for resistance to the newly imported disease powdery mildew. Infestations were traced through England to France, where the insect initially killed vineyards in the Rhone Valley before it spread rapidly throughout the remaining areas of French production and then across the rest of Europe, beginning in the 1860s. After associating dying vines with the presence of this insect on the roots, many control tactics were tested. Carbon bisulfide fumigation of infested soil was widely employed before the phylloxera resistance of American grapevines was discovered. The initial use of phylloxera-resistant vines was as fruit-producing interspecific hybrids between *V. vinifera* and native American species. The winning technology, however, was the use of resistant American vines as rootstocks, upon which more popular European cultivars were grafted as the scion. Rootstock development began in the 1870s and breeding and selection continued intensively for 50 years in many European agricultural experiment stations. Hundreds of rootstocks were developed for a multitude of specific conditions, and these can be credited with a renewal of European viticulture.

Cultivation of *V. vinifera* in California expanded with the Missionary period (1760s to 1840s) and began to flourish with population expansion in the 1850s. However, phylloxera soon spread into California and damaged vines there as it had in Europe. The importation of rootstocks into the United States, beginning in the late nineteenth century, rescued viticulture in California and expanded the use of *V. vinifera* scions across warmer parts of the United States. Many regions of the United States with winter temperatures unsuitable for survival of *V. vinifera* vines cultivated own-rooted American grape species and *V. vinifera*-American grape ("French") hybrids that possessed commercially adequate levels of resistance or tolerance to phylloxera.

Symptoms and Damage

Grape phylloxera are minute insects less than 1 mm in length, requiring magnification (e.g., a 10× hand lens) for visualization. Phylloxera galls induced on susceptible plants by the insect's feeding can be seen without magnification and take several forms. Galls on mature roots, called tuberosities, are swellings supporting one to (often) many feeding phylloxera (Fig. 285). The insect's color makes large colonies of them appear yellow. Infestations on immature feeder roots form smaller root swellings, called nodosities, which can be identified by their hooked to clublike shape (Fig. 286). Leaf galls, induced by phylloxera feeding and growing within them, are obvious as spiny pocketlike protrusions on the lower surface of the lamina, with up to several hundred developing on a single leaf (Fig. 287). Leaf galls only develop as the leaf is expanding; once fully expanded, leaves do not produce additional galls.

Leaf galling phylloxera do not severely impact vine survival but can reduce shoot growth and grape yield when injury is severe. Nodosities in the absence of tuberosities cause very limited vine injury. The most significant phylloxera damage by far is caused by tuberosities: a large population of these can lead to vine death. Phylloxera feeding wounds on mature or immature roots can allow invasion by secondary fungal pathogens, including species of *Fusarium*, *Alternaria*, *Rhizoctonia*, *Macrophominia*, and *Phaeoacremonium*. In some cases, these fungi can spread radially from infection sites around the phloem parenchyma, girdling the root. Experimental data on the role that such opportunistic fungal infections play in vine damage are limited, although there is some evidence of their significance.

Damage in vineyards of *V. vinifera* cultivars begins with a few declining or dying vines in a small epicenter where the insect first invades. Damage widens from these foci (Fig. 288), with downwind spread showing up as discrete damage sites. Vine death may occur rapidly over 2 or 3 growing seasons, or slowly over many years. Conditions detrimental to phylloxera populations (e.g., sandy soil, adverse temperatures, flooding) tend to contain or slow losses.

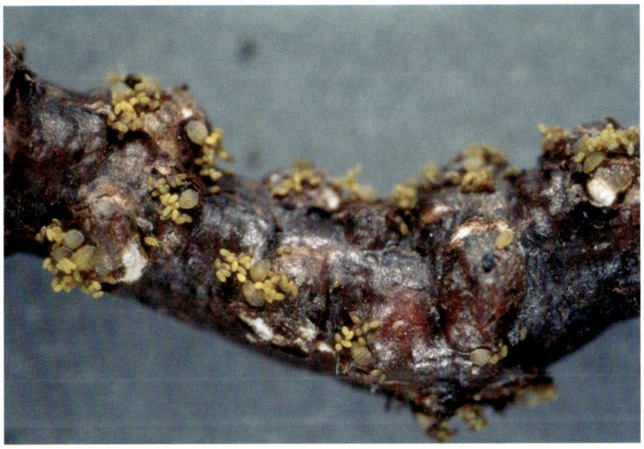

Fig. 285. Mature adults, nymphs, and eggs of grape phylloxera upon a root exhibiting tuberosities. (Courtesy J. Granett)

Fig. 286. Nodosities caused by the feeding of grape phylloxera upon fine roots of a *V.* x *labruscana* 'Concord' vine. (Courtesy R. N. Williams)

Life Cycle and Population Biology

The life cycle of grape phylloxera (Fig. 289) is predominantly asexual (only females are active) with periodic sexual forms. The asexual life cycle includes an egg, four nymphal instars (immature developmental stages) as the insect grows, and finally the reproductive adult female. This life cycle is common both on the surface of root galls and inside leaf galls. The first instar, called the "crawler," is mobile on roots or canes and can disperse in wind. Its function is to establish a feeding site, whereupon a gall forms after the crawler settles and starts to feed. Subsequent instars and adults are morphologically similar and tend not to move from their feeding sites. Overwintering on roots is thought to be by dormant 1st instars called hibernants.

The sexual generation was described in the 1870s but has not been verified to be functional in most parts of the insect's current range. The sexual cycle is produced when a root generation under stressful conditions gives rise to winged adults (alates). The alates, after growing to adulthood on roots underground, emerge from this habitat and disperse by flying or crawling. Without feeding or mating, they lay three to five male and female eggs. These eggs hatch and develop to the adult state without moving or feeding. As adults, the females are mostly sessile whereas males crawl toward them and mate, after which a female lays a single egg. It is thought that the crawler from

Fig. 287. Leaf galls caused by phylloxera on the lower lamina of a *Vitis* interspecific hybrid. (Courtesy D. M. Gadoury)

Fig. 288. Infrared photo of a vineyard with vines dead and dying from phylloxera feeding, showing the expansion of damage from individual, initial foci. (Courtesy J. Granett)

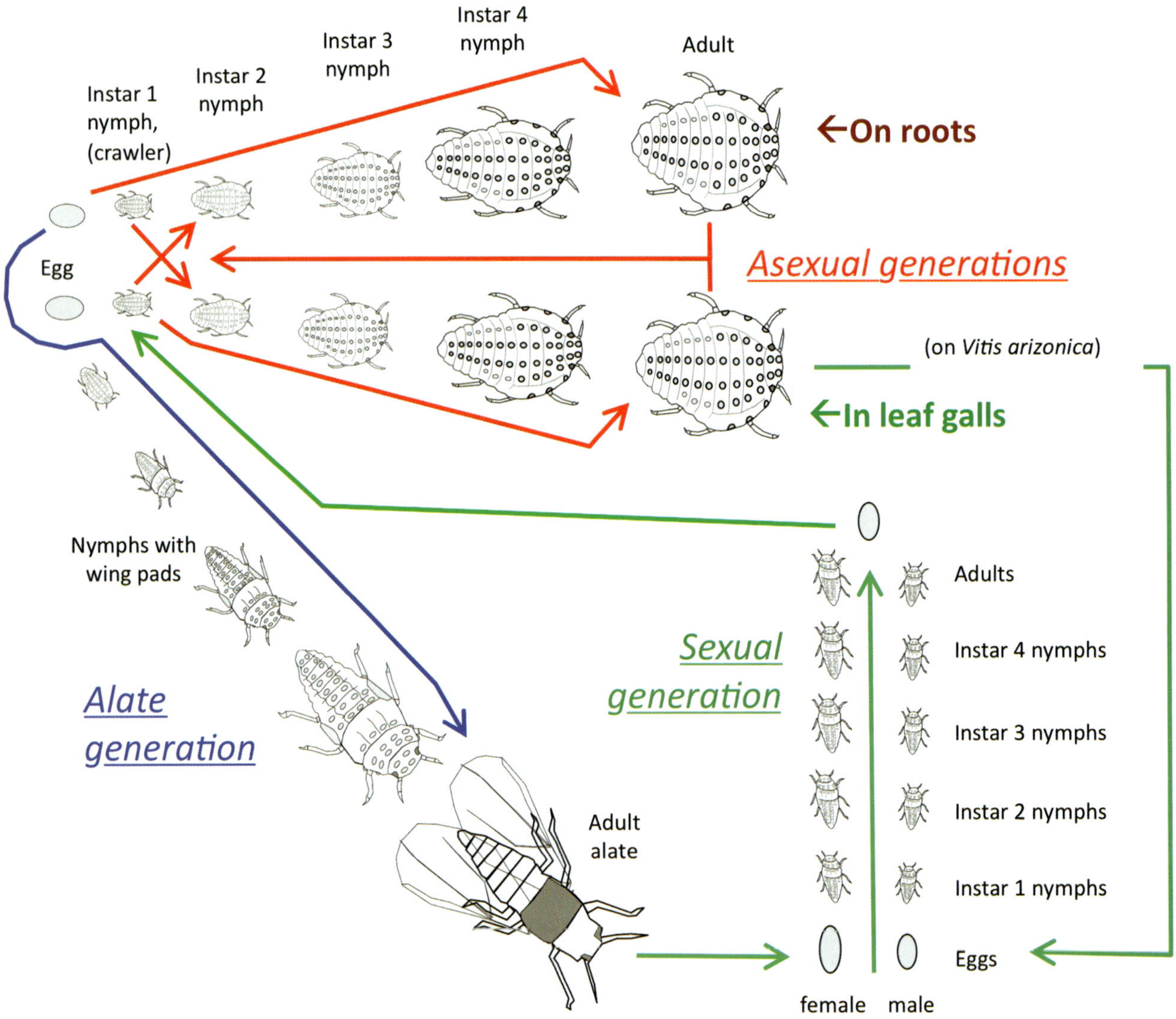

Fig. 289. Life cycle of grape phylloxera. (Courtesy J. Granett)

this egg initiates leaf galls primarily, although they may also initiate root galls. In the southwestern United States, *V. arizonica* supports nonwinged sexual adults arising from leaf galls.

Overwinter populations of grape phylloxera on susceptible vine roots are very low and rise exponentially in spring. In California, populations reach a peak in the midsummer around veraison, at which time the population falls rapidly and recovers only after harvest. In eastern Ontario, Canada (and presumably, similar climatic regions in the eastern United States), populations do not decline at veraison.

Populations are constrained by the quality of gall formation and temperature. Root gall formation is influenced by the level of susceptibility of the own-rooted cultivar or the rootstock upon which it is grafted (resistance may slow gall formation, which also increases generation time) and by soil type (clay soils enhance growth of phylloxera population whereas sandy soils inhibit them). Galls do not form below 18°C, although the minimum threshold for feeding is below that on both roots and leaves. The optimal temperature for phylloxera development is between 20 and 30°C. Developmental time from egg to adult is at least 15 days, dependent on the constraints discussed above. Adults may live a month or more and lay several hundred eggs over their lifetime. In warm climates there may be more than 10 generations per year, whereas in cool climates there may be only two or three. Soil temperatures above 32°C and standing water are both lethal to phylloxera. No biological control agents are known to effectively regulate populations on roots or leaves, though many species of arthropod predators have been observed.

Humans have been responsible for the dispersal of grape phylloxera on a local to international scale, through the movement of infested plants or soil adhering to equipment or boots. The insect also can be transported up to several kilometers independent of human aid. First instar nymphs emerging from the soil or leaf galls can be blown by wind to previously uninfested vines, and flood waters are also believed to have moved phylloxera to uninfested vineyards. Although alates fly, they have not been implicated in the spread of infestations.

Rootstocks and Other Controls

The resistance of many American *Vitis* species to forming tuberosities prevents their damage by phylloxera. If these plants are used as rootstocks with *V. vinifera* cultivar scions, damage is completely prevented. Rootstocks with pure American parentage (no partial *V. vinifera* background) have not succumbed to phylloxera after more than a century of use, even though there have been observations of high phylloxera populations on immature roots (nodosities) of strongly resistant selections. However, rootstocks of mixed parentage that includes *V. vinifera* have sometimes failed after decades of intensive and previously successful use. In these cases, phylloxera have become able to produce high populations on tuberosities. When a vineyard fails as a result of damage by grape phylloxera, the grower must replant with a strongly resistant rootstock. Local viticultural extension workers and grape nurseries often have the expertise to advise on regional choices for rootstocks. However, because rootstocks are sold as rooted materials with no foliage and cannot be identified by visual inspection, the buyer must choose sources reputable for the quality of their planting material and proper identification.

A variety of rootstocks are now available with unquestioned resistance to grape phylloxera based on their pure American parentage. Among such interspecific hybrid rootstocks are: *V. berlandieri* × *V. riparia* selections, including SO4, Teleki 5C and 420A Mgt; hybrids between *V. riparia* and *V. rupestris*, including 3309 Couderc, 101-14 Mgt, and Schwarzmann; *V. berlandieri* × *V. rupestris* hybrids, including 110 Richter and 1103 Paulsen; and many others. The names of these rootstocks often include the name or abbreviation of the breeder or place of origin; Teleki, for example, was a prominent Hungarian rootstock breeder. Examples of rootstocks that are selections of a single American species include Riparia Gloire and Rupestris du Lot (also known as St. George), selections of *V. riparia* and *V. rupestris*, respectively.

A number of interspecific hybrid rootstocks include *V. vinifera* in their parentage. An example of a failed rootstock in this category is AXR#1 (also known as Ganzin 1 or ARG1). This rootstock was initially bred in 1876 and found considerable use in Europe, but it failed there and in South Africa in the first decades of the twentieth century. AXR#1 was planted in experimental vineyards in California beginning in 1905, where over the next half century it was found to be excellent for the prevailing conditions and was recommended for primary use in the premiere coastal valley regions. However, once it was widely planted, more aggressive strains of phylloxera were selected, which were capable of feeding on and damaging AXR#1. By the early 1980s, it was clear that such strains, then termed biotype B, were destroying vineyards on this rootstock; the resulting damage caused $1 billion in losses over the next decade, necessitating massive replantings on rootstocks with greater levels of resistance. Other commercial *V. vinifera* hybrid rootstocks are 41B, Fercal, Georgikon 28, VR O39-16, Freedom, and Harmony. They are recommended in appropriate situations for specific exceptional attributes, although growers are cautioned that their phylloxera resistance may not be adequate to prevent damage from this pest.

In eastern North America and parts of Europe, cultivars of the native American *Vitis labrusca* crossed with *V. vinifera* (which forms the cultigen *V.* × *labruscana*), such as 'Concord' and 'Niagara', and complex interspecific hybrids between various American species and *V. vinifera* have long been in use. These varieties often exhibit sufficient resistance to withstand grape phylloxera and can be cultivated on their own roots without grafting. However, the market for such grapes is limited and these cultivars represent only a small proportion of worldwide grape production.

Other tactics for managing grape phylloxera are less successful than host plant resistance. Quarantines have limited the spread of phylloxera from infested zones of Australia; however, although useful, quarantines are at best a delaying tactic. In regions newly infested by phylloxera, growers can slow vineyard decline by balancing the crop load, irrigating well, and fertilizing. Although experiments have shown some insecticides to be efficacious with regard to population control, little work has demonstrated rejuvenated vineyard health at a level that would economically justify the treatment costs. Insecticides have been most successful as a delaying agent when replanting an infested vineyard, but to date, they have not prevented later infestation and damage. (Several insecticides provide effective seasonal control of leaf galling on non-*vinifera* cultivars susceptible to this form of attack). Amendments to adjust and balance soil microbial activity or root health are sold for control of phylloxera damage. These are generally developed without objective, replicated, and controlled evaluations, although they are sometimes supported by grower testimonials.

Selected References

Davidson, W. M., and Nougaret, R. L. 1921. The grape phylloxera in California. U. S. Dep. Agric. Bull. 903.

Granett, J., Walker, M. A., Kocsis, L., and Omer, A. D. 2001. Biology and management of grape phylloxera. Annu. Rev. Entomol. 46:387-412.

Omer, A. D., and Granett, J. 2000. Relationship between grape phylloxera and fungal infections in grapevine roots. Z. Pflanzenkrankh. Pflanzenschutz 107:285-294.

Ordish, G. 1987. The Great Wine Blight. Sidgwick & Jackson, London, UK.

Pongracz, D. P. 1983. Rootstocks for Grapevines. David Philip, Cape Town, South Africa.

Riley, C. V. 1876. The grape phylloxera (*Phylloxera vastatrix* Planchon). Pages 30-87 in: 6th Annual report of the noxious, beneficial, and other insects of the State of Missouri. Regan & Carter, Jefferson City, MO,

(Prepared by J. Granett, M. A. Walker, and G. Loeb)

Thrips

In California vineyards, thrips are a minor problem on wine and raisin grapes. On table grapes, two species of thrips, Western flower thrips, *Frankliniella occidentalis* (Pergande) and European grape thrips, *Drepanothrips reuteri* Uzel, are a concern because they can injure clusters and cause shoot stunting. In the eastern United States, *F. tritici* (Fitch), sometimes called the eastern flower thrips, and *D. reuteri* are considered potential pests. In Chile, *F. cestrum* Moulton, also called flower thrips, and *D. reuteri* are important pests of table grapes. In Europe, *D. reuteri* is widely distributed and is primarily a pest of new shoot growth and foliage in the summer. It has been reported from Switzerland, Italy, France, Spain, Greece, Algeria, and Egypt.

Western flower thrips and grape thrips cause stunted shoot growth in early spring. The adult flower thrips are attracted to the blooming clusters and produce spotting on berries when ovipositing in newly developing fruit. This consists of a small dark scar at the oviposition puncture site surrounded by whitish tissue, sometimes referred to as a halo spot. Varieties such as Almeria, Calmeria, and Italia are normally affected.

Western flower thrips nymphs cause scarring on the berries by rasping the tissue as they feed. On Thompson Seedless, Western flower thrips cause scarring in a characteristic starfish pattern at the stylar end of the berry (Fig. 290) when nymphs feed under persisting calyptras. This scarring, which occurs only when the caps fail to shed normally, renders the clusters unmarketable for fresh consumption. Bloom thinning sprays of gibberellin accentuate the scarring, probably by both stimulating the production of scar tissue and increasing berry growth, which increases the visibility of the scar. The damage can be reduced greatly by controlling the nymphal thrips population from late bloom through the shatter stage. Eliminating bloom thinning sprays also reduces the amount of scarring. Secondary damage can occur when the berry skin splits at the site of halo spots or when scarring during ripening leads to bunch rot.

Drepanothrips reuteri is host-specific for grapes and a common inhabitant of the foliage. In California, the European grape thrips spends the winter in the soil as an adult; infestations are much greater in sandy soils than heavy loam or clay soils. This pest is mainly responsible for damage to the foliage, which often is severe during the summer. Affected leaves are curled and bronzed and fail to grow to normal size; internodes

Fig. 290. Typical starfish pattern of berry scarring caused by western flower thrips on cv. Thompson Seedless (Sultanina). (Photo by A. N. Kasimatis; reprinted, by permission, from Jensen et al., 1981)

Fig. 291. European grape thrips injury to cv. Thompson Seedless (Sultanina) foliage. (Courtesy R. H. Gonzalez)

Fig. 292. European grape thrips injury to cv. White Malaga berries. (Reprinted, by permission, from Jensen et al., 1981)

are shortened and scarred (Fig. 291). These symptoms might be confused with Eutypa dieback or boron deficiency. Populations of the causal pest often have decreased by the time damage becomes evident.

European grape thrips also sometimes cause troublesome fruit scarring. Nymphs and, to a lesser extent, adults injure fruit when berries are approximately 3 mm in diameter. White cultivars such as Thompson Seedless, Malaga, and Moliniere are most sensitive. Severely scarred White Malaga berries split and crack as they grow (Fig. 292). Exposed fruit on vines with a light canopy of foliage are most susceptible to injury.

Selected References

Bailey, S. F. 1942. The grape or vine thrips, *Drepanothrips reuteri*. J. Econ. Entomol. 35:382-386.

Gonzalez, R. H. 1983. Manejo de plagas de la vid. Cienc. Agric. 13. Departamento de Sanidad Vegetal, Universidad de Chile, Santiago.

Jensen, F. L. 1973. Timing of halo spotting by flower thrips on table grapes. Calif. Agric. 27:6-8.

Jensen, F., and Luvisi, D. 1973. Flower thrips nymphs involved in scarring of Thompson Seedless grapes. Calif. Agric. 27:8-9.

Jensen, F. L., Flaherty, D. L., and Luvisi, D. A. 1981. Thrips. Pages 176-186 in: Grape Pest Management. D. L. Flaherty, F. L. Jensen, A. N. Kasimatis, H. Kido, and W. J. Moller, eds. Publ. 4105. Division of Agricultural Sciences, University of California, Berkeley.

Yokoyama, V. Y. 1977. *Drepanothrips reuteri* on Thompson Seedless grapes. Environ. Entomol. 6:21-24.

Yokoyama, V. Y. 1977. *Frankliniella occidentalis* and scars on table grapes. Environ. Entomol. 6:25-29.

(Prepared by C. Gispert)

Part III. Disorders Caused by Abiotic Factors

Chimeras

Chimeras are noticeably different plant parts that arise on a grapevine as a result of a spontaneous genetic change in meristematic tissue and its subsequent proliferation. The term *sport* or *bud sport* is often used in reference to a noticeably different phenotype that comes from a clearly traceable origin, such as one bud, arm, or cane. However, sports also can be encountered as entire vines that develop originally from a single bud with a genetic change (Fig. 293). Sports may appear very different from the original cultivar, such as exhibiting white berries rather than black, or the changes may be barely observable, such as a small percentage increase in berry size. Sports are usually stable and can be propagated asexually just like any other grapevine genotype. They occasionally are selected as new cultivars. For example, it is likely that Pinot Gris and Pinot Blanc are sports of Pinot Noir.

Because of their genetic origin and the lack of any sign indicating a fungal or bacterial pathogen, sports are sometimes incorrectly assumed to be of viral origin. Indeed, the appearances of some sports can resemble the symptoms of certain virus diseases. However, although they can be propagated through buds or cuttings, sports do not transfer their appearance when they are grafted onto healthy plants. This lack of graft transmissibility readily distinguishes their defining phenotypic characters from symptoms of virus infection.

Four common types of sports occur in grapes. Variegation, or different tissue colors appearing side by side, is a common sport reported from many varieties. Leaves (Fig. 294), shoots, inflorescences, or fruit (Fig. 295) may all be affected. Even an individual cluster or berry may show different colors (Fig. 296). A spot within a leaf, a fruit cluster, a single shoot, a major portion of the vine, or the entire vine may show the mosaic pattern of variegation, depending on the position of the mutation within the bud or meristem and the stage of development of the vine when this genetic change occurred. Variegation or mosaic symptoms on shoots and leaves also can be a symptom of grapevine fanleaf degeneration disease.

Fasciation is another common sport in grapevines. The shoots, petioles, or pedicels of the clusters develop with a flattened shape rather than their normal cylindrical form. This can appear on any part of the vine. This sport occurs more frequently in some cultivars than in others and is especially common in Petite Sirah (Fig. 297.). Genetic fasciations frequently have been confused with fasciations induced by *Grapevine fanleaf virus* infection.

In the witches'-broom sport (also called "bull vine"), all lateral buds on the sport grow for a short time after they are initiated, developing into a bushy growth (Fig. 298). Other buds on the same vine can produce normal growth. Affected leaves are smaller and remain green longer into the fall than normal leaves. The growth of the witches' broom sport is often more vigorous than that of the typical part of the vines. Witches'-

Fig. 293. Leaf variegation on a new vine apparently propagated from a bud with this mutation. (Courtesy P. Cousins)

Fig. 294. Leaf variegation on a vine of cv. Riesling. (Photograph by W. McFadden-Smith, Ontario Ministry of Agriculture and Food; Queen's Printer of Ontario, © 2009. Reproduced with permission.)

Fig. 295. Variegation of an entire cluster, resulting from a bud sport or chimera. (Courtesy P. Cousins)

Fig. 296. Variegation involving individual berries within a cluster. (Courtesy P. Cousins)

Fig. 297. Genetic fasciation on shoots of a cv. Petite Sirah vine. (Courtesy A. C. Goheen)

Fig. 298. Witches' broom sports, characterized by the proliferating growth of multiple, sterile lateral buds. (Courtesy P. Cousins)

broom shoots usually are not fertile; thus, their presence in a vineyard reduces yield and they should be removed by pruning.

Leaf distortion is a fourth sport occasionally found in commercial vineyards. Affected leaves are misshapen and smaller than normal leaves, sinuses of the leaf blade are reduced, and the color of the blade indicates that some chlorophyll-bearing tissues do not develop (Fig. 299). The shoot is not greatly reduced in size.

Undesired sports can be removed by pruning back to normal tissue. This practice effectively eliminates most sports in grapevines, although fasciations can be more difficult to control. Avoid collecting propagating material from sports by inspecting vineyards during the growing season, as many sports are not easily identified during the dormant season.

Fig. 299. 'Sauvignon Blanc' shoot with leaf distortion chimera (bottom) and normal shoot (top). (Courtesy A. C. Goheen)

Selected References

Dermen, H. 1947. Histogenesis of some bud sports and variegations. Proc. Am. Soc. Hortic. Sci. 50:51-73.

Flaherty, D. L., Christensen, P. T., Lanini, T., Marois, J., and Wilson, L. T. 1992. Grape Pest Maanement. Division of Agriculture and Natural Resources, University of California, Davis, CA.

Hocquigny, S., F. Pelsy, V. Dumas, S. Kindt, M. C. Heloir, and D. Merdinoglu. 2004. Diversification within grapevine cultivars goes through chimeric states. Genome 47:579-589.

(Prepared by J. VandenHeuvel and P. Cousins)

Environmental Stress

Drought

The first response of grapevines subjected to water stress is the cessation of shoot elongation. Stressed plants are small, do not fill their trellis with leaves, and tend to have loose clusters with small berries. Fruit set may be reduced on vines stressed before or soon after bloom. Under prolonged and severe drought conditions, basal leaves first show marginal necrosis and then senesce and abscise (Fig. 300). Periderm tends to form early on the shoots of stressed vines. Absent irrigation, the entire vineyard often shows signs of stress. In irrigated vineyards, stress may be observed only on vines growing in isolated areas where the soil has very low water-holding capacity, or in areas where root growth is restricted by phylloxera, nematodes, or physical impediments to rooting depth.

Excess Water

Prolonged flooding after growth begins kills roots by depriving them of soil oxygen. Vine growth is restricted, and plants may show symptoms resembling those of drought or simply a general decline and eventual collapse. Roots killed by asphyxiation may appear black or slightly purple, and typical odors of anaerobic decay (e.g., reduced sulfur aromas) may be present. If water is in abundant supply but not asphyxiating, as when the water table is stable and near the surface, shoot growth may be excessive and prolonged. Shoots on excessively vigorous vines may fail to mature and may be killed by the first low temperature in fall or early winter.

Hail

Hail injures vines by contact. Early in the season, hail may break shoots, leaves, or clusters (Fig. 301) or damage portions of the internode (Fig. 302). These "hail pecks" can heal over to resemble small galls or insect feeding wounds. Hail later in the season can cause tattered leaves and split fruit. Berries injured during the early stages of their development either abscise or shrivel and turn brown. The shriveled berries sometimes resemble black rot mummies in regions where this disease occurs. Fruit injured later in the season may become infected with *Botrytis cinerea* and other pathogens. Fruit injured after *veraison* typically rot.

Heat

Soon after exposure to damaging high temperatures, nonlignified portions of shoots wilt and discolor (Fig. 303). Subsequently, the pith may dry and separate, as on vines struck by lightning. Shoots then turn brown and desiccate. Succulent tissues such as shoot tips, young leaves, and tendrils are most susceptible to heat injury. Berries also can be damaged by excessive temperatures (see Shriveled Fruit section).

Lightning

Lightning damage to vines can be difficult to diagnose because of the variable and general nature of the vine response. In some cases, the only response is bronzing and necrosis of leaves, which is reminiscent of chemical burns. In other cases, the entire vine or vine row can collapse and die. The pith of shoots may dry and separate (Fig. 304). Lightning strikes are indicated by the pattern of injury in the vineyard. Entire single rows are often damaged when lightning hits trellis wires, with necrosis sometimes evident on individual vines where they contacted the wire. In nontrellised vineyards, a single vine or small area of vines is usually affected. Evidence of lightning on the trellis or vine stakes, such as discolored wire or shattered posts/stakes, sometimes helps to confirm the diagnosis more than observations of the vines themselves.

Salt Toxicity

Grapevines are moderately sensitive to salinity (chloride), and vines growing in a saline environment may show injury.

Fig. 302. Internode damage caused by hail injury. (Courtesy T. E. Martinson)

Fig. 300. A cv. Cabernet Sauvignon vine with drought stress symptoms. (Courtesy A. N. Kasimmatis)

Fig. 303. Heat-injured (left) and uninjured (right) shoot tips on a cv. Chenin Blanc vine. (Courtesy L. P. Christensen)

Fig. 301. Early season vine damage caused by hail injury. (Courtesy T. E. Martinson)

Fig. 304. Dried and separated pith of a grape shoot struck by lightning. (Courtesy W. Gärtel)

The primary symptom is marginal necrosis of mature leaves (Fig. 305). The expression of salt toxicity can vary markedly in vineyards because of differential susceptibility of scions and differential ability of rootstocks to exclude the chloride ion. Vines can also be injured by windblown sea spray. Sprinkle-irrigated vineyards are injured occasionally by salt applied to the leaf surfaces, with irrigation water containing more than 3 meq of sodium or chloride especially problematic. Salt toxicity from sprinkler irrigation is most likely during periods of high evaporation, when sprinkler rotation is slow, and when the new spring growth is young and tender.

Spring Freeze

After buds begin to swell, their tissues become much less tolerant of low temperatures. Malformation of tissues within the expanded bud, resulting in distorted leaves that develop from them after growth begins (Fig. 306), can occur at higher temperatures than those that would cause similar injury in mid-winter. Such malformed leaves could be confused with symptoms of Phomopsis cane and leaf spot. Following bud break, young and succulent tissues are most susceptible to injury from spring freezes. Soon after thawing, frozen tissue collapses and turns brown (Fig. 307). Cold damage is often restricted to low-lying portions of the vineyard or to vine growth near the ground. Vines damaged by spring freezes may produce little or no crop depending on the timing, number, and severity of freeze events, and therefore regrowth can be extremely vigorous.

Wind and Sand

Vines growing in windy regions can exhibit stunted growth or structural damage. Vines subjected to high winds may have broken shoots, which results in unusual branching patterns. When strong winds blow late in the season, fruit may be injured and become prone to bunch rot. Blowing sand can cause fewer shoots, stunted growth, partial defoliation, and/or malformed leaves. Such damage is usually restricted to a few vines on the windward side of the vineyard.

Winter Injury

Buds and phloem are the vine tissues most sensitive to cold damage during dormancy. Frequently the damage is restricted to low-lying areas of the vineyard or to parts of the vineyard where vine maturity the previous season was restricted by excessively vigorous growth or the effects of disease development, such as premature defoliation from downy or powdery mildew. Bud injury can be detected while vines are still dormant by sectioning buds that have been thawed for more than 24 hours. Injured shoot primordia appear dark brown to black rather than the normal pale green (Fig. 308). Shoot growth on vines with significant bud injury is sparse and irregular. Many buds fail to grow, and secondary and tertiary buds from injured nodes typically begin developing later than healthy primary buds. In some cases the shoot primordium survives but the basal leaf primordia are injured. These leaves are often very small, can be malformed, and are frequently rugose with irregular, patchy chlorosis (Fig. 309). These symptoms can be confused with those of virus diseases or with herbicide injury.

Fig. 305. Salt toxicity on a cv. Chardonnay leaf, expressed as marginal necrosis developing into general leaf scorch. (Courtesy A. N. Kasimmatis)

Fig. 307. A young grape shoot killed by spring frost. (Courtesy T. Martinson)

Fig. 306. Leaf malformation caused by cold injury to buds while they were swollen but yet to break. (Courtesy W. Gärtel)

Fig. 308. Cross section through a winter-inured node showing the dead primary shoot bud. (Courtesy T. J. Zabadal)

Fig. 309. Malformed and mottled leaf developing after cold injury. (Courtesy R. C. Pearson)

Phloem injury is also visible soon after the tissue thaws. Because temperatures are lowest near the ground, phloem injury is often found only on the trunk near the soil or snow line. On such vines, shoot growth may appear normal until the shoots collapse during hot weather when the injured trunk cannot supply adequate water to meet the demand. Vines with cold injury to the trunk often have a large number of suckers growing more vigorously than shoots above the injured tissues. Vines with winter-injured trunks may survive one or two years but grow so poorly that they fail to survive the next harsh winter. These vines often have large cracks in the trunk that form as the tissues dehydrate. Winter-injured trunks often serve as sites for crown gall formation.

Selected References

McCarthy, M. G., Jones, L. D., and Due, G. 1992. Irrigation principles and practices. Pages 104-128 in: Viticulture, Vol. 2: Practices. B. G. Coombe and P. R. Dry, eds. Winetitles, Adelaide, Australia.

Ross, D. S., and Wolf, T. K. 2008. Grapevine water relations and irrigation. Wine Grape Production Guide for Eastern North America. T. K. Wolf, ed. Natural Resource, Agriculture and Engineering Service, Ithaca, NY.

Williams, L. E., Dokoozlian, N. K., and Wample, R. 1994. Grape. Pages 85-133 in: Handbook of Environmental Physiology of Fruit Crops. B. Schaffer and P. C. Anderson. eds. CRC Press, Boca Raton, FL.

Zabadal, T. J., Dami, I. E., Goffinet, M. C., Martinson, T. E., and Chien, M. L. 2007. Winter injury to grapevines and methods of protection. Mich. State Ext. Bull. E2930.

(Prepared by J. Vanden Heuvel and P. Cousins)

Nutritional Disorders

Nutritional disorders in grapevines are manifested by changes in the shape, color, chemical composition, performance, and attainable age of individual organs or the whole vine. In addition to specific symptoms, the appearance of the whole vine, multiple vines, or the entire vineyard may also aid in the diagnosis. Soil and petiole analysis can confirm nutrient imbalance. These data, combined with knowledge of the soil, the sensitivity of the grape cultivar, and the environmental conditions, increase the accuracy and significance of the diagnosis. Interpretation of soil and/or petiole nutrient concentrations data as well as optimal ameliorative efforts are best determined using recommendations tailored to the individual growing region.

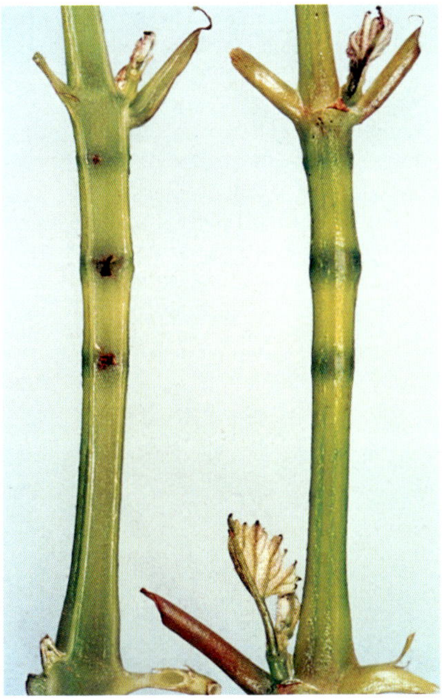

Fig. 310. Longitudinal section of an internode showing internal necrosis due to boron deficiency (left) and external symptoms (right). (Courtesy W. Gärtel)

Boron

Boron deficiency can have dramatic effects on vine growth and fruiting. It is observed especially on strongly acid (pH 3.5–4.5) soils and less frequently on neutral and alkaline soils (pH 7–8.5). Dry soil in the rooting area effectively hinders the absorption of boron. Vineyards in areas with very high rainfall and those irrigated with water devoid of boron are also susceptible, especially in easily leached, sandy soils.

The first symptoms of boron deficiency appear on tendrils near the shoot tip before bloom. Dark, knotty bulges form and become necrotic. The distal portions dry up, and flower clusters die. During the period of rapid shoot growth, younger internodes swell slightly at one or several places and the pith becomes necrotic (Fig. 310). Usually, the part of the shoot distal to the swelling dies. Leaves have short, thick petioles sometimes showing longitudinal lesions or necrotic caverns. Their blades are misshapen and show interveinal chlorosis or necrosis (Fig. 311). Buds formed on boron-deficient vines may produce short, bushy, branched, sterile shoots the following season. Boron deficiency also affects the development of berries and clusters. Only a few seeded berries set, and small, seedless berries develop (sometimes called a "pumpkin and peas" or "hens and chicks" symptom), as also can occur with some virus diseases on certain cultivars (Fig. 312). Roots remain short, are thickened, and swell into knots that break open longitudinally.

An oversupply of boron can also be detrimental, affecting the development of all aboveground parts of the vine. The younger leaves can be severely malformed. Necrosis develops on the tips of the serrations of older leaves and progresses from the margin to interveinal areas (Fig. 313). Elongation of the main shoots decreases in favor of lateral shoot growth, producing vines that look weak and bushy.

Calcium

Calcium deficiency appears occasionally on quartz gravel soils that are strongly acid (pH < 4.5). A narrow necrotic border appears at the margin of affected leaves and advances incrementally toward the point of attachment of the petiole. Dark-brown pimples up to 1 mm in diameter may develop on the primary bark of the internodes. The growing clusters dry up starting from the tip, which resembles severe stem necrosis.

Iron

Iron deficiency (also called iron chlorosis, lime chlorosis, and lime-induced chlorosis) is especially common in regions with high-pH soils, and is often noted first in wet sections of the vineyard. The loss of chlorophyll starts between the small leaf veins. Fading begins at the leaf margins but progresses interveinally (Fig. 314). Leaves may dry and fall, and set may be reduced. In cold, wet soils, transient iron deficiency symptoms may develop in the spring. *V. labrusca* and related cultivars tend to be more susceptible to iron deficiency than *V. vinifera*.

Magnesium

Magnesium deficiency occurs on light, acid soils with a low magnesium content, sandy soils with a relatively high potassium content, and calcareous (high calcium carbonate) soils. Large or frequent applications of potassium (which competes with magnesium for uptake) or ammonium (which acidifies the

Fig. 311. Necrotic shoot tip, interveinal chlorosis, and swelling of internodes caused by boron deficiency. (Courtesy W. Gärtel)

Fig. 312. Cluster with one normal seeded berry and many small, seedless berries of equal size, as a result of boron deficiency. (Courtesy W. Gärtel)

Fig. 313. Chenin Blanc leaves with marginal necrosis caused by excess boron. (Courtesy P. Christiansen)

Fig. 314. Interveinal chlorosis and necrosis caused by iron deficiency. (Courtesy R. M. Pool)

Fig. 315. Magnesium-deficient leaf with the green margins and interveinal chlorosis that distinguish this disorder from deficiency of other nutrients. (Courtesy W. Gärtel)

soil) can induce magnesium deficiency symptoms on soils that are not otherwise deficient in this element.

Magnesium deficiency takes two forms: leaf necrosis dominates early in the season, whereas interveinal yellowing is the primary symptom during summer and fall. The first symptoms usually appear before bloom and may be small, brown-green spots near the margins and in the interveinal tissues of young, growing leaves, which retain their green margins (Fig. 315). The elliptic or oval necrotic spots form chains that run within a few millimeters of the leaf margin and parallel to it. In early to late summer, a brightening of the tissue appears between the main veins and, with increasing yellowing, proceeds from the leaf margin toward the attachment of the petiole in the form of a wedge between the primary and secondary veins (Fig. 316). Magnesium deficiency at this time can be differentiated from deficiencies of other nutrients such as manganese, potassium, zinc, and boron by its striking straw yellow chlorosis pattern and its appearance first on basal or sometimes middle leaves of the shoot. Symptoms of magnesium deficiency are sometimes confused with the yellow mosaic syndrome produced upon infection by *Grapevine fanleaf virus*.

Fig. 316. Interveinal yellowing progressing inwards from the margins, resulting from magnesium deficiency. (Courtesy L. Cheng)

Manganese

Manganese deficiency is observed mainly on alkaline, sandy soils high in humus and on manganese-poor lime soils. In early summer the leaves at the base of shoots start to pale, and shortly afterward small, yellow spots appear in the interveinal tissue. The spots show a mosaiclike arrangement and are bordered by the smallest green veins, with only a very small seam along first- and second-order veins remaining green (Fig. 317). The symptoms are more severe on leaves exposed to the sun than on shaded leaves. Leaves are not malformed as with zinc deficiency. Advanced manganese deficiency affects the growth of shoots, leaves, and berries, and the maturation of the clusters is delayed. Symptoms of manganese deficiency in vineyards where soils contain lime are frequently obscured by the simultaneous yellowing due to lime-induced chlorosis.

In highly acid soils or soils rich in the element, manganese can be in excess and toxic (see Acidic Soil Sickness).

Nitrogen

When nitrogen becomes limiting, vine foliage turns pale green and then yellow (Fig. 318). Young shoots, petioles, and cluster stems become pink or red. Shoot growth is greatly re-

Fig. 318. Pale green discoloration caused by nitrogen deficiency (left) compared with healthy leaf on the right. (Courtesy T. Bates)

Fig. 317. Symptoms of manganese deficiency expressed as interveinal chlorosis beginning as chlorotic islands. (Courtesy W. Gärtel)

Fig. 319. Spring symptoms of potassium deficiency, including chlorotic leaf margins and necrotic sports. (Courtesy W. Gärtel)

Fig. 320. Advanced spring symptoms of potassium deficiency, including marginal and interveinal necrosis accompanied by downward rolling of leaf margins. (Courtesy W. Gärtel)

Fig. 321. "Black leaf" symptom of potassium deficiency seen in late summer on leaves that receive direct sunlight. (Courtesy L. Cheng)

duced. Sometimes light-brown islands of dead tissue appear between the main veins of the basal leaves, and in extreme cases the leaf blades may wilt and abscise. Berries may be small. No specific patterns of organ malformation are associated with nitrogen deficiency. In commercial plantings, the first symptoms may appear after berry veraison because nitrogen is translocated from the leaves near the clusters into the berries.

During long periods of cool, wet weather, vines may develop "cool-weather chlorosis," which can be confused with nitrogen deficiency. Low temperatures reduce chlorophyll synthesis, and the chlorosis disappears as soon as temperatures rise. Mechanical injuries and root damage from soil pests such as nematodes and phylloxera may hinder the absorption and transport of nutrients and lead to symptoms that also may be confused with nitrogen deficiency.

Excess nitrogen increases growth; internodes become long and large, and leaf blades become deep green, thick, and sometimes cupped. Shoot growth may be excessive and prolonged. Grapevines are more tolerant of high nitrogen when other micro- and macronutrients are in adequate supply and when the trellis system allows for adequate leaf exposure to light. Excessive rates of foliar urea can cause phytotoxicity that presents as foliar burn.

Phosphorus

Phosphorus deficiency is an uncommon nutritional problem in grape production. Thus, most reports are based on symptoms obtained in nutrient culture. In such cases, shoot and root growth are reduced and leaves are small and dark green. Leaf margins turn down without rolling. Under severe deficiency, leaves develop red, punctiform spots. Only a few experimental results report crop increases achieved through phosphorus fertilization. Rather, a history of excess phosphorus fertilization is more common. Excess phosphorus may induce zinc and iron deficiencies. Phosphorus deficiency may be one component of the nutrient imbalance brought about by low soil pH called acidic soil sickness (see Acidic Soil Sickness).

Potassium

The symptoms of potassium deficiency vary with the stage of development of the leaf blades when the potassium content

Fig. 322. Symptoms of "acidic soil sickness" on various cultivars grown in low-pH soil (bottom) compared with leaves from those cultivars grown in the same soil to which lime was applied (top). (Courtesy T. Bates)

Fig. 323. Leaf with interveinal chlorosis, reduced size, blade asymmetry, and opened petiolar sinus caused by zinc deficiency. (Courtesy W. Gärtel)

Fig. 324. Loose clusters of seeded berries that vary in size, from a zinc-deficient vine. (Courtesy W. Gärtel)

in the tissues falls below the critical level. During the early part of the growing season, the leaves lighten in color in areas, and a few necrotic spots appear along the margin of young blades (Fig. 319). During dry weather, necrotic areas varying in form, number, and size develop sporadically in the interveinal tissue. Leaf margins dry and roll up or down, and blades become distorted and ruffled (Fig. 320). In late summer, the surface of older leaves at the base of shoots, receiving direct sunlight, becomes violet brown to dark brown ("black leaf"), especially near the clusters. Black leaf begins interveinally but can progress to completely cover the upper leaf surface (Fig. 321). The leaf browning is especially pronounced on heavily cropped vines because, after veraison, maturing berries become a potassium "sink." Potassium deficiency symptoms are more common in dry years and are exacerbated in soils high in magnesium.

Acidic Soil Sickness (Säureschäden)

"Acidic soil sickness" is used to describe leaf symptoms on vines growing in extremely acid (pH 3.4–4.5) soils low in calcium and magnesium (Fig. 322). The symptoms are associated with deficiencies of calcium, magnesium, and phosphorus. Excess potassium, aluminum, and especially manganese are also involved.

Shortly after bloom the margins of older leaves turn yellow or light brown. Brown spots develop along the margin and coalesce into larger, elongated, rust-brown areas with irregular outlines. Red cultivars may have bright red spots. As in calcium deficiency, the damaged area dies slowly. In dry weather, acidic soil sickness progresses rapidly. Frequently, the symptoms appear together with those of magnesium deficiency. Basal leaves may abscise. In early summer on acid soils, a bewildering variety of combinations and transitions of these symptoms can be observed. The clusters on grapevines with acidic soil sickness rarely mature fully, and the poorly matured canes do not tolerate low winter temperatures.

Zinc

Zinc may be deficient in soils with low zinc levels such as very sandy soils or where topsoil has been removed. Zinc availability can also be affected by high phosphorus, which precipitates zinc as insoluble zinc phosphate, or in high-pH soils. The first symptoms of zinc deficiency are small leaf blades with opened petiolar sinuses and sharp teeth (Fig. 323). The leaf blades are commonly asymmetric, one half of the leaf being larger than the other. The interveinal areas turn light green to yellow in a mosaic pattern and can become reddish in black and red cultivars. The leaf veins also become clear, with narrow borders of green. The more advanced chlorotic areas can become necrotic. The intensity and distinctness of the symptoms vary somewhat among cultivars. Zinc deficiency reduces yields because vines produce fewer seeds and smaller berries (Fig. 324.).

Selected References

Bates, T. R., and Wolf, T. K. Nutrient management. Winegrape Production Guide for Eastern North America. T. K. Wolf, ed. Natural Resource, Agriculture and Engineering Service, Ithaca, NY.

Christensen, L. P., Kasimatis, A. N., and Jensen, F. L. 1978. Grapevine nutrition and fertilization in the San Joaquin Valley. Publ. 4087. Division of Agricultural Sciences, University of California, Berkeley.

Dami, I., Bordelon, B., Ferree, D. C., Brown, M., Ellis, M. A., Williams, R. N., and Doohan, D. 2005. Midwest Grape Production Guide. Ohio State Univ. Ext. Bull. 919.

Davenport, J. R., and Horneck, D. A. 2011. Sampling guide for nutrient assessment of irrigated vineyards in the inland pacific northwest. Pacific Northwest Ext. Publ. PNW622.

(Prepared by J. VandenHeuvel and P. Cousins)

Shriveled Fruit

Physiological disorders of the fruit that reduce organoleptic qualities can occur during growth and ripening. Although progress has been made on describing the morpho-anatomical and compositional changes that occur with these disorders, determining precise causal factors and associated control practices has been challenging.

Dehydration

Dehydration is caused by water transpiration from the berry, which results in a slow shriveling and a concentration of sugars. Symptomatic berries are generally scattered throughout the cluster (Fig. 325) and are usually noted toward the end of ripening in warmer climates, but can be found in cooler regions as well. The resulting impacts of dehydration (reduced berry weight due to water loss) are sometimes considered desirable as solutes are concentrated in the berry.

Fig. 325. Shriveled, dehydrated berries scattered throughout an otherwise healthy cv. Cabernet Sauvignon cluster. (Courtesy M. Keller)

Bunch Stem Necrosis

Bunch stem necrosis is generally identified by the appearance of patches of dead (necrotic) tissue on the rachis or peduncle (Fig. 326), which results in a shriveling of the berries since the supply of water and nutrients to them is impeded by the necrotic patches. Symptoms are usually most visible at the tip of the cluster or on a shoulder, although entire clusters can be affected. The berries generally remain sour, although the impact on fruit composition is dependent on when the rachis becomes necrotic; late-season onset of the symptoms can concentrate the sugars in the berries. The exact cause of bunch stem necrosis is unknown although it is not thought to be caused by pathogens. Widely fluctuating temperatures favor its appearance, as do low temperatures and high humidity around bloom and excessive rainfall after veraison. The incidence of bunch stem necrosis has also been correlated with the concentration or ratio of nitrogen, magnesium, potassium, and calcium in the vine.

Sunburn

Excess solar radiation and high temperatures can cause damage to the berry skin, causing the grapes to shrivel (Fig. 327). Damaged berries are often on sun-exposed shoulders and can appear raisinlike and lighter in color, although the physical appearance depends on the cultivar and stage of development. Early-season sunburn generally appears as a brown discoloration on the sun-exposed side of the cluster in both white and red cultivars, whereas postveraison damage results in fruit with less color and a shiny appearance. This type of injury occurs in warm climates, but is also common in cooler climates when a later-season leaf removal in the cluster zone is followed by sunny days. Sunburn can be limited or avoided in hotter climates by reducing or avoiding management practices that expose them to direct sunlight such as shoot thinning and leaf removal, or by foliar application of a kaolin-based particle film. In cooler climates, timing canopy management practices that result in sudden fruit exposure so that they occur during cooler, cloudier days and/or earlier in the season may reduce sunburn incidence.

Fig. 326. Symptoms of bunch stem necrosis on a cv. Cabernet Sauvignon cluster. (Courtesy M. Keller)

Fig. 327. 'Cabernet Sauvignon' berries shriveled due to sunburn. (Courtesy M. Keller)

Fig. 328. A cv. Cabernet Sauvignon cluster with symptoms of berry shrivel (sour shrivel, sugar accumulation disorder, or *Traubenwelke*), compared with a healthy cluster immediately above it. (Courtesy M. Keller)

Berry Shrivel

Grape berries suffering from berry shrivel (also called sour shrivel, sugar accumulation disorder, or Traubenwelke) closely resemble those suffering from bunch stem necrosis; however, in contrast to bunch stem necrosis, there are no necrotic patches on the rachis or peduncle. Symptoms can impact the entirety of the cluster or only its tip. Affected berries are flaccid and waxy; the outer portion shrivels while the flesh begins to break down, sugar fails to accumulate, the pH remains low, and yield is reduced. Not all clusters on a vine are symptomatic (Fig. 328). Berry shrivel occurs under a wide variety of conditions and its cause is currently unknown.

Selected References

Bondada, B. R., and Keller, M. 2012. Not all shrivels are created equal – Morpho-anatomical and compositional characteristics differ among different shrivel types that develop during ripening of grape (*Vitis vinifera* L.) berries. Am. J. Plant Sci. 3:879-898.

Krasnow, M. N., Matthews, M. A., Smith, R. J., Benz, J., Weber, E., and Shackel, K. A. 2010. Distinctive symptoms differentiate four common types of berry shrivel disorder in grape. Calif. Agric. 64:155-159.

Shellie, K., and Glenn, D. M. 2008. Wine grape response to foliar particle film under differing levels of preveraison water stress. HortScience 43:1392-1397.

(Prepared by J. VandenHeuvel and P. Cousins)

Pesticide Toxicity

If fungicides, insecticides, herbicides, or growth regulators are applied improperly, they may injure grapevines and elicit a phytotoxic response. Phytotoxicity may result from an excessive application rate, the application of incompatible pesticide mixtures, application at a sensitive stage of vine development, application during or before unfavorable environmental conditions, or the application to unusually sensitive cultivars. The addition of adjuvants or oils that promote absorption across the waxy cuticle of the leaves and berries, or mixing with other products that might be formulated with such an agent, can trigger a phytotoxic response to certain products that otherwise would be harmless. Herbicide injury to grapevines may occur through absorption by roots or by leaves from either a direct application or from the drift of spray droplets or volatiles. Herbicide drift from offsite applications can be a particularly pernicious source of injury. Preplant soil fumigants may injure a new vineyard if plants are set before residues of the fumigant have been adequately reduced.

Phytotoxicity symptoms vary depending on the chemical, its concentration, and the stage of vine development at the time of exposure. Some of these symptoms are presented in Table 7, although this listing is far from exhaustive. Diagnosing the specific cause of herbicide injury is a discipline unto itself, and whereas several photographs are provided herein for reference (Fig. 329-348), growers and practitioners are encouraged to consult an appropriate specialist on this topic as warranted.

Young, unfolding, or expanding leaves are especially sensitive to chemical injury. Stunted shoots, malformed leaves, chlorotic leaves, or necrotic leaves may be observed. Berries sprayed during the early stages of their development are highly sensitive to injury, which may appear as russeting, scarring, or even cracking. Phytotoxicity symptoms caused by some pesticides and growth regulators used in grape production (Figs. 349-360) may be easily confused with symptoms of certain diseases (e.g., paraquat injury with black rot on leaves, glyphosate injury with Eutypa dieback on newly expanding shoots in the spring, and endosulfan injury with Rotbrenner) or with abiotic disorders such as magnesium deficiency. Injury may be associated with specific environmental parameters, such as high temperatures during or shortly after applications of sulfur or dinocap, or those that delay drying of spray solutions (cool weather, high humidity) and thereby promote absorption of pesticide droplets that contain materials intended to remain on plant surfaces. Injury from herbicide drift is often associated with applications during windy conditions, higher temperatures that promote the evaporation of spray water droplets and/or volatilization of the active ingredient, or temperature inversions that allow small spray particles to remain suspended in the air for extended periods.

Diagnosis of phytotoxicity is frequently possible only after a thorough analysis of the grower's spray record during the current season and in previous years, including accompanying weather records, an examination of neighboring crops and weeds within the vineyard, observations on the location of injured vines within the vineyard, and determination of the injured organs on individual vines. Injury symptoms in newly set vineyards may be related to previous crops and the pesticides used on those crops. Because of the complexity of diagnosing phytotoxic responses, it is frequently advisable to consult more than one specialist.

Selected References

Bolay, A., and Caccia, R. 1979. Effets de traitements cupriques sur le rougissement précoce du feuillage du cépage Merlot au Tessin. Rev. Suisse Vitic. Rev. Suisse Vitic., Arboric., Hortic.11:205-211.

Bovey, R., Gärtel, W., Hewitt, W. B., Martelli, G. P., and Vuittenez, A. 1980. Virus and Virus-like Diseases of Grapevines. Payot Lausanne, Switzerland.

Clore, W. J., and Bruns, V. F. 1953. The sensitivity of the Concord grape to 2,4-D. Proc. Am. Soc. Hortic. Sci. 61:125-134.

Doster, M., and Sall, M. A. 1984. Phytotoxicity to grapevines of fenarimol and triadimefon. Am. J. Enol. Vitic. 35:97-99.

Haeseler, C. W., and Petersen, D. H. 1974. Effect of cupric hydroxide vineyard sprays on Concord grape yields and juice quality. Plant Dis. Rep. 58:486-489.

Kuck, K. H., and Scheinpflug, H. 1986. Biology of sterol-biosynthesis inhibiting fungicides. Pages 65-96 in: Chemistry of Plant Protection. Vol. I. G. Haug and H. Hoffmann, eds. Springer-Verlag, Berlin.

Pearson, R. C. 1986. Fungicides for disease control in grapes, advances in development. Pages 145-155 in: Fungicide Chemistry, Advances and Practical Applications. M. B. Green and D. A. Spilker, eds. American Chemical Society, Washington, DC.

Shaulis, N. J., Crowe, D. E., and Rogers, R. A. 1978. The relation of the phytotoxicity of glyphosate to its injury-free use in vineyards. Part I. Glyphosate studies as a basis for injury-free use. Proc. Northeast. Weed Sci. Soc. 32:246-253.

Suit, R. F. 1948. Effect of copper injury on Concord grapes. Phytopathology 38:457-466.

Taschenberg, E. F., and Braun, A. J. 1960. Evaluation of phaltan for the control of grape diseases. Plant Dis. Rep. 44:560-565.

Tweedy, B. G. 1981. Inorganic sulfur as a fungicide. Residue Rev. 78:43-68.

Weaver, R. J. 1970. Some effects on grapevine of exogenous plant regulators and herbicides. Pages 247-254 in: Virus Diseases of Small Fruits and Grapevines. N. W. Frazier, ed. Division of Agricultural Sciences, University of California, Berkeley.

(Prepared by W. F. Wilcox)

TABLE 7. Symptoms of Pesticide Injury to Grapevines

Pesticide	Symptoms	Remarks
Herbicides		
Synthetic auxins (e.g., 2,4-D; dicamba; fluoxypyr; triclopyr)	Leaves typically are distorted. They may be strap-shaped with greatly opened sinuses, cupped, elongated, and show various forms of enations. Young shoots may become twisted and distorted. Berry set may or may not be affected. Severe injury may result in necrosis of the new growth. (Figs. 329–337)	Grapes are highly sensitive to these herbicides. Injury is commonly associated with drift (spray particles and volatiles of some materials and formulations), as these materials are used commonly on many other crops and for control of broad leafed weeds in non-agricultural settings such as roadsides, ditch banks, etc.
Glyphosate	*Year of application*: Leaves are arrow-shaped and very rugose, with or (more frequently) without interveinal chlorosis (Fig. 338). Internodes may be shortened. Apical dominance may be broken, resulting in growth of numerous lateral shoots. *Year after application*: Early shoot growth is stunted, with severely strapped leaves (Fig. 339) that can be confused with symptoms of Eutypa dieback (Fig. 113).	Injury is usually due to translocation of glyphosphate applied to ground suckers or to drift from material applied under the trellis. Glyphosate moves with the photosynthates and may affect one or several shoots on a vine. Late summer or fall applications may produce severe symptoms the following spring.
Simazine	Basal leaves are usually most affected. Chlorotic patches originate at leaf margins. Leaves are not deformed (Fig. 342).	Symptoms are caused by translocation from roots. Injury is usually associated with application of high rates to coarse soils low in organic matter and/or excessive application of water.
Diuron	Primary symptom is yellow veins on otherwise normal leaves (Fig. 343).	Most common symptoms are caused by translocation from roots. Injury is more common where root growth is shallow, soil is coarse, or excessive water is applied and is frequently observed in mulched vineyards. More unusual is contact injury, which can resemble simazine injury.
Paraquat	Injury generally results from contact. Discrete spots initially are cleared, chlorotic areas (Fig. 345) and develop into necrotic spots (Fig. 346). These spots sometimes can be confused with black rot leaf lesions (Fig. 38). Treated suckers show a general burn.	Injury is usually from vineyard application. Rarely, paraquat applied near ground level is absorbed and translocated to the canopy, resulting in a general chlorosis (Fig. 347). Limited translocation may also lead to necrosis along the leaf veins (Fig. 348).
Growth Regulators		
Gibberellic acid	*First year*: Clusters are elongated, with long pedicels, many shot berries, and retained ovaries. Rachis may be woody and twisted (Fig. 349). *Second year*: Bud break is reduced and irregular. Flower clusters are greatly reduced in number and size. Set is normal, but very few berries develop (Fig. 350).	Injury usually occurs from misapplication to sensitive cultivars. Most seedless grapes tolerate gibberellic acid.
Fungicides		
Sulfur	Injury on foliage generally appears as interveinal bleaching that becomes necrotic (Figs. 351 and 352). Defoliation may occur. Injured fruit have brown to black scar tissue, which may eventually crack, depending on the stage of berry development at the time of exposure.	Injury to *V. vinifera* cultivars is generally associated with exposure at temperatures over 30°C, and often is more severe under humid conditions. Some cultivars and rootstocks of other *Vitis* spp. or interspecific hybrids are highly susceptible to sulfur injury even at lower temperatures.
Sulfur dioxide	Bleaching, the most common type of sulfur dioxide injury in stored table grapes, is most pronounced around the pedicel (Fig. 353) or any break in the epidermis. Tissue under the bleached area dries out, collapses, and forms a depression. Changes in color are usually accompanied by objectionable changes in taste.	
Copper	Injury ranges from slight bronzing of leaves to foliar necrosis (Fig. 354) and defoliation. Fruit may exhibit black necrotic areas where tissue has been killed.	Cu^{++} ions are the toxic moiety. Injury is most likely with formulations or spray solutions that allow excessive amounts of copper to persist in the ionic form (e.g., low pH, insufficient lime). Injury also is exacerbated by conditions that maintain the ions in solution and promote their absorption, such as cool slow-drying and highly humid environments.
Dinocap	Injury ranges from stunted, malformed leaves with necrotic sectors (Fig. 355) to large, necrotic burn areas (Fig. 356) on leaves exposed to the sun. Berry symptoms range from circular black spots to russeted areas where spray residues have dried.	Leaf malformation occurs when dinocap is applied to young, unfolding or expanding leaves of susceptible cultivars. Foliar burn is generally associated with exposure to temperatures over 30°C and is generally independent of cultivar. Emulsifiable concentrate formulations are more phytotoxic than wettable powder formulations.
Trizoles (propiconazole, tebuconazole, difenoconazole, etc.)	Most commonly, a growth-regulator effect, with shortened internodes and leaves that are small, thick, dark green, and puckered (Fig. 357). In a few specialized cultivar × product interactions, mild to severe interveinal chlorosis and necrosis may occur (Fig. 358)	Injury is usually associated with excessive rates of application and may be exacerbated by adjuvants that promote absorption into plant tissues.
Phthalimides (captan, folpet, etc.)	Leaves may be misshapen, puckered, and rugose, or show a general burning and develop necrotic blotches; berries are scarred (Fig. 359).	Injury is usually associated with circumstances that promote absorption of the chemicals across the waxy cuticle of the leaves and berries. These include the incorporation of oils or certain adjuvants in the spray solution, sometimes as a component of another pesticide product applied concurrently, or applications made during slow-drying conditions. Foliar burning caused by folpet also has been associated with droughty soils and high temperatures.
Insecticides and Miticides		
Endosulfan	A rapid necrosis appears on portions of leaves that receive the heaviest spray deposit (Fig. 360). Leaves may drop if injury is severe.	Injury is most severe on cv. Concord and the interspecific hybrid cultivars Baco noir, Cascade, Chancellor, Colobel, and is associated with hot weather at the time of application.

Fig. 329. Rugose, fan-shaped leaf with typical symptoms of 2,4-D drift injury. (Courtesy A. Senesac)

Fig. 330. 2,4-D injury to cv. Flame Tokay shoots. (Courtesy A. N. Kasimatis)

Fig. 331. Poor berry fill resulting from 2,4-D drift injury. (Courtesy A. Senesac)

Fig. 332. Malformed young growth resulting from triclopyr drift injury. (Courtesy J. Roncoroni)

Fig. 333. Stem malformation resulting from drift of triclopyr. (Courtesy A. Senesac)

Fig. 334. Necrosis of new shoot growth caused by drift of triclopyr applied adjacent to the vineyard. (Courtesy R. M. Dunst)

Fig. 335. Stem malformation resulting from drift of fluroxypyr. (Courtesy A. Senesac)

Fig. 336. Upward leaf curling caused by fluroxypyr injury. (Courtesy A. Senesac)

Fig. 337. Dicamba injury to cv. Thompson Seedless (Sultanina) foliage. (Courtesy A. N. Kasimatis)

Fig. 338. Foliar symptoms of glyphosate injury. (Courtesy M. Norton)

Fig. 339. Severe glyphosate injury resulting from application the previous autumn. (Courtesy B. Hanson)

Fig. 340. Symptoms of injury caused by sulfometuron introduced in runoff water. (Courtesy B. Hanson)

Fig. 341. Norflurazon injury. (Courtesy A. Senesac)

Fig. 342. Simazine injury. (Courtesy R. M. Pool)

Fig. 343. Diuron injury to cv. Concord leaves. (Courtesy R. C. Pearson)

Fig. 344. Injury caused by topical application of oryzalin; new growth appears normal. (Courtesy A. Senesac)

Fig. 345. Early symptoms of paraquat injury, including discrete necrotic spots surrounded by a yellow halo. (Photograph by W. McFadden-Smith, Ontario Ministry of Agriculture and Food; Queen's Printer of Ontario, © 2014. Reproduced with permission.)

Fig. 346. Paraquat injury to a cv. Petite Sirah leaf. (Courtesy A. N. Kasimatis)

Fig. 347. Leaf injured by translocated paraquat. (Courtesy A. N. Kasimatis)

Fig. 348. Injury to leaf veins caused by slight translocation of paraquat. (Courtesy A. Senesac)

Fig. 349. Effect of excess gibberellic acid on cv. Riesling grapes, within the year of treatment. (Courtesy R. C. Pearson)

Fig. 350. Shoots of cv. Chenin Blanc with very small clusters one year after treatment with gibberellic acid. (Courtesy A. N. Kasimatis)

Fig. 351. Sulfur injury to foliage of cv. Sauvignon Blanc. (Courtesy A. N. Kasimatis)

Fig. 352. Sulfur injury to foliage of an interspecific hybrid. (Courtesy W. F. Wilcox)

Fig. 353. Sulfur dioxide injury to cv. Emperor grapes in storage, characterized by bleached and sunken tissue , particularly around the attachment to the pedicel. (Courtesy J. Harvey)

Fig. 354. Severe copper injury to cv. Concord foliage. (Courtesy R. C. Pearson)

Fig. 355. Stunted, malformed leaves resulting from dinocap injury. (Courtesy R. C. Pearson)

Fig. 356. Dinocap injury on a cv. Concord leaf associated with exposure to the sun. (Courtesy R. C. Pearson)

Fig. 357. Etaconazole injury to foliage of the interspecific hybrid cultivar Aurore, including stunted lateral shoots with small, downward-cupped leaves. (Reprinted, by permission, from Pearson, 1986; copyright 1986 American Chemical Society)

Fig. 358. Injury to cv. Concord leaves caused by difenoconazole applied with an absorption-promoting surfactant. (Courtesy R. M. Dunst)

Fig. 359. Berries injured by captan. (Courtesy W. Gärtel)

Fig. 360. 'Chancellor' leaves injured by endosulfan. (Courtesy G. L. Jubb, Jr.)

Part IV. Grapevine Disease Management

Grapevine Fungicides

Fungicides are valuable tools for grapevine disease management and their effective deployment is a critical component of the production system in most commercial vineyards. To utilize these various materials most efficiently, it is necessary to understand how they function, their individual strengths and weaknesses, and both the risk of and techniques for managing the development of resistant pathogen strains.

Fungicides are often categorized by their number of action sites in target fungi (multiple or single), their systemicity (movement properties within the plant), their chemical makeup, their biochemical mode of action, and/or their physical mode of action. Which category is used for reference often depends on the person communicating and the nature of the communication, and it is not unusual to hear these categorical terms used interchangeably. For example, thiophanate-methyl can properly be referenced as a single-site, systemic, methyl benzimidazole carbamate (MBC), mitosis-inhibiting, and/or protective plus postinfection fungicide, as appropriate to the context.

Fungicide Properties

Systemicity. The potential for fungicide movement within a plant or plant organ is largely dependent upon the material's chemical composition and ranges from negligible to fully systemic. Nonsystemic or "contact" fungicides are intended to adhere to the outer surfaces of the vine (leaves, clusters, stems, etc.) and prevent the initiation of infection, usually by inhibiting spore germination. Because they do not enter the plant tissues, they have no postinfection or "curative" activity against pathogens once they have breached the cuticle, and therefore must be present on the plant before infection occurs in order to control disease. Some may also inhibit sporulation by the pathogen as it emerges from the host tissue following infection and lesion formation, although this effect typically is modest for contact fungicides. Because these fungicides do not move within the plant and are restricted to the surface of sprayed organs, thorough coverage of plant tissue is essential for effective disease control. Furthermore, such materials are subject to physical degradation by the elements (e.g., removal by rain, breakdown in light) and do not generally protect new tissues that have developed since their previous application, although limited quantities are sometimes redistributed from treated to untreated tissues by rainwater. Typically, such materials require relatively frequent reapplication to maintain effective control during periods suitable for new infections to occur, particularly during periods of rapid vine growth. Most fungicides commercialized before 1970 belong to this category.

Systemic fungicides are absorbed into plant tissues, allowing them to provide activity against pathogens for a limited time after infection has occurred. Some of these materials also are maintained to varying degrees in or near the cuticle at the tissue surface, allowing them to inhibit spore germination and/or interfere with the initial stages of the infection process. However, there is considerable variation in the degree of systemicity among such materials. "Locally systemic" fungicides are able to penetrate into plant tissues but subsequent movement is limited, with little transport beyond the particular leaf or berry into which they are absorbed. Such materials are less subject to environmental degradation than contact fungicides, typically allowing longer application intervals, but they similarly have no activity on newly emerging tissues unless redistributed to them in the vapor phase.

"Limited systemic" fungicides are absorbed into plants and transported through the xylem with water movement and more locally via apoplastic movement (through and along cell walls). As result, limited systemics move acropetally (upward in the plant), with redistribution through the tissues as they move. These properties provide increased spatial uniformity and persistence in fungicide activity, although the need for good spray coverage is still not negated. Relatively few fungicides move in such a manner, the most important group for grape producers being the phenylamides (e.g., metalaxyl/mefenoxam).

"Fully systemic" fungicides also move with water, both apoplastically and symplastically (within cells), resulting in both acropetal and basipetal (downward) movement. Such movement has been observed only with phosphonate/phosphite type fungicides, such as fosetyl-Al and the phosphorous acid salts.

Fungicides can also be transported through translaminar movement and vapor-phase redistribution. The action of a fungicide in penetrating through one side of a leaf blade and accumulating at biologically active concentrations at the other side is often referred to as "translaminar" systemicity, and can be observed for some weakly systemic fungicides such as the strobilurins. Vapor-phase redistribution may occur when fungicides vaporize from the plant surface and are reabsorbed by nearby tissues. Several individual fungicides exhibit significant vapor-phase redistribution, which can be very useful within dense vine canopies but becomes less important when canopy structure is open and/or there is substantial air movement through vines. The practical contribution of this phenomenon is difficult to quantify in the vineyard and published studies examining it under field conditions are rare.

Biochemical mode of action. A fungicide's biochemical mode of action encompasses the mechanism of its toxicity to target fungi and nontarget organisms. This mechanism determines both the point in the infection cycle where the fungicide is most effective and the risk that pathogens will develop resistance to it. The biochemical mode of action is the criterion used to group fungicides into classes for resistance management purposes, as a fungal individual resistant to one member of a given class is generally resistant to all other members of the same class (cross-resistant). Based upon the number of fungal target sites affected, fungicides are generally categorized as either single-site or multisite inhibitors.

Multi- vs. single-site fungicides. Multisite fungicides tend to be (i) older synthetic materials developed from the 1930s through the 1960s or inorganic minerals such as sulfur and copper, which interfere with multiple enzyme systems; or (ii) other materials (e.g., oils, inorganic salts, calcium polysulfide, hydrogen peroxide) that kill or inhibit fungi in a nonspecific manner such as asphyxiation, desiccation, and oxidation. Such materials often are active against a relatively wide spectrum of fungi and are at very little risk for resistance development. However, this same lack of biochemical specificity often confers an associated risk of activity against nontarget organisms, including the treated plant, and is the reason that many materials in the first group (e.g., captan, chlorothalonil) cause plant injury when they are applied in a manner that promotes their absorption by treated vine tissues, such as in tank mixes with oil or certain surfactants. Lack of target specificity can also be an impediment to regulatory approval for use of a material, and is a major reason that few modern synthetic fungicides are multisite inhibitors.

In contrast, single-site fungicides inhibit fungi by binding to a specific fungal enzyme, often at a very specific site (the so-called "lock and key" model). These generally have far fewer nontarget effects than the multisite materials and comprise the vast bulk of the newest fungicides, which must demonstrate minimal environmental and toxicological impact as a requirement for registration. However, because of the high specificity of single-site fungicides, oftentimes only relatively minor genetic changes in the pathogen are necessary for a slight alteration of the target site that subsequently makes the organism resistant to the material (i.e., the specific fungicide "key" no longer fits the slightly altered "lock").

Physical mode of action. The physical mode of action, or at what temporal point in the infection process a fungicide exerts its activity against pathogens, is determined by both the biochemical mode of action and the systemicity of the material. There are three general physical modes of action—preventive, postinfection, and postsymptom—with narrower subcategories within them sometimes described. Individual fungicides and fungicide groups exhibit different relative strengths within the context of each.

In a broad sense, preventive activity refers to the ability of a fungicide to forestall disease symptoms if applied prior to the start of a potential infection event. More narrowly, it sometimes refers to a fungicide's ability to prevent infection by inhibiting spore germination or ingress of the fungal germling. Fungicides with strictly preventive activity are generally active on or within the outer waxy cuticle of green tissues or the surface of woody tissues. Systemic fungicides also may prevent infection in this manner if sufficient residue is present near the plant surface, but often provide preventive activity in a broader sense when residual concentrations within the plant stop the pathogen from proliferating after it has penetrated the cuticle and passed into the epidermal cells or beyond.

"Postinfection activity" describes the ability of a fungicide to arrest pathogen growth when applied after spore germination and penetration into plant tissues but before symptoms are expressed. This is often referred to as "curative" activity in the jargon of fungicide researchers and industry specialists. However, producers and advisors often interpret this latter term as they would within the context of human and animal medicine, to denote the eradication of visible symptoms from diseased plants and their restoration to full health following treatment. Unfortunately, this is beyond the promise of the vast majority of fungicides with postinfection activity.

The term "presymptom" activity is sometimes used to refer to a specific manifestation of postinfection activity: when applied too long after the initiation of infection to completely prevent symptom development, some fungicides nevertheless cause the resulting lesion to form abnormally and produce few if any spores. This is also a form of "antisporulant" activity. Unless the target pathogen is growing primarily upon the surface of the infected tissues (e.g., powdery mildew), any fungicide that provides postinfection activity must by definition be absorbed into those tissues (have some degree of systemicity) in order to be able to exert it.

As implied, postsymptom activity refers to the effect provided by a fungicide when it is applied after disease symptoms have developed. In the general and broadest sense, this refers to a fungicide's ability to halt or severely restrict the production of new spores when it is applied to existing lesions. Fungicides with postsymptom activity typically inhibit key metabolic processes involved in spore production (e.g., QoIs, phenylamides, benzimidazoles) or are directly toxic to new spores when contact is made after the spores are produced (e.g., copper, EBDCs, captan). A narrower and relatively uncommon form of postinfection activity is eradication, the ability of a material to kill or severely damage an active fungal colony upon treatment, thereby halting or severely restricting any further development from it. With a few specific exceptions, such activity on grapes is largely confined to several materials that kill active powdery mildew colonies on contact, resulting from the unique and largely superficial nature of that pathogen upon infected host surfaces and its consequent physical exposure to such agents. In the case of most postsymptom fungicide applications, noticeable damage has already occurred to the plant tissue by the time of treatment and it is not possible to restore such tissue to a "healthy" state with the spray. Rather, the basic benefit of postsymptom activity is a reduction of secondary inoculum production and subsequent disease spread. If the pathogen is truly eradicated or kept in check continually, it may be possible to maintain some physiological function of the infected organ if a significant amount of uncolonized tissue remains, although typically it will be physically disfigured.

Fungicide Resistance

A number of fungicides that once were highly effective have lost their efficacy against specific pathogens in certain regions of the world because those pathogens have developed resistance to them. It is likely that this phenomenon will increase in importance into the future, as modern fungicides are almost always more prone to resistance development than traditional, multisite contact products. The biochemical, single-site specificity of most newer materials, responsible for the limited nature of their effects on nontarget organisms (now a key factor for successful pesticide registration by government regulatory agencies), also greatly increases the probability that individuals will arise within the targeted pathogen population who possess a minor genetic variation that renders them resistant to the fungicide. In some cases (e.g., benzimidaoles, QoIs), this requires only a single amino acid substitution at one site on one gene, thereby preventing the fungicide from binding to its target site on the fungus. Simply put, pathogens treated with single-site materials may become resistant relatively simply and quickly. Thus, implementing effective resistance management strategies has become a requirement for the sustainable use of modern fungicides.

Within an agricultural context, fungicide resistance is a classic example of evolution, i.e., it results from the selection of specific individuals from within the entire pathogen population, whose genetic structure enables them to better survive than other individuals when exposed to a particular fungicide or fungicide group. When such individuals subsequently reproduce preferentially and their proportion of the population increases to the point that the fungicide (or group thereof) no longer provides an acceptable level of disease control when applied properly, a condition termed "practical resistance" is reached. The risk of this occurring is a function of both the fungicide itself (the biochemical basis of the resistance response) and the individual disease involved.

Diseases at greatest risk are those with a capacity for rapid spread, because they are caused by pathogens (i) with a potential for developing multiple generations per year, thereby allowing resistant individuals to preferentially reproduce and increase their proportion of the population quickly; and (ii) which also produce large numbers of spores that can be widely dispersed (e.g., on air currents), allowing the resistant biotype to spread spatially. Both downy and powdery mildew are prime examples of such diseases. In contrast, diseases at least risk are those with a single annual cycle and those caused by pathogens with limited potential for movement; Phomopsis cane and leaf spot is a prime example on both accounts.

Resistance to a fungicide is said to be qualitative when individuals within the pathogen population are either sensitive to the field dose or are virtually immune to even many multiples of it. In such cases, shifts in the makeup of the population can occur quickly ("disruptive" shifts), with resistant phenotypes becoming predominant after just a few years of use and control failures occurring suddenly. Common examples in viticulture include development of resistance to the phenylamides among some populations of *Plasmopara viticola* (downy mildew); to the QoI fungicides among populations of *P. viticola* and *Erysiphe necator* (powdery mildew); and to the benzimidazoles among populations of *E. necator* and *Botrytis cinerea*.

In contrast, resistance is said to be quantitative when individuals poorly controlled by one dose of a fungicide may be controlled by incremental increases in that dose or by the substitution of a field dose of a related material with greater intrinsic activity. In such cases, the distribution of sensitivities within the pathogen population shifts incrementally over time following repeated exposure to fungicides with this biochemical mode of action, with progressively greater proportions of the individuals within the population requiring progressively higher doses of the material for a given level of control ("displacement" shifts). A well-characterized example of this phenomenon is found among *E. necator* populations that have developed resistance to the DMI fungicides.

Although resistance development is a complex process, basic resistance management strategies are relatively simple. These are designed to limit both the selection of resistant individuals and their subsequent reproduction and spread, and include:

- Limit the number of selection events, i.e., restrict the total number of seasonal applications of an at-risk fungicide or related products with the same biochemical mode of action. Many fungicide labels in the United States and other countries now provide a numerical designation that indicates the resistance group to which that material has been assigned on the basis of its biochemical mode of action (Table 8).

TABLE 8. Grape Fungicide Active Ingredients and Associated Resistance Management Groups

Active ingredient[a]	Group name[b]	Group[c]	Active ingredient[a]	Group name[b]	Group[c]
ametoctradin	QxI[d]	45	mancozeb	dithiocarbamate	—[h]
azoxystrobin	QoI[e]	11	maneb	dithiocarbamate	—[h]
benomyl	MBC[f]	1	mandipropamid	CAA[k]	40
boscalid	SDHI[g]	7	mefenoxam (metalaxyl-M)	phenylamide	4
captan	phthalimide	—[h]	metalaxyl	phenylamide	4
carbendazim	MBC[f]	1	metconazole	DMI[j]	3
chlorothalonil	chloronitrile	—[h]	metiram	dithiocarbamate	—[h]
copper (various compounds)	inorganic	—[h]	monopotassium (dihydrogen potassium) phosphate	inorganic salt	—[l]
cyazofamid	QiI[i]	21	myclobutanil	DMI[c]	3
cymoxanil	cyanoacetamide-oxime	27	oil (botanical, petroleum)	oil	—[l]
cyproconazole	DMI[j]	3	penconazole	DMI[j]	3
cyprodinil	anilinopyrimidine (AP)	9	phosphorous acid and salts	phosphonate	33
dichlofluanid	sulfamide	—[h]	polyoxin	polyoxin	19
diethofencarb	N-phenyl carbamate	10	potassium bicarbonate	inorganic salt	—[l]
difenoconazole	DMI[j]	3	procymidone	dicarboximide	2
dimethomorph	CAA[k]	40	propiconazole	DMI[j]	3
epoxyconazole	DMI[j]	3	proquinazid	azanaphthalene	13
famoxadone	QoI[e]	11	pyraclostrobin	QoI[e]	11
fenamidone	QoI[e]	11	pyrimethanil	anilinopyrimidine (AP)	9
fenarimol	DMI[j]	3	quinoxyfen	azanaphthaline	13
fenbuconazole	DMI[j]	3	spiroxamine	amine (morpholine)	5
fenhexamid	hydroxyanilide	17	sulfur	inorganic	—[h]
fenpyrazamine	hydroxyanilide	17	tebuconazole	DMI[j]	3
ferbam	dithiocarbamate	—[h]	tetraconazole	DMI[j]	3
fludioxonil	phenylpyrrole	12	thiabendazole	MBC[f]	1
fluopicolide	benzamide	43	thiophanate-methyl	MBC[f]	1
fluopyram	SDHI[g]	7	thiram	dithiocarbamate	—[h]
flusilazole	DMI[j]	3	tolyfluanid	sulfamide	—[h]
flutriafol	DMI[j]	3	triadimefon	DMI[j]	3
fluxapyroxad	SDHI[g]	7	triadimenol	DMI[j]	3
folpet	phthalimide	—[h]	trifloxystrobin	QoI[e]	11
fosetyl-Al	phosphonate	33	triflumizole	DMI[j]	3
iprodione	dicarboximide	2	zineb	dithiocarbamate	—[h]
iprovalicarb	CAA[k]	40	ziram	dithiocarbamate	—[h]
kresoxim-methyl	QoI[e]	11	zoxamide	benzamide	22

[a]Common names of active ingredients used on grapes internationally. Not all active ingredients are registered in all countries. Trade names for products containing a given active ingredient may vary among countries and manufacturers or distributors.

[b]Name based upon chemical relatedness, as compiled from various sources by industry Fungicide Resistance Action Committee (FRAC).

[c]FRAC-proposed resistance group as shown on product labels in the United States and other participating countries. Active ingredients in the same group are considered cross-resistant to one another and should not be rotated or combined for resistance management purposes.

[d]QxI = Quinone x [unknown] inhibitor.

[e]QoI = Quinone outside inhibitor.

[f]MBC = methyl benzimidazole carbamate.

[g]SDHI = succinate dehyrdogenase inhibitor; previously referenced as carboxin.

[h]— = multisite inhibitor, little to no resistance risk.

[i]QiI = Quinone inside inhibitor.

[j]DMI = Demethylation inhibitor.

[k]CAA = carboxylic acid amide.

[l]— = biochemical activity uncertain, little to no resistance risk.

- Limit the size of the pathogen population from which resistant individuals might be selected, thereby limiting the potential number of resistant survivors that might subsequently reproduce and spread. Simply put, the chance of selecting resistant individuals from a treated population is proportional to the population's size. In practical terms, this means avoid using a fungicide at risk of resistance development as a "rescue" treatment in a vineyard with a significant outbreak of a disease that is controlled by that material.
- Limit the subsequent reproduction of resistant individuals that have survived exposure to the at-risk fungicide. This can be accomplished several ways. (i) Utilize appropriate cultural practices to limit disease development (pathogen reproduction). (ii) Rotate with effective, unrelated fungicides. The fewer sequential applications of an at-risk fungicide before alternating with an effective unrelated material, the less opportunity for reproduction of resistant individuals before they are controlled by the alternative. (iii) Apply an at-risk material in combination with another unrelated fungicide, either through tank mixing or use of a prepackaged product containing two or more ingredients active against the target pathogen. However, it is critical to note that resistance management efforts dependent upon rotation and/or mixture with unrelated fungicides are only as effective as the companion materials themselves; a companion material with poor activity against the target pathogen or applied at a relatively low dose will have only a limited effect on slowing the reproduction of any resistant individuals that might survive exposure to the at-risk component of the program.
- An additional strategy appropriate to fungicides subject to quantitative resistance development (e.g., DMI materials) is to reduce the proportion of the pathogen population resistant to (surviving) an application by increasing the activity of that application. This can be accomplished either by increasing the rate of the product within legal limits or by substituting a related fungicide with greater intrinsic activity.

Specific Fungicides and Groups

Individual active ingredients used on grapes in various countries and their associated resistance-management groups are provided in Table 8. Selected properties of important materials and groups are provided below. Single-site inhibitors, at risk for resistance development, are identified with an asterisk (*); multisite inhibitors, at little risk for resistance development, are identified with a "plus" sign (+).

***Amines (morpholines).** Amine (also known as morpholine) fungicides include fenpropimorph, fenpropidine, spiroxamine, and tridemorph. They selectively control various powdery mildew species, including that which affects grapevines, but have no significant activity against other diseases. Although approved for use on grapevines in many other countries, they are not registered on this crop in the United States.

Like DMIs, these fungicides disrupt cell membrane function by inhibiting ergosterol biosynthesis (see section on DMIs below), but do so at a different step of the synthetic pathway (delta14 reductase and delta8-delta7 isomerase in sterol biosynthesis). Similar to the DMIs, they are inhibitors of germ tube elongation and mycelial growth with good postinfection activity against powdery mildews.

Although a shift in sensitivity to morpholines has been documented in Europe for populations of the wheat powdery mildew fungus, the development of practical resistance has not been a problem for grapevine powdery mildew as it has for the DMIs. There appears to be little cross-resistance between these two classes of fungicides, which have been placed in different resistance groups.

***Anilinopyrimidines (APs).** The AP fungicides (cyprodinil and pyrimethanil) have been used for Botrytis control since the 1990s, but have little activity against other grape diseases at commercial use rates. They appear to act by interfering with the synthesis of a specific amino acid (methionine) and the secretion of enzymes involved in pathogenesis, and inhibit both germ tube elongation following spore germination and mycelial growth following infection. They are absorbed by plant tissues in a locally systemic mode and provide both preventive and postinfection activity.

The risk of the Botrytis fungus developing resistance to the AP fungicides is high, and isolates resistant to either cyprodinil or pyrimethanil are cross-resistant to the other. Practical resistance has been documented to occur relatively rapidly if these materials are used intensively and with little rotation, whereas their utility has been maintained since their introduction in many vineyards where the annual use of all APs is strictly limited and other recommended resistance management practices are employed.

***Azanaphthalenes.** Quinoxyfen and proquinazid (currently not sold in the United States) are active only against powdery mildews, including that of grapevine. They function by disrupting signal transduction in these fungi, although the exact mechanism of action remains unknown. They do not stop spore germination but act by inhibiting appressorium formation, thereby stopping the fungus from successfully penetrating and infecting the plant. As a result, they are good protectant fungicides but lack postinfection or postsymptom activity. Furthermore, quinoxyfen and proquinazid have strong vapor activity, allowing some degree of redistribution within the vine canopy to help protect tissues formed since the previous application or otherwise lacking adequate spray coverage.

Resistance to azanaphthalenes has been documented for isolates of grapevine powdery mildew. Resistant isolates appeared to have a "tolerance" (quantitative resistance) phenotype, with cross-resistance between proquinazid and quinoxyfen.

***Benzimidazoles.** See MBCs.

***Benzamides.** Fluopicolide and zoxamide are oomycete-specific fungicides used for the control of downy mildew on grapes. Although they are chemically similar and inhibit cytoskeleton formation in fungi, they do have different modes of action and are considered to be in different resistance groups (not cross-resistant). Zoxamide inhibits beta-tubulin formation during cell mitosis whereas fluopicolide inhibits the formation of spectrinlike proteins essential for cytoskeleton function. Both inhibit zoospore motility and germination as well as mycelial growth and are locally systemic, lending them both protective and limited postinfection activity. Resistance to these relatively new fungicides has not yet been documented, although they are considered to be at risk for its development.

***CAAs (carboxylic acid amides).** These fungicides, which include dimethomorph, iprovalicarb, and mandipropamid, exert their activities through inhibition of an enzyme involved in cellulose biosynthesis. Thus, they are only active against oomycetes (e.g., downy mildew), whose cell walls contain cellulose, and not against true fungi (causes of all other grape fungal diseases), whose cell walls are composed primarily of chitin. They provide activity in both preventive (inhibition of zoospore germination) and postinfection (mycelial growth) modes, in addition to inhibiting the development of sporangia. However, they have a high affinity for waxes, which limits their absorbance into plant tissues (especially berries), thereby limiting the scope of their postinfection effects.

The resistance risk for grape downy mildew is considered moderate, although it has been documented in a few locations in Europe, with resistant isolates characterized as having an "immunity" (qualitative resistance) phenotype and exhibiting cross-resistance among mandipropamid, dimethomorph and iprovalicarb.

+Chloronitriles. Chlorothalonil (not labeled for use on grapes in the United States) is the only member of this group. Its properties are similar to those of the phthalimides discussed below.

+Coppers. Copper exerts activity in the form of its cupric ion (Cu^{++}), which binds to amino acid residues involved in providing protein shape and function. Thus, at sufficient concentration it is detrimental to multiple enzymes of numerous organisms, including plants. As a fungicide, it was originally developed to control downy mildew as the active ingredient in the "Bordeaux mix" of copper sulfate plus lime, in which cupric ions produced from dissociation of the former compound combine with the hydroxyl ions produced in aqueous suspensions of lime to form copper hydroxide. Copper hydroxide is a weakly soluble compound that typically provides a "slow release" of cupric ions at a dosage detrimental to microbial pathogens on plant surfaces but not to the host itself. Today, manufactured formulations of copper hydroxide and several other weakly soluble "fixed" copper compounds can be purchased, all acting on the same principle. Most *V. vinifera* cultivars are tolerant of copper when it's applied using standard formulations and rates, whereas green tissues of some native American cultivars and interspecific hybrids express varying levels of injury following treatment. Highly acidic spray water and conditions that promote absorption of applied sprays (e.g., high relative humidity and other slow-drying conditions) increase the danger of phytotoxicity, sometimes even on otherwise tolerant cultivars.

Although most active against downy mildew, copper provides measurable suppression of numerous grape diseases, and is often the most effective chemical tool available for "organic" management of diseases other than powdery mildew. However, copper's lack of selectivity, nondegradable nature, and immobility in soils has resulted in accumulation to toxic levels at some sites after heavy use over many years. This has led some agencies and organizations to limit its use in systems that emphasize a sustainable or ecological approach to production.

Coppers are similar to the organic (carbon-based) contact fungicides in many respects; residues are intended to remain on treated surfaces, and fungicidal activity is preventive through inhibition of spore germination. However, copper products do provide some antisporulant activity against active downy and powdery mildew colonies when applied postsymptom.

***Dicarboximides.** The dicarboximides (iprodione, procymidone, vinclozolin) were introduced for Botrytis control in the late 1970s and early 1980s. These fungicides bind to a protein (MAP/histidine-kinase) involved in osmotic signal transduction, leading to a disruption of cell walls and membranes. Dicarboximides inhibit both spore germination and mycelial growth, but are more active against the latter. Thus, they provide both preventive and postinfection activity, including a strong inhibition of sporulation from active lesions. Several studies have shown that at least some of these materials provide improved control when applied with an oil or surfactant to enhance absorption into fruit, thereby enhancing their internal activities.

Because dicarboximides were the only highly active Botrytis fungicides available between the time that MBC resistance became widespread and release of the AP fungicides, they were used intensively in some regions for over a decade without access to an effective rotational or mixing partner. Thus, practical resistance developed in a number of locations. This resistance is typically quantitative and several studies have shown that the frequency of the most resistant isolates declines between seasons, probably because the mutated target protein that confers resistance also makes these individuals less fit than their wild-type counterparts. Practically, the use of dicarboximides in many vineyards has continued to be successful within rotational programs. Limited use of the dicarboximides at full rates is key for maintaining the utility of these fungicides for Botrytis control.

+Dithiocarbamates. The dithiocarbamates are a large and important class of broad spectrum, multisite inhibitor (contact) fungicides. They are sometimes split into two chemical groups, the older dimethyldithiocarbamates (e.g., ferbam, thiram, ziram) and the more common ethylenebisdithiocarbamates or EBDCs (maneb, mancozeb, zineb, metiram). Although exhibiting similar fungicidal properties, this chemical distinction is important within certain contexts, e.g., some U.S. processors impose special restrictions pertaining to the use of EBDC fungicides on grapes that they will purchase, which do not apply to the dimethyldithiocarbamates.

The dithiocarbamates have the same general properties as other broad-spectrum contact fungicides. They function as protectants by inhibiting spore germination, and therefore must be present before an infection event (rain, typically) occurs. Although formulated with "inert" ingredients to increase adherence to treated surfaces, they are nevertheless subject to removal by rain and must be reapplied regularly during periods of significant precipitation. The risk of resistance is low.

***DMIs (demethylation inhibitors).** This large and important group of broad-spectrum fungicides was first introduced in the 1970s, and various compounds have been used on grapes in different countries throughout the world since shortly thereafter. They act by inhibiting a particular step (demethylation) in the biosynthesis of ergosterol, a specific sterol necessary to maintain cell membrane integrity in most fungi (animal cell membranes contain an analogous but different material, cholesterol). Following spore germination, on which the DMIs have no effect, susceptible fungal germlings exposed to these materials are unable to synthesize additional ergosterol as they attempt to grow and colonize their plant host. The resulting deficient membranes cause cells to leak, and the nascent fungal colonies die. Note that the downy mildew pathogen (*Plasmopara viticola*), like most other members of its family (Pythiaceae) in the oomycetes, does not synthesize ergosterol for its growth, and is therefore not affected by inhibitors of this process. Although the DMI fungicides are sometimes called "sterol inhibitors," "sterol biosynthesis inhibitors," or "ergosterol biosynthesis inhibitors," two other unrelated fungicide groups (the hydroxyanilides and the amines or morpholines) also act by inhibiting sterol biosynthesis, although at different enzyme target sites. Thus, the three types of sterol biosynthesis inhibitors comprise three distinct groups with respect to their biochemical modes of action and associated resistance management considerations, and therefore should be referenced independently to avoid potential ambiguity.

DMI fungicides are absorbed rapidly by treated plant tissues, and are therefore not subject to removal by rain soon after their application. They exhibit varying degrees of local systemicity, with those prone to greater movement providing correspondingly shorter periods of residual activity within treated tissues. They sometimes show stronger activity in leaves than in fruit, likely the result of differential uptake and/or movement in the two organs. Because they operate against developing mycelia within plant tissues, postinfection activity is the strength of the DMI fungicides; residual preventive activity typically is significant, but may be limited. Individual DMIs can vary significantly in their intrinsic activities against specific pathogens and their associated abilities to provide commercial levels of disease control, particularly after long use of these products and attendant population shifts in favor of less sensitive individuals.

Throughout more than three decades of use, multiple pathogens on multiple crops in multiple locations have developed practical resistance to DMIs. Nevertheless, this phenomenon is not universal, and the DMIs continue to be a very important group of fungicides in grape production throughout the world. On grapes, reported resistance problems have been confined to powdery mildew, where reductions in control have conformed to a typical quantitative model and have been addressed with significant success by increasing dosages or substituting alternative DMIs with greater intrinsic activity against this pathogen than those that failed. Nevertheless, continuous and conscientious adherence to resistance management recommendations will be necessary to maintain the utility of these compounds into the future.

***Hydroxyanilides.** Fenhexamid and the newly introduced fenpyrazamine are the only two members of this group. At registered use rates, they provide commercial levels of control only against Botrytis. The hyrdoxyanalides are sterol biosynthesis-inhibiting fungicides, although their biochemical mode of action is different from that of both the DMIs and the morpholines, with no cross-resistance among these three groups. As with the DMIs, the hyrdoxyanalides do not inhibit spore germination but do affect germ tube elongation and mycelial growth. They provide good residual protective activity and, like other Botrytis fungicides, are used most effectively in a preventive control program. However, contrary to initial concepts concerning these materials, they also have been shown to exert locally systemic and postinfection activities against Botrytis, comparable to those of the AP and dicarboximide fungicides.

The resistance risk for this group is considered to be moderate. Isolates showing variable levels of quantitative resistance have been detected in grapes and several other crops, and tests have shown them to be more difficult to control than isolates with baseline levels of sensitivity, particularly in a postinfection mode. Isolates exhibiting qualitative resistance due to a single-point mutation also have been reported, although less frequently. However, no reports of practical resistance have been published, and there is some indication that resistant isolates may be less fit than typical baseline isolates.

+Inorganic salts. Dilute solutions of inorganic salts such as potassium bicarbonate and monopotassium (dihydrogen potassium) phosphate can be used to control powdery mildew. They provide virtually no residual preventive activity against spore germination or mycelial growth, but kill live fungal structures (spores, germlings, and visible colonies) through desiccation when applied both postinfection and postsymptom, probably due to osmotic shock. Although the waxy cuticle on leaves and clusters largely protects underlying plant cells from the same fate, sufficient absorption to cause plant injury can occur at high concentrations. Young powdery mildew colonies are more severely affected by postsymptom applications than are older, well-established colonies. Because of the absence of residual ("forward") activity, these materials are most effective when applied at relatively short spray intervals, thereby maximizing the number of contact episodes per unit of time. Complete coverage of infected plant parts is essential for effective control. Although some products are marketed otherwise, the inorganic salts provide little if any practical control diseases other than powdery mildew, owing to the subcuticular or subepidermal growth of all other grape pathogens and their resultant protection from contact with the spray solution.

***MBCs (methyl benzimidazole carbamates).** Developed in the late 1960s, materials in this group, which include benomyl, carbendazim, and thiophanate-methyl, were the first systemic fungicides with a broad spectrum of activity. They inhibit mitosis and hence cell division by binding to a specific site of a fungal protein (beta-tubulin) that composes microtubules, which are essential to that process. MBC fungicides do not bind to the beta-tubulin of oomycetes, and therefore do not control diseases caused by this group of microorganisms (e.g., downy mildews). In susceptible fungi, MBCs inhibit germ tube elongation, mycelial growth, and sporulation from lesions that do form, and can be effective against sensitive fungal populations when used in either a preventive or a postinfection mode.

The development of practical resistance to the MBCs within numerous populations of the fungi causing powdery mildew and Botrytis bunch rot has significantly limited the utility of these fungicides in vineyard disease management programs internationally. Resistance is of a qualitative (immune) nature and has proved to be stable over many years. However, MBCs are still effective against these diseases in select locations, and they remain generally effective against various canker diseases, which reproduce relatively slowly. When using MBC fungicides, it is crucial to rigidly implement appropriate resistance management strategies in order to maintain their utility.

Diethofencarb, an N-phenyl carbamate, has specific utility for the control of MBC-resistant Botrytis isolates. It exhibits 'negative cross resistance' with the MBC fungicides, binding only to beta-tubulin strands with the mutation that prevents binding of MBC fungicides. Thus, it controls only MBC-resistant but not MBC-sensitive fungi. It is not registered for use on grapes in the United States.

As diethofencarb also inhibits microtubule formation and mitosis in fungi that are susceptible to it, it shares many of the same systemic, protectant, and postinfection characteristics of the MBC fungicides. Similarly, the resistance risk for diethofencarb also is high, and Botrytis isolates resistant to both MBCs and diethofencarb can be found.

+Oils. Some highly purified oils derived from both botanical and petroleum sources can be used as fungicides. They have good efficacy against powdery mildew and although some products claim activity against additional diseases, control of them is typically erratic at best. Oils are most active when applied directly to germlings and mycelia of the powdery mildew fungus in both postinfection and postsymptom modes, and are among the most effective products available for eradication of visible colonies. As with the inorganic salts, thorough spray coverage is critical for the successful use of oils, since their activity is largely the result of direct contact with the fungus. Unlike the salts, however, these materials also provide up to several days of residual preventive activity, although much of this is lost following even moderate rainfall. Oils must be used with caution, as applications with or near those of certain insecticides and fungicides designed to remain on treated plant surfaces can promote absorption of these latter materials, causing plant injury; even when used alone, application of oils at high temperatures can result in burning.

***Phenylamides.** This fungicide class contains metalaxyl and its active enantiomeric form, mefenoxam (metalaxyl-M), in addition to benalaxyl and oxadixyl. These materials are highly active against the downy mildew pathogen and other sensitive oomycetes, but do not affect true fungi. They function by binding to and inhibiting an enzyme (RNA polymerase) essential for the synthesis of cellular proteins, the basic machinery of life. Unlike most fungicides, they are highly soluble in water and are highly systemic, being readily transported acropetally (upward) throughout the plant in the xylem. Although they do not inhibit spore germination, they are very active against mycelial growth and spore formation, providing residual (preventive) activity as well as activity in both the postinfection and postsymptom modes; however, postsymptom use is discouraged since it can promote the development of large populations of resistant individuals very quickly. Metalaxyl has been shown to provide significant activity through the vapor phase, with the vapors emanating from sprayed tissues providing protection of unsprayed neighboring clusters inoculated with downy mildew sporangia.

Qualitative resistance to these materials develops readily and quickly, and they became ineffective soon after their introduction in many European regions with high downy mildew pressure and intensive spray programs. Where still effective, it is essential that the number of seasonal applications be strictly limited in order to preserve their utility. Although they are often formulated with a contact fungicide for resistance management purposes, the labeled use rates of some such commercial products provide the mixing partner at a dose that is likely to provide only modest control of resistant isolates if used at the spray intervals typically recommended for phenylamides. In such cases, additional applications of the mixing partner alone or supplementing its rate in mixture may be needed to limit the development of isolates resistant to the phenylamides.

***Phosphonates (phosphites).** Discussions of these fungicides are subject to potential confusion over nomenclature on several bases. The terms phosphonate and phosphite are synonymous within the context of fungicidal activity and are often used interchangeably. These materials are the salts and esters of phosphonic and phosphorous acid, respectively; the latter two terms refer to the same molecule as it shifts back and forth between two slightly different configurations (isomers). The discovery of activity of phosphonates against oomycetes was made in the 1970s, and since then several forms have been used to control diseases caused by oomycetes on a number of crops, including grapes. Commercial products in this class consist of either fosetyl-Al, which quickly breaks down into phosphonate and aluminum within treated plants, or mixtures of various phosphorous acid salts. These materials should not be confused with phosphoric acid (phosphate, PO_4^{3-}), the standard form of phosphorus provided in commercial fertilizers, which has no fungicidal activity beyond that of a general osmoticum. Conversely, phosphite (HPO_3^{2-}) has fungicidal activity but is a poor fertilizer, as plants are able to utilize very little phosphorus in this form.

Although the biochemical mode of action is not known, phosphites are toxic to a number of oomycete species and this is almost certainly the means by which they control grapevine downy mildew. Whereas there is some evidence that they may also elicit natural defense responses within some treated plants, which could provide additional activity against diseases caused by true fungi or bacteria, control of nonoomycete diseases has been notoriously erratic on grapes and only occasionally significant under field conditions.

Phosphonate products vary in their absorption characteristics, likely owing in part to differences in their formulation chemistries, as polar molecules such as the phosphite ion are poorly absorbed through the plant cuticle; an esterified form such as fosetyl-Al is more readily absorbed. Nevertheless, multiple phosphite products and fosetyl-Al have provided good control of grapevine downy mildew at labeled use rates. The systemicity exhibited by the phosphonates is unique, inasmuch as they are transported readily in both the phloem and xylem, both acropetally and basipetally. They provide excellent postinfection activity against downy mildew and greatly reduce subsequent sporulation when applied to nascent lesions too late after infection to prevent them from expanding. The phosphonates also provide significant postsymptom activity, reducing further sporulation when applied to active lesions, although not eradicating them. However, the duration of residual preventive activity provided by these materials following application is limited by their degree of systemicity, declining more quickly in older leaves from which they apparently are exported (source) than in newer ones to which they likely arrive (sink). The risk of resistance development is poorly understood but appears to be moderate, as reduced disease control due to quantitative (rate-dependent) resistance has been documented for downy mildews on other crops.

+Phthalimides. These materials, which include captan and folpet, are multisite toxicants. They exert their activity through a nonspecific reaction with thiol groups, including several thiol-containing enzymes involved in respiration. This nonselectivity renders them damaging to plant cells as well as those of fungi, and they typically cause plant injury under conditions conducive to their absorption across the waxy cuticle of leaves and berries, e.g., when applied in combination with a nonpolar solvent such as oil or certain liquid insecticide formulations. Thus, they function strictly as protectant fungicides, by inhibiting spore germination. Their resistance risk is low.

***Polyoxins.** Polyoxin-D was originally developed in Japan as a naturally derived fermentation product from soil-inhabiting *Streptomyces* species for the control of Rhizoctonia diseases, but has recently been developed for use on grapevine for the control of Botrytis and powdery mildew. In the United States, it received registration as a "reduced-risk" biopesticide.

Polyoxin-D inhibits chitin synthesis in fungi and affects spore germination and mycelial growth. It has locally systemic properties and provides both protective and limited postinfection activity. No resistance to polyoxin-D has been documented for powdery mildew or Botrytis at this time and its resistance risk is considered moderate.

***Quinone inside inhibitors (QiIs).** Cyazofamid is a QiI fungicide that is oomycete-specific and is therefore used on grapevines only against downy mildew. Like the QoI fungicides (see below), cyazofamid also targets complex III of mitochondrial respiration, but at a different site. Thus, there is no cross resistance between QiI and QoI fungicides. Resistance of *P. viticola* isolates to cyazofamid has not yet been detected in nature; however, resistance due to a single target site mutation has been characterized in lab mutants, and the fungicide is considered at moderate to high risk for resistance development. Limiting seasonal applications of QiI fungicides and/or tank mixing with multisite fungicides is recommended for prolonging the utility of this fungicide group.

Cyazofamid inhibits zoospore formation, germination, and motility; oospore production; and mycelial growth, but it has limited systemic and translaminar properties. Because the fungicide remains largely on the outside of the plant, it functions primarily in a protective mode.

***Quinone outside inhibitors (QoIs).** These fungicides function by inhibiting a specific step ("complex III") in fungal mitochondrial respiration, the process necessary to convert a food source into usable energy. The individual QoIs vary significantly in their biological range of activity and include the broad-spectrum strobilurins (azoxystrobin, kresoxim-methyl, pyraclostrobin, trifloxystrobin) and the chemically distinct, oomycete-specific imidazole fungicides fenamidone and famoxadone. All QoIs exhibit some limited local systemicity, but are confined primarily to the waxy cuticle of treated tissues, although they do exhibit translaminar activity as well. Kresoxim-methyl and trifloxystrobin exhibit measurable vapor-phase redistribution, which contributes to their effectiveness against powdery mildew.

Although QoI fungicides inhibit both spore germination and mycelial growth, their physical placement causes them to function primarily as protective fungicides against most sensitive grape pathogens. Whereas postinfection applications directed against powdery mildew can provide excellent control, against other diseases this approach typically allows the formation of pathogen lesions following the infection and treatment events, albeit with greatly reduced sporulation. The risk of qualitative resistance has proven to be high, having been documented in multiple viticultural regions for downy mildew and in select regions for powdery mildew after just a few years of use. Therefore, resistance management strategies should be strictly adhered to when using these materials. Individual materials have caused injury to specific cultivars of specific crop plants, including certain native American and interspecfic hybrid grapes, but not *V. vinifera*.

***SDHIs (succinate dehydrogenase inhibitors).** SDHI fungicides inhibit mitochondrial respiration at complex II (succinate dehydrogenase). Old members of this group (e.g., carboxins) had a narrow spectrum of activity and were used primarily for treating seeds of agronomic crops. However, a new generation of SDHIs, represented by materials such as boscalid, fluopyram, fluxapyroxad, and penthiopyrad was introduced in the early years of the twentieth century for use on horticultural crops, where they provide activity against a much broader spectrum of pathogens. On grapevines, they provide good control of Botrytis bunch rot and powdery mildew, but have little significant activity against any other major disease. SDHIs show local to limited systemicity and are effective against sensitive pathogens when applied both preventively and postinfection.

The SDHIs are considered to be at moderate to high risk for resistance development, and *B. cinerea* isolates exhibiting qualitative resistance have been detected in commercial vineyards. The proliferation of new products containing SDHI fungicides makes it essential that growers carefully adhere to recommended resistance-management practices, including seasonal limits on the use of all products containing them, in order to sustain the utility of these materials.

+Sulfur. Elemental sulfur is used on grapes more commonly than any other fungicide in the world, and is probably the oldest pesticide known to humankind. It is sold in various formulations based largely on particle diameter: dusts (about 25 μm); wettable powders (0.1–25 μm, averaging 7.6 μm); dry, flowable, wettable granules (1–8 μm); and liquid flowables (average of 3.5 μm). In general, tenacity on leaves (resistance to removal by rain), vaporization rate, and efficacy per unit weight applied are all inversely proportional to particle size.

Despite more than two millennia of agricultural experience with sulfur, surprisingly little is known about its biochemical mode of action. It is believed that sulfur is absorbed by permeable fungal cells in its elemental state (S^0), either through direct contact with solid particles or in its vapor phase. Once taken into the cytoplasm it appears to interfere with mitochondrial respiration at multiple sites, through a variety of postulated mechanisms. It is a strong inhibitor of spore germination but also is toxic to fungal mycelium. Powdery mildew fungi are particularly sensitive to sulfur, and this disease is its primary target in grape production. Sulfur is also used for early-season Phomopsis control in some Mediterranean climates, although it has not been effective for that purpose under high disease pressure in the wet springtime climate of New York State.

Normally, applied sulfur remains on the surface of treated tissues or is absorbed into the waxy cuticle, which increases its resistance to washing off. This surface orientation of the material, combined with its inhibition of spore germination and multiple biochemical action sites, is typical of a contact fungicide that provides only preventive activity. Indeed, sulfur has traditionally been considered a protectant and provides very effective control in this mode; and as with most contact fungicides, it causes injury to grapevine tissues when absorption across the cuticle and into the plant cells is promoted, such as through application with or near an oil spray. However, its activity against fungal mycelia and the surface nature of *E. necator* result in sulfur providing strong postinfection activity against powdery mildew as well. Furthermore, postsymptom applications significantly reduce subsequent spore formation from treated colonies and sometimes eradicate them completely, although this latter effect is often most pronounced with young colonies and requires extremely thorough spray coverage of the infected tissues. Trials demonstrating postinfection and postsymptom properties have been conducted with small-particle sprayable formulations, which might improve the opportunity for direct contact of the S^0 particles with developing mycelia relative to that provided by dust applications, although this question has not been addressed experimentally.

Volatilization increases with increasing temperature, which can affect both disease control and the potential for phytotoxicity. Threshold levels of 15–18°C are quoted in some literature as the minimum temperature necessary for significant volatilization of sulfur products and reliable powdery mildew control with them. However, field experience often contradicts this claim, as do controlled-environment experiments from labs in both the United States and Australia, showing little reduction in the efficacy of sprayable formulations on plants maintained at a temperature of 15°C relative to those at 20–30°C. Temperatures below 15–18°C are also suboptimal for powdery mildew development, and the practical effect of reduced volatility at lower temperatures appears to have been overstated for at least some situations. Supraoptimal levels of volatility at temperatures above the upper threshold for powdery mildew development (32°C) not only hasten dissipation of sulfur residues but also increase the potential for phytotoxicity, which is further increased under humid versus arid conditions. In general, use at temperatures above 35°C is not recommended. Some native American and interspecific hybrid cultivars become injured even at moderate temperatures and should not be treated with sulfur.

Selected References

Avenot, H. F., and Michailides, T. J. 2010. Progress in understanding molecular mechanisms and evolution of resistance to succinate dehydrogenase inhibiting (SDHI) fungicides in phytopathogenic fungi. Crop Prot. 29:643-651.

Bartlett, D. W., Clough, J. M., Godwin, J. R., Hall, A. A., Hamer, M., and Parr-Dobrzanski, B. 2002. The strobilurin fungicides. Pest Manag. Sci. 58:649-662.

Blum, M., Waldner, M., and Gisi, U. 2010. A single point mutation in the novel PvCesA3 gene confers resistance to the carboxylic acid amide fungicide mandipropamid in *Plasmopara viticola*. Fungal Genet. Biol. 47:499-510.

Erickson, E. O., and Wilcox, W. F. 1997. Distributions of sensitivities to three sterol demethylation inhibitor fungicides among populations of *Uncinula necator* sensitive and resistant to triadimefon. Phytopathology 87:784-791.

Forster, B., and Staub, T. 1996. Basis for use strategies of anilinopyrimidine and phenylpyrrole fungicides against *Botrytis cinerea*. Crop Prot. 15:529-537.

Gubler, W. D., Ypema, H. L., Ouimette, D. G., and Bettiga, L. J. 1996. Occurrence of resistance in *Uncinula necator* to triadimefon, myclobutanil, and fenarimol in California grapevines. Plant Dis. 80:902-909.

Hilber, U., and Hilber-Bodmer, M. 1998. Genetic basis and monitoring of resistance of *Botryotinia fuckeliana* to anilinopyrimidines. Plant Dis. 82:496-500.

Hoffman, L. E., and Wilcox, W. F. 2003. Factors influencing the efficacy of myclobutanil and azoxystrobin for control of grape black rot. Plant Dis. 87:273-281.

Köller, W., ed. 1992. Target Sites of Fungicide Action. CRC Press, Boca Raton, FL.

Latin, R. 2011. A Practical Guide to Turfgrass Fungicides. American Phytopathological Society, St. Paul, MN.

Leroux, P. 2004. Chemical control of *Botrytis* and its resistance to chemical fungicides. Pages 195-222 in: *Botrytis:* Biology, Pathology, and Control. Y. Elad, B. Williamson, P. Tudzynski, and N. Delen, eds. Kluwer, Dordrecht, Netherlands.

Leroux, P., and Clerjeau, M. 1985. Resistance of *Botrytis cinerea* Pers. and *Plasmopara viticola* (Berk. & Curt.) Berl. and de Toni to fungicides in French vineyards. Crop Prot. 4:137-160.

Lyr, H. 1995. Modern Selective Fungicides: Properties, Applications, Mechanisms of Action. Fischer Verlag, New York.

Ma, Z., and Michailides, T. 2005. Advances in understanding molecular mechanisms of fungicide resistance and molecular detection of resistant genotypes in phytopathogenic fungi. Crop Prot. 24:853-863.

Ohshima, T., Komyoji, T., Mitani, S., Matsuo, N., and Nakajima, T. 2004. Development of a novel fungicide, cyazofamid. J. Pestic. Sci. 29:136-138.

Steffens, J. J., Pell, E. J., and Tien, M. 1996. Mechanisms of fungicide resistance in phytopathogenic fungi. Curr. Opin. Biotechnol. 7:348-365.

Szkolnik, M. 1981. Physical modes of action of sterol-inhibiting fungicides against apple diseases. Plant Dis. 65:981-985.

Toffolatti, S., Maffi, D., Serrati, L., and Vercesi, A. 2011. Histological and ultrastructural studies on the curative effects of mandipropamid on *Plasmopara viticola*. J. Phytopathol. 159:201-207.

Williams, J. S., and Cooper, R. M. 2004. The oldest fungicide and newest phytoalexin – a reappraisal of the fungitoxicity of elemental sulfur. Plant Pathol. 53:263-279.

Wong, F. P., and Wilcox, W. F. 2001. Comparative physical modes of action of azoxystrobin, mancozeb, and metalaxyl against *Plasmopara viticola* (grapevine downy mildew). Plant Dis. 85:649-656.

Young, D. H., and Slawecki, R. A. 2001. Mode of action of zoxamide (RH-7281), a new Oomycete fungicide. Pestic. Biochem. Physiol. 69:100-111.

Ziogas, B. N., Markoglou, A. N., and Malandrakis, A. A. 2003. Studies on the inherent resistance risk to fenhexamid in *Botrytis cinerea*. Eur. J. Plant Pathol. 109:311-317.

(Prepared by W. F. Wilcox and F. P. Wong)

Spray Technology for Grapevines

Acceptable disease control in commercial vineyards is generally dependent on the efficacious use of fungicides. Traditionally, the vast majority of research and education relative to this aspect of disease management has focused on determining the optimum timing of fungicide applications and the properties of the materials used. However, implicit to these foci is the assumption that efficacious materials applied at the correct times will be deposited efficiently on their intended targets at the intended dosages. Unfortunately, this assumption is not always correct.

Good application of pesticides to grapevines is a perennial challenge as a result of many interrelated factors, including variations in trellis design, canopy structure, sprayer design, and operator skill (Table 9). Management of the spraying system has assumed much greater importance in recent years owing to the dual emphases on (i) increasing pest management efficiency (maximized control with minimized inputs of materials, labor, and equipment); while (ii) reducing the potential negative impacts of pesticide use, e.g., through improved operator safety and minimized environmental pollution.

Poor spray coverage is one of the most common factors contributing to inadequate disease control in commercial viticulture. The challenge of every operator is to uniformly apply the spray material to both sides of the targeted leaves and/or grape clusters throughout the entire canopy. This is best accomplished through combining an optimum volume of water with an optimum size of droplets, which will both reach and adhere to the targeted leaves or fruit. Optimum spray volume is determined by canopy size and spray classification. If air is used to deliver the droplets, its volume must also match that of the canopy. Forward speed must be slow enough to allow time for the spray cloud to gain adequate canopy penetration without unduly prolonging the application process. The best specific means by which to accomplish these goals varies among individual sprayers and canopies.

Nozzles and droplets

A cloud of spray droplets can be created by one of three methods. In the majority of sprayers, hydraulic pressure of 0.8–1.8 MPa (1 MPa = 10 bar = 145 psi), created by a high-pressure pump, delivers pesticide solution to full or hollow cone nozzles. The liquid breaks up into a fine spray of droplets, with a volume median diameter (VMD) <200 µm, by means of a swirl plate and a nozzle tip. The size of the swirl plate and nozzle tip determines flow rate and droplet size. Modern design includes a one-piece swirl plate/tip, color coded by size for ease of selection, that is easier to replace than traditional ceramic two-piece swirl plate and tip nozzles; the newer design also does not crack within the nozzle cap, owing to a plastic case surrounding the ceramic tip.

Alternatively, a low-pressure pump (0.20–0.25 MPa) may deliver a stream of liquid to an air-shear nozzle, where it then is cut into fine droplets by a high-speed air stream passing over a shear plate. Droplet size is determined by the ratio of liquid to air at the shear nozzle, and is inversely proportional to the air velocity. Air-shear nozzles tend to provide finer atomization than hydraulic nozzles, with spray droplet diameters typically in a range of 50–150 µm. Although these fine droplets can provide excellent coverage of the spray target in some situations, they can also result in increased losses due to off-target drift, which can be further exacerbated by evaporation of the droplets, particularly in hot weather.

The third method of droplet creation is the controlled droplet applicator (CDA). Liquid is fed onto a high-speed spinning disc or cage, then centrifugal forces spread the liquid and throw droplets from the periphery. Approximately 95% of these droplets are the same size, as determined by flow rate and the speed of the spinning disc or cage.

A droplet with twice the diameter of another has four times the surface area and eight times the volume. Thus, eight small droplets having the same total volume as the larger one can cover twice the surface area of the larger. Similarly, for the same volume of liquid, halving the diameter of the droplets increases their number eight fold. For example, a single 200-µm droplet can be dispersed into eight droplets of 100 µm diameter, 64 droplets of 50 µm diameter, etc.

The objective when spraying grapevines is to deliver many small droplets that reach and adhere to the surface of the intended target. Very small or fine droplets (less than 150 µm) provide excellent spray coverage, but are also very prone to drift and evaporation; such droplets are created by using a nozzle tip with a small orifice or operating at high pressures, as is typical for low-volume spray systems. The challenge is to provide a spray comprised of fine to medium-sized droplets (100–250 µm), which will not only penetrate the canopy and thoroughly cover their target but also reduce the number of drift-prone fine droplets that never reach their target.

Spray classification tables are commonly provided in nozzle catalogues, allowing operators to select the correct nozzle on the basis of its output at a given forward speed and pressure as well as the target characteristics. Nozzles are given a spray quality classification based on the resulting droplet size and the relative need for coverage versus drift avoidance, e.g., fine for contact fungicides and insecticides; medium for postemergence herbicides and systemic products; and coarse for preemergence herbicides.

Drift

Pesticide drift is an important and costly problem facing applicators. Drift can result in damage to susceptible, off-target crops; environmental contamination to watercourses; and a lower-than-intended dosage deposited on the target crop, thereby reducing the effectiveness of the pesticide application. Pesticide drift also affects neighboring properties, often leading to concern, acrimony, and even litigation. Throughout the world, legislation is being directed at pesticide applicators in an attempt to reduce drift.

There are two types of drift: airborne and vapor. Airborne drift, which is often very noticeable (Fig. 361), comprises spray

TABLE 9. Inter-related Factors that Impact Spray Delivery

Sprayer	Application	Target	Weather	Operator
Design, tank capacity, number of rows treatable	Water volume, nozzle size, and pressure	Varietal susceptibility, required degree of control, growth habit	Wind speed	Skill
Droplet size	Nozzle orientation	Canopy structure	Wind direction	Attitude
Fan size	Forward speed	Area	Temperature	Management
Air volume, direction, and velocity	Plume penetration	Distance from sprayer (e.g., single-versus alternate-row application)	Humidity	Time available
Power requirement	Drift potential	Growth stage	Inversion layer	Safe practices

droplets carried outside the treatment area by wind currents during application, often to long distances. Vapor drift results from the volatilization of spray materials, due to evaporation and sublimation of the water carrier and active ingredients during or after application, and the subsequent aerial distribution of the resulting vapors. The type and amount of drift depends upon the type of sprayer being used; atmospheric conditions such as humidity, temperature, and wind; and the product being applied.

Drift reduction. Drift is influenced by many interrelated factors, such as machine design, droplet size, canopy structure, and sprayer settings. Although small water droplets are advantageous in terms of spray volume applied and target coverage, the potential for drift imposes a practical limit to minimum droplet size. Droplets <150 μm generally pose the greatest hazard and droplets <50 μm have insufficient momentum to penetrate the canopy or impact the target, and therefore are likely to remain suspended in the air and evaporate. Traditional airblast sprayers give the greatest cause for concern as they produce many small droplets, which often are moved off-target owing to the high volume of air created by the large fan.

Modern nozzle technology, such as air induction (or AI) nozzles, produces larger droplets with a volume median diameter (VMD) >350 μm versus conventional hollow cone nozzles with a VMD of 150 μm. Although larger droplets are desirable in terms of drift reduction, they normally are impractical for pest control, as they roll off the leaf or berry following impact. However, air induction nozzles create air bubbles within the large water droplets, which then collapse on contact with the target, dissipating the energy of propulsion and dispersing the liquid. Recent vineyard research has shown promising results using air induction nozzles with airblast sprayers. Air induction nozzles, when orientated properly, can reduce drift by at least 50% relative to standard systems.

If drift is to be reduced, a management strategy that includes operator training, correct use of a well-calibrated sprayer, buffer zones, alertness to changes in weather conditions, and a well-planned spraying regime must be implemented. There are many adjustments that may be made to the sprayer in order to minimize drift:

- The orientation of the nozzles may be adjusted to counteract the directional airflow created by a rotating fan. Most airblast sprayers have an uneven airflow on each side of the sprayer as a function of fan direction. Directing nozzles downward to counteract the upward air flow and vice versa on the opposite side of the sprayer improves deposition considerably while reducing drift.
- An endplate may be fitted in situations where only one side of the sprayer is required to be in use. A shroud can be used to block the air outlet on one side of the sprayer, thus ensuring that all spray droplets are emitted only in the airstream toward the canopy. This can be particularly useful when spraying the perimeter three rows in a vineyard block, reducing off-target drift by nearly 50% during that time.
- Adjusting air speed and volume can improve deposition and reduce drift considerably. In a well-managed canopy of moderate vigor, using an axial-fan sprayer to produce an air speed of 8–10 m/sec that displaces an air volume of around 3,000 m^3/h is preferable to the common usage of large fans that displace air volumes of up to 50,000 m^3/h at speeds as high as 80 m/sec. In order to accommodate variations in crop canopies during different portions of the season, some modern sprayers are fitted with adjustable pitch propellers to provide a variable airflow. However, a number of other techniques also can be used to adjust airflow and mitigate drift. These include regulating the power take off (PTO) speed of the tractor; fitting an adjustable-speed hydraulic motor to drive the sprayer fan; and adjusting airflow via the fan air intake or outlet. For example, fitting an air-inlet restrictor such as a circular plate surrounding a hole sized according to canopy volume (Fig. 362), or fitting an adjustable louver on the air outlet (Fig. 363), has been shown to reduce drift by 75% by restricting air volume and speed to a degree that the spray plume is maintained within the canopy of the target row. Alternatively, reducing tractor PTO speed by 25% reduced drift by 75% at a distance of 25 m in New York trials, illustrating a simple yet effective approach to air volume reduction.

Sprayer Design

Traditional designs: The airblast sprayer. Airblast sprayers are commonly used to apply pesticides to grapevines. They comprise a tank containing dilute pesticide solution, a pump and filtration system, and a manifold fitted with nozzles. A fan is used to create a high volume of air into which the pesticide solution is dispensed through nozzles and subsequently projected toward the target as a large radial plume of droplets. The airblast from the fan moves these droplets into the canopy, displacing the air within and depositing them onto the target surfaces. Unfortunately, this basic design is fraught with potential inefficiencies. Poor coverage of the inner leaves and grape clusters can be a major problem, as leaves tend to "shingle" (lie flat against each other) because of the high wind speed created by the fan. The airblast can also deposit the pesticide far from its intended target, providing environmental pollution rather than effective pest control. Recent developments in sprayers include designs that seek to minimize such problems by directing the spray plume into the target area via flexible air tubes, small

Fig. 361. Spray droplets not being deposited on the target surface, resulting in airborne drift of the applied pesticide. (Courtesy A. J. Landers)

Fig. 362. Restrictor fitted to the sprayer air intake, limiting air outflow volume and resultant spray drift. (Courtesy A. J. Landers)

fans producing converging turbulent air, or recycling tunnels. Growers with traditional airblast sprayers can improve deposition and reduce drift via simple modifications such as deflecting airflow into the canopy, adjusting nozzles to provide better directional orientation of the spray, and selecting the correct nozzle for the intended use (spray classification).

Conventional airblast sprayers use a single, axial flow fan mounted directly behind the sprayer to direct the air in an upward and outward direction (Fig. 364). Resulting air speeds have been recorded up to 80 m/sec, and are commonly found in the range of 40–45 m/sec. Axial fans are designed to move large volumes of air at low pressures through a shroud or former, which turns the air 90° through an outlet; nozzles are situated radially within the outlet. It has been suggested that deflecting this air into the canopy at an angle more than 90° relative to forward travel may prevent shingling (layering) of the vine leaves, improve penetration, and therefore increase uniformity of deposition within the canopy. The theory behind this approach is that diverting the angle of entry of the spray plume away from perpendicular allows it to travel a longer distance through the canopy, thereby increasing the probability that droplets will become deposited within. However, the theory has not been borne out by experimental evidence, e.g., test results in Italian vineyards indicated that orientating the air jets 20° to the rear of perpendicular actually produced poorer spray deposition on the vine targets than did a perpendicular orientation.

Air assistance aimed at directing and improving penetration of the spray into the canopy has been proved effective; however, over- and undersupply of air energy can result in inadequate deposition, poor control of the target pest or disease, and off-target drift. Early- to midseason sprays are frequently applied at full fan speed, resulting in a dense plume of pesticide rising into the air above the vineyard or heading rapidly toward a sparsely arranged group of small leaf targets. Some growers try to utilize this excess air by driving down only alternate rows and attempting to treat two rows on each side of the sprayer at the same time; however, research has shown that deposition on the second row distant from the sprayer is erratic and can be affected by even modest wind blowing across the vineyard during the spray operation. Indeed, deposition efficiencies are often erratic even on the target row next to the sprayer when the air volume produced by the fan is excessive, as it blows many droplets off target. In experiments, decreasing airspeed from 13 to 9 m/sec and air volume from 7,000 to 3,000 m^3/h increased spray deposition throughout the season, indicating that airflow rates often used in practice could be reduced substantially. Many traditional airblast sprayers are capable of producing air volumes of 20,000 m^3/h.

Fig. 363. Electronically adjustable louver fitted to the air outlet of a sprayer to reduce airflow and spray drift. (Courtesy A. J. Landers)

Forward speed of the tractor also affects airflow, penetration, and deposition. There is an optimum forward speed that depends on canopy density and air volume from the sprayer fan. In general, driving at speeds above approximately 6–7 km/h reduces penetration into the canopy. The desire to increase forward speed for improved work capacity and timeliness of operation must be tempered with speeds that allow time for the spray plume to penetrate the canopy.

To improve output, many growers have purchased multirow sprayers that direct the spray plume horizontally into each side of the canopy. Depending on machine design, multirow airblast sprayers deliver pesticides to two to four rows of vines via a folding frame or gantry system. In research trials, spraying both sides of the canopy at the same time has increased deposition by up to 50% relative to single-sided spraying while improving uniformity of distribution within the canopy and limiting losses of spray onto the ground.

Grape canopies can be highly variable according to their seasonal development stage, cultivar, pruning and training system, and many other cultural and environmental factors. For efficient pesticide use, both air and liquid application volumes must change throughout the season to match the developing canopy. Although some operators change application volume per hectare in response to seasonal canopy development, frequently no concomitant changes are made to airflow components (speed, volume, and direction) nor forward tractor speed. Ideally, fan output should match the canopy size during different portions of the season if deposition is to be improved and drift reduced.

Modern designs of canopy sprayers. *Tower sprayer.* Conventional airblast sprayers can be modified with the addition of a "grape tower" or volute (Fig. 365). The tower, mounted onto the air outlet of the axial fan, directs the air horizontally toward the canopy. A vertical spray boom is mounted on each side of the air outlets and the resulting spray cloud penetrates the canopy much better than the upward-directed plume created by the traditional airblast sprayer. Drift is also reduced considerably.

Directed air ducts. A number of sprayers direct the air to the target via adjustable outlets mounted around the canopy. Each outlet incorporates a hydraulic or air-shear nozzle affixed within a stream of air. The outlets may be mounted on either side of the sprayer (Fig. 366) or fitted to a frame or gantry that allows two to three rows to be sprayed. The outlet angle is adjustable, depending upon growth stage, shape, and density of

Fig. 364. A conventional airblast sprayer. (Courtesy A. J. Landers)

the canopy. The converging air streams provide good leaf turbulence that limits shingling, resulting in excellent penetration with minimum drift.

Tangential blowers. A few sprayers are available that use a hydraulically driven tangential or crossflow fan to blow air radially into the canopy. A line of hydraulic nozzles (Fig. 367) or spinning discs produce droplets that are then carried by the air stream. The position of the fans is adjustable, allowing a more directed spray plume than from a traditional airblast sprayer.

Multihead fans. These sprayers use multiple, small, axial fans to create individually directed airblasts, driven by either hydraulic or electric motors; in front of each small fan are hydraulic nozzles or a spinning cage (Fig. 368). Hydraulic nozzle versions allow operators to change droplet size by changing nozzle size or system pressure. The spinning cage system is a controlled droplet application (CDA) technique, with droplet size determined by the speed of cage rotation. Most CDA applications are made at reduced volumes of water per hectare with good success.

Electrostatic sprayers. Electrostatic charging of spray droplets has been a great success with certain industrial spray applications (e.g., paints), handheld sprayers for agricultural use, and some aerial applications to field crops. A number of manufacturers offer electrostatic charging for vineyard application, although this technique has provided mixed results in terms of pest and disease control. A high voltage/low current charge is applied to the spray solution, usually at or near the nozzle outlet. The spray cloud is attracted to the grounded plant because the negatively charged droplets attract free positive charges to the surface of the vine, a process known as induction. The corresponding free negative charges are repelled down the plant to the soil. The degree of attraction depends upon many factors, such as how the spray is directed toward the vine and the relative humidity of the air. To aid penetration of the canopy, an air stream is often provided. This system utilizes very low water volumes and extremely fine (< 50 μm) water droplets. The attendant advantages of this combination are good coverage of impacted targets and a large area sprayed per tank of solution, whereas the disadvantage is a relatively high potential for poor deposition due to evaporation and drift.

Tunnel sprayers. Tunnel sprayers straddle the canopy and employ a relatively inert material (plastic, stainless steel, etc.,) to form a three-sided "tunnel" that encases both sides and the top of the vine. A vertical spray boom, fitted with hydraulic

Fig. 365. A conventional airblast sprayer modified through addition of a "grape tower" or volute. (Courtesy A. J. Landers)

Fig. 367. A tangential-blower sprayer. (Courtesy A. J. Landers)

Fig. 366. A directed air-duct sprayer. (Courtesy A. J. Landers)

Fig. 368. A multihead fan sprayer. (Courtesy A. J. Landers)

nozzles, is enclosed within this unit on each of its two sides (Fig. 369). The spray is directed into the canopy and any droplets that pass through are captured by the inside tunnel walls and recirculated, via filters, into the spray tank. Recycling spray solution that would otherwise impact primarily on nontarget surfaces has been shown to reduce the total amount of pesticide applied per hectare by an average of 35% in vineyards, and allows the sprayer to travel greater distances per tank load. The reduction in pesticide use is greatest in the early prebloom season (up to 85% recycling) versus later in the season when the canopy is in full foliage. In vineyards, a reduction in drift of 90% relative to airblast sprayers has been documented for tunnel sprayers, which represent the best drift-reducing spray technology currently available. Operation on very uneven or sloping terrain can be problematic.

Canopy sensors. Drift problems increase dramatically when a space occurs within the row because of the absence of a target to intercept the spray plume (e.g., missing vine or arm). Vineyard sprayers can be fitted with ultrasonic sensors that detect the presence or absence of the target canopy via sound waves. Sensors are linked to an onboard computer, which automatically switches electric solenoid valves on or off to operate individual nozzles, thereby applying spray only where vine canopy is detected. Advantages include reduced drift and ground deposition, less pesticide applied per hectare, and increased vineyard area treated per tank load. Trials with these units show pesticide savings of up to 30% and a reduction in drift up to 50%, in direct proportion to the size and number of gaps in the canopy wall.

Small sprayers for trial plots and small vineyards. A number of size-appropriate sprayers are available for smaller vineyards and trial plots. Portable sprayers include handheld and knapsack machines. They are simple in design and construction, leading to easy maintenance, but their work rate is relatively low compared with field sprayers.

Handheld sprayers, such as those using a CDA spinning disc, are very lightweight and useful for weed control or spot spraying (e.g., individual clusters) on a small scale. They are often battery operated with a small tank of up to 10-L capacity. The spinning disc creates uniform droplets resulting in low application volumes.

Knapsack sprayers are larger in capacity, around 20 L, and pressure can be provided by hand via a lever or powered by a small battery or combustion engine. For proper operation, it is necessary to provide a constant flow of liquid to the nozzle at a constant pressure and walking speed. Better-quality units have both a pressure gauge and a spray-management valve to maintain uniform output. They can be used for both canopy and herbicide spraying, and multiple nozzles can sometimes be fitted to a wider boom to improve coverage relative to a single nozzle.

Knapsack mistblowers also are available and comprise a small fan driven by a gasoline engine with a tank and nozzle assembly. The airflow from the fan is emitted via a tube and a nozzle provides the droplets. The resulting spray is blown into the canopy (Fig. 370) and gives better penetration and deposition than provided by a traditional knapsack sprayer.

Dry pesticide applicators or dusting machines. Dusting machines are used to apply dry, pulverized sulfur particles to the canopy. The machine, attached to or pulled by a tractor, mixes the sulfur dust with a stream of air, which is then blown into the canopy. Designs vary from simple mounted devices, driven by the tractor PTO to larger machines with independent engines. A hopper (70–130 kg capacity) holds the sulfur dust and an agitator prevents the particles from coagulating, so that a constant volume of material is metered (by an augur, on larger machines) into a fan-provided airstream, typically moving at a speed of 45–70 m/sec. The air/powder mix leaves the machine via independently adjustable outlet heads mounted on each side of the unit.

For machine longevity, it is essential that all traces of sulfur dust be removed between applications, as residues corrode metal parts if allowed to accumulate. Dust particles (approximately 25 µm diameter) drift just as do very fine water droplets, so great care must be taken to prevent this from occurring.

Operating the Sprayer

The use of pesticides in agriculture introduces occupational hazards and health risks to personnel involved in handling and application tasks. The main hazard to operator health is direct chemical exposure, particularly during mixing and loading. Both engineering controls (e.g., induction bowls and closed transfer systems) and personal protective equipment (coveralls, gloves, face shield, etc.) reduce exposure to pesticides. Sprayer operators are also at risk from splashes of dilute pesticide while washing out the sprayer and from spray drift. Time taken to measure and decant pesticides carefully and for adequate washing of the pesticide containers and sprayer tank after use sometimes seems costly and may encroach upon the time available for spraying; however, if the operator hurries these tasks, the chance of personal contamination increases.

A number of nontraditional methods of mixing and loading pesticides into sprayers have been developed with the goal of improving operator safety. When loading a sprayer, closed transfer systems allow concentrated pesticide to be moved from the original shipping container to the sprayer mix tank with minimal or no applicator contact. Induction bowls, hoppers attached to the side of the sprayer, allow pesticides to be added

Fig. 369. A tunnel sprayer, which captures and recirculates spray solution that would otherwise be lost to drift. (Courtesy A. J. Landers)

Fig. 370. A knapsack mistblower. (Courtesy A. J. Landers)

to the tank without the applicator climbing onto the sprayer. Pesticides are poured into the induction bowl and water rinses the bowl and carries the pesticide to the spray tank.

Direct injection of pesticides allows them to be mixed directly with water in the sprayer plumbing system rather than in the main spray tank. The pesticide is pumped from its container and mixed with the water either in a manifold or at the main water pump. Only clean water is held in the main tank of the sprayer, negating the need to cleanse the inside of the tank after spraying.

Cleaning sprayers and containers presents potential hazards to the operator. To address this danger, tank rinse systems have been devised, comprised of a clean-water supply tank mounted to the sprayer and one or more rotating discs or nozzles mounted inside the main sprayer tank. These systems are designed for infield rinsing of the sprayer, reducing time, quantity of rinsate, and the risk of operator contamination.

Reducing pesticide loss and resulting ground contamination from dripping nozzles is another consideration. Check-valves installed at each nozzle prevent dripping by using a spring-loaded rubber diaphragm to close off the flow of liquid once the system pressure drops. Contamination also can occur when operators change or unclog nozzles during an application, so rubber gloves and spare nozzles should be carried on the tractor to facilitate replacement. Low-drift nozzles also reduce contamination of the tractor and sprayer.

Providing adequate wash water is essential for operator safety and often is legally mandated. A simple container with a hand-operated valve can be mounted on the side of the sprayer to provide clean water for hand washing and personal hygiene.

Minimizing operator contamination while driving the tractor and sprayer is essential. The use of an enclosed tractor cab reduces contamination and provides a more comfortable work environment. Carbon filtration systems are used to remove pesticide odor and pesticide-laden mist from the fresh air inside the tractor or self-propelled sprayer cab. Preventing contamination of the cab interior also requires the removal of protective clothing worn while mixing and loading before entering the cab. Protective clothing should be kept in a locker at the fill site or placed in a simple compartment mounted to the side or front of the tractor or sprayer.

Inspection of Sprayers

In many countries, there is growing concern among the general public regarding pesticides in the environment and on food in particular. The understanding of and skills associated with the correct application of pesticides is provided via education and operator training legislation. There is concern, however, that while training and certification may ensure the operator is competent, the application machine may or may not be operating to an acceptable standard. Across the world, surveys of the mechanical condition and the accuracy of sprayers show that many are substandard in these regards.

In many western European countries for example, sprayer inspections occur every 3–5 years and most are mandatory, depending on local regulations. The objectives are to maximize the efficiency of application by assuring the mechanical integrity of the equipment, thereby reducing both pesticide costs and environmental pollution, while improving operators' knowledge of application technology and sprayer management.

A five-point test sheet is used to check each sprayer, in particular for mechanical safety, with attention paid to wear and corrosion. Ancillary items such as the hydraulic, pneumatic, and electrical systems are inspected for leaks and/or inadequate connections, and a check of the visual condition of pipes and cables is made. Components of the liquid delivery system, such as the tank and fittings, are checked under pressure. The pressure adjustment valve and gauge must work properly and there should not be any fluctuations in pressure. A uniform nozzle output is required, and spray distribution is compared along the length of the manifold. Each test takes 2–3 hours, depending on the model, size, and complexity of the sprayer. A report is given to the grower detailing the components tested and drawing attention to any critical defects.

Assessing Spray Coverage

A number of methods can successfully illustrate localized spray-droplet deposition. Artificial targets, such as water-sensitive cards whose surface-area changes color beneath individual droplets (Fig. 371), can be used to show where spray droplets have impacted within specific areas of interest, e.g., upper versus lower leaf surfaces or the immediate cluster zone. They can easily be attached to leaves with paper clips or wire staples, providing the canopy is dry and the user wears rubber gloves to prevent other moisture from reacting with the cards, and then removed after a test spray with clean water. Spray coverage can be quantified by means of image analysis software that determines the relative area of each card that has reacted to the droplets. Alternatives include high-quality photographic (chromocote) paper cut into 50- × 25-mm strips and attached to the target. A food-coloring dye mixed into water is sufficient to produce bright coloration on these surfaces when they are sprayed with droplets of the solution, allowing quantification as above.

Dyes mixed in water provide an excellent tracer for visual assessments. Pyranine, a fluorescent tracer, can also be used but requires ultraviolet ("black") light for visualization of deposited droplets. Alternatively, leaves and grapes sprayed with this material can be collected and rinsed with deionized water to provide a sample for assessment in a fluorimeter, thereby providing a quantitative measure of spray penetration and deposition. Spray products based upon kaolin clay are safe to use as a tracer, dry rapidly, and are highly visible in daylight (Fig. 372). However, these products do not adhere as well to the grape berry surface as they do to the leaves, which may lead the observer to underestimate the degree of deposition on fruit surfaces.

Monitoring sprayer output to determine where materials are going and to adjust nozzle orientation can be accomplished by means of a vertical patternator, comprised of a series of vertical collecting trays situated at regular intervals (typically, 25 cm) above the ground (Fig. 373). When the sprayer is operated, the spray plume impacts on the trays, whence it is conveyed by tubes into a row of measuring cylinders corresponding to each position (Fig. 374). The resulting distribution graph can then be used to adjust the nozzle orientation to better direct the spray plume into the canopy. The vertical patternator can also be used to conduct repeatable tests that quantify the effects on spray deposition and drift of changes in airflow, nozzle orientation, water volume, etc.

Fig. 371. A yellow water-sensitive card stapled to a leaf to assess spray coverage; purple coloration has resulted from a reaction with the applied water droplets. (Courtesy A. J. Landers)

Fig. 372. Grape leaves treated with a suspension of kaolin clay to assess spray coverage. (Courtesy A. J. Landers)

Fig. 373. A vertical patternator device used for monitoring relative sprayer output at different heights. (Courtesy A. J. Landers)

For simple visualization, growers can construct a similar device by obtaining a pole approximately 5 m high or by joining two appropriate lengths of 50-mm × 100-mm lumber end to end, and affixing a roll of paper tape (such as used for printing in an office calculator or cash register) along the length of the edge facing the sprayer. The pole is then placed between two vines within one or more rows of interest and a mixture of clean water and food coloring is sprayed while traveling between two rows as usual. The spray stains the paper where it hits, demonstrating where the spray plume is traveling, how much is going above the target row, and the relative output between the left and right side of the sprayer.

Automated Precision Technology

Field crop growers have long been able to utilize precision farming techniques to apply fertilizers and pesticides and to measure crop yields by means of GIS maps and GPS-based navigation systems. Research is underway to develop a fully automatic vineyard sprayer that drives along the rows of vines, monitoring the canopy volume and any absence of vines. Sensors monitor the dimensions of the canopy and then alter both

Fig. 374. Measuring cylinders of a vertical patternator device, allowing quantification of spray deposited upon collecting trays at different heights and conveyed by the attached tubing. (Courtesy A. J. Landers)

airflow from the fan and liquid flow (application rate) according to the immediate canopy density. Patches of diseases or insect activity that may have been identified previously through scouting and their exact location recorded on a handheld GPS device can be targeted using a GPS device or a transponder mounted on the sprayer.

Airflow can also be adjusted automatically according to wind direction across the vineyard. A sonic anemometer with a GPS locator, fitted to the tractor, can determine wind speed and direction. The air outlet of the sprayer could automatically increase air output to move spray into the canopy against the wind and reduce airflow when spraying in the direction of the wind.

Electronic controllers can be used to adjust sprayer output according to forward speed and desired application rate. Changing products during a spray operation, e.g., in response to a change in cultivars, can be carried out using techniques where pesticide is injected directly into the water flow on the sprayer.

Autonomous spraying using robotic sprayers is also under development. Robots are used in many industries and offer the advantage of removing the spray operator from an unpleasant and potentially dangerous situation.

Selected References

Backer, G. 2007. Lasting plant protection procedures by innovative technology. Pages 34-46 in: Proc. Pflanzenschutz im Wein und Obstbau: International ATW Symposium on Sustainable Production, Measuring Berry Ripening and Global Climate Change. KTBL, Darmstadt, Germany.

Balsari, P., Marucco, P., and Tamagnone, M. 2001. Assessment of the most suitable setup of mist blowers in vineyards. Pages 116-125 in: Proc. Pflanzenschutz im Wein und Obstbau: 6. Internationales ATW Symposium Technik im Weinbau. KTBL, Darmstadt, Germany.

Balsari, P., Marucco, P., and Tamagnone, M. 2005. Assessment of spray application in vineyards employing a new model of cross flow sprayer. Pages 101-106 in: Proc. Pflanzenschutz im Wein und Obstbau: International ATW Symposium on Plant Protection for Viticulture and Fruit Cultivation. KTBL, Darmstadt, Germany

Doruchowski, G., Roettele, M., Herbst, A., and Balsari, P. 2013. Drift evaluation tool to raise awareness and support training on the sustainable use of pesticides by drift mitigation. Comput. Electron. Agric. 97:27-34.

Ganzelmeier, H., and Rautmann, D. 2000. Drift, drift reducing sprayers and sprayer testing. Pages 1-10 in: Aspects of Applied Biology 57: International Advances in Pesticide Application. Assoc. Applied Biologists, Welsbourne, UK.

Gil, E., 2006. The Spanish perspective on pesticide application issues on international standards and regulatory demands. Pages 51-62 in:

Aspects of Applied Biology 77: International Advances in Pesticide Application. Assoc. Applied Biologists, Welsbourne, UK.
Gil, E., Landers, A., Gallart, M., and Llorens, J. 2013. Development of two portable patternators to improve drift control and operator training in the operation of vineyard sprayers. Span. J. Agric. Res. 11:615-625.
Hall, F. R. 1993. Application to plantation crops. Pages 187-213 in: Application Technology for Crop Protection. G. A. Matthews and E. C. Hislop, eds. CAB International, Wallingford, UK.
Koch, H., 2007. How to achieve conformity with the dose expression and sprayer function in high crops. Pflanzenschutz-Nachrichten Bayer 60:71-84.
Landers A. J. 2002. The answer is blowing in the wind. Pages 177-184 in: Aspects of Applied Biology 66: International Advances in Pesticide Application. Assoc. Applied Biologists, Welsbourne, UK.
Landers, A. J. 2008. Innovative technologies for the precise application of pesticides in orchards and vineyards. Pages 343-348 in: Aspects of Applied Biology 86: International Advances in Pesticide Application. Assoc. Applied Biologists, Welsbourne, UK.
Landers, A. J. 2008. Spray drift mitigation. Pages 196-215 in: Wine Grape Production Guide for Eastern North America. A. K. Wolf, ed. Natural Resource, Agriculture and Engineering Service, Ithaca, NY.
Landers, A. J. 2010. Effective Vineyard Spraying: A Practical Guide For Growers. Effectivespraying.com, Keuka Park, NY.
Pergher, G., and Gubiani, R. 1995. The effect of spray application rate and airflow rate on foliar deposition in a hedgerow vineyard. J. Agric. Eng. Res. 61:205-216.

(Prepared by A. J. Landers)

Certification and International Regulation of Planting Materials

Grapevines have had a long and complex historical relationship with humanity and the development of civilization. They have been selected for superior fruit and horticultural characteristics and subsequently propagated through vegetative means since the beginnings of recorded history. While the existence of grapevine virus diseases and their debilitating effects on vegetative growth and fruit production and quality were recognized long before, they have been documented extensively only from the mid-twentieth century to the present day. Concomitantly, researchers have developed sanitary selection procedures and collaborated with regulators on the development of certification programs as key virus disease management tools.

Grapevine Certification and Sanitary Selection

Grapevine certification schemes worldwide share a common goal of providing standard procedures to produce foundation stock that test negative for select graft-transmissible diseases. The actual procedures and protocols vary widely depending upon the target diseases, the diseases endemic to the production region, the technical and financial resources available, and the expectations of the industry served by the program.

The quality and reliability of any certification scheme is dependent upon the techniques used for detection of pathogens. Technologies for the detection of grapevine viruses and other graft-transmissible pathogens have undergone rapid development since the turn of the twenty-first century and continue to do so. Progress in this area has been enormous; in contrast, regulatory programs for certification and registration of plant materials have been historically slow to change. New technology is often not included in regulations and formal certification schemes until many years after development. For example, at this writing, the U.S. Federal grapevine quarantine program specifies only biological indexing to indicator hosts as a detection tool. Although laboratory tests are widely used in grapevine clean stock programs, the regulations and quarantine requirements underlying these programs are infrequently updated to include newer technologies. This is partially because changes in regulation may require extensive pest risk assessments, and data comparing testing technologies is often inadequate for such decision making.

The most reliable testing programs use a combination of tests, including biological indexing, serological tests, and PCR (polymerase chain reaction) assays, to screen for target pathogens. Next-generation sequencing has recently emerged as an important diagnostic tool of graft-transmissible agents (GTAs).

Biological indexing. Two types of biological indicators—herbaceous and woody plants—are used in grapevine clean stock programs. Herbaceous indexing is performed in a greenhouse in the spring and involves extracting grape tissues of candidate accessions into a buffer solution and rubbing this onto leaves of sensitive indicator plants, e.g., *Chenopodium quinoa*, *C. amaranticolor*, *Nicotiana clevelandii*, and *Cucumis sativus*, among others. After an incubation period of 10–14 days, the presence of certain mechanically transmissible viruses induces characteristic symptoms.

Woody indexing requires two years (minimum) to complete. Grapevine indicators that are especially sensitive to specific virus diseases, e.g., *Vitis rupestris* St. George, *V. vinifera* cv. Cabernet Franc, LN33, and Kober 5BB, among others, are grafted with buds from a candidate accession. After a four-week incubation in a greenhouse, grafted buds are assessed for viability. Indicator plants with viable grafts then spend two weeks in a screenhouse to become acclimated to ambient conditions before transplanting into the field, where they are periodically inspected for canopy symptoms over two growing seasons. To assess for wood symptoms (e.g., stem pitting on *Vitis rupestris* St. George), indicator plants are uprooted, the bark removed to expose the underlining woody cylinders, and symptoms recorded.

Although biological indexing is labor intensive, time consuming, and expensive, this procedure is mandated in most regulatory programs to ensure a high level of confidence in the final diagnosis. In particular, the use of biological indicators is required by many quarantine test schemes where protocols are designed to exclude exotic viruses and other GTAs not known to occur in the growing region where imports are destined. In addition, woody indexing is still required to ascertain the presence of unknown causal pathogens or those insufficiently characterized for laboratory assays (e.g., GTAs of vein mosaic, vein necrosis, rugose wood complex, and some of the leafroll-associated viruses).

Serological tests. Serological methods utilize antibody reactions for known GTAs, usually viruses or bacteria. Antiserum is produced in an animal (typically, a rabbit) by immunizing it with a purified preparation of a pathogen, such as a virus. The animal reacts to the foreign proteins by producing specific antibodies to react against them. The antibodies are purified from the blood serum drawn from the animal and the resulting antiserum is used in diagnostic procedures to test for the presence of the original vaccinating agent. The most commonly used serological test is the enzyme-linked immunosorbent assay (ELISA).

Many grapevine clean stock programs incorporate testing for viruses by ELISA. It is relatively simple to run, has a high throughput, and results are obtained in just one or two days. Test samples should consist of the appropriate tissue(s) collected at the proper time of year; these should be kept cool while en route to the laboratory and maintained in sound condition.

One limitation of ELISA testing is that antisera are not available in sufficient quality and quantity for all of the known grapevine viruses and, in some cases, the virus itself is not yet characterized. Therefore, if a grape plant is infected with a virus for which antibodies are lacking, the ELISA cannot detect it. This is the current situation for some of the leafroll disease agents, which are only detected by indexing on the indicators *V. vinifera* cv. Cabernet Franc or Pinot Noir.

Molecular tests. Molecular assays have been developed that target the genetic material (genome) of plant pathogens, specifically directed to the nucleotide sequences unique to the pathogen. One sensitive molecular method currently used is polymerase chain reaction (PCR) for DNA viruses and reverse transcriptase (RT)-PCR for RNA viruses.

The PCR process involves the selective amplification of a small part of a pathogen's genome. If the unique sequence is present, even in minute quantity, the amplification steps in the reaction mixture cause its exponential increase and accumulation and subsequent detection. It is this amplification process that makes PCR such a sensitive assay.

PCR is used to detect grape pathogens because each GTA has its own unique genetic code. In the past decade, scientists have identified specific genetic markers for many individual pathogens, and PCR testing is currently available for most of the viral and bacterial pathogens of grapevines. This is an active area of research and recent sequencing technology has expanded the number of available tests significantly. A looming issue with respect to regulation and certification is that we are now able to sequence the virome of a grapevine and develop PCR tests for detection of viruses that may or may not be of economic importance for grape production.

The California Grapevine Registration and Certification Program

In the New World, the oldest grapevine certification program is the California Department of Food and Agriculture (CDFA) Grapevine Registration and Certification (R&C) Program. Other state programs in the United States are modeled after it and several use its clean stock as initial source materials for their own certification or clean stock programs. The California program also has historic ties to clean stock programs of Canada and Australia, where similar techniques and protocols have been adopted.

By the early 1950s, California had experienced serious virus disease problems with infected planting stocks. Hewitt listed major grapevine virus diseases known or suspected at that time: "White Emperor disease" (leafroll); yellow mosaic (caused by *Grapevine fanleaf virus*, [GFLV]); fanleaf; and rough bark (later renamed corky bark). He remarked:

> "... the job is to produce certified healthy vines as a source of cuttings for the grower. This calls for a definite program on an industry-wide basis. Such programs have been proposed to the industry, and I am happy to note that efforts to implement controlled methods of introducing only healthy vines and cuttings in California vineyards have already been started."

By 1956, California had issued "Regulations for Registration of Grapevines Inspected and Tested for Virus Diseases," thereby creating the California Grapevine R&C Program. The program has since evolved into its present form.

The current program is described in some detail to illustrate typical protocols used for grapevine clean stock programs. This program differs from many European programs in two key areas. First, it is entirely voluntary; nurseries are free to participate or not under this system. Second, it does not include cultivar or clonal evaluation, a key aspect of many European grapevine clean stock programs. Selections entering the program are normally chosen for their potential horticultural merit and become available for performance evaluations once the phytosanitary requirements of the program have been met. These evaluative efforts are managed separately from the regulatory program, which focuses on target diseases and pathogens coupled with management protocols intended to reduce the chance of re-infection of clean stock.

Foundation Plant Services (FPS), formerly known as Foundation Plant Materials Service, is a service unit in the College of Agricultural and Environmental Sciences at the University of California, Davis. This unit performs the disease testing required by the CDFA R&C program and maintains the foundation stocks that supply clean material to nursery participants.

Each grapevine in the FPS collection is assigned a registration category according to its health and cultivar identification status. "Registered" foundation grapevines are those propagated from sources that met requirements mandated by the CDFA R&C program guidelines, and have been identified professionally to cultivar. "Provisional" foundation grapevines are those that have passed the disease tests (CDFA R&C program), but whose cultivar identification is incomplete. "Nonregistered" grapevines are those that either have tested positive for targeted diseases, are of unknown disease status, or whose cultivar identity is questionable. Provisional and nonregistered selections are distributed only to customers who have received notification of the grapevines' status and are willing to assume the associated risks. All nonregistered grapevines are maintained with physical separation from the elite foundation block, awaiting therapy or identification.

The CDFA R&C program includes provisions for three levels of certified stock (Fig. 375). Propagation materials derived from foundation grapevines in the FPS Foundation vineyard is referred to as "California Foundation Stock." Participants in the CDFA R&C program use the California Foundation Stocks to establish their own vineyards, referred to as "California Registered increase blocks." These blocks are inspected annually and virus tested as required by CDFA inspectors. Cuttings taken from California Registered increase blocks are used to produce "California Registered Stock" and identified by a purple California Registered Stock tag. Plants produced by own-rooting California Registered Stocks or by grafting California Registered Stock (scions) to California Registered Stock (rootstocks) qualify as "California Certified Grapevines." These plants are identified by blue Certified Stock tags and are sold for commercial plantings. CDFA regulations govern all Grapevine R&C Program responsibilities, which includes: eligibility requirements; planting and maintenance requirements; inspection and testing procedures; and approval, suspension, and cancellation of certification, application, and fees.

All woody and herbaceous index tests prescribed in CDFA R&C regulations are used to screen grape materials before qualifying as foundation stock at FPS. In addition to the biological indexing, ELISA and PCR tests are performed at FPS on every accession destined for inclusion in the National Foundation Vineyard at Russell Ranch (Table 10). Furthermore, during their first two growing seasons, the new accessions in the foundation vineyard are tested twice by ELISA for grapevine leafroll-associated viruses; three nematode-transmitted viruses, namely *Arabis mosaic virus* (ArMV), GFLV, *Tomato ringspot virus* (ToRSV), *Grapevine fleck virus* (GFkV), and *Grapevine virus A* (GVA). These accessions continue to be monitored regularly thereafter, as one-fifth of the foundation grapevines at least 3 years old are ELISA-tested each year, so that over a 5-year period the entire foundation vineyard is retested. This 20% retesting procedure is repeated for the life of the foundation vineyard. As an additional precaution, the foundation grapevines are periodically re-indexed biologically and retested by ELISA and PCR. Lastly, the foundation vineyards are visually inspected for disease symptoms in the spring and summer of each year. Any grapevine exhibiting suspicious symptoms is tested immediately with laboratory-based assays and as warranted by woody host indexing, and is removed if found to be infected.

The most recent revision of the CDFA R&C Program was made official July 1, 2010. This new version of the regulations was finalized after three years of public meetings between CDFA regulators, researchers, viticulturists, and commercial nurseries.

Grapevine Registration & Certification In Europe

There is over a century of experience in grapevine selection for both clonal and sanitary purposes in the European Union (EU). Efforts were initiated first in Germany near the end of the nineteenth century. The commitment to improve grapevine

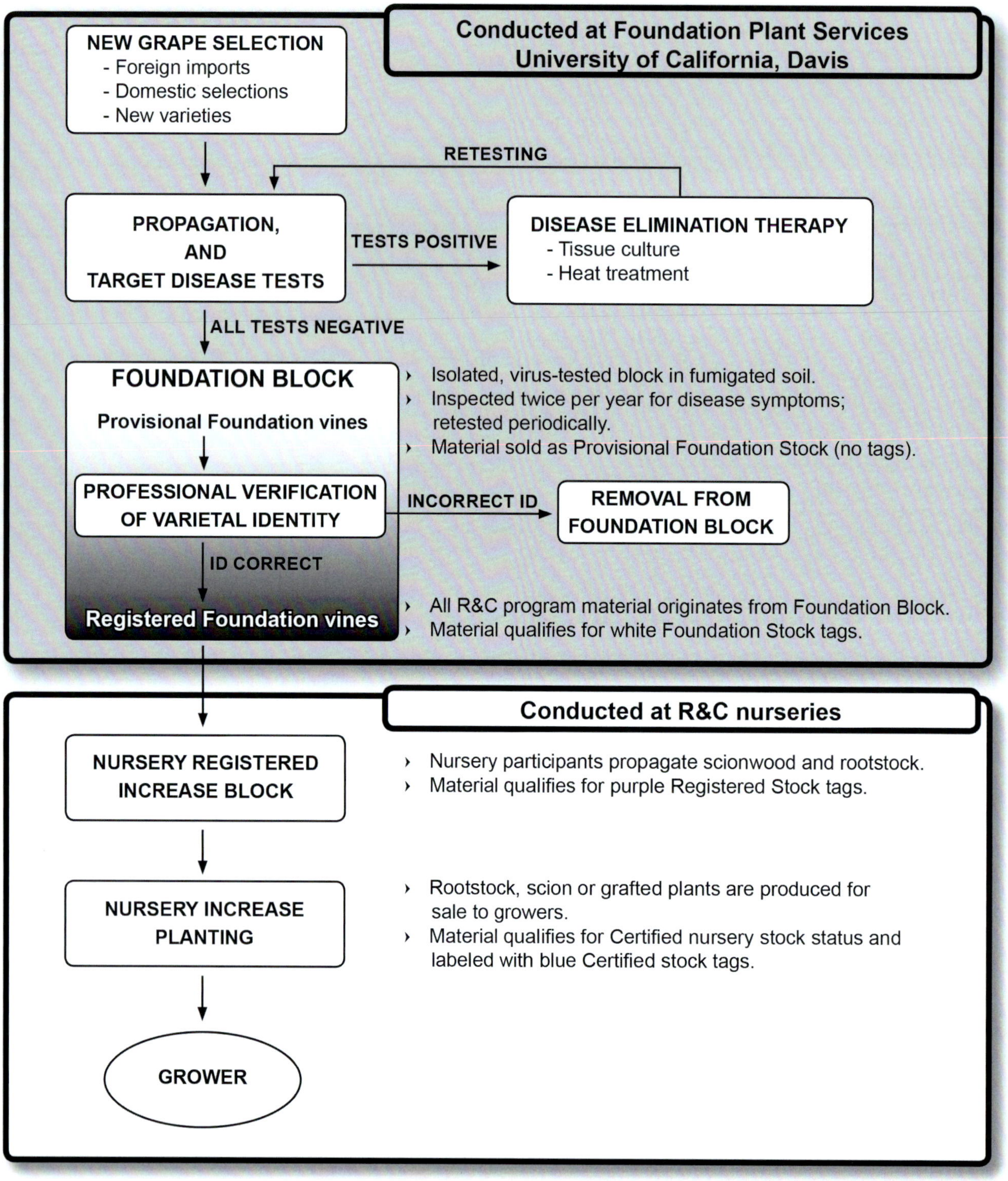

Fig. 375. Flowchart of the California Grapevine Registration and Certification (R&C) program. (Courtesy D. Golino)

sanitary conditions grew stronger in the 1960s as a consequence of two events: first, the development of new knowledge on the viral nature of some serious disorders and attendant potential to procure virus-free shoot tips from infected plants through thermotherapy; and second, the alarming sanitary deterioration of the crop in many regions, which was in part counterbalanced by the programs of individual member states (France, Germany, and Italy). These events prompted the EU Council to issue directives for the improvement of the EU's viticulture industry. The first Directive 68/193 on the "Marketing of Vegetatively Propagated Material of Grapevines" categorized propagative materials as "basic," "certified," and "standard," and defined the sanitary characteristics of mother vineyards destined for their production.

Whereas participation in the California certification system and other North American programs is voluntary, in Europe it is mandatory. Even so, in the European program, the acceptance of the "standard" category allows for the propagation of materials lacking significant phytosanitary guarantees. In reality, the "standard" category was a consequence of the lack of clones suitable for certification when the program was initiated, and of the lengthy and costly procedures required to procure and increase such. However, the same directive contained the auspices for a progressive improvement of the sanitary sta-

TABLE 10. Tests Used to Qualify Vines for the National Clean Plant Network Foundation Block at Foundation Plant Services[a]

Group	Pathogen	Symbol	ELISA	qPCR	PCR	Herbaceous Indexing	Woody Indexing
Nepoviruses	*Grapevine fanleaf virus*	GFLV	√√	√√	√	√√	**St. George**
	Tomato ringspot virus	ToRSV	√√	√√	√	√√	
	Tobacco ringspot virus	TRSV		√√	√	√√	
	Arabis mosaic virus	ArMV	√√	√√	√	√√	
	Blueberry leaf mottle virus	BLMV		√√	√	√√	
	Raspberry ringspot virus	RpRSV		√√	√	√√	
	Tomato black ring virus	TBRV		√√	√	√√	
	Grapevine deformation virus	GDefV		√√	√	√√	
	Artichoke Italian latent virus	AILV			√	√√	
Sadwavirus	*Strawberry latent ringspot virus*	SLRSV		√√	√	√√	
Closterovirids	*Grapevine leafroll-associated virus 1*	GLRaV-1	√	√√	√		**Cab. franc**
	Grapevine leafroll-associated virus 2	GLRaV-2	√	√√	√		**Cab. franc**
	Grapevine leafroll-associated virus 2 RG	GLRaV-2RG		√√	√		
	Grapevine leafroll-associated virus 3	GLRaV-3	√	√√	√		**Cab. franc**
	Grapevine leafroll-associated virus 4	GLRaV-4	√ gen.	√√	√		**Cab. franc**
	Grapevine leafroll-associated virus 4 strain 5	GLRaV-4 strain 5	√ gen.	√√	√		**Cab. franc**
	Grapevine leafroll-associated virus 4 strain 6	GLRaV-4 strain 6	√ gen.	√√	√		**Cab. franc**
	Grapevine leafroll-associated virus 4 strain 9	GLRaV-4 strain 9	√ gen.	√√	√		**Cab. franc**
	Grapevine leafroll-associated virus 4 strain De	GLRaV-4strain De		√√	√		**Cab. franc**
	Grapevine leafroll-associated virus 4 strain Pr	GLRaV-4 strain Pr			√√		**Cab. franc**
	Grapevine leafroll-associated virus 4 strain Car	GLRaV-4 strain Car	√ gen.	√√	√		**Cab. franc**
Vitiviruses	*Grapevine leafroll-associated virus 7*	GLRaV-7		√√	√		**Cab. franc**
	Grapevine virus A	GVA		√√	√		Kober 5BB
	Grapevine virus B	GVB		√√	√		**LN33**
	Grapevine virus D	GVD		√√	√		
	Grapevine virus E	GVE		√√			
	Grapevine virus F	GVF		√√			
Foveavirus	*Grapevine rupestris stem pitting-associated virus*	GRSPaV		√√	√		St. George
Maculaviruses	*Grapevine fleck virus*	GFkV	√√	√√	√		**St. George**
	Grapevine redglobe virus	GRGV			√√		
Marafiviruses	*Grapevine Syrah virus-1*	GSyV-1		√	√√		
	Grapevine vein feathering virus	GVFV			√√		
	Grapevine asteroid mosaic virus	GAMV			√√		
DNA Viruses	Grapevine red blotch-associated virus	GRBaV		√√	√		
	Grapevine vein clearing virus	GVCV			√√		
Phytoplasma	Universal detection	Phyto		√√	√		
Pierce's Disease	*Xylella fastidiosa*	PD		√√	√		

[a] √ = test is available; √√ = FPS primary test; √ gen. = ELISA using generic antibody which detects GLRaV-4 and GLRaV-4 strains 5, 6, 9, and Car in a single test; qPCR = real-time RT-PCR with cyber green or TaqMan probe; PCR for DNA viruses and RT-PCR for RNA viruses. Bold font face (Woody Indexing column) = FPS primary biological test.

tus toward production of "certified" materials. Enforcement toward that endeavour came in Directive 2002/11, issued to modify Directive 68/193, which abolished the production of rootstocks in the "standard" category. Directive 68/193, successively modified and integrated by Directives 71/140 and 77/629, was an innovative regulation of great social and economic importance because it included concepts for certification and sanitary status accepted as valid in all EU member states. It defines the sanitary requirement of current EU certification as follows: "In the vineyards producing basic material, harmful virus diseases, notably fanleaf and leafroll, must be eliminated. Vineyards producing materials of other categories must be kept free from plants showing symptoms of virus diseases." By-laws generated by these Directives were promulgated in EU member states and national certification schemes were implemented.

Directive 2005/43/EC further modified Directive 68/193 to state that the presence of harmful organisms, which reduce the usefulness of the propagation material, shall be at the lowest possible level. This "lowest possible level" denotes the absence of: (i) infectious degeneration (GFLV and ArMV); (ii) leafroll (GLRaV-1 and -3); and (iii) GFkV (only for rootstocks). Besides the bizarre tolerance of the presence of GFkV in the scions, apparently ignoring the possibility of the virus translocating to the rootstock of a grafted vine, no mention was made of *Grapevine leafroll-associated virus 2* (GLRaV-2) and its RG strain, which is unanimously recognized as most insidious inducer of graft incompatibility in grapevines, nor of any of the viruses of the rugose wood complex.

The sanitary provisions of the EU directives fail to ensure an acceptable sanitary status of propagative material in any category. Furthermore, they also fail to account for recent scientific discoveries in grapevine virology, where much progress has been made in better understanding the etiology, epidemiology, and diagnosis of grapevine viruses and viruslike diseases. In an attempt to address this issue, some European countries have enacted their own measures. Although national certification schemes implemented in EU member states are inspired by and more or less conform to EU directives, they vary. Some differences pertain to the viticultural aspects of clonal selection. In Germany, for example, the selection process requires a longer time period compared with France and Italy. Additional differences reside in sanitary requirements which, in most instances, are stricter than those in the EU directives. For example, in 2008 an association of Italian grapevine nurserymen and breeders (Moltiplicatori Italiani Viticoli Associati, MIVA) signed an agreement, endorsed by the Ministry of Agriculture, Food, and Forestry Policies, whereby GLRaV-2 and the rugose wood-associated agents *Grapevine virus A* (GVA) and *Grapevine virus B* (GVB) were added to the list of viruses whose absence from nursery productions must be certified. Portugal and France also require all selections to test negative for rugose wood diseases, but this is not required in Germany. In Spain, assays are limited to *Rupestris* stem pitting and corky bark. And, contrary to all other countries, France and Germany require the absence of fleck in the rootstocks, but not in scion cultivars.

In addition to phytosanitary requirements, EU registration and certification of source materials require identification as true-to-type and verified as to clonal origin. The identification of clones is regulated by EU Directive 72/169 and outlined in a Resolution of the Office International de la Vigne et du Vin. Thus, in the EU, clonal and sanitary selection involves interdisciplinary activity requiring the joint effort of viticulturists, virologists, and wine-grape technologists.

Although other differences exist among programs relative to the maintenance, propagation, and distribution of certified materials, certain steps are commonplace to all schemes. Registered clones (primary sources or nuclear stocks) are maintained by the organization or individual who owns them ("obtenteur" or conservation breeder) and undergo an initial multiplication in specialized facilities. The propagating material is distributed subsequently to nurseries for the establishment of certified mother blocks and used for the production of certified budwood, rooted cuttings, or grafted plants for commercial plantings. An officially authorized organization is responsible for the sanitary status, the origin of materials, and quantities of certified plants produced prior to the issuance of certification labels.

Regional and Global Regulation

In 1995, the World Trade Organization (WTO) was formed and member nations agreed to honor its charter. Included in these international codes is an Agreement on the Application of Sanitary and Phytosanitary Measures, known as the SPS Agreement. This agreement contains provisions allowing member nations to enforce regulations that might otherwise violate the terms of free trade when those provisions are "necessary to protect human, animal, or plant life or health." It is under these provisions that most national importation and quarantine regulations are allowed. It is incumbent on member states to ensure that these restrictions are not disguised trade barriers. The WTO administers disputes and provides resolutions in cases where member countries feel that the SPS agreement is being used unfairly or arbitrarily.

The WTO/SPS authorizes the setting of international trade standards. In cases involving plant health, the Food and Agriculture Organization (FAO) of the United Nations is the responsible organization, working under the terms of the International Plant Protection Convention (IPPC), a treaty signed by 107 contracting countries. At this time, the process involves establishing guidelines and definitions, as well as coordinating efforts of regional plant protection organizations to establish consistent regional standards. The regional plant protection organizations are being asked to make the first efforts to harmonize standards, because geographically contiguous areas often share common exotic pest and pathogen concerns and frequently work in concert to establish both internal and external control programs.

According to international guidelines, the national regulation programs—either through mandatory certification programs or official control programs for target diseases for each commodity—could allow classification of endemic economically important diseases as regulated nonquarantine pests. Regional, national, state, or local regulations might also serve this purpose. By establishing strict regulations, only imported nursery stock meeting high standards of freedom from specific domestic diseases could enter the regulated area.

North America. The North American Plant Protection Organization (NAPPO) is a regional organization with members from the national plant protection organizations of Canada, the United States, and Mexico. It is one of many regional organizations whose primary responsibility is to develop regional plant protection standards designed to protect the member states from the entry and establishment of pests, while facilitating trade. In addition, NAPPO participates with other regional plant protection groups within both the Western Hemisphere and at the global level to develop international standards. "Guidelines for the Importation of Grapevines into a NAPPO Member Country, RSPM #15, Part 1: Viruses and Virus-like Pests, Viroids, Phytoplasmas and Bacteria" is a document developed by a committee of experts and signed on 20 October 2002. This document serves as the initial regional guideline for the development of harmonized North American Standards for grapevine nursery stock.

Canada has a formal national certification program for grapevines, which is voluntary. The Canadian Plant Protection Export Certification Program (PPECP) for grapevine nursery stocks, established in 1997, was developed primarily for meeting the requirements of foreign importers of Canadian material. The plant material produced under this scheme can also be used within Canada. Nursery stocks accepted for PPECP purposes belong to one of three categories (elite, foundation, certified) derived from nuclear stocks, defined as: "Original mother plants tested in Canada for the viruses of concern by the Centre for Plant Health (Sidney, BC) of the Canadian Food Inspection Agency (CFIA) or a laboratory approved by it." The pathogens certified absent under PPECP are viruses (12 nepoviruses plus 10 "graft-transmissible agents," which include two leafroll-associated viruses [GLRaV-1 and GLRaV-3] and some unspecified rugose wood-associated agents), four phytoplasmas, two bacteria (*Xylella fastidiosa* and *Xanthomonas ampelinus*), and eight fungi.

Mexico has no national grape certification program, but the majority of grape nursery stock comes from California and must meet California certification standards to be imported. The U.S. operates under a system of voluntary state certification programs which, combined with strict quarantine regulations, have resulted in high-quality grape nursery stock with a minimum of regulatory infrastructure. There is no national standard for grapevine nursery stock, although a voluntary national standard is being developed under the National Clean Plant Network organized in 2009. However, as regional organizations like NAPPO and international agencies like the FAO work to harmonize standards for the movement of plant materials internationally, a more formal and coordinated national and international system is being discussed within the U.S., to insure that the industry is protected from nonquarantined but damaging diseases that can be carried in planting stocks.

Argentina and Chile. Certification procedures with the characteristics of clean stock programs were established in Argentina in 2001 and in Chile in 2007. In both countries, nuclear stocks, which do not undergo clonal selection, are certified for the absence of a limited number of viruses. In Argentina, these are GFLV; GLRaV-1, -2, and -3; GFkV; rugose wood; vein mosaic; and vein necrosis. In Chile, they are GFLV; GLRaV-1, -2, and -3; GVA; and GVB.

South Africa. A Vine Improvement Association was established in 1986 with the mission to promote the interests of the South African wine industry through plant improvement and certification of vine propagation material. In 1992, a certification scheme for wine grapes was licensed under the Plant Improvement Act No. 53 of 1976 of the South African Department of Agriculture, Forestry, and Fisheries. The South African certification protocol encompasses clonal selection and requires, for both rootstocks and scions, freedom from the following virus diseases: fanleaf, fleck, leafroll, corky bark, stem pitting/grooving, and "Shiraz disease" (no known etiologic agent). It also requires visual freedom from crown gall (*Agrobacterium vitis*) and bacterial blight (*Xylophilus ampelinus*), *Pythium* spp.,

Phythophthora spp., and a group of nematodes and insects, among which are the pseudococcid mealybugs (*Pseudococcus longispinus* and *Planococcus ficus*), known vectors of some viruses associated with leafroll and rugose wood.

New Zealand and Australia. Grapevine certification is a private undertaking in these countries. In New Zealand, the certification scheme is supported by the New Zealand Winegrowers Board, which has developed a Grafted Grapevine Standard and an associated certification program with the objective of minimizing the chances that infected material be released to the industry. Certified mother plants are true-to-type, clonally selected, and free from GLRaV-3, the only virus taken into account by the certification protocol.

The Australian Standard for Grapevine propagation material is a clean stock program announced by the nongovernment organization Standards Australia with the publication "AS 5588-2013 Grapevine propagation material." As specified, "The objective of the standard is to provide the purchaser of grapevine propagation material with assurance as to the origin, varietal identity, physical specifications, and health status of grapevine propagation material." To meet the requirements of the standard, vines must undergo hot-water treatment targeting the agent of crown gall (*A. vitis*) and some fungi of trunk diseases, and must test negative for GLRaV-1, GLRaV-3, GVA, and GVB.

The European Union. The EU adopted a very strict regulation concerning the introduction of grapevine germplasm from areas outside it. According to Directive 2000/68, the importation of plants or plant parts (except fruits) of *Vitis* spp. is forbidden. However, under certain conditions, member states can allow the importation of germplasm after its testing and indexing under strict quarantine conditions. For example, Italian legislation (DM 31.01.1996), in agreement with the European directives, requires the following procedure concerning *Blueberry leaf mottle virus*, the grapevine yellows phytoplasmas (including that causing *flavescence dorée*), *Peach rosette mosaic virus*, *Tobacco ringspot virus*, *Tomato ringspot virus*, *Xylella fastidiosa*, *Xylophilus ampelinus*, Ajinashika disease, Grapevine stunt, and Summer mottle. First, therapeutic treatments must be completed according to the technical guidelines FAO/IBPGR (International Board for Plant Genetic Resources), to be followed by indexing and laboratory analysis in proper facilities.

Current Trends, Future Prospects and Challenges

The high value of grapevine plant materials, the competitive nature of the grapevine nursery business, and rapid globalization of the grapevine and wine community have resulted in some fundamental changes in the availability of clean grapevine selections, cultivars, and clones. Recent years have seen a worldwide increase in patented grapevine cultivars and rootstock selections, the introduction of trademarked clones of traditional cultivars, and the development of proprietary programs in which valuable selections are marketed exclusively and cannot be obtained by all growers. Some sales contracts for grapevine nursery stock even limit the rights of growers to subsequently propagate selections. Governments are having increasing difficulties funding public programs for importation, certification, and distribution of grapevine stock, making the few remaining programs increasingly international in influence. The technology involved in both grapevine identification and plant disease detection has changed radically in the past decade and is likely to continue doing so. All of these factors will have a profound effect on the future of international grapevine quarantine, clean stock, and certification programs.

Continuing work on the development of harmonized international standards for grapevine nursery stock is likely to increase trade among the grape-growing regions of the world. This could result in more open competition for the nursery industries and lower prices for growers, but also increases the potential for importation of damaging pests and diseases and a resulting degradation of crop quality and productivity. As efforts are made to harmonize grapevine certification protocols, high standards are essential to ensure that no viticultural region is compromised by the introduction and spread of diseases. In the United States, there is a now a National Clean Plant Network for Grapevines that has the responsibility to produce the top tier (G1) plants that would serve as the starting material for state certification programs. There is also an effort to develop a grapevine certification standard that would be agreed to by all states producing certified stocks, essentially creating a national standard. These efforts include an attempt to harmonize certification standards with those of trading partners (e.g., EPPO and/or NAPPO standards) where possible. Within this context lies the challenge of determining which microbes are relevant and actionable from a regulatory point of view, in order to maintain the delicate balance of protecting local and national industries from introduced pests and diseases while reaping the benefits of expanded trade.

Selected References

Alley, L., and Golino, D. 2001. The origins of the grape program at Foundation Plant Materials Service. Pages 222-230 in: Proc. ASEV 50th Anniversary Annual Meeting. ASEV Publications, Davis, CA.

Anonymous, 1991. Programme type pour la réalisation de la sélection clonale de la vigne. Le Bulletin de l'OIV 64:715-716.

Anonymous. 1994. Pathogen-tested material of grapevine varieties and rootstocks. OEPP/EPPO Bull. 24:347-367.

Boidron, R. 1995. Clonal selection in France: Methods, organization and use. Pages 1-7 in: International Symposium on Clonal Selection. ASEV Publications, Davis, CA.

Bovey, R. 1981. Aspects de la sélection sanitaire de la vigne. Pages 293-301 in: 3rd International Symposium on Clonal Selection in Vines. CNR, Conegliano, Italy.

Frison, E. A., and Ikin. R., eds. 1991. Technical Guidelines for the Safe Movement of Grapevine Germplasm. International Plant Genetic Resources Institute (IPGRI), FAO, Rome.

Garnsey, S. M., and Cambra, M. 2003. Enzyme-linked immunosorbent assay (ELISA). Pages 169-192 in: Graft-transmissible Diseases of Grapevines: Handbook for Detection and Diagnosis. G. P. Martelli, ed. FAO Press, Rome.

Goheen, A. C. 1989. Virus diseases and grapevine selection. Am. J. Enol. Vitic. 40:67-72.

Golino, D. 2000. Trade in grapevine plant materials: Local, national and worldwide perspectives. Pages 216-221 in: Proc. ASEV 50th Anniversary Annual Meeting. ASEV Publications, Davis, CA

Golino, D. A. 2003. Emerging grapevine diseases. Pages 136-138 in: Extended Abstracts 14th Meeting, International Council for the Study of Virus and Virus-like Diseases of the Grapevine (ICVG). http://web.pppmb.cals.cornell.edu/fuchs/icvg/data/2003Session12.pdf.

Golino, D. A., and Wolpert, J. A. 2003. Vine selection and clones. Pages 8-11 in: Wine Grape Varieties in California. Publ. 3419. Division of Agriculture and Natural Resources, University of California, Berkeley.

Hewitt, W. B. 1951. Virus and virus-like grape diseases. Wines Vines 32:14-15.

Hewitt, W. B. 1968. Viruses and virus diseases of the grapevine. Rev. Appl. Mycol. 47:433-455.

Kreith, M., and Golino, D. A. 2003. Regulatory framework and institutional players. Pages 19-38 in: Exotic Pests and Diseases: Biology and Economics for Biosecurity. D. Sumner, ed. Iowa State Press, Ames.

Martelli, G. P. 1993. Use of herbaceous hosts. Pages 157-162 in: Graft-transmissible Diseases of Grapevines: Handbook for Detection and Diagnosis. G. P. Martelli, ed. FAO Press, Rome.

Martelli, G. P., Savino, V., and Walter, B. 1993. Indexing on *Vitis* indicators. Pages 137-155 in: Graft-transmissible Diseases of Grapevines: Handbook for Detection and Diagnosis. G. P. Martelli, ed. FAO Press, Rome.

Martelli, G. P., and Walter, B. 1998. Virus certification of grapevines. Pages 261-276 in: Plant Virus Disease Control. A. Hadidi, R. K. Ketarpal, and H. Koganezawa, eds. American Phytopathological Society, St. Paul, MN.

Rowhani, A., Uyemoto, J. K., Golino, D. A., and Martelli, G. P. 2005. Pathogen testing and certification of *Vitis* and *Prunus* species. Annu. Rev. Phytopathol. 43:261-278.

Walter, B. ed. 1997. Sanitary Selection of the Grapevine: Protocols for Detection of Viruses and Virus-like Diseases. Les Colloques 86, INRA Editions, Paris, France.

(Prepared by D. Golino, V. Savino, G. Martelli, and M. Fuchs)

Appendix I

Common Names and Causal Agents of Grapevine Diseases and Disorders

FUNGI AND OOMYCETES

Angular leaf scorch
Pseudopezicula tetraspora Korf, Pearson & Zhuang
Angular leaf spot
Mycosphaerella angulata Jenkins
(syn. *Cercospora brachypus* Ell. & Ev.)
Anthracnose (Bird's-eye rot, Black spot)
Elsinöe ampelina (de Bary) Shear
(syn. *E. viticola* Raciborski)
(syn. *Sphaceloma ampelinum* deBary)
Armillaria root rot (Oak root rot)
Armillaria mellea (Vahl:Fr.) P. Kumm.
Berry Rots (Miscellaneous) and Raisin Molds
Alternaria rot
Alternaria alternata (Fr.) Keissl.
(syn. *A. tenuis* Nees)
Aspergillus rot
Aspergillus niger van Tiegh.
Blue mold
Penicillium spp.
Cladosporium rot
Cladosporium herbarum (Pers.:Fr.)
Rhizopus rot
Rhizopus arrhizus Tesher
R. stolonifer (Ehrenb.:Fr.) Lind
Other berry rots and secondary invaders, raisin molds and rots
Alternaria geophila Deszew.
Ascochyta sp.
Aspergillus aculeatus Iizuka
Aspergillus flavus Link
Aspergillus ochraceous Wilhelm
Aspergillus wentii Wehmer
Candida sp.
Chaetomium elatum Kunze
Cladosporium cladosporioides (Fres.) de Vries
Cladosporium oxysporium Berk. & Curt.
Fusarium moniliforme Sheldon
Hormiscium sp.
Hormodendrum spp.
Monilia sp.
Penicillium brevicompactum Diercks.
Penicillium cyclopium West.
Penicillium frequentaus West.
Penicillium stoloniferum Thom
Saccharomyces cerevisiae Kreger-van Rij
Sclerotinia sclerotiorum (Lib.) de Bary
Stemphylium botryosum Walker
Torula sp.

Bitter rot
Greeneria uvicola (Berk. & M. A. Curtis) Punith.
(syn. *Melanconium fuligineum* (Scribn. & Viala) Cav.)
Black foot disease
Campylocarpon fasciculare Schroers, Halleen and Crous
Campylocarpon pseudofasciculare Halleen, Schroers and Crous
Cylindrocarpon destructans (Zinssm) Sholten
Cylindrocarpon liriodendri (Cooke & Harness) Wollew.
Cylindrocarpon macrodidymum Halleen, Schroers and Crous
Cylindrocarpon obtusisporum (Cooke & Harness)
Black rot
Guignardia bidwellii (Ellis) Viala and Ravaz
(syn. *Phyllosticta ampelicida* [Engleman] Van der Aa)
Botryosphaeria dieback (Black dead arm, Botryosphaeria canker, Diplodia cane dieback)
Botryosphaeria dothidea (Moug. : Fr.), Ces. & De Not.
(syn. Fusicoccum aesculi Corda)
Diplodia corticola A.J.L. Phillips, A. Alves & J. Luque
Diplodia mutila (Fr. : Fr.) Fr.
Diplodia seriata De Not.
Dothiorella americana J.R. Úrbez-Torres, F. Peduto & W.D. Gubler
Dothiorella iberica A.J.L. Phillips, J. Luque & A. Alves
Dothiorella neclivorem W.M. Pitt & J.R. Úrbez-Torres
Dothiorella vidmadera W.M. Pitt, J.R. Úrbez-Torres & F.P. Trouillas
Dothiorella vinea-gemmas W.M. Pitt & J.R. Úrbez-Torres
Lasiodiplodia crassispora T.I. Burgess & Barber
Lasiodiplodia missouriana J.R. Úrbez-Torres, F. Peduto & W.D. Gubler
Lasiodiplodia theobromae (Pat.) Griff. & Maubl.
Lasiodiplodia viticola J.R. Úrbez-Torres, F. Peduto & W.D. Gubler
Neofusicoccum australe (Slippers, Crous & M.J. Wingf.) Crous, Slippers & A.J.L. Phillips
Neofusicoccum luteum (Pennycook & Samuels) Crous, Slippers & A.J.L. Phillips
Neofusicoccum macroclavatum T.I. Burgess, Barber & Hardy
Neofusicoccum mediterraneum Crous, M.J. Wingf. & A.J.L. Phillips
Neofusicoccum parvum (Pennycook & Samuels) Crous, Slippers & A.J.L. Phillips
Neofusicoccum viticlavatum (Van Niekerk & Crous) Crous, Slippers & A.J.L. Phillips
Neofusicoccum vitifusiforme (Van Niekerk & Crous) Crous, Slippers & A.J.L. Phillips

Spencermartinsia viticola (A.J.L. Phillips & J. Luque) A.J.L. Phillips, A. Alves & Crous
Spencermartinsia westrale W.M. Pitt, J.R. Úrbez-Torres & F.P. Trouillas
Sphaeropsis porosa (Van Niekerk & Crous) A.J.L. Phillips & A. Alves
Botrytis bunch rot and blight
Botrytis cinerea Pers.
(syn. *Botryotinia fuckeliana* (*de Bary*) Whetzel)
Brulure
Anthostomella pullulans (de Bary) Bennett
Cladosporium leaf spot
Cladosporium viticola Cesati
Cercospora leaf spot
Phaeoramularia dissiliens (Duby) Deighton
Dematophora root rot
Rosellinia necatrix Prill.
(syn. *Dematophora necatrix* Hartig)
Downy mildew
Plasmopara viticola (Berk. & Curt.) Berl. & DeToni
Esca
Cause undetermined; associated with:
Fomitiporia punctate (Fr.) Murrill
(syn. *Fomitiporia mediterranea* M. Fischer)
(syn. *Phellinus punctatus* [P. Karst.])
(syn. *Stereum hirsutum* [Willd. :Fr.] Pers.)
other *Fomitiporia* spp. and *Phellinus* spp.
Esca proper = Esca + Grapevine leaf stripe disease
(see Grapevine leaf stripe disease)
Eutypa dieback
Eutypa lata (Pers.:Fr.) Tul. & C. Tul.
(syn. *E. armeniacae* Hansf. & Carter)
(syn. *Libertella blepharis* A. L. Smith [syn. *Cytosporina* sp.])
Eutypa sp.
Grapevine leaf rust
Phakopsora euvitis Y. Ono
(syn. *Aecidium meliosmae-myrianthae* P. Hennings & Shirai)
Grapevine leaf stripe disease
Phaeoacremonium aleophilum W. Gams, P.W. Crous, M.J. Wingfield &L. Mugnai
(syn. *Togninia minima* (Tul. & C. Tul.) Berl.)
Phaeoacremonium spp.
Phaeomoniella chlamydospora (W. Gams, Crous, M.J. Wingf. & Mugnai) Crous & W. Gams
Leaf blight (Pseudocercospora leaf blight, Isariopsis leaf spot)
Pseudocercospora vitis (Lév) Speg.
(syn. *Isariopsis clavispora* (Berk. & Curt.) Sacc.)
(syn. *Mycosphaerella personata* Higgins)
Leaf blotch
Briosia ampelophaga Cav.
Leaf spots (miscellaneous)
Asperisporium minutulum (Sacc.) Deighton
Phaeoramularia heterospora (Ell. & Gall.) Deighton
Macrophoma rot
Botryosphaeria dothidea (Moug. ex Fr.) Ces. & de Not.
(syn. *B. ribis* Grossenbacher & Duggar)
(syn. *Fusicoccum aesculi* = *Macrophoma* sp.)
Phomopsis cane and leaf spot
Phomopsis viticola (Sacc.) Sacc.
Phymatotrichopsis root rot (formerly, Phymatotrichum root rot; Texas root rot; cotton root rot)
Phymatotrichopsis omnivora [Duggar] Hennebert
(syn. *Phymatotrichum omnivorum* [Shear] Duggar)
(syn. *Ozonium omnivorum* Shear)
Phytophthora crown and root rot
Phytophthora cactorum (Leb. & Cohn) Schroet.
Phytophthora cinnamomi Rands
Phytophthora cryptogea Pethyb. & Laff.
Phytophthora drechsleri Tucker
Phytophthora parasitica Dastur
Powdery mildew
Erysiphe necator Schw.
(syn. *Uncinula necator* (Schw.) Burr)
(syn. *Oidium tuckeri* Berk.)
Ripe rot
Colletotrichum gloeosporioides (Penz.) Penz. & Sacc.
(syn. *Glomerella cingulata* (Stonem.) Spauld. & Schrenk)
Colletotrichum acutatum J.H. Simmonds
(syn. *Glomerella acutata* Guerber & Correll)
Rotbrenner
Pseudopezicula tracheiphila (Müll.-Thurg.) Korf & Zhuang
(syn. *Pseudopeziza tracheiphila* Müll.-Thurg)
(syn. *Phialophora tracheiphila* [Sacc. & Sacc.] Korf)
(syn. *Botrytis tracheiphila* Sacco & Sacc.)
Septoria leaf spot
Septoria ampelina Berk. & Curt.
Sour rot
Aspergillus niger Tiegh.
Aspergillus carbonarius (Bainier) Thom
Hanseniaspora uvarum (Niehaus) Shehata et al.
(syn. *Kloeckera apiculata* (Reess) Janke)
Candida krusei (Castellani) Berkhout
(syn. *Issatchenkia orientalis*)
Candida zemplinina (Kroemer & Krumbholz) Meyer & Yarrow
(syn. *C. stellata*)
Metschnikowia pulcherrima Miller and van Uden
Pichia kluyveri (Bedford)
Pichia membranaefaciens Hansen
Saccharomycopsis crataegensis Kurtzman and Wickerham
(syn. *S. vini, Endomycopsis vini*)
Acetobacter spp. (bacteria)
Gluconobacter spp. (bacteria)
Verticillium wilt
Verticillium dahliae Kleb.
White rot
Pilidiella diplodiella (Speg.) Crous & Van Niekerk
(syn. *Coniella diploidella* [Speg.] Petrak & Sidow)
(syn. *Coniothyrium diplodiella* [Speg.] Sacc.)
(syn. *Phoma diplodiella* Speg.)
Zonate leaf spot
Cristulariella moricola (Hino) Redhead
(syn. *C. pyramidalis* Waterman & Marshall)
(syn. *Sclerotium cinnamomi* Sawada)

BACTERIA

Bacterial blight
Xylophilus ampelinus (Panagopoulos 1969) Williams et al
(syn. *Xanthomonas ampelina* Panagopoulos)
Crown gall
Agrobacterium vitis (Ophel and Kerr 1990)
Pierce's disease
Xylella fastidiosa Wells et al. 1987

PHYTOPLASMAS

Australian grapevine yellows
Australian grapevine yellows phytoplasma ('*Candidatus* Phytoplasma australiense')
Tomato big bud phytoplasma ('*Candidatus* Phytoplasma australasia')
Buckland Valley grapevine yellows phytoplasma
Bois noir (= *Vergilbungskrankheit*)
Bois noir phytoplasma ('*Candidatus* Phytoplasma solani'-related strains; stolbur phytoplasmas group)

Flavescence dorée
Grapevine flavescence doree phytoplasma (‘*Candidatus* Phytoplasma vitis’-related strains; Elm yellows phytoplasmas group)
North American grapevine yellows
‘*Candidatus* Phytoplasma pruni’-related strains (peach X-disease phytoplasmas group)
‘*Candidatus* Phytoplasma asteris’-related strains (Aster yellows phytoplasmas group)
Palatinate grapevine yellows
Palatinate grapevine yellows phytoplasma (Elm yellows phytoplasmas group)

VIRUSES AND VIRUSLIKE AGENTS

Corky bark disease (see also Rugose Wood complex)
Genus *Vitivirus*, *Grapevine virus B* (GVB)
Enation disease
Unknown graft-transmissible agent
Fanleaf degeneration (see also Infectious degeneration)
Grapevine fanleaf virus (GFLV)
Fleck complex
Asteroid mosaic
Genus *Marafivirus*, Grapevine asteroid mosaic-associated virus (GAMaV)
Fleck
Genus *Maculavirus*, *Grapevine fleck virus* (GFkV)
Grapevine redglobe virus
Genus *Maculavirus*, Grapevine redglobe virus (GRGV)
Rupestris vein feathering
Genus *Marafivirus*, Grapevine rupestris vein feathering virus (GRVFV)
Grapevine angular mosaic
Grapevine angular mosaic virus (GAMV)
Grapevine berry inner necrosis
Genus *Trichovirus*, *Grapevine berry inner necrosis virus* (GBINV)
Grapevine leafroll disease
Genus *Ampelovirus*
Grapevine leafroll-associated virus 1 (GLRaV-1)
Grapevine leafroll-associated virus 3 (GLRaV-3)
Grapevine leafroll-associated virus 4 (GLRaV-4)
Genus *Closterovirus*, *Grapevine leafroll-associated virus 2* (GLRaV-2)
Genus *Velarivirus*, *Grapevine leafroll-associated virus 7* (GLRaV-7)
Grapevine graft incompatibility
Genus *Closterovirus*, *Grapevine leafroll-associated virus 2* strain Red Globe (GLRaV-2 RG)
Grapevine line pattern
Grapevine line pattern virus (GLPV)
Grapevine little berry
Genus *Nepovirus*, *Tomato ringspot virus* (ToRSV)
Grapevine red blotch
Grapevine red blotch-associated virus (GRBaV)
Grapevine Roditis leaf discoloration
Possible interaction of *Carnation mottle* virus (CarMV) and *Grapevine virus B*
Grapevine yellow mottle
Alfalfa mosaic virus (AMV)
Grapevine yellow speckle disease
Grapevine yellow speckle viroid 1 (GYSVd-1)
Grapevine yellow speckle viroid 2 (GYSVd-2)
Grapevine yellow vein
Genus *Nepovirus*, *Tomato ringspot virus* (ToRSV)
Infectious degeneration (Infectious malformation syndrome, Yellow mosaic syndrome)
Genus *Nepovirus*
Arabis mosaic virus (ArMV)
Cherry leafroll virus (CLRV)
Grapevine Anatolian ringspot virus (GARSV)
Grapevine chrome mosaic virus (GCMV)
Grapevine deformation virus (GDefV)
Grapevine fanleaf virus (GFLV)
Raspberry ringspot virus (RpRSV)
Tomato black ring virus (TBRV)
Strawberry latent ringspot virus (SLRSV)
Infectious malformation (see Infectious degeneration)
Kober stem grooving disease (see also Rugose Wood complex)
Genus *Vitivirus*, *Grapevine virus A* (GVA)
LN33 stem grooving disease (see Rugose Wood complex)
Rugose Wood complex (LN33 stem grooving disease; see also Corky bark disease, Kober stem grooving disease, and Rupestris stem pitting disease)
Genus *Vitivirus*
Grapevine virus A (GVA) (association)
Grapevine virus B (GVB) (association)
Grapevine virus D (GVD) (association)
Grapevine virus E (GVE) (association)
Grapevine virus F (GVF) (association)
Rupestris stem pitting disease
Genus *Foveavirus*, *Grapevine rupestris stem pitting-associated virus* (GRSPaV)
Vein banding
Interaction of *Grapevine yellow speckle viroid 1* or *-2* and *Grapevine fanleaf virus*
Vein mosaic
Unidentified graft-transmissible agent
Vein necrosis
Unidentified graft-transmissible agent (some genetic variants of *Grapevine rupestris stem pitting-associated virus* are associated)
Virus-induced grapevine decline
Genus *Nepovirus*
Peach rosette mosaic virus (PRMV)
Tobacco ringspot virus (TRSV)
Tomato ringspot virus (ToRSV)
Yellow mosaic (see Infectious degeneration)

NEMATODES (PARASITIC)

Citrus Nematode
Tylenchulus semipenetrans Cobb
Dagger Nematodes
Xiphinema americanum Cobb
Xiphinema diversicaudatum (Micoletzky) Thorne
Xiphinema index Thorne & Allen
Xiphinema vuittenezi Luc, Lima, Weischer & Flegg
Xiphenema spp.
Needle Nematodes
Longidorus africanus Merny
Longidorus spp. (pathogenicity not determined)
Ring Nematode
Mesocriconema xenoplax (Raski) Loof & de Grisse (syn. *Criconemoides xenoplax* Raski)
Root-knot Nematodes
Meloidogyne arenaria (Neal) Chitwood
Meloidogyne hapla Chitwood
Meloidogyne incognita (Kofoid & White) Chitwood
Meloidogyne javanica (Treub) Chitwood
Meloidogyne spp.
Root-lesion Nematodes
Pratylenchus brachyurus (Godfrey) Filipjev & Schuurmans Stekhoven
Pratylenchus vulnus Allen & Jensen
Pratylenchus zeae Graham
Pratylenchus spp.

ABIOTIC DISORDERS

Chimeras (sports)
- Spontaneous genetic change in meristematic growing points and their subsequent proliferation
  - Variegation
  - Fasciation
  - Witches' broom
  - Leaf distortion

Environmental Stress
- Drought, excess water, hail, heat, lightning, salt toxicity, spring freeze, wind and sand, winter injury

Nutritional disorders
- Acidic soil sickness (*Säureschäden*)
  - Deficiencies of calcium, magnesium, and phosphorous; excess potassium, manganese, and aluminum
- Boron deficiency
- Boron toxicity
- Calcium deficiency
- Iron deficiency (iron chlorosis, lime chlorosis, lime-induced chlorosis)
- Magnesium deficiency
- Manganese deficiency
- Manganese toxicity (see Acidic soil sickness)
- Nitrogen deficiency
- Nitrogen excess
- Phosphorous deficiency
- Potassium deficiency (black leaf)
- Zinc deficiency

Shriveled fruit
- Dehydration
- Bunch stem necrosis
  - Cause unknown; associated with nutrient supply or imbalances
- Sunburn
- Berry shrivel (sour shrivel, sugar accumulation disorder, *Traubenwelke*)
  - Cause unknown

Pesticide toxicity

Appendix II

Equivalent Names (English and Other Languages) of Grapevine Diseases and Disorders

Common Name Causal Factors	Other names in English	French	German	Italian	Spanish
Acidic soil sickness	Acid soil Sour soil	Sol acide	Saurer Boden Säureschäden	Danni da eccesso di acidità del terreno	Suelo ácido
Alternaria rot *Alternaria alternata*		Alternariose	Alternariafäule	Alternariosi	Pudrición por *Alternaria* Alternariosis
Angular leaf scorch *Pseudopezicula tetraspora*				Disseccamento angolare fogliare	Quemadura angular de la hoja
Angular leaf spot *Mycosphaerella angulata*				Maculatura angolare fogliare	Mancha angular de la hoja
Anthracnose *Elsinoë ampelina*	Bird's-eye rot Black spot	Anthracnose	Anthraknose Schwarzer Brenner	Antracnosi	Antracnosis Ojo de pájaro
Armillaria root rot *Armillaria mellea*	Oak root rot Mushroom root rot	Pourridié à armillaire Pourridié agaric	Wurzelfäule durch Hallimasch Heckenschwamm Honigpilz	Marciume radicale fibroso	Pudrición de raíces por *Armillaria*
Australian grapevine yellows Australian grapevine yellows phytoplasma ('*Candidatus* Phytoplasma australiense') Tomato big bud phytoplasma ('*Candidatus* Phytoplasma australasia') Buckland Valley grapevine yellows phytoplasma	Grapevine yellows				
Aspegillus rot *Aspergillus niger*				Marciume da *Aspergillus*	Aspergilosis Pudrición por *Aspergillus*
Bacterial blight *Xylophilus ampelinus* (*Xanthomonas ampelina*)	Bacterial necrosis	Nécrose bactérienne Maladie d'Oléron	Bakteriennekrose	Mal nero	Tizón bacteriano Necrosis bacteriana
Bitter rot *Greeneria uvicola* (*Melanconium fuligineum*)		Pourriture amère	Bitterfäule	Marciume amaro	Pudrición amarga
Black foot disease *Cylindrocarpon* spp. *Campylocarpon* spp.				Piede nero	Pie negro
Black leaf	Potassium deficiency			Carenza di potassio	Deficiencia de potasio Fiebre de primavera
Black rot *Guignardia bidwellii* (*Phyllosticta ampelicida*)		Black-rot Pourriture maculée	Schwarzfäule	Marciume nero	Pudrición negra

Common Name Causal Factors	Other names in English	French	German	Italian	Spanish
Blue mold *Penicillium* spp.	Penicillium rot	Pourriture amère Pourriture acide (sur grappe)	Grünfäule Speckfäule durch *Penicillium glaucum* und *Penicillium* spp.	Marciume verde-azzurro	Moho azul Pudrición por *Penicillium*
Bois noir Bois noir phytoplasma (‘*Candidatus* Phytoplasma solani’-related strains)	Black wood disease *Vergilbungskrankheit*	Bois noir	Vergilbungskrankheit Schwarzholzkrankheit	Legno nero	Madera negra
Botryosphaeria dieback *Botryosphaeria dothidea* (*Fusicoccum aesculi*) *Diplodia* spp. *Dothiorella* spp. *Lasiodiplodia* spp. *Neofusicoccum* spp. *Spencermartinsia* spp. *Sphaeropsis porosa*	Black dead arm Botryosphaeria canker Diplodia cane dieback			Deperimento da Botryosphaeria Braccio nero morto	Brazo muerto negro Cancrosis de la madera
Botrytis bunch rot and blight *Botrytis cinerea* (*Botryotinia fuckeliana*)	Gray mold	Pourriture grise	Botrytis-Traubenfäule und Botrytis-Triebfäule Graufäule	Muffa grigia	Pudrición gris de racimos y tizón de cañas Moho gris Piel suelta
Bunch stem necrosis Physiological disorder	Grape peduncle necrosis Stem necrosis Stalk necrosis Water berry	Dessîchement de la rafle Flétrissement pédonculaire	Stiellähme	Disseccamento del rachide da disordini fisiologici	Palo negro Necrosis del raquis Desecación del escobajo Baya acuosa
Cercospora leaf spot *Phaeoramularia dissiliens*		Cercosporiose		Cercosporiosi	Cercosporiosis Mancha foliar por *Cercospora*
Citrus nematode *Tylenchulus semipenetrans*		Nématode des citrus Nématode du dépérissement des agrumes	Citrus nematode	Nematode degli agrumi	Nemátodo de los cítricos
Cladosporium leaf spot *Cladosporium viticola*		Cladosporiose		Maculatura da *Cladosporium*	Cladosporiosis Moho verde Mancha foliar por *Cladosporium*
Cladosporium rot *Cladosporium herbarum*		Cladosporiose		Marciume da *Cladosporium*	Cladosporiosis Moho verde Pudrición por *Cladosporium*
Corky bark *Grapevine virus B*	Rugose wood complex	Ecorce liégeuse Maladie de l’écorce liégeuse	Korkrindenkrankheit der Rebe	Suberosi corticale	Corteza corchosa
Crown gall *Agrobacterium vitis*		Broussin parasitaire	Mauke Grind Krebs des Weinstocks	Tumore batterico	Agalla de la corona Tuberculosis
Dagger nematodes *Xiphinema* spp.		Nématode poignard Nématode à dague	Dolchnematode	Nematode daghiforme	Nemátodo daga
Dematophora root rot *Rosellinia* (*Dematophora*) *necatrix*	White root rot Rosellinia root rot	Pourridié de la vigne Aubernage Blanc de raciness Champignon blanc	Wurzelschimmel Wurzelfäule Wurzelpilze	Marciume radicale bianco	Pudrición blanca de raíces Pudrición lanosa de las raíces
Downy mildew *Plasmopara viticola*		Mildiou	Peronospora Falscher Mehltau Blattfallkrankheit	Peronospora	Mildiú Mildiú lanoso Mildeo velloso Niebla Añublo

Common Name Causal Factors	Other names in English	French	German	Italian	Spanish
Esca Cause undetermined; associated with *Fomitiporia punctata* (*Fomitoporia mediterranea, Phellinus punctatus, Stereum hirsutum*), other *Fomitiporia* and *Phellinus* spp.	Apoplexy	Esca Apoplexie	Apoplexie Schlaganfall der Rebe	Mal dell'esca	Yesca Esca
Esca proper (= Esca + Grapevine leaf stripe disease; see individual disease listing for causal factors)	Apoplexy Black measles Black wood streaking Esca Grapevine leaf stripe disease Leaf stripe disease (= Young esca) Measles Petri disease	Esca Apoplexie	Apoplexie Schlaganfall der Rebe	Mal dell'esca	Yesca Esca
Eutypa dieback *Eutypa lata (E. armeniacae), Eutypa* sp.	Dying arm (dead arm)	Eutypiose	Eutypose Holzhartfäule	Eutipiosi	Eutipiosis Brazo muerto
Fanleaf degeneration *Grapevine fanleaf virus*	Infectious degeneration	Court-noué Dégénérescence infectieuse	Reisigkrankheit der Rebe	Arricciamento Degenerazione infettiva	Hoja de abanico Virus del entrenudo corto infeccioso de la vid
Fasciation Chimera (genetic disorder)		Fasciation	Verbänderung Fasziation	Fasciazione (origine genetic)	Fasciación
Flavescence dorée Grapevine flavescence doree phytoplasma ('*Candidatus* Phytoplasma vitis'-related strains	Grapevine yellows	Flavescence dorée Maladie du Baco 22A	Goldgelbe Vergilbung der Rebe	Flavescenza dorata	Flavescencia dorada
Fleck *Grapevine fleck virus*		Marbrure	Marmorierung	Maculatura infettiva (Fleck)	Punteado
Grapevine leaf rust *Phakopsora euvitis*		Rouille de la vigne	Rebenrost	Ruggine	Roya
Grapevine leaf stripe disease *Phaemoniella chlamydospora, Phaeoacremonium aleophilum* (*Togninia minima*) *Phaeoacremonium* spp.	Black measles Black wood streaking Grapevine leaf stripe disease Leaf stripe disease (= Young esca) Measles Petri disease	Esca Apoplexie	Apoplexie Schlaganfall der Rebe	Mal dell'esca	Yesca Esca
Grapevine leafroll disease *Grapevine leafroll-associated virus-1, -2. -3, -4,* and *-7*	Leafroll	Enroulement	Blattrollkrankheit der Rebe	Accartocciamento fogliare	Enrollamiento foliar
Grapevine yellows Various phytoplasmas	Australian grapevine yellows *Bois nor* (*Vergilbungs-krankheit*) *Flavescence dorée* North American grapevine yellows Palatinate grapevine yellows	Panachure Jaunisse de la vigne	Vergilbungen	Giallume da fitoplasmi	Amarillez de la vid
Lime-induced chlorosis	Iron deficiency Lime chlorosis		Eisenmangelchlorose	Clorosi ferrica da eccesso di calcare	Chlorosis férrica, caliza
Kober stem grooving *Grapevine virus A*	Rugose wood complex	Cannelures sur Kober		Scanalatura del Kober (sindrome del legno riccio)	
Leaf blight *Pseudocercospora vitis* (*Isariopsis clavispora, Mycosphaerella personata*)	Isariopsis leaf spot Pseudocercospora leaf blight			Maculatura fogliare	Tizón foliar

Common Name Causal Factors	Other names in English	French	German	Italian	Spanish
Leaf blotch *Briosia ampelophaga*				Maculatura necrotica fogliare	Mancha foliar
Lesion nematodes *Pratylenchus* spp.		Nématode des lesions racinaires	Läsionenälchen	Nematode delle lesion	Nemátodo de las lesions
Macrophoma rot *Botryosphaeria dothidea (B. ribis; Fusicoccum aesculi = Macrophoma* sp.)	Botryosphaeria rot			Marciume da *Botryosphaeria*	Pudrición por *Botryosphaeria*
Needle nematodes *Longidorus* spp.		Nématode aiguille	Nadelälchen	Longidoride	Nemátodo aguja
North American grapevine yellows '*Candidatus* Phytoplasma pruni'-related strains '*Candidatus* Phytoplasma asteris'-related strains	Grapevine yellows				
Palatinate grapevine yellows Palatinate grapevine yellows phytoplasma	Grapevine yellows				
Phomopsis cane and leaf spot *Phomopsis viticola*	Phomopsis rachis and fruit rot Phomopsis canker	Excoriose	Phomopsis	Escoriosi	Escoriosis Mancha foliar y de sarmientos
Phymatotrichopsis root rot *Phymatotrichopsis omnivora* (*Ozonium omnivorum*)	Cotton root rot Ozonium root rot Phymatotrichum root rot Texas root rot			Marciume radicale da *Phymatotrichopsis*	Pudrición Texana
Phytophthora crown and root rot *Phytophthora* spp.	Root and crown rot	Pourriture du collet	Wurzelfäule Kragenfäule	Marciume radicale e del colletto da *Phytophthora* spp.	Pudrición del cuello
Pierce's disease *Xylella fastidiosa*		Maladie de Pierce	Pierce'sche Krankheit	Malattia di Pierce	Enfermedad de Pierce
Pin nematodes *Paratylenchus* spp.		Nématode épingle	Pinnematode	Nematode spilliforme	Nemátodo alfiler
Powdery mildew *Erysiphe (Uncinula) necator*	Oidium	Oïdium	Oidium Mehltau Schimmel Äscherich Echter Mehltau	Oidio Mal bianco	Cenicilla Oídio Mildeo polvoso Peste ceniza
Reniform nematode *Rotylenchulus* spp.		Nématode réniforme		Nematode reniforme	Nemátodo reniforme
Rhizopus rot *Rhizopus* spp.		Pourriture à *Rhizopus* Pourriture acide		Marciume nero	Pudrición por *Rhizopus* Moho sal y pimiento
Ring nematode *Criconemoides* (*Mesocriconema*) *xenoplax*			Ringnematode	Nematode ad anelli	Nemátodo anillado
Ripe rot *Colletotrichum gloeosporioides* (*Glomerella cingulata*), *Colletotrichum* (*Glomerella*) *acutatum*	Improperly referenced as "anthracnose" in some international English-language literature			Antracnosi	Antracnosis
Root-knot nematodes *Meloidogyne* spp.		Nématode à galles Nématode cécidogène Nématode des racines noueuses	Wurzelgallenälchen	Nematode galligeno	Nemátodo del nudo Nemátodo agallador Nemátodo nodulador
Rotbrenner *Pseudopezicula* (*Pseudopeziza*) *tracheiphila*		Rougeot parasitaire Brenner	Roter Brenner Sang Brand	Rossore parassitario	Quemadura roja
Rupestris speckle Physiological disorder		Moucheture du rupestris		Picchettatura fogliare di *Vitis rupestris*	Manchado rupestris

Common Name Causal Factors	Other names in English	French	German	Italian	Spanish
Rupestris stem pitting *Grapevine rupestris stem pitting-associated virus*	Rugose wood complex	Maladie du bois strié Cannelures sur Rupestris		Butteratura del tronco di *Vitis rupestris* (sindrome del legno riccio)	
Septoria leaf spot *Septoria ampelina*	Melanose	Mélanose infectieuse	Melanose	Septoriosi	Septoriosis Mancha foliar por *Septoria*
Sour rot Acetic acid bacteria (*Acetobacter* spp., *Gluconobacter* spp.) *Hanseniaspora uvarum* (*Kloeckera apiculata*) and other yeast species *Aspergillus* spp.		Pourriture acide	Sauerfäule	Marciume acido	Pudrición ácida
Spiral nematodes *Helicotylenchus* spp.		Nématode spiralé	Spiralälchen	Nematode a spirale	Nemátodo espiral
Stubby-root nematode *Paratrichodorus christiei*		Rattle Nématode des racines tronquées	Borstenwurzelälchen	Nematode delle radici tronche	Nemátodo atrofiador de raíces
Stunt nematodes *Tylenchorhynchus* spp.		Nématode du rabougrissement			Nemátodo atrofiador
Tar spot *Rhytisma vitis*				Croste nere	Mancha de chapopote
Vein mosaic Some strains of *Grapevine rupestris stem pitting-associated virus*		Mosaïque des nervures	Adernmosaik	Mosaico delle nervature	Mosaico de las nervaduras
Verticillium wilt *Verticillium dahliae*		Flétrissement à *Verticillium* Verticilliose	Verticillium-Welke	Verticilliosi	Verticilosis Amarillez foliar Marchitamiento por *Verticillium*
Virus-induced grapevine decline					
Tomato ringspot virus	Grape yellow vein Grapevine little berry	Taches annulaires de la tomate	Tomatenringflecken	Maculatura anulare del pomodoro	Decaimiento por el virus de la mancha anular del tomate
Tobacco ringspot virus *Peach rosette mosaic virus*	Grapevine decline	Taches annulaires du tabac	Tabakringflecken	Maculatura anulare del tabacco	Decaimiento por el virus de la mancha anular del tabaco
White rot *Pilidiella (Coniella, Coniothyrium, Phoma) diplodiella*		Rot blanc Coître	Weissfäule	Disseccamento parassitario del rachide Carie bianca Marciume bianco degli acini	Pudrición blanca por *Coniella*
Yellow speckle *Grapevine yellow speckle viroid-1* and *-2*		Moucheture jaune	Gelbsprenkelung	Picchettatura gialla	Moteado amarillo
Zonate leaf spot *Cristulariella moricola*	Target spot			Maculatura zonata	Mancha foliar zonada

(Prepared by M. Fuchs, G. Hill, P. Cortesi, and B. Latorre)

Glossary

A—acre
a.i.—active ingredient
C—Celsius or centigrade
cm—centimeter (1 cm = 0.01 m = 0.3937 in.)
F—Fahrenheit
ft-c—foot-candle
g—gram (1 g = 0.03527 oz)
gal—gallon (1 gal liquid [U.S.] = 3.785 L)
h—hour
ha—hectare (1 ha = 2.471 acres)
in—inch (1 in. = 2.540 cm)
kg—kilogram (1 kg = 2.205 lb)
L—liter (1 L = 1.057 quarts liquid [U.S.])
lb—pound (1 lb = 453.59 g)
m—meter (1 m = 39.37 in.)
mg—milligram (1 mg = 0.001 g)
min—minute
ml—milliliter (1 ml = 0.001 L)
mm—millimeter (1 mm = 0.001 m = 0.03937 in.)
µg—microgram (1 µg = 10^{-6} g)
µm—micrometer (1 µg = 10^{-6} m)
nm—nanometer (1 nm = 10^{-9} m)
oz—ounce (1 oz = 28.35 g); fluid ounce (1 fl oz [U.S.] = 29.57 ml)
ppm—parts per million (= mg/L, µg/ml)
sec—second

abaxial—directed away from the stem of a plant; pertaining to the lower surface of a leaf (*see* adaxial)
abiotic—pertaining to the absence of life, as diseases not caused by living organisms
abscise—to separate from a plant, as leaves, flowers, and fruits do when they fall
abscission—the shedding of leaves or other plant parts as the result of physical weakness in a specialized layer of cells (the abscission layer), which develops at the base of the structure
acervulus (pl. acervuli)—erumpent, cushionlike fruiting body bearing conidiophores, conidia, and sometimes setae
acid—having a pH less than 7
acid precipitation—precipitation of low pH because of the presence of nitric and sulfuric acid formed by the combination of air pollutants (NO_x and SO_2) with water
acid rain—*see* acid precipitation
acropetal—upward from the base to the apex of a shoot of a plant (*see* basipetal); in fungi, producing spores in succession in the direction of the apex, so that the apical spore is the youngest
acute—with reference to symptoms, developing suddenly (*see* chronic)
adaxial—directed toward the stem of a plant; pertaining to the upper surface of a leaf (*see* abaxial)
adventitious—arising from other than the usual place, as roots from a stem rather than branches of a root
aeciospore—a dikaryotic spore produced in the aecium of a rust fungus; in heteroecious rust fungi, the spore stage that infects the alternate host
aecium (pl. aecia; adj. aecial)—the fruiting body of a rust fungus in which the first dikaryotic spores (aeciospores) are produced
aerial—occurring in air
aerobic—living only in the presence of oxygen
aflatoxin—a chemical by-product of *Aspergillus flavus* and *A. parasiticus,* harmful to humans and other animals
agar—a jellylike material derived from algae and used to solidify liquid culture media; a medium containing agar
alate—winged (syn. apterous)
albino (n. albinism)—white or light-colored; having a marked deficiency in pigmentation
alkaline—having basic (nonacidic) properties; having a pH greater than 7
alkaloid—any of various nitrogen-containing ring compounds produced by plants and having physiological effects in animals
allele—any of one or more alternative forms of a gene
allelopathy (adj. allelopathic)—the ability of one species to inhibit or prevent the growth of another species through the production of one or more toxic substances
alternate host—one of two kinds of plant on which a parasitic fungus (e.g., a rust fungus) must develop to complete its life cycle
alternative host—a plant other than the main host that can be colonized by a parasite, but which is not required for completion of the developmental cycle of the parasite
amphigynous—having the antheridium surrounding and penetrated by the ooginial stalk (*see* oogonium)
anaerobic—living in the absence of oxygen
anamorph (adj. anamorphic)—the imperfect state, or asexual form in the life cycle of certain fungi, producing asexual spores (e.g., conidia) or no spores (*see* holomorph; teleomorph)
anastomosis (pl. anastomoses)—fusion of branches of the same or different structures (e.g., hyphae) to make a network
annual—a plant that completes its life cycle and dies within one year (*see* biennial; perennial)
antagonist—an organism or substance that counteracts or limits the action of another
anterior—situated toward the front or head (*see* posterior)
anther—the pollen-bearing portion of a flower
antheridium (pl. antheridia)—the male sexual organ (male gametangium) of certain fungi
anthesis—the period of the opening of a flower, during which pollination can occur
anthracnose—disease caused by acervuli-forming fungi (in the order Melanconiales), characterized by sunken lesions and necrosis
antibiotic—any of various chemical compounds produced by microorganisms and killing or inhibiting the growth of other living organisms
antibody—a protein formed in the blood of warm-blooded animals in response to the presence of an antigen
antigen—any foreign chemical (normally a protein) that induces antibody formation in warm-blooded animals
antiserum (pl. antisera)—blood serum containing an antibody
apex (pl. apices; adj. apical)—the tip of a root or shoot, containing the apical meristem
aphid—any of numerous species of small, sucking insects of the family Aphididae (order Homoptera), which produce honeydew and injure plants when present in large populations

apical meristem—terminal zone at the tip of shoots and roots containing the apex proper and the youngest primordial organs (*see* apex)
aplerotic—with reference to oospores, not filling the oogonium
apothecium (pl. apothecia)—open, cuplike or saucerlike, ascus-bearing fruiting body (ascocarp) of ascomycetous fungi, often supported on a stalk
appressed—closely flattened down or pressed against a surface (syn. adpressed)
appressorium (pl. appressoria)—a swollen, flattened portion of a fungal filament that adheres to the surface of a higher plant, providing anchorage for invasion by the fungus
apterous—wingless (syn. alate)
arbuscule (adj. arbuscular)—a branched haustorial structure of certain endomycorrhizal fungi, formed within living cells of the root
arm—viticulturally, a branch more than one year old that is attached to a trunk and that bears canes or spurs
arthropod—any member of the phylum Arthropoda, which consists of animals with articulated bodies and limbs, including insects, arachnids, and crustaceans
ascocarp—sexual fruiting body (e.g., apothecium, ascostroma, chasmothecium, perithecium, pseudothecium) of ascomycetous fungi, producing asci and ascospores (syn. ascoma)
ascogenous—pertaining to ascus-producing hyphae
ascogonium (pl. ascogonia)—a specialized cell that gives rise to hyphae that produce asci
ascoma (pl. ascomata)—*see* ascocarp
Ascomycetes (adj. ascomycetous)—a class of fungi, also called the Ascomycotina, characterized by the production of sexual spores (ascospores) within an ascus
ascospore—a sexual spore borne in an ascus
ascostroma (pl. ascostromata)—a fruiting body of ascomycetous fungi containing bitunicate (double-walled) asci in locules (cavities), usually dark and containing multiple locules, but sometimes containing only a single locule (*see* pseudothecium)
ascus (pl. **asci**)—a saclike structure containing ascospores (typically eight) and usually borne in a fruiting body
aseptate—lacking cross-walls (septa); nonseptate; coenocytic
asexual—vegetative; lacking sex organs, gametes, or sexual spores (as in the imperfect or anamorphic stage of a fungus)
asexual reproduction—any type of reproduction not involving the union of gametes and meiosis
asexual stage—the imperfect (anamorphic) stage of certain fungi
attenuate—to narrow; to weaken; to decrease in virulence or pathogenicity
autoecious—with reference to rust fungi, producing all spore forms on one species of host plant (*see* heteroecious)
autotroph—any of various organisms (e.g., plants capable of photosynthesis) that synthesize their nutritive substances from inorganic molecules
avirulent—nonpathogenic, unable to cause disease (*see* virulent)
axenic—culture in the absence of living bacteria or other organisms; pure culture
axil—the angle formed by a leaf petiole and the stem
axillary—pertaining to or placed within an axil

bacilliform—shaped like a short rod with rounded ends
backcross—to cross (mate) an offspring with one of its parents
bactericide—a chemical or physical agent that kills bacteria
bacterium (pl. bacteria)—a prokaryotic, microscopic, single-celled organism having a cell wall and increasing by binary fission
bark—viticulturally, brown periderm (a protective tissue) or, more generally, all the tissues outside the cambium (including the phloem and periderm) on woody organs
basidiocarp—the sexual fruiting body of fungi in the class Basidiomycetes (syn. basidioma)
Basidiomycetes (adj. basidiomycetous)—a class of fungi, also called the Basidiomycotina, characterized by the formation of external basidiospores on basidia
basidiospore—a haploid (1*n*) sexual spore produced on a basidium
basidium (pl. basidia; adj. basidial)—a structure on which basidiospores are produced externally
basipetal—downward from the apex toward the base of a shoot of a plant; developing in the direction of the base, so that the apical part is oldest (*see* acropetal)
berm—in agricultural usage, a ridge of soil upon which a row of a perennial crop is established
berry—a soft, simple fleshy fruit derived from the ovary of a flower
bicellular—two-celled
biennial—a plant that produces seed and dies at the end of its second year of growth (*see* annual; perennial)
binucleate—having two nuclei per cell (*see* multinucleate; uninucleate)
bioassay—any test (assay) using a living organism
biocide—a compound toxic to all forms of life
biological control—exploitation of the natural competition, parasitism, or antagonism of organisms for the management of pests and pathogens (syn. biocontrol)
biotic—pertaining to living organisms; with reference to disease, caused by a living organism
biotroph—an organism that can live and multiply only on another living organism (syn. obligate parasite) (*see* necrotroph)
biotype—a subdivision of a species, subspecies, or race based on some identifiable physiological trait, such as a particular virulence pattern
bitunicate—having two walls, as for the asci of certain Ascomycete fungi
blade—the flat, expanded portion (lamina) of a leaf
blight—sudden, severe, and extensive spotting, discoloration, wilting, or destruction of leaves, flowers, stems, or entire plants
blotch—an irregularly shaped, usually superficial spot or blot
bloom—flowering, as indicated by the shedding of caps (calyptra); the waxy coating on grape berries that gives a frosted appearance, particularly on dark-colored varieties
botryose—shaped like a bunch of grapes
bract—a reduced leaf associated with a flower or inflorescence; a modified leaf from the axil of which a flower arises
bursa—extension or flap of cuticle at the side of the male nematode sex organ, used for orienting the body during mating

calcareous—rich in calcium carbonate (lime)
callose—a polysaccharide encrusting the sieve plates between conductive phloem cells, especially during the fall acclimation and winter periods
callus—specialized tissues that form over a wound or cut in a plant (cork cambium may form, and the cells produced will gradually seal the wound)
calyptra—*see* cap
calyx—outermost whorl of organs of a flower
cambium (pl. cambia)—sheath of meristematic cells separating the xylem and phloem in stems and roots, which divide to produce secondary xylem (to the inside) and phloem (to the outside)
cane—viticulturally, a mature woody shoot, which attains this designation from leaf fall its first year through its second year of growth
canker—a plant disease characterized in woody plants by the death of cambium tissue and malformation or loss of bark, typically associated with a slowly-expanding lesion; a lesion caused by such a disease
canopy—viticulturally, the mass of leaf-bearing shoots, measured in height, width, or density
cap—calyptra, or corolla, of a grape flower, formed by petals interlocking along their edges and at anthesis separating along their bases, thus falling as a unit
cap stem—*see* pedicel
capsid—the protective layer of protein surrounding the nucleic acid core of a virus; the protein molecules that make up this layer (syn. coat protein)
carbohydrate—any of various chemical compounds composed of carbon, hydrogen, and oxygen, such as sugars, starches, and cellulose
carpel—the ovule-bearing structure of a flower in angiosperms
catenulate—formed in a chain, with reference to the spores of certain fungi
causal agent—the organism or agent that produces a given disease
chasmothecium (pl. chasmothecia)--a spherical ascocarp that is closed at maturity, characteristic of powdery mildew fungi (formerly clestothecium, cleistothecia)
chimera, **chimaera**—a plant or an organ consisting of two or more genetically different tissues
chlamydospore—a thick-walled or double-walled, asexual resting spore formed from hyphal cells (terminal or intercalary) or by transformation of conidial cells, functioning as an overwintering stage

chlorophyll (adj. chlorophyllous)—any of a group of green pigments found in chloroplasts and critical to photosynthesis
chloroplast—a disk-like structure containing chlorophyll, in which photosynthesis occurs in green plants
chlorosis (adj. chlorotic)—failure of chlorophyll development, caused by disease or a nutritional disturbance; fading of green plant parts to light green, yellow, or white
chromosome—the structure that contains the genes of an organism (in eukaryotes, chromosomes are in the nucleus and can be visualized with an optical microscope as threads or rods during meiosis and mitosis; in bacteria, the chromosome is usually a single circle of DNA and cannot be visualized with an optical microscope)
chronic—with reference to symptoms, slow-developing, persistent, or recurring (*see* acute)
circulative—with reference to viruses, passing through the gut and circulating in the body of an insect vector before being transmitted to a host
circulative transmission—virus transmission in which the virus must accumulate within or pass through the lymphatic system of an insect vector before it can be transmitted to a plant
cirrhus (pl. cirrhi), cirrus (pl. cirri)—a curled, tendril-like mass of exuded spores, held together by a slimy matrix
clamp connection—a bridge- or buckle-like hyphal protrusion in basidiomycetous fungi, formed at cell division and connecting the newly divided cells
clavate—club-shaped (syn. claviform)
cleistothecium (pl. cleistothecia)—*see* chasmothecium
clone—vegetatively (asexually) propagated plant or group of such plants derived from a single original source plant and therefore representing the same genetic entity
closterovirids—long (1,400-2,200 nm) flexuous plant viruses
cluster—common name for the inflorescence and subsequent fruiting bunch of the grapevine
coalesce—to grow together into one body, as enlarging lesions may join together to form a single spot
coenocytic—having multiple nuclei embedded in cytoplasm without cross walls; nonseptate
collenchyma—a specialized tissue of living cells whose non-lignified cell walls are differentially thickened to provide organs mechanical support with great flexibility
colonize—with reference to plant pathogens, to infect and ramify through plant tissue
colony—a microorganism growing in mass, especially as a pure culture
compound winter bud—viticulturally, an overwintering bud at the node of a cane (also called an eye)
conidiogenous—producing and bearing conidia
conidiophore—a simple or branched hypha on which conidia are produced
conidiomata—fruiting structures within which conidia are produced, e.g., pycnidia and acervuli
conidium (pl. conidia)—an asexual, nonmotile fungal spore that develops externally or is liberated from the cell that formed it
cordon—heavy stem trained horizontally from the trunk and on which fruiting canes or their shortened spurs are positioned
coremium—*see* synema
cork—a waterproofing layer of cells produced by a cork cambium that originates in the outermost phloem in stems and in the pericycle tissue of roots
cortex (adj. cortical)—the region of parenchyma tissue between the epidermis and the phloem in stems and roots; the region beneath the rind of a sclerotium
cotyledon—the primary embryonic leaf within the seed, in which nutrients for the new plant are stored (monocots have one cotyledon; dicots have two)
coulure—excessive drop (shatter) of flowers and fruitlets
crawler—first-instar nymph of grape phylloxera, scale insects, and related groups
crown—the junction of the root and stem of a plant, usually at the soil line; in grafted woody plants, the rootstock portion of the plant near the soil surface
cultivar (abbr. **cv.**)—a cultivated variety of a plant species, resulting from deliberate genetic manipulation or selection and having recognizable characteristics such as color; shape of flowers, fruits, and seeds; height; and form (syn. variety)
cultural practices—the methods by which plants are grown (application of nutrients, irrigation, cultivation, and so forth)
culture—the growth and propagation of microorganisms on nutrient media; the growth and propagation of living plants
cuticle (adj. cuticular)—the water-repellent, waxy layer of epidermal cells of plant parts, such as leaves, stems, and fruit; the noncellular outer layer of an insect or a nematode
cutting—in grapevine propagation, a bud-bearing piece of shoot or cane induced to produce adventitious roots and develop new shoots
cymbiform—boat-shaped
cyst—in fungi, a resting structure in a protective membrane or shell-like enclosure; in nematodes, the egg-laden carcass of a female nematode; in bacteria, a specialized cell enclosed in a thick wall, often dormant and resistant to environmental conditions
cytoplasm—the living protoplasm in a cell, not including the nucleus

defoliation—loss of leaves from a plant, whether normal or premature
degree Brix—a unit of measurement representing the total soluble solids content of the juice in grapes (approximately, percent sugar)
dehiscent—opening by breaking into parts
deliquesce—to become liquid after maturing
demicyclic—pertaining to the life cycle of rust fungi (e.g., many species of *Gymnosporangium*) that lack the urediniospore (repeating) stage (*see* macrocyclic; microcyclic)
desiccate—to dry out
denticulation—small, tooth-like projection
diagnostic (n. diagnosis)—pertaining to a distinguishing characteristic important for the identification of a disease or disorder
diapause—a period of hormonally controlled quiescence, especially in immature insects
dichasium (pl. dichasia)—branch of a grape inflorescence in which a central flower is subtended by two equal lateral flowers or, if more complex, by two equal lateral flowering branches
dichotomous—branching, often successively, into two more or less equal arms
dicot—a dicotyledon, or flowering plant having two cotyledons (*see* monocot)
dieback (v. die back)—progressive death of shoots, leaves, or roots, beginning at the tip
differential host—a plant host that develops disease symptoms that distinguish between strains or races of a plant pathogen (syn. differential cultivar)
digitate—having lobes radiating from a common center
dikaryon (adj. dikaryotic)—an organism having two sexually compatible haploid nuclei per cell, which divide simultaneously, in the reproductive phase called dikaryophase
dilution end point—the stage of a serial dilution of cells or preparations at which growth or infection from a standard sample of the suspension no longer occurs
dimorphic—having two distinct shapes or forms
dioecious—having male (staminate) and female (pistillate) flowers on separate plants
diploid—having two complete sets of chromosomes ($2n$ chromosomes) (*see* haploid; polyploid)
disbudding—viticulturally, the removal of unwanted buds from a cutting or a graft
disease—abnormal functioning of an organism
disease cycle—the succession of events and interactions between a host, a parasite, and the environment, from initial infection through pathogenesis to overseasoning, until another infection occurs
disinfest—to kill pathogens that have not yet initiated disease or other contaminating microorganisms that occur in or on inanimate objects, such as soil or tools, or on the surface of plant parts, such as seed
dispersal—the spread of infectious material (inoculum) from diseased to healthy plants (syn. dissemination)
dissemination—*see* dispersal
distal—away from the point of attachment or origin or away from the main body (*see* proximal)
DNA—deoxyribonucleic acid, the double-stranded, helical molecule that contains genetic code information, consisting of repeating units, or nucleotides, each of which is composed of deoxyribose (a sugar), a phosphate group, and a purine (adenine or guanine) or a pyrimidine (thymine or cytosine) base

dolipore septum—a type of cross-wall in fungi in the class Basidiomycetes, characterized by distinctive swellings and membranes in association with the septal pore

dominant—pertaining to a phenotypic trait that is expressed in hybrid progeny of diploid organisms even if contributed by only one of the parents (*see* recessive)

dormancy (adj. dormant)—a condition of suspended growth and reduced metabolism of an organism, generally induced by internal factors or environmental conditions as a mechanism of survival

dormant—resting; living in a state of reduced physiological activity

dorsal—situated toward the back or top of the body (as opposed to **ventral**)

drop—viticulturally, the dropping of unpollinated flowers and fruitlets from a grape cluster, typically about one to two weeks after bloom (syn. shatter)

dwarfing—underdevelopment of a plant or plant organs caused by disease, inadequate nutrition, or unfavorable environmental conditions; growth control of a scion variety obtained by grafting onto certain rootstocks

echinulate—having small spines projecting from the cell wall

ecology—the study of the interactions between individual organisms, between groups of organisms, and between organisms and their environment

ectoparasite—a parasite that feeds from the exterior of its host (*see* endoparasite)

ectotrophic—pertaining to fungal development primarily over the root surface

edema, **oedema**—blisters produced on leaves or other plant parts under conditions of high moisture and restricted transpiration (syn. intumescence)

effuse (adj.)—stretched out, especially with reference to a film-like growth

egg mass—a group of eggs held together by a gelatinous matrix

ELISA—*see* enzyme-linked immunosorbent assay

ellipsoidal—elliptical in plane section

embryo—an organism in the early stages of development, such as a young plant in the seed or a nematode before hatching from the egg

embryo sac—central structure in an ovule of a flower, containing an egg cell and accessory cells that facilitate egg fertilization and embryo development

enation—an abnormal outgrowth from the surface of a stem or leaf

encapsidate—to enclose viral nucleic acid in a protein coat

encyst—to form a cyst or protective covering

endemic—native to a particular place; pertaining to a low and steady level of natural disease occurrence

endocarp—the inner layer of a fruit wall (*see* pericarp)

endoconidium (pl. endoconidia)—a conidium produced inside a hypha or conidiophore

endodermis—single cell layer making up a tissue surrounding the vascular core of young roots and which serves to regulate perfusion of water and mineral nutrients into the xylem

endogenous—arising from inside (*see* exogenous)

endoparasite—a parasite that lives and feeds inside its host (*see* **ectoparasite**)

enzyme—a protein that catalyzes a specific biochemical reaction

enzyme-linked immunosorbent assay (ELISA)—a serological test in which the sensitivity of the reaction is increased by attaching an enzyme that produces a colored product in response to one of the reactants

epidemic—an increase of disease in a population; a general and serious outbreak of disease (*see* epiphytotic)

epidemiology (adj. epidemiological)—the study of factors influencing the initiation, development, and spread of infectious disease; the study of disease in populations

epidermis (adj. epidermal)—the surface layer of cells of leaves and other plant parts

epinasty—abnormal downward curling of a leaf, leaf part, or stem

epiphyllous—located on the upper surface of a leaf

epiphyte—an organism growing on a plant surface, but not as a parasite

epiphytotic—widespread and destructive outbreak of a plant disease

eradicant—a chemical used to eliminate a pathogen from a host or an environment

eradication—control of plant disease by eliminating the pathogen after it has become established or by eliminating the plants that carry the pathogen

eriophyid—describing mites belonging to the family Eriophyidae, commonly known as gall, rust, bud, and blister mites

erumpent—bursting or erupting through the surface of a substrate

ethylene—a plant hormone influencing vegetative growth, fruit ripening, abscission of plant parts, and senescence of flowers

etiolation—elongation of stems caused by low levels of light intensity

etiology—the study of the causes of diseases

eukaryote (adj. eukaryotic)—any of various organisms whose cells contain a membrane-bound nucleus and other organelles, including all higher plants, animals, fungi, and protists (*see* prokaryote)

evapotranspiration—total amount of water removed from a cropping ecosystem through the combined activities of evaporation and transpiration

exogenous—originating from the outside (*see* endogenous)

extracellular—not occurring or located within a cell

exudate—a liquid excreted by or discharged from diseased tissues, roots, leaves, and fungi

eye—*see* compound winter bud

f. sp.—see *forma specialis,*

facultative—capable of changing lifestyle, e.g., from saprophytic to parasitic or the reverse

facultative parasite—an organism that is normally saprophytic but is capable of living as a parasite

facultative saprophyte—an organism that is normally parasitic but is capable of living as a saprophyte

fallow—cultivated land kept free from a crop or weeds during the normal growing season

fasciation—malformation of shoots or floral organs manifested as enlargement and flattening, as if several parts were fused

fascicle (adj. fasciculate)—a small group, bundle, or cluster

fastidious—with reference to prokaryotic organisms, having special requirements for growth and nutrition

feeder root—a fine root that absorbs water and dissolved nutrients

female—botanicaly, having flowers with pistils but no stamens

fiber—elongated, narrow, thick-walled cell in xylem or phloem

filament (adj. filamentous)—a thin, flexible, threadlike structure; the stalk-like, anther-bearing portion of a stamen

filamentous—threadlike (syn. filiform)

filiform—long, needlelike (syn. filamentous)

flaccid—wilted, lacking turgor

flagellum (pl. flagella)—a hairlike, whiplike, or tinsel-like appendage of a motile cell, bacterium, or zoospore, providing locomotion

fleck—minute discolored spot in green tissue

flexuous—having turns or windings; capable of bending

focus (pl. **foci**)—a small area of diseased plants within a larger plot or field

foliar—pertaining to leaves

frass—excrement of an insect

forma specialis (pl. *formae speciales*)—a taxonomic group within a plant-pathogenic fungus species, defined by its host range (members of different *formae speciales* infect different groups of plants), abbr. f. sp.

fructification—any of the various spore-bearing organs formed by macro- and microfungi

fruit set—stage of berry development immediately after pollination when the fertilized ovary begins to enlarge

fruiting body—any of the various complex, spore-bearing structures of fungi

fruiting zone—basal region of the shoot containing the flowering clusters and subsequent bunches of fruit

fumigant (v. fumigate)—a gas or volatile substance used to kill or inhibit the growth of microorganisms or other pests

Fungi Imperfecti—a group of fungi lacking a sexual stage; a group comprising the asexual stages of fungi in the classes Ascomycetes and Basidiomycetes (syn. Deuteromycetes)

fungicide (adj. fungicidal)—a chemical or physical agent that kills or inhibits the growth of fungi

fungistat (adj. fungistatic)—a compound that inhibits growth or sporulation of a fungus without killing it

fusiform—spindle-shaped, tapering at each end

gall—an abnormal swelling or localized outgrowth, often roughly spherical, produced by a plant as a result of attack by a fungus, bacterium, nematode, insect, or other organism (syn. tumor)
gametangium (pl. gametangia)—a cell containing gametes or nuclei that act as gametes
gamete—sex cell
gelatinous—resembling gelatin or jelly
gene—a unit located on a chromosome and controlling the transmission of a heritable characteristic
genetic—pertaining to heredity or heritable characteristics
genetically modified organism (GMO)—an organism possessing a gene from another species, having been the subject of genetic engineering
geniculate—bent, like a knee
genome—the complete genetic information of an organism or virus
genotype—the genetic constitution of an individual or group; a class or group of individuals sharing a specific genetic makeup (*see* phenotype)
genus (pl. genera)—a taxonomic category that consists of species that are closely related structurally or phylogenetically (the genus or generic name is the first name in a Latin binomial)
germ tube—a hypha resulting from an outgrowth of the spore wall and cytoplasm after germination
germinate (n. germination)—to begin growth (as of a seed, spore, sclerotium, or other reproductive body)
germling—the organism produced by germination of an algal or fungal spore
germplasm—bearer of heredity material (often loosely applied to cultivars and breeding lines)
giant cell—an enlarged, multinucleate cell formed in a root by repeated nuclear division without cell division, induced by secretions of certain sedentary plant-parasitic nematodes
girdle—to destroy vascular tissue of the woody portion of a plant in a circular pattern, either mechanically or through the feeding action of a pathogen or parasite
girdling (syn, ringing)—viticulturally, the removal of a complete ring of bark from a shoot, cane, or trunk, temporarily interrupting the downward translocation of photosynthates and other metabolites
globose—nearly spherical
graft transmission—transmission of a pathogen from one host plant to another through fusion of living tissue from the diseased host with that of a healthy host
gram-negative bacteria—bacteria that are stained red or pink after treatment with Gram's stain
gram-positive bacteria—bacteria that are stained violet or purple after treatment with Gram's stain
gravid—containing an egg or eggs; capable of depositing eggs
guttation—exudation of watery, sticky liquid from hydathodes, especially along leaf margins

haploid—having a single complete set of chromosomes (*see* diploid; polyploid)
hardiness—the ability to withstand stress
hardwood cutting—in grapevine propagation, a piece of cane with buds induced to produce adventitious roots
haustorium (pl. haustoria)—a specialized branch of a parasite formed inside a host cell to absorb nutrients
head—viticulturally, upper potion of a vine consisting of trunk and arms
herbaceous—pertaining to primary, soft, nonwoody tissue of a plant or plant part; having the characteristics of an herb
herbicide—any of various chemicals that kill plants or inhibit their growth (e.g., a weed or grass killer)
hermaphrodite (adj. hermaphroditic)—having both male and female reproductive organs in a single individual
heteroecious—with reference to rust fungi, passing different stages of the life cycle on alternate hosts (*see* autoecious)
heterokaryon (adj. heterokaryotic)—a cell with genetically different nuclei
heterothallism (adj. heterothallic)—a fungal life cycle in which sexual reproduction can occur only in the presence of genetically different mycelia (*see* homothallism)
heterozygous—having the same form (allele) of a gene on different chromosomes (*see* homozygous)
holomorph—a fungus having sexual and asexual forms in its life cycle, considered in all its forms (*see* anamorph; teleomorph)
homothallism (adj. homothallic)—a fungal life cycle in which sexual reproduction occurs in a single, self-fertile thallus (*see* heterothallism)
homozygous—having the same form (allele) of a gene on homologous chromosomes (*see* heterozygous)
honeydew—a sugary ooze or exudate, often secreted by aphids and other sap-feeding insects; a characteristic symptom of ergot
host plant—a living plant attacked by or harboring a parasite or pathogen, from which the invader obtains part or all of its nourishment
host range—the range of plants on which a pathogen or parasite feeds
hull—the outer coat of a seed
hyaline—transparent or nearly so; translucent; colorless
hybrid (v. hybridize)—the offspring of two individuals of different genotypes
hydathode—an epidermal leaf structure specialized for the secretion or exudation of water; the opening at the terminus of a leaf vein
hymenium—the continuous, spore-bearing layer of a fungal fruiting body
hyperplasia (adj. hyperplastic)—an abnormal increase in the number of cells, often resulting in the formation of galls or tumors
hypersensitive—extremely or excessively sensitive, often with reference to an extreme reaction to a pathogen
hypertrophy (adj. hypertrophic)—an abnormal increase in the size of cells in a tissue or organ, often resulting in the formation of galls or tumors
hypha (pl. hyphae; adj. hyphal)—a tubular filament of a fungal thallus or mycelium, the basic structural unit of a fungus
hypophyllous—on the undersurface of a leaf

icosahedral (n. icosahedron)—having 20 faces, like polyhedral virus particles (*see* isometric)
immune—not capable of being infected by a given pathogen
immunogenic—able to induce the production of antibodies
imperfect state—the anamorph, or asexual form in the life cycle of certain fungi, in which asexual spores (e.g., conidia) or no spores are produced (*see* perfect state)
in vitro—in glass, on artificial media, or in an artificial environment; not within a living host
in vivo—within a living organism
inclusion body—a structure developed within a plant cell as a result of infection by a virus, often useful in identifying the virus
incubation period—time elapsed between infection by a pathogen and the appearance of symptoms (*see* latent period)
index (n. **indexing**)—viticulturally, to determine the presence of a pathogen in a potential source of propagation material through grafting onto a sensitive indicator plant and observing for symptom
indicator plant—a plant in which specific or distinctive symptoms develop in reaction to a pathogen or to certain environmental conditions, used to detect or identify the pathogen or determine the effects of the environmental conditions
infection—the process in which an organism enters, invades, or penetrates and establishes a parasitic relationship with a host plant
infection court—a site in or on a host plant where infection can occur
infection cushion—an organized mass of hyphae formed on the surface of a plant and producing numerous infective hyphae
infection focus—the initial site of infection, generally with reference to a population of plants
infection peg—the specialized, narrow hyphal strand located on the underside of an appressorium and penetrating host cells (syn. **penetration peg**)
infectious—with reference to disease, capable of spreading from plant to plant
infective—able to attack a host and cause infection; with reference to vectors, carrying or containing a pathogen and able to transfer it to a host plant
infest (n. **infestation**)—to attack as a pest (used especially for insects and nematodes); to contaminate, as with microorganisms; to be present in large numbers
inflorescence—a flower or flower cluster
initial inoculum—*see* primary inoculum
injury—damage caused by transitory interaction with an agent such as an insect, a chemical, or an unfavorable environmental condition

inoculate (n. inoculation)—to place inoculum in an infection court; to insert a pathogen into healthy tissue
inoculum (pl. inocula)—a pathogen or its parts, capable of causing infection when transferred to a favorable location
inoculum density—a measure of the number of propagules of a pathogenic organism per unit of area or volume
inoperculate—referring to an ascus or sporangium that opens by an irregular apical split to allow the discharge of its spores
intercalary—inserted within (e.g., located along a hypha, as opposed to located at the end of the hypha)
intercalary meristem—zone of active cell division in elongating tissues where cells divide and expand in the direction of organ extension
intercellular—between or among cells
internode (adj. internodal)—the portion of a stem between two successive nodes
interveinal—between veins (of a leaf)
intracellular—through or within cells
intumescence—blisters produced on leaves or other plant parts under conditions of high moisture and restricted transpiration (syn. edema)
isolate—(n.) a culture or subpopulation of a microorganism separated from its parent population and maintained under controlled conditions; (v.) to remove from soil or host material and grow in pure culture
isometric—with reference to virus particles, having an icosahedral structure and thus appearing approximately round

juvenile—an immature form that appears similar to the adult stage but is usually smaller and is not sexually mature, as in nematodes or in insect species that undergo gradual metamorphosis

knot—a gall or localized abnormal swelling

lamina—the broad, expanded portion, or blade, of a leaf (*see* **petiole**)
larva (pl. larvae)—immature, wormlike or caterpillar stage in the life cycle of insects that undergo complete metamorphosis
latent—present but not manifested or visible, as a symptomless infection
latent bud—viticulturally, a non-growing bud, originally axillary and usually located on wood more than one year old, capable of developing into a shoot
latent infection—infection unaccompanied by visible symptoms
latent period—time elapsed between infection and the appearance of symptoms and subsequent production of new inoculum (although sometimes used synonymously with *incubation period*); the time elapsed after the acquisition of a pathogen by a vector but before the pathogen can be transmitted
lateral bud—bud formed in the axil of a leaf
lateral shoot—viticulturally, a shoot produced from a lateral or axillary bud of a leaf and developing in the same year as the bud
layer—viticulturally, a long cane from a vine placed on the ground and induced to root in order to replace an adjacent missing vine; the new vine derived from this process
leaf spot—a leaf lesion typically restricted in development after reaching a limited, characteristic size
leafhopper—any of various species of mobile insects with sucking mouthparts in the order Hemiptera
lenticel—a natural non-stomatal opening in the surface of a stem, tuber, fruit, or root, permitting gas exchange
lenticular—lens-shaped (convex on both faces)
lesion—a localized diseased area or wound
life cycle—the cyclical stages in the growth and development of an organism
lignin—a complex organic substance or group of substances that impregnate the cell walls of xylem vessels and certain other plant cells, constituting wood
lobe—a rounded projection of an organ
local lesion—a small, restricted lesion, often the characteristic reaction of a differential cultivar to a specific pathogen, especially in response to mechanical inoculation with a virus
locule (adj. locular)—cavity, especially in a fungus stroma
lumen (pl. lumina)—the central cavity of a cell or other structure

macerate—to cause disintegration of tissues by separation of cells; to soften by soaking
macroconidium (pl. macroconidia)—the larger of two kinds of conidia formed by certain fungi (*see* microconidium)
macrocyclic—pertaining to the life cycle of rust fungi that typically exhibit all five spore stages (pycniospores, aeciospores, urediniospores, teliospores, and basidiospores) in the course of their development (*see* demicyclic; microcyclic)
male—botanically, having flowers with stamens but no pistils
male-sterile—referring to monoecious plants that produce no viable pollen
manual transmission—the spread or introduction of inoculum to an infection court by hand
mating types—compatible strains, usually designated + and – or A and B, both of which are necessary for sexual reproduction in heterothallic fungi and oomycetes
matrix—the material within which an organism or organ is embedded
mechanical injury—injury of a plant part by abrasion, mutilation, or wounding
mechanical transmission—the spread or introduction of inoculum to an infection court by human manipulation, accompanied by physical disruption of host tissues (wounding)
medium (pl. media)—a mixture of organic or inorganic chemical compounds and water providing the nutrients needed for the growth of a microorganism *in vitro;* with reference to higher plants, a mixture of fertilizers and other components in which plants are grown
meiosis—nuclear division in which the number of chromosomes per nucleus is halved, i.e., the diploid state is converted to the haploid state (*see* mitosis)
melanin (adj. melanoid)—a brown black pigment
meq—milliequivalents, a measure of quantity based on the amount of a substance that will react with or supply one mole of $H+$ ions in an acid-base reaction or one mole of electrons in a redox reaction, multiplied by 1000
meristem (adj. meristematic)—undifferentiated plant tissue characterized by frequent cell division, producing cells that subsequently differentiate into specialized tissues
meristem culture—aseptic culture of a plant or plant part from a portion of the meristem
mesophyll—the central, internal, nonvascular tissue of leaves, consisting of the palisade and spongy mesophyll
messenger RNA (abbr. mRNA)—a form of RNA that carries information to direct the synthesis of protein
metabasidium—the part of the basidium in which meiosis occurs
metabolite—any chemical participating in metabolism; a nutrient
microclimate—weather conditions on a small scale, as at the surface of a plant or within a field
microconidium (pl. microconidia)—the smaller of two kinds of conidia formed by certain fungi (*see* macroconidium)
microcyclic—pertaining to the life cycle of rust fungi that produce only teliospores and basidiospores (*see* demicyclic; macrocyclic)
microflora—the combination of all microorganisms in a particular environment
microorganism—an organism small enough that individuals can be seen only with the aid of a microscope (syn. microbe)
micronutrient—nutrient required in small quantities for normal growth and reproduction, often toxic at higher levels
microsclerotium (pl. microsclerotia)—a microscopic, dense aggregate of darkly pigmented, thick-walled hyphal cells
middle lamella—the layer, consisting largely of pectic substances, between the walls of adjacent plant cells
midrib—the central, thickened vein of a leaf
migratory—moving from place to place on a plant or from plant to plant when feeding (*see* sedentary)
mildew—a thin coating of mycelial growth and spores on the surfaces of infected plant parts
millerandage—a disorder characterized by production of small, seedless, green grape berries (shot berries)
mitochondrion (pl. mitochondria)—a cellular organelle outside the nucleus, functioning in respiration
mitosis—nuclear division in which the number of chromosomes per nucleus remains the same (*see* meiosis)
MLO—mycoplasmalike organism (*see* phytoplasma)
mold—any microfungus with conspicuous, profuse, or woolly superficial growth (mycelium or spore masses) on various substrates, commonly growing on damp or decaying matter and on the surface

of plant tissues, especially with reference to economically important saprobes
monoecious—having female (pistillate) and male (staminate) flowers on the same plant; hermaphroditic
mollicute—any of a group of prokaryotic organisms bounded by a flexuous membrane and lacking a cell wall (*see* phytoplasma; spiroplasma)
molt—to shed a cuticle or body encasement during a phase of growth
monocot—a monocotyledon, or flowering plant having only one cotyledon, such as species of grasses, orchids, palms, and bananas (*see* dicot)
monoculture—the cultivation of plants of the same species in close proximity, with few or no other plant species present
monoecious—having male and female reproductive organs on a single individual
monogenic—determined by a single gene (*see* polygenic)
morphology (adj. morphological)—the study of the form of organisms; the form and structure of organisms
mosaic—a disease symptom characterized by nonuniform coloration, with intermingled normal, light green and yellowish patches, usually caused by a virus (often used interchangeably with mottle)
motile—capable of self-propulsion by means of flagella, cilia, or amoeboid movement
mottle—a disease symptom characterized by light and dark areas in an irregular pattern, usually caused by a virus (often used interchangeably with mosaic)
mucilaginous—viscous, slimy
mulch—a layer of material, such as organic matter or plastic film, applied to the surface of the soil for purposes such as retention of water and inhibition of weeds
multinucleate—having more than one nucleus per cell (*see* binucleate; uninucleate)
multiseptate—having many septa (cross-walls)
mummy (v. mummify)—viticulturally, a dried and shriveled grape berry
muriform—having transverse and longitudinal septa
mutation—an abrupt heritable or genetic change in a gene or an individual as a result of an alteration in a gene or chromosome or an increase in chromosome number
mycelium (pl. mycelia; adj. mycelial)—the mass of hyphae constituting the body (thallus) of a fungus
mycoparasite—a fungus that attacks another fungus
mycoplasmalike organism (abbr. MLO)—*see* phytoplasma
mycorrhiza (pl. mycorrhizae; adj. mycorrhizal)—a symbiotic association between a nonpathogenic or weakly pathogenic fungus and the roots of a plant (*see* ectomycorrhiza; endomycorrhiza)
mycotoxin—any poisonous compound produced by a fungus

necrosis (adj. necrotic)—death of cells or tissue, usually accompanied by blackening or browning
necrotroph—a parasite that typically kills and obtains its energy from dead host cells (*see* biotroph)
nectary—viticulturally, one of the five swellings from the base of the ovary of a grape flower, the function of which is not known
nematicide—an agent, usually a chemical, that kills nematodes
nematode—any of various unsegmented roundworms (animals), parasitic on plants or animals, or free living in soil or water
neopvirus—any of a group of nematode-transmitted polyhedral viruses
node (adj. nodal)—the enlarged portion of a shoot at which leaves or buds arise
nodosity—gall, usually bent and club-shaped, formed on feeder roots by phylloxera
nodule—a small knot or irregular, rounded lump
noninfectious disease—a disease caused by an abiotic agent and not capable of being transmitted from plant to plant
nonpersistent transmission—transmission mode in which a virus is acquired and transmitted by a vector after a short feeding time and is retained by the vector for only a short time (syn. stylet-borne transmission)
nonseptate—lacking cross-walls, coenocytic
nucleotides (abbr. nt)—the building blocks of DNA or RNA, the order of which comprise the genetic sequence
nymph—the juvenile stage in the life cycle of an insect with incomplete metamorphosis, superficially resembling the adult

obclavate—shaped like an upside-down club
obligate parasite—an organism that can grow only as a parasite in association with its living host plant (syn. biotroph)
obovoid—shaped like an upside-down egg
obtuse—rounded or blunt; pertaining to an angle greater than 90°
oedema—*see* edema
oogonium (pl. oogonia)—female gametangium of the oomycetes, containing one or more gametes
oomycete (adj. oomycetous)—an organism of the phylum Oomycota (or Oomycetes), formerly classified as fungi but now included in the kingdom Straminopiles, typically having nonseptate mycelium, asexual sporangia and zoospores, and sexual oospores
oospore—the thick-walled, sexually derived resting spore (produced within an oogonium) of an oomycete
ooze—a mass of bacterial cells mixed with host fluids
operculate—referring to an ascus or sporangium that opens by an apical lid to allow the discharge of its spores
organelle—any of various membrane-bound structures contained within cells and having a specialized function (e.g., mitochondria and chloroplasts)
ostiole (adj. ostiolate)—a pore; the opening in the papilla or neck of a perithecium, pseudothecium, or pycnidium, through which spores are released
ovary—the female reproductive structure of organisms; in plants, the enlarged basal portion of a pistil, containing the ovules and developing into the fruit
overseason—to survive or persist from one planting season to the next
oversummer—to survive or persist over the summer
overwinter—to survive or persist through the winter
oviposit—to deposit or lay eggs with an ovipositor, as for insects
ovule—an enclosed structure that, after fertilization, becomes a seed; an egg contained within an ovary
own-rooted vine—asexually propagated, non-grafted vine whose root and shoot systems are of the same genotype
ozone—a highly reactive form of oxygen (O_3), injurious to plants

palisade parenchyma—the tissue located beneath the upper epidermis of leaves, composed of elongate, tubular cells arranged upright in the manner of posts in a palisade fortification

palmate—having lobes or veins radiating from a common point
papilla (pl. papillae, adj. papillate)—a small, blunt projection
paragynous—having the antheridium attached to the side of the oogonium
paraphysis (pl. paraphyses)—a hairlike cell within a fungal fruiting structure
parasexual—pertaining to the recombination of genetic characters without sexual processes
parasite (adj. parasitic)—an organism that lives in intimate association with another organism, on which it depends for its nutrition, but not necessarily a pathogen
paratracheal parenchyma—parenchyma cells that line the outer side of the dead, hollow xylem vessels; may contribute to vessel plugging by forming tyloses (q.v.)
parenchyma (adj. parenchymatous)—the soft tissue of living plant cells with undifferentiated, thin, cellulose walls
parthenogenesis (adj. parthenogenetic)—reproduction by the development of an unfertilized egg
pasteurization—the process by which a material, usually a liquid, is treated with heat to eliminate selected harmful microorganisms
pathogen (adj. pathogenic)—a disease-producing organism or agent
pathogenesis—the initiation and development of disease
pathogenicity—the ability to cause disease
pathology—the study of diseases
pathotype—a subdivision of a pathogenic species characterized by its pattern of virulence or avirulence to differential host varieties
pathovar (abbr. pv.)—a subdivision of a plant-pathogenic bacterial species, defined by its host range (a pathovar of a bacterial species is analogous to a *forma specialis* of a fungal species)
PCR—polymerase chain reaction, a technique used to isolate and amplify a particular DNA sequence, often for sequence analysis or detection of specific microorganisms, e.g., for disease diagnosis
pedicel—a small, slender stalk; a stalk bearing an individual flower, fruit, inflorescence, or spore

peduncle—the stalk or main stem of an inflorescence; part of an inflorescence or a fructification
penetration—the initial invasion of a host by a pathogen
penetration peg—the specialized, narrow hyphal strand located on the underside of an appressorium and penetrating host cells (syn. infection peg)
perennial—a plant that survives for several to many years (*see* annual; biennial)
perfect—sexual; capable of sexual reproduction
perfect flower—a flower possessing both stamens and pistils
perfect state—the teleomorph, or sexual form in the life cycle of certain fungi (*see* imperfect state)
pericarp—the outer layer of a seed or fruit (*see* endocarp)
pericyclic—pertaining to the outer part of the phloem of a stem or root
periderm—*see* bark
perithecium (pl. perithecia)—the flask-shaped or subglobose, thin-walled fruiting body (ascocarp) of some ascomycetous fungi, containing asci and ascospores, and having a pore (ostiole) at its apex, through which ascospores are expelled or released
peritrichous—having hairs or flagella distributed over the whole surface
persistent transmission—transmission mode in which a virus is acquired and transmitted by a vector after a relatively long feeding time and remains transmissible for a prolonged period while in association with the vector (syn. circulative transmission)
pest—any organism that damages plants or plant products
pesticide—any of various chemicals used to control pests
petiole—the stalk portion of a leaf (*see* lamina)
pH—the negative logarithm of the effective hydrogen ion concentration, a measure of acidity (pH 7 is neutral; values less than pH 7 are acidic; values greater than pH 7 are alkaline)
phenology– the study of the natural progression of annual growth and development
phenological system--viticulturally, an organization of arbitrary growth stages describing development applicable to vine physiology and management
phenotype—the physical properties of an organism produced by the interaction of its genotype with the environment
phialide—the end cell of a conidiophore with one or more open ends, through which a basipetal succession of conidia develops
phialospore—a conidium produced on a phialide
phloem—the food-conducting, food-storing tissue in the vascular system of roots, stems, and leaves
photosynthate—a chemical product of photosynthesis
photosynthesis—the manufacture of carbohydrates from carbon dioxide and water in the presence of chlorophylls, using light energy and releasing oxygen
phyllody—a disorder in which floral organs are transformed into leaf-like structures
phyllotaxy—viticulturally, the arrangement of leaves, tendrils, and clusters along the grapevine shoot
physiological race—a group of individuals belonging to the same species and differing from other members of the species in their behavior or other characteristics but not in morphology; physiological form
phytoalexin—any of various substances produced by higher plants in response to chemical, physical, or biological stimuli and inhibiting the growth of certain microorganisms
phytopathology—the study of plant diseases (syn. plant pathology)
phytoplasma—any of various plant-parasitic pleomorphic mollicutes (prokaryotes lacking cell walls) found in phloem tissue and not capable of growth on artificial nutrient media (previously called mycoplasmalike organisms [*see* MLOs])
phytotoxic (n. phytotoxicity)—harmful to plants, usually with reference to chemicals
plasmid—a self-replicating piece of DNA that is stably inherited in an extrachromosomal state
pigment—a colored compound, such as chlorophyll, in the cells of plants or fungi
pinnate—featherlike; having parts arranged along two sides of an axis
pistil—the ovule-bearing organ of a plant, consisting of the ovary and its appendages (e.g., the style and stigma)
pith—parenchymatous tissue occupying the center of a stem
plant pathology—the study of plant diseases (syn. phytopathology)
plasmodium (pl. plasmodia)—a naked multinucleate mass of protoplasm moving and feeding in amoeboid fashion
pleomorphic—able to assume various shapes and perhaps sizes; having a life cycle in which a succession of distinctly different forms occurs
plumule—a rudimentary shoot of the plant embryo
polar—located at one end (pole) of the cell
pollen—male sex cells produced by the anthers of flowering plants or the cones of seed plants
pollination—the transfer of pollen from anther to stigma or from a staminate cone to an ovulate cone
polygenic—pertaining to or governed by many genes (*see* **monogenic**)
polyploid—having three or more complete sets of chromosomes (*see* haploid; diploid)
pomace—skins, seeds, stems, and other waste remaining after extraction of juice from grapes
positive-sense RNA—RNA that can serve directly as messenger RNA
posterior—situated toward the back or tail (*see* anterior)
predispose (n. predisposition)—to make prone to infection and disease
primary bud—viticulturally, the central and largest bud in a compound winter bud, which usually develops into a fruiting shoot in the following growing season
primary inoculum—inoculum, usually from an overwintering source, that initiates disease in the field, as opposed to inoculum that spreads disease during the season (syn. initial inoculum)
primordium (pl. primordia, adj. primordial)—the initial structure from which a plant part is formed
procambium—a strand of meristematic provascular tissue whose cells differentiate into primary xylem and phloem
prokaryote (adj. prokaryotic)—any of various organisms lacking internal membrane-bound organelles and a distinct nucleus, such as bacteria and mollicutes (*see* eukaryote)
promycelium (pl. promycelia)—in rust and smut fungi, a germ tube issuing from a teliospore and bearing basidiospores
propagative virus—a virus that multiplies within its arthropod vector
propagule—any part of an organism capable of independent growth
prophyll—small, scale-like structure borne at the first one or more nodes of a stem
protectant—an agent, usually a chemical, applied to a plant surface to prevent infection, in advance of exposure to a pathogen
protein—any of numerous nitrogen-containing organic compounds containing combinations of amino acids
protoplasm—the living contents of a cell
protoplast—the living contents of a cell, exclusive of the cell wall
proximal—situated toward or near the point of attachment or main body (*see* distal)
pseudoparenchymatous—referring to fungal tissue that consists of closely packed, roughly isodiametric or oval cells resembling the parenchyma cells of higher plants
pseudothecium (pl. pseudothecia)—a perithecium-like fruiting body containing asci and ascospores dispersed rather than in an organized hymenium; an ascostroma with a single locule (cavity) containing bitunicate asci
pulvinate—cushionlike in form
pupa (pl. pupae; v. pupate)—the quiescent stage between the larva and the adult stages of certain insects
pustule—a small, blisterlike elevation of the plant epidermis, formed as spores emerge
pv.—*see* pathovar
pycnidiospore—a spore (conidium) produced in a pycnidium
pycnidium (pl. pycnidia)—an asexual, globose or flask-shaped fruiting body of certain imperfect fungi, producing conidia
pycniospore—a haploid, sexually derived spore formed in the pycnium of a rust fungus (syn. spermatium)
pycnium (pl. pycnia)—the globose or flask-shaped haploid fruiting body of rust fungi, bearing receptive hyphae and pycniospores (syn. spermagonium)
pyriform—pear-shaped

quarantine—legislative control of the transport of plants or plant parts to prevent the spread of pests or pathogens
quiescent—dormant, inactive

race—a subgroup or biotype within a species or variety, distinguished from other races by virulence, symptom expression, or host range, but not by morphology
rachis—the elongated main axis of an inflorescence
radicle—the part of the plant embryo that develops into the primary root
ray tissue—radially oriented xylem or phloem tissue differentiating from the vascular cambium that provide radial transport and carbohydrate storage in secondarily thickened stems and roots
receptacle—the structure of a flower that bears the reproductive organs
receptive hypha—in rust fungi, the part of a pycnium (spermagonium) that receives the nucleus of a pycniospore (spermatium)
recessive—pertaining to a phenotypic trait that is expressed in the progeny of diploid organisms only if contributed by both parents (*see* dominant)
reniform—kidney-shaped
resinosis—exudation of resin
reservoir—an organism in which a parasite or pathogen of another organism lives and multiplies without causing damage
resistant (n. resistance)—possessing properties that prevent or impede disease development (*see* susceptible)
respiration—a series of chemical reactions that make energy available through oxidation of carbohydrates and fat
resting spore—a spore, often thick-walled, that can remain alive in a dormant state for an extended period, later germinating and capable of initiating infection
reticulate—having netlike markings
rhizomorph—a macroscopic rope-like strand of compacted vegetative tissue formed by certain fungi
rhizosphere—the microenvironment in the soil immediately around plant roots
ribosome—a subcellular protoplasmic particle, made up of one or more RNA molecules and several proteins, involved in protein synthesis
rind—the firm outer layer of a rhizomorph or other organ
ring spot—a disease symptom characterized by yellowish or necrotic rings enclosing green tissue, as in some plant diseases caused by viruses
RNA—ribonucleic acid, any of several nucleic acids composed of repeating units of ribose (a sugar), a phosphate group, and a purine (adenine or guanine) or a pyrimidine (uracil or cytosine) base, transcribed from DNA, and involved in translation to proteins
rogue—to remove and destroy individual plants that are diseased, infested by insects, or otherwise undesirable
root cap—a group of cells that protects the growing tip of a root
root hair—a threadlike, single-celled outgrowth from a root epidermal cell
rootstock—viticulturally, a rooted cutting possessing desirable belowground properties to which a fruiting (scion) cultivar is grafted to produce above-ground growth
rooting—a young vine produced from a rooted cutting grown for one season
rosette—a disease symptom characterized by short, bunchy growth habit due to a shortening of internodes with no comparable reduction in leaf size
rot—softening, discoloration, and often disintegration of plant tissue as a result of fungal or bacterial infection
rotation—the practice of programmatically utilizing unrelated groups of a pesticide in order to reduce selection pressure favoring resistant phenotypes of the target organisms; the practice of growing different kinds of crops in succession in the same field
rugose—wrinkled, roughened
russet—a brownish, roughened area on the skin of fruit, resulting from cork formation
rust—any of several diseases caused by a specialized group of fungi in the class Basidiomycetes, some of the spores of which are a rusty color

sanitation—destruction or removal of infected or infested plants or plant parts; decontamination of tools, equipment, containers, work space, hands, etc.
saprobe—*see* saprophyte
saprophyte (adj. saprophytic)—an organism that obtains nourishment from nonliving organic matter (syn. saprobe)
sapwood—outer part of xylem, in which active translocation of water and solutes takes place
satellite virus—a virus that accompanies another virus and depends on it for its multiplication
scab—a roughened, crustlike diseased area on the surface of a plant organ
scald—a necrotic condition in which tissue is usually bleached and has the appearance of having been exposed to high temperature
scion—fruiting variety that is grafted or budded onto a rootstock
sclerenchyma (adj. sclerenchymatous)—tissue made up of thick-walled plant cells
sclerotium (pl. sclerotia)—a vegetative resting body of a fungus, composed of a compact mass of hyphae with or without included host tissue, usually with a darkened rind
scorch—a symptom that suggests the action of flame or fire on the affected tissue, often at the margins of leaves
secondary bud—viticulturally, the second most developed of the three buds in a compound winter bud, adjacent to the leaf scar of the cane
secondary infection—infection resulting from the spread of infectious material from a primary or secondary infection during the same growing season
secondary inoculum—inoculum produced by an infection established previously in the same growing season
secondary organism—an organism that multiplies in already-diseased tissue but is not the primary pathogen
secondary root—a branch of a primary root
sedentary—remaining in a fixed location (*see* migratory)
seed—a ripened ovule consisting of an embryo and stored food enclosed by a seed coat
seedborne—carried on or in seed
selective medium—a culture medium containing substances that inhibit or prevent the growth of certain unwanted species of microorganisms
senesce (adj. senescent; n. senescence)—to decline with maturity or age, often hastened by stress from environment or disease
sepal—one of the modified leaves composing a calyx
septate—having septa (cross-walls)
septum (pl. septa; adj. septate)—a dividing wall; in fungi, a cross-wall
serology (adj. serological)—a method using the specificity of the antigen-antibody reaction for the detection and identification of antigenic substances and the organisms that carry them
serration—sharp teeth-like projections pointing forward on a leaf margin
sessile—with reference to leaves, leaflets, flowers, florets, fruit, ascocarps, basidiocarps, etc., lacking a stalk, petiole, pedicel, stipe, or stem; with reference to nematodes, permanently attached, not capable of moving about
seta (pl. setae)—a bristle- or hairlike structure, usually deep yellow or brown and thick-walled
sexual spore—a spore produced during the sexual cycle of a fungus
shatter—*see* drop
shelling—abscission of flowers before or at bloom, occurring before drop (shatter)
shoot—viticulturally, the succulent, green, current-season growth from a bud, including the leaves
shoot removal—deshooting; removal of unwanted shoots on the trunk of a vine below the head
shoot thinning—removal of excessive or unwanted shoots from the head, cordon, or arms of a vine when the shoots are still small, usually 15-40 cm long
shoot-tip culture—propagation of a plant from a growing shoot tip, often done in vitro
shot berry—small, seedless grape berry that does not ripen
shot hole—a symptom in which small lesions on a leaf fall out, giving the leaf the appearance of being hit by buckshot
sieve element—a tube-shaped living cell in the phloem, functioning in the transport of dissolved organic substances in the plant (syn. **sieve tube element**)
sign—an indication of disease from direct observation of a pathogen or its parts on a host plant (*see* symptom)
sinuous—having curves, bends, or turns
sinus—a cleft or indentation between the lobes of a leaf blade
slime molds—saprophytic organisms in the class Myxomycetes forming vegetative amoeboid plasmodia and spores

soft rot—a mushy disintegration of fleshy fruit or vegetables as a result of bacterial or fungal infection
softwood cutting—in grapevine propagation, a piece of (green) shoot induced to produce adventitious roots
soil drench—the application of a solution or suspension of a chemical to the soil, especially a pesticide to control soilborne pathogens
soilborne—carried on or beneath the soil surface
solarization—a disease management practice in which soil is covered with polyethylene sheeting to increase heating by sunlight and thus control soilborne plant pathogens
sooty mold—black, nonparasitic, superficial fungal growth, often on honeydew produced by aphids and other phloem-feeding insects
sorus (pl. sori)—a compact fruiting structure, especially with reference to spore masses of rust and smut fungi
sp. (pl. spp.)—species (the singular and plural abbreviations are used only after a genus name; the singular form is used to refer to an undetermined species; the plural form is used to refer to two or more species without naming them individually)
spermagonium (pl. spermagonia)—a structure in which male reproductive cells are produced; in rust fungi, the pycnium, a globose or flask-shaped haploid fruiting body composed of receptive hyphae and spermatia (pycniospores)
spermatium (pl. spermatia)—a male sex cell; a nonmotile male gamete; a haploid male gamete; in rust fungi, a pycniospore
spicule—the copulatory organ of a male nematode
spikelet—the spike-like appendage comprised of one or more reduced flowers and associated bracts; a unit of inflorescence in grasses; a small spike
spiroplasma—any of various spiral-shaped plant-pathogenic mollicutes (prokaryotes lacking cell walls)
sporangiophore—the sporangium-bearing body of a fungus or oomycete
sporangiospore—a nonmotile, asexual spore borne in a sporangium
sporangium (pl. sporangia)—a saclike fungal structure, the entire contents of which are converted into an indefinite number of asexual spores
spore—the reproductive structure of fungi, oomycetes, and some other organisms, containing one or more cells; a bacterial cell modified to survive an adverse environment
sporidium (pl. sporidia)—the basidiospore of rust fungi, smut fungi, and other members of the class Basidiomycetes
sporocarp—a spore-bearing fruiting body
sporodochium (pl. sporodochia)—a superficial, cushion-shaped, asexual fruiting body consisting of a cluster of conidiophores
sporulate—to produce spores
spot—a small necrotic area occurring as a disease symptom on a leaf, flower, or stem
stamen (adj. staminal)—the male structure of a flower, composed of a pollen-bearing anther and a filament, or stalk
staminate flower—a male flower
stele—the central cylinder of vascular tissue, especially with reference to roots
stem pitting—a viral disease symptom characterized by depressions on the stem
sterigma (pl. sterigmata)—a small, usually pointed projection that supports a spore
sterile—unable to reproduce sexually; free of living microorganisms
sterile fungus—a fungus that is not known to produce any kind of spores
sterilization (adj. sterilized)—the total destruction of living organisms by various means, including heat, chemicals, and irradiation
stigma—the portion of a flower that receives pollen and on which pollen germinates
stipe—a stalk
stippling—small dots or speckles on leaves, where chlorophyll is absent
stipule—a small leaflike appendage at the base of a leaf petiole, usually occurring in pairs
stolon—a slender, horizontal stem that grows close to the soil surface (syn. runner); in fungi, a hypha that grows horizontally along a surface
stoma, stomate (pl. stomata; adj. stomatal)—a structure composed of two guard cells and the opening between them in the epidermis of a leaf or stem, functioning in gas exchange
strain—a distinct form of an organism within a species or virus, differing from other forms of the species biologically, physically, or chemically
striate (n. striations)—marked with delicate lines, grooves, or ridges
stroma (pl. stromata)—a compact mass of mycelium (with or without host tissue) that supports fruiting bodies or in which fruiting bodies are embedded
stunting—reduction in the height plants resulting from a progressive reduction in the length of successive internodes or a decrease in their number
stylar scar—small corky area remaining at the apex of a berry after the style dries and falls off following pollination
style—the slender part of many pistils, located between the stigma and the ovary, and through which the pollen tube grows
stylet—the stiff, slender, hollow feeding organ of plant-parasitic nematodes or sap-sucking insects, such as aphids and leafhoppers
stylet-borne transmission—virus transmission in which a virus is acquired and transmitted by a vector after a short feeding time and is retained by the vector for only a short time (syn. nonpersistent transmission)
subepidermal—located or occurring just below the epidermis
subgenomic RNA—a piece of viral RNA, shorter than the entire genome of the virus, found in cells infected by the virus and sometimes encapsidated
substrate—the substance on which an organism lives or from which it obtains nutrients; a chemical substance acted upon, often by an enzyme
sunscald—injury of plant tissues burned or scorched by direct sunlight
suscept—a susceptible plant
susceptible (n. susceptibility)—prone to develop disease when infected by a pathogen (*see* resistant)
symbiosis (adj. symbiotic)—a mutually beneficial association of two different kinds of organisms
symbiont—either of the organisms in a symbiotic association
symptom—an indication of disease by a reaction of the host, e.g., canker, leaf spot, wilt (*see* sign)
symptomatology—the study of disease symptoms
symptomless carrier—a plant that, although infected with a pathogen (usually a virus), has no obvious symptoms
syncytium (pl. syncytia)—a multinucleate structure in root tissue formed by the dissolution of common cell walls induced by secretions of certain sedentary plant-parasitic nematodes, e.g., cyst nematodes
synergism (adj. synergistic)—interaction in which the total effect of the interacting factors is greater than the additive effects of the factors acting individually
synnema (pl. synnemata)—compact or fused, generally upright conidiophores, with branches and spores forming a headlike cluster (syn. coremium)
systemic—pertaining to a disease in which the pathogen (or a single infection) spreads generally throughout a plant; pertaining to chemicals that spread internally through a plant

taproot—a primary root that grows vertically downward and from which smaller lateral roots branch
taxonomy (adj. taxonomic)—the science of naming and classifying organisms
teleomorph—the perfect state, or sexual form in the life cycle of certain fungi (*see* anamorph; holomorph)
teliospore—a thick-walled resting spore produced by some fungi, notably rust and smut fungi, from which a basidium is produced
telium (pl. **telia**)—the fruiting body (sorus) that produces teliospores of a rust fungus
tendril—twining organ borne opposite certain leaves on grapevines
tertiary bud—viticulturally, the least developed of the three buds in a compound winter bud, farthest from the leaf scar of the cane
thallus—the vegetative body of a fungus
thermal inactivation point—the lowest temperature at which heating for a limited period (usually 10 min) is sufficient to cause a virus to lose its infectivity or an enzyme its activity
tissue culture—an *in vitro* method of propagating healthy cells from plant tissues
tissue—a group of cells, usually of similar structure, that perform the same or related functions

titer—a measure of the concentration of a substance in a solution

tolerance (adj. tolerant)—the ability of a plant to endure an infectious or noninfectious disease, adverse conditions, or chemical injury without serious damage or yield loss; with reference to pesticides, the amount of chemical residue legally permitted on an agricultural product entering commercial channels, usually measured in parts per million (ppm)

tomentose—covered with a dense mat of hairs

torulose—cylindrical bud having swellings at intervals

toxicity (adj. toxic)—the capacity of a substance to interfere with the vital processes of an organism

toxin—any poisonous substance of biological origin

trabecular (pl. trabeculae)—bar of wall-thickening material extending across the lumen from one tangential wall to the other of a conducting cell of xylem, as in certain virus-infected grapevines

transgenic—possessing a gene from another species, having been the subject of genetic engineering

translocation—the movement of water, nutrients, chemicals, or food materials within a plant

translucent—clear enough to allow the transmission of light

transmit (n. transmission)—to spread or transfer, as in spreading an infectious pathogen from plant to plant or from one generation of plants to another

transovarial passage—the passage of a virus through the eggs or offspring of its vector and then to the next generation of host plants

transpiration—loss of internal water by evaporation from leaf surfaces and through stomata

trichome—epidermal cell or a cell group protruding from the epidermis on plants to form an elongate structure, such as hairs on bud scales and leaf surfaces

triturate—to grind, as with a mortar and pestle

truncate—ending abruptly, as though an end had been cut off

trunk—viticulturally, the vertical above-ground axis of a grapevine

tuberosity—gall formed on a mature root by the feeding of phylloxera

tumor—an abnormal swelling or localized outgrowth, often roughly spherical, produced by a plant as a result of attack by a fungus, bacterium, nematode, insect, or other organism (syn. gall)

tumorigenic—able to incite tumor formation

turgid—swollen, inflated; plump or swollen as a result of internal water pressure

turgidity—the state of being turgid

tylosis (pl. tyloses)—a balloonlike extrusion of parenchyma cells into the lumina of contiguous vessels, partially or completely blocking them

ultrastructure—the submicroscopic structure of a macromolecule, cell, or tissue

unicellular—one-celled (*see* multicellular)

uniflagellate—having one flagellum

uninucleate—having one nucleus per cell (*see* multinucleate)

unitunicate—having one defined wall or cover, as for the asci of many Ascomycete species

urediniospore, **urediospore**, **uredospore**—the asexual, dikaryotic, often rusty-colored spore of a rust fungus, produced in a uredinium; the repeating stage of a heteroecious rust fungus, in which it is capable of infecting the host plant on which it was produced

uredinium (pl. uredinia), uredium (pl. uredia)—the fruiting body (sorus) in which urediniospores of a rust fungus are produced

vacuole—a generally spherical organelle within a plant, fungal, or oomycete cell, bound by a membrane and containing dissolved materials, such as metabolic precursors, storage materials, and waste products

variegation—a pattern of two or more colors in a plant part, as in a green and white leaf

variety (adj. varietal)—a plant type within a species, resulting from deliberate genetic manipulation and having recognizable characteristics such as color; shape of flowers, fruits, and seeds; height; and form (syn. cultivar)

vascular—pertaining to fluid-conducting tissues (xylem and phloem) in plants

vascular bundle—a strand of conductive tissue, usually composed of xylem and phloem (*see* vein)

vascular cylinder—the cylinder of vascular tissue in stems and roots (syn. **stele**)

vascular wilt disease—a xylem disease that disrupts normal uptake of water and minerals, resulting in wilting and yellowing of foliage

vector—a living organism (e.g., insect, mite, bird, higher animal, nematode, parasitic plant, human) able to carry and transmit a pathogen; in genetic engineering, a cloning vehicle, or self-replicating DNA molecule, such as a plasmid or virus, used to introduce a fragment of foreign DNA into a host cell

vegetative—pertaining to the somatic or asexual parts of a plant, which are not involved in sexual reproduction

vegetative propagation—asexual reproduction; in plants, the use of cuttings, bulbs, tubers, or other vegetative parts to grow new plants

vein—a small vascular bundle in a leaf

vein banding—a symptom of virus disease in which regions along veins are darker green than the tissue between veins

vein clearing—disappearance of green color in or around leaf veins

veinlet—small branch of a vein ending in the mesophyll of a leaf

véraison—a phonological stage characterized by the loss of green color in grape berries beginning to ripen

vermiform—worm-shaped

vertical resistance—resistance thought to be conferred by a single gene or a few genes and completely protecting a host, but only against specific races or strains (genetic variants) of a pathogen

verticillate—whorled

vesicle—the thin sac in which zoospores are differentiated and from which they are released; the bulbous head terminating the conidiophores of *Aspergillus;* a structure formed by an endomycorrhizal fungus within living cells of the root

vessel—the water-conducting structure of xylem tissue, with pit openings in end walls

viable (n. viability)—alive; capable of germination (of seeds, spores, sclerotia, etc.); capable of growth

virescence—a state or condition in which normally white or colored tissues (e.g., flower petals) become green

virion—a complete virus particle

viroid—any of various small, unencapsidated (naked), circular, single-stranded RNA molecules capable of causing disease in plants

virome—the collection or community of all the viruses existing in a given organism, such as a plant

viroplasm—an inclusion in a virus-infected cell, where viral components are synthesized and virus particles are assembled

virulence—degree or measure of pathogenicity; relative capacity to cause disease

virulent—pathogenic, having the capacity to cause disease (*see* avirulent)

viruliferous—virus-laden, usually with reference to insects or nematodes as vectors

virus—any of numerous submicroscopic, intracellular, obligate parasites consisting of a core of infectious nucleic acid (either RNA or DNA) usually surrounded by a protein coat

water soaking—a disease symptom in which plant tissues or lesions appear wet, dark, and usually sunken and translucent

whorl—a circular arrangement of like parts

wild type—the phenotype characteristic of most individuals of a species under natural conditions

wilt—drooping of leaves and stems due to a lack of water (inadequate water supply or excessive transpiration); a vascular disease in which normal water uptake is interrupted

wing—viticulturally, a large basal branch of the fruit cluster

witches'-broom—a disease symptom in which many weak shoots arise at or close to the same point, producing abnormal massed, brushlike growth

wood—lignified secondary xylem of stem and roots; also used to describe grape stems without leaves that have a brown periderm ("ripe wood")

xylem—the water- and mineral-conducting, food-storing, supporting tissue of a plant

yellows—a plant disease characterized by chlorosis and stunting

zonate—pertaining to any symptom appearing in concentric rings
zoosporangium—a sporangium (spore case) bearing zoospores
zoospore—an oomycete spore with flagella, capable of locomotion in water
zygospore—the sexual resting spore formed from the union of gametangia of fungi in the class Zygomycetes

Index